DAVID CANNADINE

Victorious Century

The United Kingdom, 1800–1906

PENGUIN BOOKS

PENGUIN BOOKS

UK | USA | Canada | Ireland | Australia
India | New Zealand | South Africa

Penguin Books is part of the Penguin Random House group of companies whose addresses can be found at global.penguinrandomhouse.com.

First published by Allen Lane 2017
Published in Penguin Books 2018
001

Set in 8.98/12.3 pt Sabon LT Std
Typeset by Jouve (UK), Milton Keynes
Printed and bound in Great Britain by Clays Ltd, Elcograf S.p.A.

A CIP catalogue record for this book is available from the British Library

ISBN: 978-0-141-01913-0

www.greenpenguin.co.uk

In memory of
Asa Briggs
and
Peter Carson

It was the best of times, it was the worst of times, it was the age of wisdom, it was the age of foolishness, it was the epoch of belief, it was the epoch of incredulity, it was the season of Light, it was the season of Darkness, it was the spring of hope, it was the winter of despair, we had everything before us, we had nothing before us, we were all going direct to Heaven, we were all going direct the other way . . .

Charles Dickens, *A Tale of Two Cities* (1859)

Men [and women] make their own history, but they do not do so freely, not under conditions of their own choosing, but rather under circumstances which directly confront them, and which are historically given and transmitted.

Karl Marx, *The Eighteenth Brumaire of Louis Bonaparte* (1852)

Contents

List of Illustrations

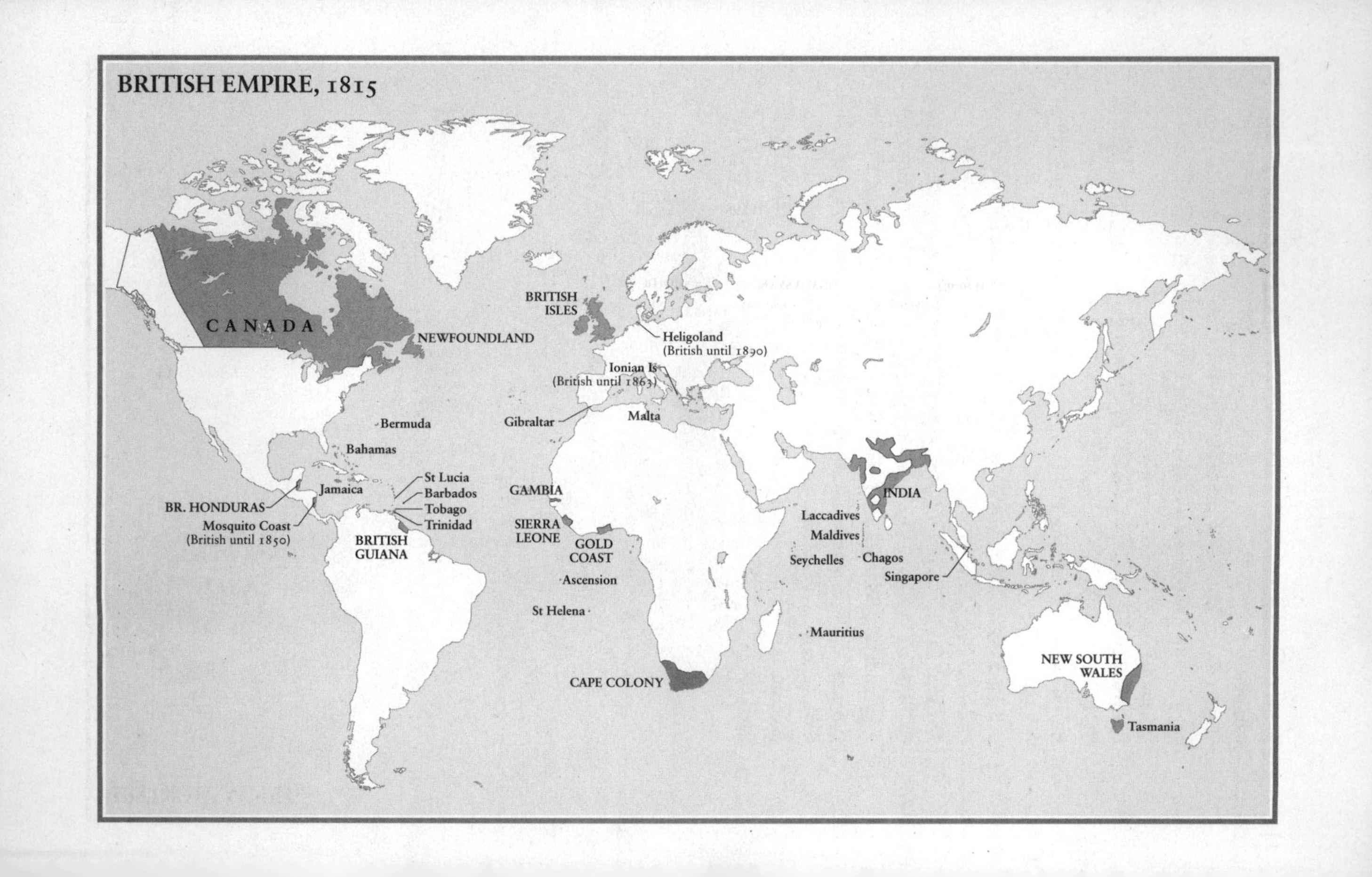
BRITISH EMPIRE, 1815
CANADA
NEWFOUNDLAND
BRITISH ISLES
Heligoland (British until 1890)
Ionian Is (British until 1863)
Gibraltar
Malta
Bermuda
Bahamas
Jamaica
St Lucia
Barbados
Tobago
Trinidad
BR. HONDURAS
Mosquito Coast (British until 1850)
BRITISH GUIANA
GAMBIA
SIERRA LEONE
GOLD COAST
Ascension
St Helena
CAPE COLONY
INDIA
Laccadives
Maldives
Seychelles
Chagos
Singapore
Mauritius
NEW SOUTH WALES
Tasmania

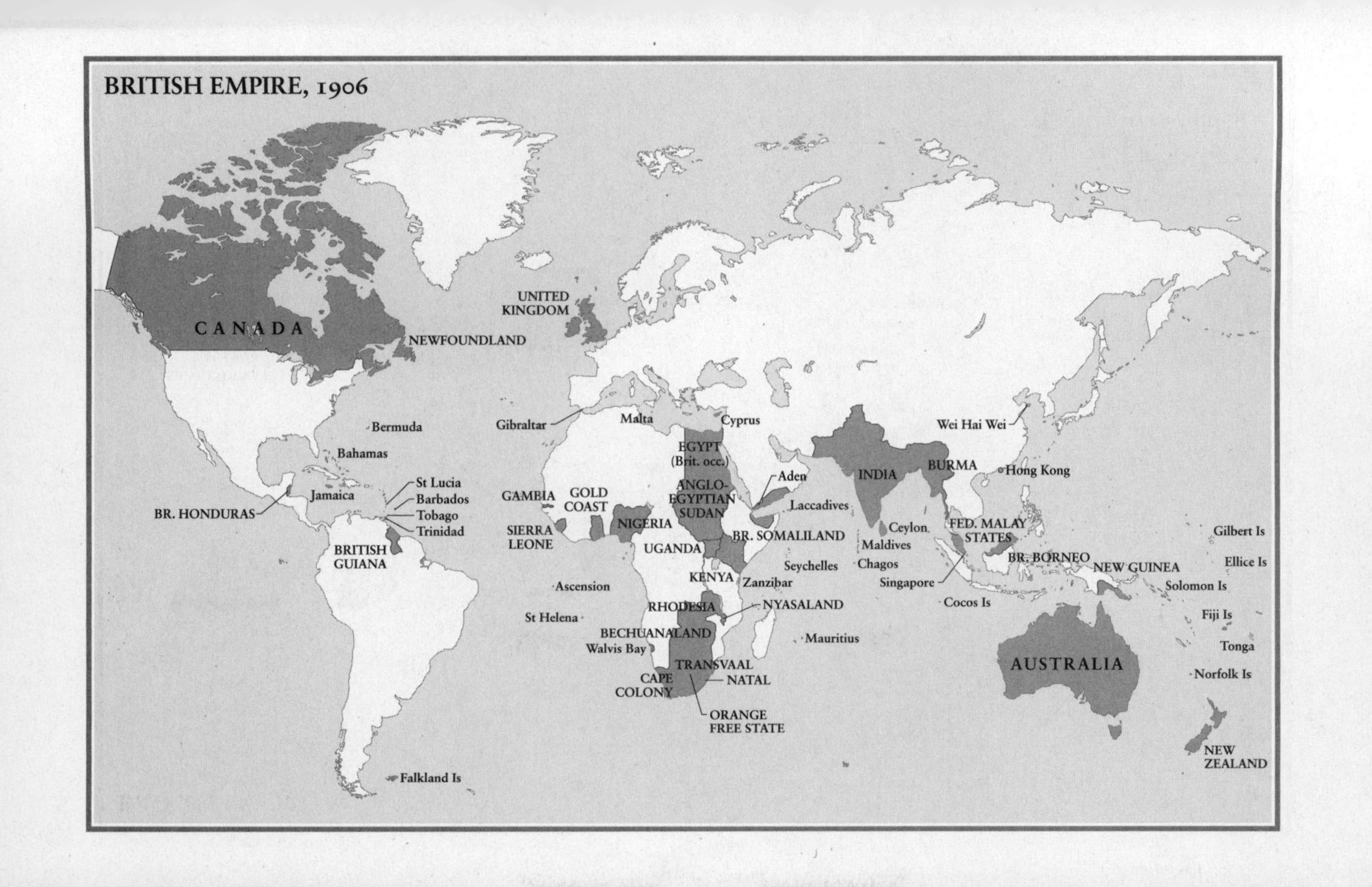
BRITISH EMPIRE, 1906
CANADA
NEWFOUNDLAND
UNITED KINGDOM
Bermuda
Bahamas
Jamaica
BR. HONDURAS
St Lucia
Barbados
Tobago
Trinidad
BRITISH GUIANA
Falkland Is
Gibraltar
Malta
Cyprus
GAMBIA
SIERRA LEONE
GOLD COAST
NIGERIA
EGYPT
(Brit. occ.)
ANGLO-EGYPTIAN SUDAN
Aden
BR. SOMALILAND
UGANDA
KENYA
Zanzibar
Ascension
St Helena
RHODESIA
NYASALAND
BECHUANALAND
Walvis Bay
TRANSVAAL
CAPE COLONY
NATAL
ORANGE FREE STATE
Seychelles
Mauritius
Laccadives
INDIA
Ceylon
Maldives
Chagos
BURMA
Wei Hai Wei
Hong Kong
FED. MALAY STATES
Singapore
Cocos Is
BR. BORNEO
NEW GUINEA
AUSTRALIA
Gilbert Is
Ellice Is
Solomon Is
Fiji Is
Tonga
Norfolk Is
NEW ZEALAND

Preface and Acknowledgements

[illegible] not know [illegible] at the time, but I was lucky [illegible] in [illegible], in many ways [illegible] of the [illegible] was [illegible] ely the [illegible] that [illegible] Britain was still a formidable [illegible] power, with a [illegible] navy and a [illegible] compulsory [illegible] national [illegible] mining, steel [illegible] engine [illegible] [illegible] p[illegible] [illegible] the [illegible] ship [illegible] and [illegible] eal, [illegible] vessels [illegible] steam engines, and [illegible] [illegible] Mason [illegible] [illegible] of the Birmingham where I grew up would have been instantly recognizable [illegible] who had been [illegible] and [illegible] in the [illegible] volume. This was especially [illegible] magnificent [illegible] buildings [illegible] Gallery, and Mason College [illegible] [illegible] 1880s [illegible] who had lived during their youth [illegible] father's parents [illegible] nineteenth-century [illegible] house in Birmingham [illegible] and [illegible] what was then known as the 'Black Country' [illegible] every year to Blackpool for [illegible] holidays. Both their houses

Preface and Acknowledgements

Although I did not know it at the time, I was lucky enough to grow up in what were in many ways the final years of the 'long' nineteenth century – namely the 1950s. At the beginning of that decade, the United Kingdom was still a formidable international power, with a large navy and a world-encompassing empire; the traditional staple industries remained coalmining, steel manufacturing, engineering and textiles; London, Liverpool and Glasgow were among the pre-eminent ports in the world, just as Cunard, Canadian Pacific, the Royal Mail Lines and P&O were some of the greatest global shipping firms; houses were warmed and factories were powered by coal, goods trains and passenger expresses were pulled by steam engines, and as a result, Holmes-and-Watson fogs were still a common occurrence. Much of the Birmingham where I grew up would have been instantly recognizable to Joseph Chamberlain, who had been by turns a reforming mayor, an advanced Liberal, and an assertive imperialist, and who will be a significant figure in the later chapters of this volume. This was especially true of the magnificent ensemble of public buildings surrounding the square that bore Chamberlain's name: the Birmingham and Midland Institute, the Reference Library (where I spent many happy hours), the Town Hall, the Council House and Art Gallery, and Mason College. Corporation Street was very much as Chamberlain had created it, the tramlines remained in many of the cobbled city-centre thoroughfares, and just a short distance away was Edgbaston, a quintessential Victorian suburb, where some of Chamberlain's descendants still lived. My four grandparents had all been born during the 1880s, and they seemed to me very old, and very Victorian; they invariably dressed in black, as if still in mourning for the Gas-Lit Gloriana who had reigned during their youth. My father's parents lived in a nineteenth-century terraced house in Birmingham with an outside lavatory, and my mother's parents, who lived in what was then known as the 'Black Country', went every year to Blackpool for their summer holidays. Both their houses seemed full of nineteenth-century novels and atlases, along with coins and stamps with images of Queen Victoria, who thus became a real presence in my life from an early age.

My four grandparents died between 1961 and 1970, breaking my personal link to Britain's nineteenth-century past. During that same decade, the United Kingdom itself began to de-Victorianize and embrace modernity, and this was vividly and literally brought home to me as much of Joseph Chamberlain's Birmingham was torn down, including many of its finest public buildings. Entire streets of the sort of houses in which my Cannadine grandparents had lived were demolished, to be replaced by offices, shops, hotels and flats. The grammar school that I attended was a creation of the 1880s, but it had left its original, inner-city home a few years before I became a pupil there, and it had been transformed into a concrete and plate-glass academy relocated on the edge of the city. But as much of Victorian Birmingham was disappearing before my very eyes, the serious study of the nineteenth-century past was simultaneously beginning in earnest; and having grown up during the last decade when much of it had still been real and visible, I was equally fortunate to live through the first decade of the twentieth century – the 1960s – in which a new generation of post-war scholars was bringing that rapidly vanishing world alive again. Pre-eminent among them was Asa Briggs: I can still recall the sense of excitement with which I read his history of Birmingham (appropriately in the Reference Library), covering the city's heroic years from 1865 to 1938; soon after I devoured *Victorian People* and *Victorian Cities*; and like many sixth-formers, I relied on *The Age of Improvement* to help me pass History A Level. Ever after, Briggs would remain an iconic and inspirational figure, as well as being in some ways a quintessential Victorian himself – in his boundless energy, his incorrigible optimism and his exceptional public-spiritedness. But by the 1960s, he was joined by many other pioneering scholars, including Robert Blake, Norman Gash, George Kitson Clark, Eric Hobsbawm, Harold Perkin, and two historians of Africa who were invariably known as 'Robinson and Gallagher', as if they were a firm of solicitors rather than a pair of academics.

So it was perhaps over-determined that as an undergraduate reading history at the University of Cambridge from 1969 to 1972, I should concentrate on nineteenth-century Britain, not only in terms of its domestic past, which in those days was deemed to consist exclusively of political and constitutional, and social and economic history, but also in its interactions with the greater world of the British Empire and Commonwealth and with what were deemed by some scholars to be its informal realms in Latin America. Thereafter, I resolved to investigate the relations between aristocratic landowners and the management of their estates bordering on large towns or the seaside, and Edgbaston, which had been developed as 'the Belgravia of Birmingham' by the Calthorpe family, seemed the obvious place to begin. Entirely by chance, I began my researches at just the time when Professor H. J. Dyos of the University of Leicester was defining and proclaiming the new sub-discipline of urban history – in part in response to the widespread demolition and destruction of large areas of Victorian cities that I had witnessed at first hand in Birmingham. Since then, scholarly approaches to the past have widened in many ways and diversified in many directions, with the rise of women's history, cultural history and the 'new' imperial history, while the IT revolution has made available many new sources, which means it is now possible to research topics that were impractical and unthinkable a quarter of a century ago. As a result, the study of the past has become a much more varied, complex, multi-layered enterprise, and that has certainly been true of the British nineteenth century, about which more scholars are writing more kinds of history than ever before.

But this welcome and prodigious expansion in academic labour also has its downside, as it is now impossible for anyone to keep up with the unrelenting outpouring of academic literature in the way that was still practicable when I was starting out in the early 1970s. So much learning, so much erudition and so much information has undoubtedly extended and enriched our knowledge in ways that were unforeseeable and unimaginable several decades ago. But this has also had a dampening and deadening effect, which may explain why, in recent decades, the focus of scholarly interest and excitement has moved backwards, to a rejuvenated eighteenth century, where Namerite torpor and Thompsonian simplicities have alike been banished, and forward to the twentieth century where the opening of official records and other archival collections means it has become a fertile and compelling field, in which new discoveries are constantly being made. By contrast, the British nineteenth century has for some time been in what Miles Taylor calls 'a state of suspended animation', and he believes it needs to be 'brought back to life'. This book is one such attempt at historical resurrection and scholarly resuscitation. As with the other volumes in the *Penguin History of Britain*, it is primarily a political history, with the politics imaginatively understood in the broadest possible way, encompassing many activities that, at first glance, may not seem conventionally 'political' at all. It also attempts to tell that history – or, more accurately, those many parallel and interlinked histories – in essentially narrative form, in the process aiming to restore to the subject something of the brio of those earlier, pioneering times. Moreover, the long years that have elapsed since I signed the contract for this book have witnessed the rise of global history as another significant sub-specialism, and this has helped to de-parochialize the nineteenth-century British (or English) past, while at the same time relocating the all too often self-enclosed history of the British Empire in a broader international context.

In writing this book, my first thanks are to those many historians who, since 1945, have written so much about the British nineteenth century, and whose works I have plundered and pillaged in the pages that follow. I have acknowledged some more specific debts in the bibliographical essay at the end of this book, but anyone who has tried their hand at any such work of panoramic synthesis knows how much they owe to the giants of an earlier time on whose shoulders they stand, and to past and present colleagues and co-workers in the same field. More particularly, I am grateful to Derek Beales, the late Christopher Platt, Ronald Hyam and Ged Martin who taught me nineteenth-century British, imperial and Commonwealth history when I was an undergraduate at Cambridge, and to the late Peter Mathias and the late Jim Dyos who respectively supervised and examined my Oxford doctoral dissertation. Since then, I have professed and practised modern British history at Cambridge, Columbia, London and Princeton Universities, and I have learned more from my doctoral students and personal assistants than I fear I have ever taught any of them. Much of this book was drafted while I was a Visiting Professor at Stern Business School, New York University, and I am grateful to Professor George Smith for having made that arrangement possible, and to him and his colleagues for having welcomed me so warmly. Earlier versions have been read by my good friends Jonathan Parry, Bill Lubenow, Stephanie Barczewski,

Michael Silvestri and Martha Vandrei, and I am much in their debt for their comments and corrections. Eve Waller brought her careful and critical eye to bear on the earlier chapters of this book, and greatly to their benefit. And as fellow historian, departmental colleague, adored spouse and best friend, Linda Colley has done more than anyone to make this book possible and to make it happen.

I am also once again beholden to my long-standing friend and long-suffering editor at the Penguin Press, Simon Winder, who has shown extraordinary forbearance in waiting so patiently for this book, and who has seen it into print with consummate professionalism and unfailing cheerfulness. He has been ably assisted by Maria Bedford, to whom I also render thanks, along with Ingrid Matts and Pen Vogler for their work in publicizing and promoting the book. I am also indebted to Cecilia Mackay for her inspired research on the illustrations, to Richard Mason for his meticulous and painstaking copy-editing, and to Dave Cradduck for compiling the index. I dedicate this book to the memories of that great historian who did so much to bring the British nineteenth century alive, for me and for so many other people, and to a great publisher who commissioned this book and the series to which it belongs. How I wish that they were both still here to receive in person my homage and my thanks.

DNC
Princeton
St George's Day 2017

Prologue

History is not just about dates and events, which are often arbitrary and accidental: it is at least as much about processes, which do not begin and end with any such tidy temporality or calendrical precision. But dates *do* matter, because important events happen at particular times, and this explains why many histories of nineteenth-century Britain begin with the Battle of Waterloo in 1815 and end with the outbreak of the First World War in 1914. Such were the defining years of Elie Halévy's pioneering but incomplete *History of the English People in the Nineteenth Century*, of David Thomson's brief and bracing volume in the post-war *Pelican History of England*, of the two volumes contributed by Sir Llewellyn Woodward and Sir Robert Ensor to the *Oxford History of England*, and of the works by Norman Gash and E. J. Feuchtwanger for the later series published by Edward Arnold. Even today, in light of the recently observed bicentennial of the Battle of Waterloo, and of the hundredth anniversary of the Battle of the Somme, the idea that the British nineteenth century was defined by the triumphant conclusion of the Revolutionary and Napoleonic Wars (which ushered in the century of Britain's global greatness), and the less sure-footed beginning of what was originally known as The Great War (after which Britain ceased, according to Sellar and Yeatman in *1066 and All That*, to be 'top nation'), has much to recommend it. In between these two massive continental conflicts, so this argument runs, the United Kingdom largely (but not entirely) avoided European military entanglements, and as a result enjoyed a remarkable period of peace and prosperity at home, and of engagement and hegemony elsewhere in the world. Hence the establishment of the so-called 'Pax Britannica', the widespread contemporary belief that God conversed in English, and the undeniable reality that more people were speaking that language, and doing so in more parts of the world, than had ever been true before.

But the very fact that so many distinguished historians have already opted for these beginning and end dates is itself one very good reason for trying to define and delineate the British nineteenth century differently. Other scholars have begun their histories in 1783, with the rise of the Younger Pitt to power, or continued on until 1918, with the close of the First World War, and these more spread-out years certainly allow for a fuller treatment of Britain's 'long' nineteenth century. But in addition, and as Jürgen Osterhammel has recently pointed out, there are many other ways of dating and defining that period, depending on the topics selected, the themes chosen, the subjects to be treated, and the part of the world to be concentrated on, while the deeper processes of change defy and deny any such precise temporal pinpoints. This book seeks to break new ground, and to offer new perspectives, by beginning the British nineteenth century with the passing of the Act of Union in 1800, which brought into being the United Kingdom of Great Britain and Ireland, and by ending it with the general election of 1906, which witnessed a landslide Liberal victory that was the last great triumph of nineteenth-century progressive politics, but also brought to power the first great reforming government of the twentieth century. In 1800, and again in 1906, some contemporaries hoped, and others feared, they were living through revolutionary times and witnessing revolutionary events. But they might all have agreed that those two dates and those two events had been and were of the first importance. Yet so far as I know, no one has attempted a history of nineteenth-century Britain bracketed and bounded in this way, which is as good a reason as any for trying to do so. Whether this enterprise has been worthwhile, 'these pages must show' (as Dickens wrote at the opening of *David Copperfield*), or hope to show.

These two dates also serve to remind us of the extraordinary dominance and unique continuity of parliament in the political culture and public life of the United Kingdom and the British Empire. Across those years, the Westminster legislature was, with all its faults, drawbacks and limitations, to which reformers and radicals often drew attention, a uniquely enduring institution of political authority, government legitimacy, popular sovereignty and national identity – in ways unmatched in Spain or France (where there were absolute monarchs, revolutions and republics), the United States (its democracy ruptured by civil war and the attempted Southern secession), Austria-Hungary (both nations' parliaments only established in 1868), Italy or Germany (neither country completely unified until 1871), Japan (without a constitution before 1889), Russia (without a Duma until 1905) and China (without a constitution before 1913). Small wonder, then, that both the permanence and the adaptability of the British constitution and the British parliament were acclaimed and envied by many commentators from overseas. They were also lauded and celebrated in Britain itself, as the franchise was peacefully and progressively extended in 1832, 1867 and 1884–85, thereby successfully fending off any potential revolutionary threat. The constitution of the United Kingdom may have been unwritten, but as Edmund Burke had earlier appreciated, that had turned out to be a huge advantage: for it could be constantly adapted and adjusted, to take account of the extraordinary and transformative changes of the nineteenth century, whereas nations with rigid and inflexible written constitutions all too often had to embrace revolution, tearing everything up and starting again, while nations with no constitution at all also had to resort to similar violent means and desperate measures to obtain one.

In terms of its institutions of government and authority, then, nineteenth-century Britain was uniquely stable among the nations of the world. To be sure, there was nothing preordained or inevitable about that stability, and fears of revolution would not be confined to the years 1800 and 1906. But the undeniable continuity of its governing structures is a very good reason for approaching the history of the United Kingdom via its parliament and its politics, which became the embodiment and expression of that stability, and for beginning and ending that history with events that were, appropriately, both parliamentary and political. Moreover, the United Kingdom was not only remarkably stable in terms of its unwritten constitution and its institutions of government, but also in geographical terms, avoiding armed invasion, enemy occupation and the forced loss of lands, which were often the fates of continental countries. In 1906 Great Britain and Ireland encompassed precisely the same political boundaries and national borders that had been established in 1800. This was not true of Germany or Italy (which were nineteenth-century creations), Austria-Hungary (which had forfeited its Italian provinces), France (which had ceded Alsace-Lorraine to Germany), or Russia (which gave up Port Arthur to

Japan), or of the United States (which was not so much losing territory as spreading across an entire continent). To be sure, there were 'invasion scares' in the United Kingdom in the 1790s and 1800s, at mid-century, and again during the 1900s; but uniquely among the great nations of the world, its boundaries remained unchanged and unchanging throughout the nineteenth century. This combination of continued and unchallenged parliamentary supremacy, and successfully preserved territorial integrity, was remarkable and important, and lent some credence to the contemporary view that Britain was unique, exceptional and providentially blessed.

Yet for all its constitutional continuity and geographical cohesiveness, the nation over which parliament and the politicians, rather than the monarch, exercised sovereign power (though George III, George IV, William IV and Queen Victoria and Prince Albert would all have contested this assertion) was a complex, contested and composite place. England and Wales had been unified since Tudor times; England and Wales, together with Scotland, had been united as crowns in 1603, and as legislatures in 1707; and the union between Great Britain and Ireland followed just short of a century later, which meant, incidentally, that in its final and fullest form, the United Kingdom was a more recently created nation than the United States of America. When the mid-Victorians described their legislature as the 'imperial parliament', they meant that it passed laws for all the four kingdoms, which were appropriately represented in the central lobby of the recently constructed Palace of Westminster, where mosaics depicted their respective patron saints: St George, St David, St Andrew and St Patrick. Moreover, the United Kingdom of Great Britain and Ireland would turn out to be in many ways an unsatisfactory and unstable creation, as England was overwhelmingly dominant in terms of population, wealth and resources, and as the three nations of what was disparagingly regarded as 'the Celtic fringe' came on occasions to resent or repudiate such an unequal and asymmetrical Union. This was especially so in the case of Ireland, which was not only the last nation to be assimilated but also the least successfully, and (with the exception of the six Ulster counties) for the shortest span of time. This, in turn, had major implications for the politics of the nineteenth century, as the Tories and Conservatives would use their customary electoral dominance in England to try to impose their will on the rest of the Union, whereas the Whigs and

Liberals would need the votes of Ireland, Scotland and Wales to enable them to dominate England.

Even as the jurisdiction of the parliament of the United Kingdom remained constant over England, Ireland, Scotland and Wales during the nineteenth century (with differing degrees of assent and acquiescence), the areas beyond Britain's shores where Westminster also claimed control and asserted its sovereign power spectacularly expanded from the time of the Napoleonic Wars onwards. In 1858 parliament finally declared its full authority over Britain's Indian Empire, having abolished the East India Company. Nine years later, it consolidated the scattered realms of British North America into what would be the senior imperial dominion of a confederated Canada. In 1900 it brought together the equally dispersed colonies of the Australian continent into another federated dominion; and nine years later, further London legislation would establish the Union of South Africa, binding together Cape Colony, Natal, the Transvaal and the Orange River Colony in what would be a final (and, as it turned out, flawed) act of nation-building within the British Empire of settlement. Across the nineteenth century, parliament also assumed the ultimate responsibility for many additional territories, especially in Asia and Africa, initially acquired from European colonial rivals, and subsequently annexed pre-emptively. So when the Liberal government took charge in December 1905, there were three cabinet ministers who were responsible to parliament for different jurisdictions of the British Empire, namely the Secretaries of State for Foreign Affairs, for India and for the Colonies (as well as the Chief Secretary for Ireland, who oversaw what was in practice the semi-colonial Hibernia); and in 1914, London would declare war against Germany on behalf of the whole of the empire, none of whose constituent parts had any say in the matter. For good or ill, indeed for good *and* ill, British history took place in more parts of the world during the nineteenth century than ever before.

But although there was much about the experience of nineteenth-century Britain and its empire that was exceptional, a great deal that went on in the United Kingdom was merely one nation's version and one imperial iteration of what was happening in many other parts of what was then described as the 'Western world'. The growth of population, industrial transformation, the expansion of cities, the construction of railways, libraries, museums and concert halls, the advent of electricity, film and the internal combustion engine – all these were widespread developments and transformative changes. So were the re-ordering of nations and political unions, the extension of the franchise, the growth of organized political parties, the advent of a new breed of charismatic politicians such as Cavour, Lincoln and Gladstone, the development of organized labour, the rise of an increasingly intrusive state, the pressures and claims of ethno-linguistic nationalism, and the allure and appeal of imperialism. Equally transformative was the expansion of the press, the enrichment of literary and cultural life, the intensification of scientific enquiry, the growth of religious scepticism and doubt, and the (limited) spread of education, all the way from the provision of elementary schools to the founding of universities. These improving trends, long-term developments and (generally) positive processes were to be found across the Western world, from the United States, via much of Europe, to parts of Russia and on to Japan. From this perspective, the history of the United Kingdom between 1800 and 1906 was far from being unusual and unique; but where it was so exceptional was that it somehow managed to remain more stable, politically and constitutionally, amidst so much change, than any other country in the world.

All this is but another way of saying that the British nineteenth century encompassed many geographies and comprised many histories that were often heading in contradictory directions. Looked at one way, it was an era of unprecedented national prosperity and global greatness, yet it was also a time of unexampled domestic misery, urban squalor and environmental degradation, and these contradictory developments undoubtedly impacted on the government and politics of the time. From one perspective, it was the triumphant century of Trafalgar, Waterloo, the abolition of the slave trade, the Great Exhibition and the Diamond Jubilee; but from another it was a hundred years debased and demeaned by the Peterloo Massacre, the Hungry Forties, the Great Famine in Ireland, the Great Rebellion in India, the 'Scramble for Africa' and the Boer War. In one guise, it was a century of extraordinary cultural enrichment and scientific creativity, which both reflected and transformed life and thought and politics: of Blake and Constable, Turner and Morris, Pugin and Waterhouse, Ruskin and Whistler; of Austen and Dickens, Carlyle and Eliot, Wordsworth and Trollope, Wilde and Tennyson, Kipling and Wells; of the Brunels and the

Stephensons, Newman and Manning, Banks and Davy, Darwin and Kelvin. But it was also the time when most people's existences were indeed nasty, mean, brutish and short, when starvation and unemployment were commonplace, when the finer things of life were simply unavailable to many ordinary men and women, and when popular protest often seemed the only option. Yet these were also the years when many of the things that we still take for granted today first came into being, among them stamps, photographs, bicycles, football, telephones, sewers, nurses, policemen, detectives, department stores, matches, museums and galleries, redbrick universities, restaurants, detective novels, bacon and eggs, golf, tennis, the National Trust and the old school tie. In some ways, even today, the British nineteenth century is still not yet over.

But although it is close enough to our own day to be instantly recognizable and comfortingly familiar, the British experience between the 1800s and the 1900s was in other ways very unlike ours in the early twenty-first century. For much of the time, it was an era of national greatness, global reach and imperial aggrandizement, which in our devolved, downsized, post-imperial, post-Brexit Britain, is experientially unknowable and imaginatively all but irrecoverable. It was a civilization underpinned by religious faith and a strict moral code, which may strike those with today's more secular mindset and liberal sensibilities as intolerant, intolerable and incomprehensible. For much of the nineteenth century, Britain witnessed continuous constitutional experimentation, which may, paradoxically, have been one of the reasons it was so stable a polity; the same would not be true during the twentieth century. The Victorians created and accumulated wealth in unprecedented abundance, as their heavy industries led the world, and as they invested prodigious sums overseas: we do not do much of this sort of thing now. It was a time when cities such as Manchester, Leeds and Birmingham reached the zenith of their fame, freedom and influence, by comparison with which today's urban jurisdictions are demoralized and largely dependent on handouts from central government. And it saw the construction of canals and railways, houses and factories, harbours and bridges, dams, sewers and irrigation schemes, ships and underwater telegraphs, all built with a vigour, energy and dynamism that the twenty-first century cannot rival. Thus regarded, nineteenth-century Britain was so unlike our own diminished times and limited horizons that it seems much further distant than the hundred-odd years that separates us from the death of Queen Victoria.

Yet this is no invitation to escapist nostalgia, for these were not the only ways in which nineteenth-century Britain was unlike our contemporary world. It was also a society which, until the early 1830s, tolerated slavery across the British Empire, and the employment of children in factories and down the mines at home; where, fifty years later, there were alleged to be 80,000 prostitutes in London, and where, according to the social investigator Charles Booth, one-third of the inhabitants of the greatest city on the globe were living in poverty, which to those with social consciences was unacceptable amidst so much plenty. It was a society where life expectancy was scarcely half of what it is today, where most people received only a minimal education, and where state pensions, a national health service and antibiotics were unknown. It was a society where women were legally subordinate to men and could not vote in parliamentary elections, where homosexuality became a crime, and where divorce was difficult to obtain and almost invariably spelt social disgrace, even for the innocent party. It was a society where there was growing inequality between the rich and the poor, which assumed the majority of people and all women were unfit for the vote, and which took for granted that those with white skin were generally – though not invariably – superior to people of colour. From this perspective, nineteenth-century Britain was not so much a place of great accomplishments and wholesome values to which we should aspire to return: rather it was a place of widespread vices and flawed accomplishments from which we should be glad that we have eventually escaped. Both views are valid; neither is the whole truth of things.

Here, then, lie the challenges and the opportunities in attempting any new treatment of nineteenth-century Britain: to describe with equal force and conviction the elements of sameness and similarity, and also the elements of strangeness and surprise; to evoke the energy, the talents, the achievements, the optimism, the excitement, the sense of limitless possibilities, while also recognizing the darker sides of life; and to do full justice to the global reach of the history that Britons made, without neglecting what was happening at home. On the one hand, and as Tristram Hunt has rightly noted, the nineteenth century was 'a terrible, fascinating and creative age, one that deserves greater appreciation than the twentieth century ever provided'. But it also seems greedy, hypocritical, snobbish, repressed, vulgar, uncaring, complacent, bigoted, intolerant, aggressive and jingoistic, even as nineteenth-century Britons also believed, and with some good cause, that they belonged to the most advanced country and the finest civilization on the globe. And all the while, they were also anxious, uncertain, doubting and insecure – about themselves, their society, their economy, their religion, their nation and their empire – and with equally good reason. Such were the many contradictions of progress, manifested by a nation that for much of the nineteenth century would undeniably be at the summit of the world, yet which was never entirely confident or convinced that it should be there, or that it would remain there. Perhaps, in the end, it was all something of an accident, or an illusion as, for a relatively short span of time, two recently united islands situated off the coast of mainland Europe briefly achieved industrial supremacy and imperial pre-eminence, and as a result came to wield for much of the nineteenth century a wholly disproportionate influence over the affairs and the territories of the world.

1

Act of Union, 1800–02

The union of England and Wales with Scotland that had been carried in 1707 had created a polity virtually coterminus with the island of Great Britain, but in 1800, the national boundaries of the United Kingdom were redefined and extended to encompass the neighbouring landmass of Ireland. The separate Irish parliament grudgingly voted itself out of existence, its members bribed with money and peerages from London. One hundred Irish MPs would in future join those already representing England, Wales and Scotland in the British legislature at Westminster; and four Irish bishops and twenty-eight Irish peers would henceforward be able to sit in the House of Lords. Accordingly, on the first day of January 1801, Ireland vanished as a separate nation, and the expanded and consolidated United Kingdom that came into being would endure, albeit with misgivings and challenges, disapproval and protest, for the whole of the nineteenth century and on, indeed, until the aftermath of the First World War. This Act of Union was driven through in Ireland itself by the Lord Lieutenant (or Viceroy), the Marquis Cornwallis, and in Britain by the First Lord of the Treasury (and de facto prime minister), William Pitt the Younger. This was the same Cornwallis who had previously commanded the British forces in America, where he had vainly attempted to subdue the rebellious colonists but had surrendered at the Battle of Yorktown in 1781. From 1786 to 1793 he had been a successful and innovative Governor General of British India, and Pitt had sent him to Ireland five years later tasked with getting the Irish elite to agree that their nation should cease to be a separate political entity. Pitt had been continuously in power since King George III had appointed him in 1783, and his long period in office was dominated by the fallout from two epochal events that took place on opposite sides of the Atlantic: the

American Revolution of 1776, and the French Revolution of 1789, and it was to the second of these wrenching traumas that the Act of Union, linking Great Britain with Ireland, was a delayed but determined response.

Indeed, determination, along with fortitude and resolve, would prove the hallmarks of Pitt's (first) administration, as evidenced by the fact that it lasted until 1801. Although he was only twenty-four years old, Pitt had been hand-picked by George III in the hope he might provide a more congenial government than the cynical and chaotic coalition of Charles James Fox and Lord North that the monarch had detested. Pitt might also restore the nation's finances and revive national morale in the aftermath of the loss of the American colonies, formally recognized at the Treaty of Paris which had been signed in September 1783, just three months before he took office. Defeat in the War of Independence had been a terrible national humiliation, for it had torn apart the so-called 'First British Empire', which Pitt's father, the Elder William, had so brilliantly extended in the Caribbean, North America and South Asia during the Seven Years War. The American revolutionaries had rejected the authority of the British crown and parliament, had proclaimed instead the seditious doctrine that all men (but not women) were 'created equal', had rent asunder the trans-Atlantic, Anglo-Saxon polity, and had rejected the traditions and hierarchies of 'Old Europe' by abolishing all hereditary titles. So it was scarcely coincidence that at just this time Edward Gibbon had begun publishing his multi-volume *Decline and Fall of the Roman Empire*, in which he chronicled the collapse of an earlier great imperium from the high peaks of civilized greatness and power into decadence, impotence and barbarism. Then came the French Revolution of 1789, which eventually overthrew the *ancien régime* of the Bourbons, and in the name of 'liberty, equality and fraternity' a radicalized citizenry abolished the monarchy, executed King Louis XVI and Queen Marie Antoinette and many of their aristocratic friends, and confiscated their properties along with those of the Catholic Church. Not since the English had beheaded their king in 1649 had the fabric of a European polity been so violently ruptured: the French Revolution was far more radical than its American precursor, and it represented an unprecedented and more immediate threat to Britain – and also to the British connection with Ireland.

A NECESSARY BUT UNCONSUMMATED MARRIAGE

To be sure, in the short run, many Britons had not recognized or foreseen this, as they had enthusiastically embraced the ideas and ideals of 1789. The Whig leader, Charles James Fox, who was the Younger Pitt's sworn enemy, called the French Revolution 'the greatest event . . . that ever happened in the world, and . . . much the best'. The poet William Wordsworth declared that 'Bliss was it in that dawn to be alive / But to be young was very heaven!' Thomas Paine, who had earlier gone to the American colonies to support their revolution, published *The Rights of Man* (1791), calling for the British to abolish monarchy, aristocracy and the House of Lords, urging that a republic was the best form of government, and that all men should be given the vote. Mary Wollstonecraft went further: for not only did she defend the French Revolution in *A Vindication of the Rights of Men* (1790), but she also advocated political equality for her own gender in *A Vindication of the Rights of Woman* (1792). At the same time, new, radical political societies, claiming thousands of members, and involving skilled working men in large numbers for the first time, sprang up in towns and cities across Britain, embracing Paine's subversive doctrines and corresponding with similar organizations in France. They were viewed with deep suspicion and distrust by the government, and such anxieties seemed amply borne out when in 1792 the French envoy to London, Chauvelin, ostentatiously received deputations from the Norwich Revolutionary Society, the Manchester Constitutional Society and the London Corresponding Society. In the summer and early autumn of 1795 protesters smashed the windows of Pitt's residence at 10 Downing Street, a crowd of about 200,000 people hurled abuse at the premier and his sovereign as they rode to the opening of parliament, and radical activists were organizing monster meetings across the country. The second half of the decade was even worse, characterized by steeply rising food prices and high unemployment, which reached a crisis point in 1800. There had been a succession of very poor harvests, there was widespread rioting and protest, and there were many rumours and fears that there would be armed insurrections, and that Britain would soon be facing its own revolution.

This domestic subversion seemed all the more sinister and ominous to the authorities because in February 1793 republican France had declared war on royal Britain, and these anxieties were further reinforced when the revolution lurched savagely towards destructive authoritarianism, as Robespierre's 'Reign of Terror' began later in the year, and 16,000 French men and women were eventually sent to the guillotine. To be sure, France had long been Britain's 'traditional' enemy, and the sequence of conflicts that had begun in the late seventeenth century, initially fought between William III and Louis XIV, amounted to what was in effect a 'second Hundred Years War' between the two cross-Channel countries. Moreover, France had joined in the American War of Independence on the side of the colonists, and it had thereby obtained revenge for the defeats that Pitt the Elder had inflicted during the Seven Years War. But revolutionary France was a different nation from royal France, and it was widely believed during the remainder of the 1790s that it sought to invade and defeat Britain, and also to subvert the established social and political order from within – and not only in Britain but in Ireland as well. The war that began in 1793 was thus not only a conflict of arms, but also of ideology, and the British authorities feared that the two would converge, as sedition and subversion spread in the Royal Navy, which was Britain's final bulwark against invasion. They duly came together in 1797, when there were mutinies in the Channel fleet anchored at Spithead off Portsmouth and at the Nore at the mouth of the Thames, and there were further disruptions within British squadrons in the West Indies and the Cape of Good Hope, and in the fleet commanded by Admiral Sir John Jervis off the coast of Spain. All were brutally suppressed. But this brought little respite, for later in the same year, the young Napoleon Bonaparte was put in charge of the ominously named 'Army of England'. It was located along the French coast either side of Boulogne, which meant Britain then faced a serious prospect of invasion and conquest by a continental enemy power.

Pitt responded to these increasing threats and challenges by building up a network of spies and informants (who may have exaggerated the potential dangers and threats to order) and by passing a series of repressive measures and instituting a succession of harsh prosecutions that were denounced by their victims as ushering in Britain's own 'Reign of Terror' (though it was scarcely comparable to what was occurring in

France). In 1794 his government suspended habeas corpus, which had guaranteed to all Britons that they would not be imprisoned without trial, and soon after, the leaders of the London Corresponding Society, the largest of the new, radical organizations, were charged with treason on the grounds that their plan to hold a convention to demand universal male suffrage was tantamount to sedition. But they were acquitted by an independent-minded London jury, and to widespread popular delight. Pitt retaliated by passing two further repressive measures in the following year: the Treasonable Practices Act declared it unlawful to 'imagine, invent, devise or intend' the death or destruction of the monarch, while the Seditious Meetings Act limited the size of public gatherings to no more than fifty people. But popular protests continued across the country, and members of the corresponding societies continued to meet, to which the government responded by passing the Combination Acts of 1799 and 1800, which effectively made any form of association illegal, whether for political purposes or in pursuit of improved pay and conditions of work. Local militias were widely deployed, and Pitt was so alarmed that he brought troops back from the continent to try to assure order. The Britain of the late 1790s was thus a nation ill at ease with itself, and whereas Pitt's repressive policies seemed to his supporters necessary to maintain order, they appeared to his critics excessive and paranoid. Either way, it was against this uncertain, febrile and anxious background that Pitt passed the Act of Union incorporating Ireland into the United Kingdom.

From this vexed and anxious late eighteenth-century British perspective, Ireland's problems were a complex and confusing amalgam of the economic, sociological, religious, political and geographical. The country was dominated by a small, landowning, Protestant ruling class, most of whom were settlers of English or Scottish ancestry, known as the 'Ascendancy', and they controlled the Irish parliament and the country's political and social life. Their forebears had conquered Ireland in the seventeenth century, but by the late 1790s many of them were absentees living in Britain, and their relations with the indigenous majority of the Irish population, who were generally poor, ill-educated Catholic peasants or agricultural labourers, and who were denied civil rights on the grounds of their religion, were far from close or cordial. There was also a small Presbyterian middle class that was a significant force among the legal and commercial circles of Belfast, and a larger

Presbyterian working class in the more economically developed parts of Ulster. These dissenters disliked the Catholics, whom they regarded as inferior and uncivilized, and also the Ascendancy families, whom they disdained as effete, snobbish and overbearing, and they were equally hostile to the established, Protestant Church of Ireland. It was not easy for the British ruling elite to deal with any of these three groups. As the Gordon Riots in London had shown as recently as 1780, there was a deep prejudice and ingrained hostility in Protestant Britain towards Roman Catholics and to 'popery'. Although equally anti-Catholic, the Presbyterian middle and working classes were at best ambivalent about the British connection, and resented the fact that, like the Catholics, they were prevented from holding certain political and administrative offices. Meanwhile, the London men of government were equally uncertain about the Ascendancy families: they might be Britain's only reliable ally as an 'Irish garrison'; but their 'alien' presence and power incurred the animosity of the Catholic majority, which meant that in any crisis they would lack popular legitimacy.

Since the passing of what was termed 'Poynings' Law' in 1494, the Irish parliament had been legally subordinated to the English (and subsequently the British) legislature, and this had been confirmed by the passing of the Declaratory Act in 1720. It was rarely invoked in practice during the eighteenth century, as the British knew that co-operation with the Ascendancy was essential; but the Anglo-Irish elite still bridled at such quasi-colonial inferiority as, indeed, did the Presbyterians and Catholics. This inferior status was further proclaimed by the presence at Dublin Castle of the Lord Lieutenant, who was the agent and representative of the British monarch and cabinet. But when France entered the American War of Independence on the side of the American colonists, the British government had been forced to transfer the troops stationed in Ireland to other parts of the empire where they were more urgently needed. Fearing a possible French invasion, the Irish Protestants established units of so-called Volunteers, who were not only drawn from Ascendancy families, but in some cases from Presbyterians and even from Catholics as well. There may have been as many as 40,000 of them. They were fully armed, and although undoubtedly hostile to the French, they were also determined to extract concessions from the British. By 1782, when the War of Independence was as good as lost, and with his government discredited, Lord North felt compelled to grant the Irish parliament autonomy from Westminster. But the Younger Pitt, who came to power the following year, feared that the newly liberated legislature lacked legitimacy, because the majority Catholic population were unable to vote or to hold public office. Accordingly, in 1793 he had pressured the Irish parliament to enfranchise Catholics possessing what was termed a 'forty-shilling freehold'. The Ascendancy was stunned and resentful, the Presbyterians annoyed, and the Catholics unreconciled because they were still denied the right to sit in parliament or hold high office. Such British interference in Irish affairs, which seemed to be undermining the legislative autonomy that had ostensibly been granted in 1782, was also widely resented.

The result was growing popular discontent, directed at London rather than Dublin, which was further fuelled by the French Revolution, and which would represent a greater threat to the metropolitan governing elite than the subversive mobilization and organization that were occurring simultaneously in Britain. In October 1791 the Society of United Irishmen was founded, with strong Catholic and some Protestant support, and led by Theobald Wolfe Tone, who had been greatly influenced and inspired by recent events in France. The original intention of the United Irishmen was to agitate for constitutional reform, but by the middle of the decade their failure to make progress in the face of government repression led them to demand complete separation from Britain and to embrace republicanism. The French National Assembly sent messages of support, Wolfe Tone was well received in Paris, and the revolutionary government sent agents to Ireland to assess the potential for a full-scale rebellion. The United Irishmen also sought to collaborate with a secret society known as the Defenders, whose members were drawn largely from the Catholic peasantry, and who resented having to pay tithes to the established Church of Ireland. Elsewhere in Europe peasants were rarely among the revolutionary vanguards of the 1790s, but the Defenders resorted to rural violence to intimidate the Protestant landlords and farmers, and embraced republicanism. In response, the Protestants established their own society in Ulster in 1795, known as the Orange Order, which sought to retaliate against the Defenders' disruptive activism. The resulting sectarian bloodshed, combined with the upsurge in anti-British agitation, the growing demands for a republic, and the undeniable evidence that the

United Irishmen and the Defenders were eager to ally with France, meant Pitt's government felt compelled to act. Between 1795 and 1796, legislation was passed imposing harsh punishments on the United Irishmen and the Defenders, banning popular assemblies, increasing the powers of magistrates, declaring the taking of oaths illegal, and suspending habeas corpus. At the same time, the Dublin Castle authorities established a new armed force named the Yeomanry, led by Protestant landlords and manned largely by their very own Protestant tenants.

The London government's anxieties that Ireland was on the brink of a massive, French-supported revolutionary uprising, which would be embraced by Catholics and Protestants alike, were well borne out when, in December 1796, a fleet of forty-three French ships, carrying 15,000 soldiers, came close to landing on the west coast of Ireland at Bantry Bay and was only prevented from doing so by stormy weather. A mere 400 French troops managed to land, and they were soon killed or captured by the local Yeomanry. The commander of the British forces, General Lake, retaliated with reprisals that were especially savage in Ulster, which only served to alienate many Protestants and Presbyterians. Meanwhile, the United Irishmen had set up a Directory in Dublin, modelled on the French revolutionary government in Paris, and in May 1798 thousands of rebels took up arms in the Irish capital and to the south and west of the country, with the cry of 'Death or Liberty'. By the end of the month, the rebels had established themselves in County Wexford, where they declared a republic and established a provisional government consisting of four Protestants and four Catholics. In June the Ulster United Irishmen, most of whom were Presbyterians, rebelled in County Antrim, and in August 1798 a French naval force landed at Killala in County Mayo on the northwest coast of Ireland. But it was too little, too late, since by then the Wexford rebels and the Ulster rebellion had both been brutally suppressed, and thousands of Protestants and Catholics had been killed in some of the bloodiest and most ferocious fighting in Ireland's history. In October British naval forces intercepted a French ship carrying Wolfe Tone to Donegal with the aim of fomenting a further rising. He was arrested, tried and sentenced to death, but committed suicide before he could be hanged. The Irish rebellion, which had been the most significant uprising in the British Isles since the Civil Wars of the mid-seventeenth century, was over, and it was put down as harshly as the naval mutinies in the previous year: many of the leaders were executed on the spot, and several hundred Defenders and United Irishmen were transported to Australia.

With widespread subversion in Britain, the threat of further rebellion in Ireland, and the risk of another French invasion on Britain's vulnerable western flank, Pitt and his colleagues determined at the end of 1798 to consolidate the two kingdoms into one single polity. The Union that eventually came into being in January 1801 was thus essentially a defensive wartime measure, in response to recent political and military events on both sides of the Irish Sea, rather than the culmination of some inevitable historical process, or the outcome of a long premeditated plan whereby the two nations were eventually bound to be conjoined. As such, it was also but one early example of the sort of constitutional rejigging and geopolitical rearrangements between governments, peoples and nations that would frequently take place elsewhere in nineteenth-century Europe. Yet in ways that could not have been foreseen in such troubled and traumatic times, this hurriedly created United Kingdom of Great Britain and Ireland would become the single most influential player on the nineteenth-century global stage, while Irish affairs would loom larger in British political, parliamentary and public life than they had done during the eighteenth century. But while the Act of Union might have made the British feel more secure in fighting subversion at home and in prosecuting the war against France abroad, it did not make Ireland any easier to govern, or the Irish people for their part any happier, and the full consolidation of the two kingdoms and the two peoples into one nation never took place. For as the consequence of legislation that was enacted in great haste, but also never fully completed, the Anglo-Irish Union was fundamentally flawed, and from these deficiencies flowed many of the problems with which Irish patriots and British public men vainly struggled during the nineteenth century. In particular, the governing elite in London could never decide whether the Irish were fellow Britons, to be conciliated as members of the Union, or a different and alien population altogether, to be coerced by Westminster and Whitehall.

In driving through the Act of Union, Pitt had been well aware that it would not be popular with the Irish people as a whole. There would be deep and abiding bitterness over the brutal suppression of the rebellions of 1798. The Anglo-Irish Ascendancy, though grateful for London's additional support, deplored the loss of its own legislature. The Presbyterians resented the way in which this enlarged British state bore down on them, especially in terms of the increased financial exactions levied during the Revolutionary and Napoleonic Wars. And the Catholic majority disliked being incorporated into a polity whose head of state took a public oath during his coronation service to uphold the Protestant religion, and they objected to having to pay tithes for the maintenance of the established Church of Ireland. Nor was the Union popular with many Britons who did not want their Protestant nation to be defiled and polluted by closer association with those whom they regarded as half-savage and indigent Catholics. Recognizing that coercion was not enough by itself, Pitt was the first (but by no means the last) British prime minister who tried to balance it with measures of conciliation, for he was eager to win over the Irish to this new arrangement. This policy had already been prefigured by his decision, in 1793, to give the vote to Catholics who could meet the appropriate property qualification; and now, in collaboration with Cornwallis and the young Chief Secretary for Ireland, Viscount Castlereagh, he sought to implement additional measures that he hoped would reconcile the Irish people to the Union. For Pitt intended to pass a companion piece of legislation, enacting what was termed 'Catholic Emancipation', which would repeal the much-resented laws preventing people of that faith from being MPs or from holding high public office. He also hoped to make further provision for state support of the Catholic Church in Ireland, for which there already existed a precedent established in 1795 when he had endowed the Catholic seminary at Maynooth in County Kildare. Aiding the education of the Catholic priesthood would be another way of reconciling the Irish to the Union.

Thus regarded, the abolition of the Dublin parliament was only one part of what Pitt and his colleagues intended to be a comprehensive package of Anglo-Irish legislation, in which coercion would be balanced by conciliation. But Pitt was unable to push through the necessary measures in London, because George III would not have them. Although by then playing a less active part in government than during the first two decades of his reign, the king was adamant that he would not countenance Catholic Emancipation because it would indeed be a violation of his coronation oath. Unwilling to coerce his sovereign, to whose initial choice and subsequent support he owed his long years in power, Pitt resigned early in 1801, with the promise of Catholic Emancipation unredeemed, and the proposals to give further funding to the Catholic Church disregarded. He was the first British prime minister to fall over what would become known as the 'Irish question'; but he would not be the last. Nor was this the only unhappy portent: for Pitt's failure to implement any measures of conciliation left many Irish people, still unreconciled after the savage suppression of the rebellion of 1798, yet more resentful at this further display of British domination. Nor would matters end there, for during the next quarter of a century, Ireland would be further punished and frequently disciplined, as coercion acts would be regularly and repeatedly passed during the 1800s, 1810s and early 1820s. Habeas corpus was suspended again in 1801 and that suspension lasted until 1804, while an Insurrection Act, passed in 1807, continued on the statute book until 1825. Meanwhile, the police force that Robert Peel established when Chief Secretary in 1814 was as much concerned with suppressing political agitation as with detecting 'normal' crime, and as the agent of an increasingly intrusive state it resembled the continental 'gendarmerie' that would have been deemed constitutionally unacceptable in Britain. As Gladstone would observe, many years later, when introducing his second Home Rule Bill in April 1893: 'The maintenance of the Union between 1800 and 1829 was really a maintenance not by moral agency but through the agency of force.'

THE POLITICAL NATION DEFINED AND LIMITED

Since it possessed a hereditary monarchy at its apex, the optimistically named, appropriately described but recently created United *Kingdom* of Great Britain and Ireland resembled Spain, Austria, Prussia and Russia and, indeed, much of the remainder of the globe: for monarchy and royal empires were the natural order of things, and the two recent, upstart, revolutionary republics established in the United States and France in 1776 and 1789 respectively were very much the exceptions that proved the rule. Yet for more than a century and a half after the death of Queen Elizabeth in 1603, England, and then Britain, had been ruled and reigned over by monarchs who had been of foreign rather than domestic origin: the House of Stuart was Scottish, the House of Orange was Dutch, and the House of Hanover was German – imported in 1714, when the House of Orange became extinct, to prevent the Protestant British throne from reverting to the (by then) Catholic Stuarts. Both George I and George II had been born and brought up in Hanover, and they were thus German first and English a long way second. But George III, who had succeeded his grandfather in 1760 (his own father, Prince Frederick, son of George II, having died in 1751), was the first Hanoverian monarch to be British born, and at the outset of what would be his very long reign he declared that he 'gloried in the name of Briton'. Moreover, British monarchs in many ways differed from their contemporary continental counterparts: for whereas most of the latter were regarded as 'enlightened despots', which meant they were generally despotic but only intermittently enlightened, the British throne was a 'limited monarchy', grudgingly sharing power with, and gradually relinquishing power to, the House of Lords and especially and increasingly the House of Commons; and this arrangement, which was described as a mixed or balanced constitution, was widely admired by those European savants and *philosophes* who regretted and resented the power that was concentrated in the hands of most continental sovereigns.

As such, the British monarchy embodied in the person of George III was no *ancien régime* fossil, for as Lord North explained to his sovereign when he insisted on resigning in 1782, having lost his parliamentary majority (as well as the thirteen American colonies), it had long been established that there were occasions when the king had no choice but to bow to the will of the Commons, even if he had no wish to do so. But although he was no despot, governing directly through ministers who merely did his bidding, and whom he could appoint and dismiss at will, George III had no wish to be coerced by the politicians. He believed his views on matters of policy ought to prevail, and he was outspoken in declaring who, among the politicians, were his 'friends' – and who were not. Hence his constant interventions during the 1760s, as he had sought to find ministers who were congenial to him personally, regardless of whether they could work with the legislature or not. Hence his determination to sustain Lord North in power during the 1770s, and his belief that as king he stood for the reassertion of the rule of parliament over the revolting American colonists. Hence his later loathing of the Fox-North coalition, which he regarded as an unprincipled alliance and unnatural combination, and his successful efforts to bring it down at the end of 1783. Hence his audacious decision to appoint the young William Pitt to be his first minister, while still only in his early twenties, and with little but his family name to recommend him. And hence his determined opposition to Catholic Emancipation, which prevented Pitt from passing the measures he wanted to complement (and, indeed, complete) the Act of Union with Ireland, and which meant Catholic Emancipation was effectively shelved for the remainder of George III's reign and beyond, with significant and harmful consequences on both sides of the Irish Sea.

Such right royal political interventions necessarily came at the price of controversy and unpopularity. For Whigs such as Edmund Burke, the king had played the major part in bringing about the political disruption of the 1760s, and until the early 1790s they regarded their monarch with deep suspicion. For the American colonists, and as embodied in the excoriating denunciation contained in the Declaration of Independence, George III was a tyrant and a monster who had subverted their liberties and sought to create a new imperial despotism. His health and mental balance were also suspect: in 1788 he had descended into what seemed to be madness, the attack lasted for several months, and for a time a regency had seemed imminent. In 1801 there were further signs of mental instability, apparently brought on by the suggestion that he should countenance Catholic Emancipation, and it was fear for the king's sanity as well as respect for his views which effectively made it impossible for the subject to be raised again. To make matters worse, and despite his own happy home life with Queen Charlotte, the king's sons were unappealing and unpopular: loose in their sexual morals, siring many illegitimate children, and running up huge debts. Of none of his male progeny was this more true than of his eldest son, the Prince of Wales and future King George IV. He sided with the Whigs against his father in 1788, secretly married a Catholic, Maria Fitzherbert, and his insolvent finances were a national scandal and a national disgrace. For these many reasons, not all of them his fault, George III was not widely esteemed during the first three decades of his reign. Only in the 1790s did he become a popular monarch, standing for decency and restraint, tradition and hierarchy, order and justice, in contrast to the violent excesses of the French Revolution and the Terror; and as he increasingly became the embodiment of his nation's determination to face down the mortal enemy across the Channel. Although he reigned for sixty years, the 1800s would prove to be George III's most popular decade, culminating in the celebrations marking his Golden Jubilee in 1809.

As the resignation of the Younger Pitt made plain, the British government was still in some ways the king's government, which was why it was deemed obligatory to hold a general election when the sovereign died, as had happened in 1727 and 1760, and as would occur again in 1820, 1830 and 1837 (but not thereafter). Nevertheless, it had increasingly been recognized since the late 1780s that it was the so-called prime minister and his cabinet who formed the effective executive branch, and who conducted the affairs of the British nation and empire more in the king's name than with the sovereign's active, day-to-day involvement. Like Lord North and Sir Robert Walpole before him, the Younger Pitt was a long-serving premier who owed his position to the sustained and sustaining favour of the king, to his mastery of the House of Commons, where for a time he was the only minister, to the backing of those in receipt of government patronage, and to those 'independent' country gentlemen MPs who were inclined to support the government of the day and who had been outraged by the Fox-North coalition. From 1783 to 1801 Pitt was not only First Lord of the Treasury but also Chancellor of the Exchequer, and his cabinets consisted of scarcely more than a handful of colleagues, of whom the most important were the Lord Privy Seal, the Lord President, the Lord Chancellor, the Home Secretary, the Foreign Secretary and the President of the Board of Trade; and (especially from 1793) those ministers overseeing Britain's military efforts, namely the Secretary of War, the Secretary at War, the First Lord of the Admiralty, the Commander-in-Chief, and the Master General of the Ordnance. As the titles of these offices suggests, the tasks of central government, as they related to both the United Kingdom and the British Empire, were essentially limited: to the raising of revenue by direct and indirect taxes and the prudent oversight of the public finances; to the upholding of the law and the maintenance of order; and to the conduct of foreign policy and the defence of the realm.

Pitt's long administration, beginning in the aftermath of the military defeat and high cost of the American War of Independence, had been active in all these realms of endeavour, as he sought to deal with the crisis in the national finances, the break-up of the transatlantic imperial structure, and Britain's damaged and isolated international position. In order to cope with the substantially increased public debt, Pitt raised a variety of ingenious new taxes, on everything from hats to windows; he deployed the Commissioners for Examining the Public Accounts to make government more efficient; and he further saved money by reducing the number of patronage positions, even as that lessened the bedrock of his support in the Commons. He also set up a complex and elaborate 'sinking fund', the purpose of which was to apply any government surplus to the reduction of the national debt, and in this he was successful until war with France resumed in 1793. He was equally eager to reform and consolidate what was left of the British Empire: he scaled up expenditure on the Royal Navy to increase the number of ships of the line, and he rehabilitated Britain's international position by concluding a trade agreement with France in 1786 and defensive alliances with the United Provinces and Prussia two years later. Thus did Pitt set in train what would later become known as the 'national revival' that soon took place in the aftermath of the loss of the American colonies, allaying those Gibbon-inspired fears that Britain was about to follow ancient Rome into irretrievable decline and imperial ruin. But with the outbreak of war against Revolutionary France, it was national survival, rather than national revival, that became the main task in hand. Hence the repressive measures in Britain and Ireland and the passing of the Act of Union, and hence Pitt's imposition of the income tax in 1799, which was proposed as a necessary but temporary expedient to increase government revenue. By then, Pitt was the leader of an embattled and beleaguered administration, leading an embattled and beleaguered nation, and of a people who were by turns anxious, unhappy and discontented, or patriotic, devoted and loyal.

During his long years of power, Pitt was the head of an executive whose party composition was never fully clear and also often changing. For over a century, British politicians had paid more than lip service to the two-party distinctions of Whig and Tory: the former had been dominant between 1714 and 1760, while the latter were in opposition; and although George III had resolved to put an end to what he regarded as the unhappy distinctions between parties, this meant in practice that the Tories had replaced the Whigs in power, and with temporary interruptions they would retain and consolidate their position, under Lord North, the Younger Pitt and Lord Liverpool, until 1827. In very general terms, the Tories supported the Anglican establishment and the residual power of the monarch, whereas the Whigs were sympathetic to religious dissent and to a more limited conception of the royal prerogative. But all such notions were modified according to the different circumstances of government or opposition, and in any case, combinations of Tories and Whigs were not at all unusual, as in the case of the coalition briefly constructed by North and Fox. The Younger Pitt had begun his political career by disavowing all party allegiance, and by insisting that he was in every sense the king's man, which at the beginning of his prime ministership he undoubtedly was. But his family inheritance was Whig, and he included other Whigs in his first administration, among them Earl Gower as Lord President and Earl Temple as Home Secretary. Yet because he incurred the lifelong enmity of the Foxite Whigs, Pitt was widely regarded as a Tory, an impression reinforced by his appointment of Lord Thurlow as Lord Chancellor, and later by Henry Dundas as Home Secretary and Lord Hawkesbury as President of the Board of Trade. In July 1794 Pitt further broadened his administration by bringing in the 'Portland Whigs', who were outraged and dismayed by the turn towards the 'Terror' in France, and who included not only their eponymous Duke, but such additional grandees as Earl Spencer as Lord Privy Seal and First Lord of the Admiralty and Earl Fitzwilliam as Lord President of the Council.

As these names suggest, the majority of Pitt's cabinet were members of the House of Lords, and this practice would continue well into the second half of the nineteenth century (prime ministers in the upper house, from Liverpool to Salisbury, would also become more common than in the age of Walpole, North and the two Pitts). In his brief second administration, of 1804–06, all ministers except Pitt himself would be peers, as had been the case when he had first taken office in 1783, and the senior posts at the royal court were also held by peers. As befitted the hierarchical society in which they ranked so superior, the members of the House of Lords were the greatest grandees in the land, unrivalled in their wealth, status and power, and many of them were richer than their sovereign. The Dukes of Bedford, Bridgewater, Devonshire, Portland and Buccleuch, the Marquis of Stafford, and the Earls Fitzwilliam,

Grosvenor and Gower each enjoyed incomes of more than £100,000 a year: partly drawn from their broad agricultural acres, but also from urban real estate development (especially in London), from the ownership of docks and harbours and, increasingly, of mineral rights. They were as much plutocrats as aristocrats, and during the late eighteenth century, many of them spent unprecedentedly on building extravagant town houses in London, and buying fabulous art collections of European Old Masters, which impoverished aristocrats in France and Italy were putting on the market. Such wealth also translated directly into political and social power: individual grandees often held high government or courtly office, and owned so-called 'rotten' boroughs, with electorates so small they could effectively nominate their own MPs; while the House of Lords collectively possessed the power to reject all measures sent up from the Commons except, by convention, money bills, and with the encouragement of George III they had brought down the Fox-North coalition. For his part, Pitt also appreciated the allure of peerages, and from 1783 to 1801 he was profligate in ennobling government supporters and borough owners.

Except in matters of finance (a major qualification, which explains why the First Lord of the Treasury and Chancellor of the Exchequer often sat in the lower house), the Commons was constitutionally and customarily subordinate to the Lords, and its composition reflected its junior and inferior status. The elder sons of peers habitually served their apprenticeships in the Commons, and younger sons might spend their whole lives there, often representing a family-controlled constituency, and so further reinforcing the client status of the lower house. Lord North, the elder son of the Earl of Guildford, was an example of the first; Charles James Fox, a younger son of Lord Holland, of the second (at least at the start of his parliamentary career). Baronets and country gentlemen, who generally ranked below the peerage in acreage, wealth and prestige, often represented their local constituencies, and although they were inclined to support the government of the day, there was no absolute guarantee they would do so. Together, the sons of peers, along with baronets and gentry, formed the majority of the Commons, so that, like the House of Lords, it was overwhelmingly dominated by what was termed 'the landed interest'; these ties of property and family would remain important even when the two houses collided on major issues, as they would do on several occasions between the 1830s and the 1890s. MPs also included rich, London-based merchants, bankers and financiers (such as the Barings), the owners of West Indian plantations, and so-called 'nabobs': men who had made a fortune in India (as had the Younger Pitt's grandfather), and who then returned home, purchased a landed estate, and got themselves returned to parliament. Of the 550 MPs (rising to 650 after the Act of Union with Ireland), the vast majority represented English constituencies, and only a tiny minority, fewer than fifty in all, spoke for Wales and Scotland. This may have been the British House of Commons, and subsequently the legislature for the whole of the United Kingdom, but the distribution of its membership was overwhelmingly weighted in favour of England.

Taken together, the court, the cabinet, the Lords and the Commons formed a small, tightly knit oligarchy of a few thousand families, and as members of this ruling elite, Whigs and Tories had at least as much in common with each other socially as they might disagree with each other politically. In no sense, then, did the polity that the Younger Pitt commanded approximate to anything like a modern-day democracy: the landed interest was dominant in government and in the legislature, while the size of the electorate, and its capacity to influence the composition and the deliberations of the Commons, was conspicuously limited. Under the Septennial Act, which had been passed in May 1716 to help secure the Hanoverian succession and consolidate the Whig ascendancy, general elections did not have to be held more frequently than every seven years (and that would remain the law throughout the eighteenth and nineteenth centuries). Moreover, many constituencies were not contested, because they had effectively been decided beforehand by the landowners who controlled them. This was especially so in the boroughs, but also in some of the counties, where the dominant families often came to terms beforehand, and split the representation on party lines, so as to avoid the needless cost of a contest. The electorates of many boroughs could be counted in single figures, most infamously in the case of Old Sarum, near Salisbury in Wiltshire. The representation of Wales and Scotland was notoriously restricted and corrupt, and very few constituencies, among them Yorkshire and the City of London, were truly open or free. The total electorate of the United Kingdom numbered only a few hundred thousand, the majority of men did not vote, and no women, although at that time there was no statute formally prohibiting their participation in the electoral process. Only 5 per cent of the adult population was enfranchised, which may have been less than during the late seventeenth and early eighteenth centuries. The vast majority of people in the United Kingdom thus had no direct personal involvement in choosing the men who governed them.

This was partly because in most seats the vote was tied to property, on the grounds that anyone who enjoyed the privilege of helping to select a member of parliament and indirectly chose a government should literally have a stake in the country in terms of some established material interest; but an additional qualification was adherence to the established religion of the state. In England, Wales and Ireland involvement in public life was only possible if certain tests of religious conformity to the state church were passed. In all three countries it was the Anglican, Protestant, Episcopalian church that was established by law and supported by the state: hence George III's claim that as the supreme governor of the church, he must uphold the Protestant religion because that was what his coronation oath required; and hence the fact that certain bishops of the Church of England, of Wales and (after 1800) of Ireland sat as of right in the House of Lords. In those three countries, as in Scotland, where the Presbyterian Church was the establishment, Catholics could not sit in either house of the Westminster parliament, nor hold high public office, even if they were as grand as the (Roman Catholic) Duke of Norfolk; while in England, Protestant dissenters could only participate in public life if they were willing, as late seventeenth-century legislation had allowed, to conform occasionally and publicly to the doctrines and liturgies of the Church of England. This also meant that the established church was a major supporter of the secular structure of power and authority in the United Kingdom: the coronation, which acclaimed the new sovereign as head of state, was a religious sacrament; and during the troubled times of the 1790s there was a significant rise in the number of Justices of the Peace who were clerics in the Church of England. In attacking the established church, radicals such as Tom Paine were also attacking the British state, and they knew what they were doing.

The early nineteenth-century United Kingdom was thus a monarchy, an oligarchy and a highly restricted polity, which was further supported by a state church, but it was also a more open and flexible society than the rigid regimes that had been widespread across Europe before the French Revolutionary armies began to overwhelm them. To begin with, there was a certain amount of social mobility, as enough people rose up the social scale in Hanoverian Britain to attract the attention and approbation of contemporaries. The great ducal dynasties such as Bedford and Devonshire may have assumed they ruled by hereditary right; but even they had come from humble origins, albeit generations earlier. Banking families such as the Hoares, and entrepreneurs such as the Peels, bought their way into land, while in some cases also keeping up with their business activities. The Younger Pitt was not alone in having had forebears of relatively lowly backgrounds: one of his successors as prime minister, Henry Addington, was of minor gentry background, and was the son of a doctor; another, George Canning, was the son of an actress. One way in which such talented outsiders could join the ruling elite was by attracting the notice of a rich and friendly patron who could get them returned to the Commons for one of his rotten boroughs, which was how Thomas Babington Macaulay and William Ewart Gladstone were able to become MPs at such an early age. Defenders of the British constitution also claimed that by such means, many of the different national interests and constituencies, such as commerce, trade, or the army or navy, were 'virtually' represented in the Commons by people who knew about these subjects, even if they were rarely represented directly. Moreover, that nebulous construct called public opinion could and did exert occasional influence on the deliberations of the government and the legislature, both via the ballot box and via popular protest: it had done so in 1733 (against the Excise Bill), in 1757 (in support of the Elder Pitt), and in 1784 (against the Fox-North coalition), and it would do so again on several occasions between 1830 and 1832 during the protracted political crisis over the Reform Bill.

So while the ruling elite of the United Kingdom was in many ways a narrow cabal and a self-recruiting and self-perpetuating oligarchy, and although in general it had no need of popular legitimacy, there was some slight and grudging recognition that it could not be wholly indifferent to public opinion. Moreover, in the case of Great Britain, popular Protestantism morphed into a broader sense of British national identity that often, although not invariably, transcended the divisions between Anglicans, Presbyterians and those who were termed old and new dissenters (the former including Quakers and Unitarians, the latter soon to include Methodists, although in 1800 they had not yet formally broken from the Church of England). Many Britons shared George III's visceral dislike of Catholics as being papists and idolaters, and thus his opposition to Catholic Emancipation. The religious delinquencies and absolutist aspirations of James II were part of the nation's collective memory, while the revolution of 1688 was widely revered as a benign event that had preserved Britain's true religion. This deeply engrained popular Protestantism was reinforced by the fact that Britain's two long-standing European enemies, namely Spain and France, who often combined antagonistically during the eighteenth century, had for most of their history been Catholic despotisms. Despite the severe limits placed on their participation in public affairs, many Britons saw themselves as being, by comparison, Protestant and also therefore free from papal interference, and this could give them a sense of fellow-feeling with the governing elite and with their sovereign. But while this shared sense of Protestant identity helped to bind together the inhabitants of Great Britain, it also created serious political difficulties when it came to dealing with what was overwhelmingly Catholic Ireland. On several occasions during the nineteenth century, of which Gladstone's two abortive Home Rule Bills were merely the most famous examples, it proved impossible for governments in London to carry measures intended to conciliate the Irish in the face of determined opposition in the Westminster parliament. It was no coincidence that MPs and peers were overwhelmingly Protestant, and popular opinion in England would remain anti-Catholic and anti-Irish for most of the nineteenth century.

Although England, Wales and Scotland could claim a shared identity on the basis of a parliamentary union and common Protestantism, there remained significant differences between the three nations that would be intensified as the nineteenth century wore on. In Wales there was a clear divide between the anglicized landed and religious elite, and the majority of the population, which would later be reinforced by the coming of Nonconformist religion and Liberal politics. In Scotland the fact that the established church was Presbyterian rather than Episcopalian was a significant difference, one indication of which was that the British sovereign changed religion when changing countries; and although the Union of 1707 resulted in the creation of a single, pan-British trading area, Scotland would retain its own particular educational and judicial system, and it would be separately administered from London. As for Ireland after 1801, it was in the anomalous position of being neither a completely subordinated colony nor fully integrated into the United Kingdom. Dublin had been downgraded from being a national capital and one of the largest cities in Europe to a provincial outpost; but during the course of the nineteenth century, the Presbyterian and Episcopalian Protestants of Belfast and much of Ulster would come to feel a growing sense of identity with the Union and the British crown. Like Scotland, Ireland would be separately administered: the Chief Secretary would be based in London and sit in the British cabinet, but the vice-regal post of Lord Lieutenant would also be retained, representing both the British crown and the government, and that individual would be supported by a quasi-colonial bureaucracy based in Dublin Castle, which grew by 40 per cent between the Union and 1829. Moreover, the Chief Secretaryship came almost at the bottom of the cabinet hierarchy, with the Lord Lieutenancy often having to be hawked around because no one wanted to do it, and across the nineteenth century most British politicians regarded the Irish with dismay or disdain. Like many peoples in other parts of the world, they were not deemed to be civilized – hostile and uncomprehending opinions that the Irish could scarcely fail to notice.

Between the governing elite at Westminster, and popular notions of collective national identity, lay the intermediate world of local communities and of the structures of power and administration holding them together – or, in increasing numbers of cases, failing to do so. Across the United Kingdom, the historic unit to which most people related was the county, presided over by the Lord Lieutenant, a resident grandee who was the sovereign's representative, but in practice was appointed by the government. Most county administration was carried on by the local elite of aristocrats and gentry, who regularly met at quarter sessions, to dispense justice and levy rates to ensure that essential services, such as the provision of roads, were adequately resourced. The county was also the strongest unit of local loyalty, as evidenced by the many regiments recruited on such a basis; to this day their flags and banners hang in many cathedrals. Separated off from the counties were those many towns and cities that were incorporated as boroughs, presided over by a mayor, aldermen and a corporation, who could levy rates for the upkeep of roads and for the provision of lighting. The provision of the poor law was related to the parish organization, and separate rates were levied to support the workhouses that were an essential part of its provision. In 1800 there were no police forces anywhere in Britain or Ireland, and public order was maintained by the magistrates, by their constables or, in dire circumstances, by sending in the local militia. Just as the structure of Commons constituencies and representation reflected the social and economic conditions of an earlier time, so many of the towns and cities that had recently expanded, such as Birmingham and Leeds, had no appropriate form of local government. As for London, it boasted a population of one million, and was the largest city in the Western world, but it was governed by a chaotic amalgam of vestries, boards and commissioners, and any attempt to impose a central unifying authority was denounced as bringing in alien and despotic continental methods.

PEOPLE, PLACES AND PRODUCTION

The social structure of the United Kingdom of Great Britain and Ireland over which the national and local ruling elites held a not wholly confident, certain or competent sway at the beginning of the nineteenth century is probably clearer to us now than it was to contemporaries then. In 1801, when the first census was taken, the population of England and Wales was heading towards nine million, and that of Scotland to two million, while Ireland had reached five million (though this was no more than an estimate, since there would be no Irish census until 1821). Since 1750, the populations of England, Wales and Scotland had each increased by approximately 50 per cent, whereas that of Ireland had grown even more rapidly, having been only two million in 1750. These figures help explain both the overwhelming dominance of England vis-à- vis Scotland and Wales, and also the resentment felt in Ireland at its under-representation in the Westminster parliament in the aftermath of the Act of Union, where the hundred MPs that had been provided for should have been doubled on the basis of its relative population. All four nations had witnessed significant – indeed, unprecedented – increases in their populations since the middle of the eighteenth century, as birth rates remained high, and death rates began to diminish. Until the middle of the nineteenth century, birth rates would continue to be high across the British Isles, and population growth would further accelerate, which was one reason why so many Britons would emigrate to the United States and to the settlement colonies of the British Empire. This would become an important safety valve for what would be deemed to be the 'surplus' population, for by the late eighteenth century there was growing concern that the alarming increase in numbers was creating a demand for food that was outstripping supply. Such was the pessimistic conclusion reached by the Reverend Thomas Malthus, in his influential *Essay on the Principles of Population*, published in 1798, which predicted demographic disaster for the nation if the numbers of people continued to grow geometrically while agricultural output only increased arithmetically; and this was precisely what would happen in Ireland during the 1840s.

At the apex of this rapidly expanding population were those great territorial aristocrats who were not only powerful figures in Whitehall and Westminster and at the royal court, and often prodigiously rich, but were also in the process of becoming an increasingly supranational elite, with territorial holdings extending and amalgamated across several (and in some rare cases, all) of the four nations that from 1800 comprised the United Kingdom. Great families such as the Fitzwilliams, Devonshires and Londonderrys were major landowners not only in England but in Ireland as well, and this would later influence their (hostile) views of Gladstone's schemes for Irish Home Rule. The Marquis of Bute possessed broad acres in his titular county in Scotland and in rich coal-bearing lands in Glamorgan in Wales, the Marquis of Stafford held similar mineral-bearing lands in Shropshire, along with a million acres in County Sutherland in Scotland, from which his family would subsequently take their ducal title, while the Marquis of Lansdowne drew rentals from estates in England, Ireland, Scotland and Wales. For dynasties such as these, their pan-British territorial amalgamations and accumulations were gradual processes, sometimes involving marriages to heiresses across several generations, but they also conveniently aligned with the political consolidations incorporating Scotland into Britain in 1707, and Ireland into the United Kingdom in 1800. Many of these titles were of relatively new creation, especially in the case of the marquisates bestowed on Londonderry, Bute and Stafford, acknowledging the growth in their territorial accumulations and, correspondingly, in their fortunes. Meanwhile, in Ireland, the Order of St Patrick had been established in 1783, modelled on the English Order of the Garter and the Scottish Order of the Thistle. It was given out to the greatest Anglo-Irish landowners in the Ascendancy, with the aim of binding them ever more closely to the British nation and crown.

As the nineteenth century began, most aristocrats possessed both an hereditary title and a landed income of at least £10,000 a year, which roughly corresponded to the ownership of 10,000 acres; and when the great wartime commanders such as St Vincent, Nelson and Wellington were awarded peerages, they were also given substantial grants of money by parliament so they could establish themselves on landed estates that would make it possible for them to support their titular dignities in an appropriate way. Such men often owned broad acres in more than one county (or country), possessed more than one country mansion, as well as a town house in one of London's smartest locations, such as Mayfair or, later, Belgravia. This enabled them to take part in the social rituals of London high society, to be resident in the capital while parliament was sitting, and also to spend time in the countryside maintaining their local social and political connections. By contrast, the country gentry, who enjoyed landed incomes of between one and ten thousand pounds a year, and acres to match, were much more circumscribed in their resources and activities, and perhaps in their vision as well. Their estates were more likely to be confined to one county; they did not benefit from extensive non-agricultural sources of income; and they would be unlikely to possess a London town house. But like those magnates who ranked above them, the landed gentry were members of the leisured classes, and did not expect or need to work. They took part in the activities of their localities, especially hunting and shooting; and they might represent their local county constituency. Some of them, like the Arkwrights at Hampton Court in Herefordshire, were relatively recent arrivals in the ranks of the landowning classes; others, like the Lygons at Madresfield in neighbouring Worcestershire, had held their lands for centuries.

At least three groups of people who did work, and who formed the richest members of a group appealingly but vaguely labelled 'the middling orders', were also likely to sink some of their fortunes in land. The first were the bankers, financiers and merchants who were such a significant occupational force in the City of London, and who accumulated substantial riches by servicing the metropolitan economy and by involving themselves in international trade (to which should be added the 'nabobs' returning from India). Until the Rothschilds appeared on the scene in the 1800s, the most famous British bankers were the Baring family, who were well established by the second half of the eighteenth century, not only in finance in London, but also as landowners in Hampshire and as patrons of the arts. A second group were those who rose to the top in the great professions, made their fortunes along the way, or received extensive sums from parliament, and again set themselves up in land. In addition to the military men already mentioned, there were successful lawyers such as Lord Thurlow or Lord Eldon, both of whom came from relatively humble backgrounds, but did so well in their chosen professions that they amassed substantial fortunes and duly translated them into broad acres. The final group, which became more prominent during the closing decades of the eighteenth century, were the industrialists and entrepreneurs who were both the agents and the beneficiaries of the quickening economic pace of those years. The Arkwrights had established themselves in Herefordshire on the proceeds of the fortune they had made in textiles in Derbyshire, while the Peels, who made their money in the same industry in Lancashire, settled as squires at Drayton in Staffordshire. Like the bankers, many entrepreneurs kept part of their fortunes in their businesses, which meant the landed elite and middle classes melded into each other, especially in England, in the counties surrounding London and in the industrializing north.

These three groups of rich so-called 'amphibians', who often linked landownership and business, in turn merged into the next layer of the middle class, who were lower down the social scale and less well off, and who inhabited parts of London, the provincial towns of England, Edinburgh and Glasgow in Scotland, and Dublin and Belfast in Ireland (there were no cities or towns of comparable size in Wales). In places ranging from the new industrial centres such as Manchester and Birmingham, to traditional county towns such as Lincoln and Norwich, there were networks of bankers, solicitors, brewers, scriveners, clergymen, insurance workers, doctors, and many others who were involved with agricultural processing and small-scale manufacture. This was the world depicted by Jane Austen, where the Bennet family trembled on the edge of landed gentility in *Pride and Prejudice*, and where the Woodhouses and the Fairfaxes rubbed shoulders with the landed Mr Knightly in *Emma*. In rural and agricultural Norfolk such real-life figures included the Gurneys in banking, the Bignolds in insurance, and the Buxtons in brewing, while in industrializing Birmingham their counterparts were the likes of Matthew Boulton and James Watt in manufacturing, Sampson Lloyd in banking, and George Cadbury in food processing. There were also self-made and self-taught civil engineers, such as John Rennie and Thomas Telford, there were academics in the university towns of Oxford, Cambridge, Glasgow, Edinburgh, Aberdeen, St Andrews and Dublin, and there were scientists and savants such as Humphry Davy, James Hutton and Francis Beaufort. Further down the social scale there were customs and excise workers, teachers, shopkeepers, publicans, gunsmiths, jewellers and hat makers; there were those involved in the luxury trades, the 'toy' trades and retailing; and together they comprised a lower middle class that was undoubtedly expanding by the late eighteenth century. Not for nothing would Napoleon soon describe (and mistakenly dismiss) the British as being 'a nation of shopkeepers'.

These different occupational categories and cohorts in turn dissolved into those socially inferior groups who were called the working classes or lower orders, who formed the overwhelming majority of the population of the United Kingdom, perhaps as much as 75 per cent of it. They were more varied in their occupations by the end of the eighteenth century than at any earlier time, and they may also have been more diverse than their counterparts anywhere else in Europe in the same period. Many of them were agricultural labourers: indeed, the overwhelming majority were thus employed in Ireland, Scotland and Wales (but not in England), and spent long hours in the fields in back-breaking toil, or tending their sheep on inhospitable hills and mountains. Some worked in fishing fleets, especially in the North Sea and across the Atlantic towards the coasts of Iceland, Greenland and Newfoundland. Some were miners: in the West Country, where tin was extracted, and in the Midlands and the north, where coal was being mined in ever greater quantities. Some worked in manufacturing, but few in factories, as most were employed in small-scale workshops. Others worked in government installations, such as dockyards, or in the royal mint, which were probably the largest units of employment at the time, or volunteered for the British army or were press ganged into the Royal Navy. Most of these were jobs for men, but thousands of women were employed in domestic service, in the great houses of the aristocracy, the lesser mansions of the gentry, and the comfortable villas and terraces occupied by members of the middle classes. Below these men and women in gainful employment were those who, because of infirmity or misfortune, could not work, for whom there was a basic provision of poor relief to keep them from starving. This dated back to Tudor times, but many varied local modifications were made by the Poor Law guardians during the years of high food prices and high unemployment that characterized the 1790s.

The late eighteenth-century United Kingdom was, then, a very unequal society in terms of the distribution of wealth, power and status. It was probably less so than the Russian Empire, but it was certainly more inequitable than the United States (provided slaves and native Americans were disregarded, as the founding fathers had done). There was a vast economic gap between the richest dukes and the most indigent paupers, which in real terms may have been wider than in our own day, and which meant that life chances significantly deteriorated the lower down the social scale a person was situated. There was an equally great distinction between those few who had power, like cabinet ministers, peers, MPs or Justices of the Peace, or who as voters were recognizably part of the political nation, and the overwhelming majority of the population who did not wield such power or who were not thus recognized. There was a further distinction between those Britons who were members of the Church of England, or Wales or Ireland, and who could play a full part in the public life of the country, and those committed dissenters or Catholics who were prohibited by law from doing so. There was an additional hierarchy of inequality built around gender: for men played a far greater part in public life than did women, men alone were in practice allowed the vote, and on marriage, the woman's property and legal rights were assumed by the husband, and the wife had virtually none. There was another imbalance in that although all Britons were supposed to be equal before the law, the reality was very different, and it became more so during the closing decades of the eighteenth century. For the law was more to do with preserving property (which favoured the rich) than with protecting people (which would have favoured the poor). By the 1780s and 1790s, the statutes prescribing the death penalty had multiplied greatly, and they were more concerned with crimes against property than against people; and although some members of the middle and upper classes went to debtors' prisons, the majority who were incarcerated were poor not rich.

Like many unequal societies, and even as it was experiencing severe strains and stresses, the late eighteenth-century United Kingdom produced significant examples of high culture. George III and his eldest son disagreed on almost everything, but they were both patrons of the arts. The king had embellished Windsor Castle, and was a major collector of books and astronomical instruments; the Prince of Wales would create the extraordinary oriental extravaganza of Brighton Pavilion and lavishly extend Carlton House, his London residence. With the support of an earlier sovereign, the British Museum had been established in 1753 as the first great public institution of its kind in the world, and the Royal Academy had followed in 1768, with George III's active encouragement. Its most illustrious early academicians included Joshua Reynolds, Thomas Gainsborough and Joseph Mallord William Turner (Joseph Wright of Derby, by contrast, refused election). The architects of the time were equally distinguished, among them the Adam brothers, Sir William Chambers and Sir John Soane. The earlier decades of the eighteenth century had witnessed the rise of two new forms of cultural activity. One was novel writing, and by the 1800s the two outstanding practitioners were both women: Jane Austen and Fanny Burney. The other was the critic and man of letters, of whom Samuel Johnson was the pioneering prototype, and James Boswell's biography of him had been published in 1791. Many country houses were significant cultural centres, with their libraries, music rooms, long galleries and gardens; while in London, Edinburgh and Dublin (at least before the Union), and the provincial towns and cities, associational life was becoming richer and more cultured. This was exemplified by the Birmingham-centred Lunar Society, whose early members included Samuel Galton, Joseph Priestley, Josiah Wedgwood, Erasmus Darwin and Boulton and Watt, and they met regularly by the light of the full moon to discuss scientific and cultural matters.

Such wide-ranging cultural activities, often combining politeness and sociability with the eager pursuit of knowledge and learning, indicated just how far the European Enlightenment had reached, both socially and geographically, in late eighteenth-century Britain. It was especially, but not exclusively, associated with the rise of a stable and prosperous middle class, which was increasingly attracted by the ideas and ideals of toleration and rationality, improvement and progress. If anything, the Enlightenment was a more pronounced phenomenon in Scotland than it was in England, and was closely associated with the ancient universities (Oxford and Cambridge, by contrast, were less intellectually engaged), with the construction of the neoclassical new town in Edinburgh (much of it the work of Robert Adam), and with such thinkers and writers as David Hume, Adam Smith, John Millar, William Robertson and John Playfair. In England, by contrast, the 1790s represented a serious challenge to shared Enlightenment values. The young Jeremy Bentham, recognizing the recent increase in the size of the incarcerated population, came up with the idea of the Panopticon, a new form of prison, a total institution in which the systematic surveillance of the inmates would be 'a new mode of obtaining power of mind over mind'. The deist and freethinker Joseph Priestley believed the French Revolution embodied the final triumph of Enlightenment values and aspirations; Joseph Banks, on the other hand, considered the revolution, with its violence and irrationality, to be the very negation of all the Enlightenment had ever stood for.

As this suggests, cultural and political life during the 1790s were for many Britons inseparably interlinked, and this was as true of those who supported the French Revolution as of those who opposed it. Tom Paine's *Rights of Man* was one of many radical works, calling for the overthrow of the political inequality that seemed to exist as much in late eighteenth-century British society as it had done across the English Channel before the Bastille had been stormed. At the very beginning of the decade, Catharine Macaulay denounced the Bourbons, the aristocracy, overweening executive government and political inequality, and called for a 'more extended and equal power of election'. The Scottish Whig, James Mackintosh, offered a further justification of the Revolution in *Vindiciae Gallicae* (1791), where he argued that French government and society were so despotic and diseased that there had indeed been a pressing need to create new institutions, where a new form of politics might flourish, based on reason. Two years later, William Godwin published his *Enquiry Concerning Political Justice*, in which he insisted that humanity must inevitably progress towards enlightenment and perfectibility, and that along the way it must abolish property, monarchy and marriage (although he would wed Mary Wollstonecraft in 1797). William Blake's *Songs of Innocence and Experience* (1794) mounted a poetic protest against what seemed to him the 'mind-forg'd manacles' of repressive religion, and such industrial abuses as child labour. Between 1794 and 1807, Paine produced instalments of a further polemic entitled *The Age of Reason*, which assailed established religion for its irrational doctrines and as the handmaid of state power, denounced church corruption, contested the legitimacy and authority of the Bible, and argued for the superiority of reason over revelation. At the same time, Erasmus Darwin was thinking the unthinkable in *Zoonomia* (1794), which anticipated the ideas of his grandson, Charles, concerning evolution and natural selection; and William Wordsworth and Samuel Taylor Coleridge published *Lyrical Ballads* (1798), which was as revolutionary in its poetry as it was in its politics.

The works of Paine, Macaulay, Mackintosh and Godwin had all been written in reply to Edmund Burke, whose *Reflections on the Revolution in France* (1791) had defended in the most extravagant rhetorical terms the organic roots and traditional structures of the *ancien régime*, had deplored the violent and destructive treatment that had been meted out to the French king and queen, regretted that this meant the age of chivalry had disappeared, and lamented that as a result 'the glory of Europe is extinguished for ever'. Soon after, the Anglican clergyman William Paley, in *Reasons for Contentment* (1792), urged that the 'labouring part of the British people' should be grateful for the lot to which providence had assigned them, and accept and appreciate the time-honoured and traditional social hierarchy and their own place and station within it. Hannah More made essentially the same point in her *Cheap Repository of Moral and Religious Tracts*, which were printed in their millions between 1795 and 1810, and which advocated the causes of purity, discipline, submissiveness, civility and patriotism, especially for the working classes, and even more so for those not in work. So did William Wilberforce in his *Practical View of the Prevailing Religious System of Professed Christians* (1797), in which he deplored the merely nominal religiosity of most Britons, recounted his own spiritual journey through which he found Christ, and also challenged men and women to centre their own lives on Jesus, so that the spiritual foundations of society might be restored and strengthened.

Two years earlier, Wilberforce had supported the Younger Pitt in his suspension of habeas corpus, and like More, he was a determined upholder of the established order of church and king. Both were concerned with preserving the spiritual and moral foundations of British society against what they regarded as the subversive, poisonous and atheist doctrines emanating from France.

During the troubled and traumatic decade of the 1790s, contemporaries had been unable to agree whether British society was essentially traditional, organic, unequal and unchanging, or whether it was pent up, frustrated, discontented and on the brink of revolution. This in turn meant that there was no shared view as to what exactly the social structure of the United Kingdom looked like. For conservatives such as Burke, More and Wilberforce, it was a seamlessly graded hierarchy of individuals, extending from the king and the greatest grandees at its apex, via the many social ranks and occupational layers that merged imperceptibly into one another, to those impoverished individuals languishing at the bottom. As such, the British nation was a traditional, time-hallowed, territorially rooted and organically evolving community, which needed to be preserved and defended from misguided revolutionaries. A second way of seeing the social structure was as being divided into three collective groups: the landowners, who enjoyed unearned incomes in the form of rentals; the middling orders, who derived their incomes from professional fees or the profits of their businesses; and the lower orders, who made their living by working for wages. This view of society, and the analysis underpinning it, had been popularized by Adam Smith in *The Wealth of Nations* (1776), and it appealed especially to those among the 'middling sorts' who from the 1790s onwards increasingly wished to assert their position more fully in British politics and society. The third way of envisaging the social structure of the United Kingdom was one that appealed much more to the likes of Paine, Godwin, Wollstonecraft and their radical supporters, who saw it as being deeply divided and bitterly polarized between those who possessed power and property, and those who did not, and who wished to overturn this inequitable arrangement by revolutionary force. None of these pictures was entirely accurate. But all three would remain powerful throughout the nineteenth century – and, indeed, on through the twentieth and into our own time.

Such were the vernacular versions of the social order; but how much did the ruling elite of the United Kingdom really know of the country over which they ruled, and for which they legislated, as the eighteenth century turned into the nineteenth? The answer is: more than their predecessors, but nothing like as much as any government knows now. Such figures as Gregory King and Joseph Massie had attempted to calculate the size of the population and describe the nature of the social structure: but their efforts were based on a great deal of guesswork, and they were usually confined to England. The first census was conducted in 1801, but it excluded Ireland and was full of inaccuracies. So when Malthus published his *Essay on the Principles of Population* in 1798, no one could be sure whether the population of the United Kingdom was growing (as he insisted), declining or staying constant, or whether the patterns and trends differed between the four nations. There were no reliable statistics concerning the distribution of landownership, the size and range of people's incomes, or the number of people living in particular towns and cities and counties. There was growing talk about the increased importance of the 'middle classes', but no one knew how many of those sort of people there actually were, or what they were worth. There was widespread concern during the 1790s and 1800s about poverty, but again, there were no national statistics available, since most aspects of local government were carried out independently of Westminster and Whitehall. Insofar as there was any systematic gathering of information, it was undertaken by private individuals, the most famous being Sir John Sinclair, who produced his *Statistical Account of Scotland* in twenty-one volumes between 1791 and 1799; but it would not be until the 1830s and 1840s that governments began to gather systematic information about the condition of the United Kingdom, and even then much of it would be biased and highly impressionistic.

This in turn meant that the Younger Pitt and his generation cannot have known they were living through the early stages of transformative changes to which in retrospect the names 'agricultural revolution' and 'industrial revolution' would be misleadingly but memorably applied. By the mid-eighteenth century, England was already an unusual place because, although agriculture was the largest employer of labour, the service sector was very well developed, and the proportion of the population living in towns and cities was higher than almost anywhere else in Europe. This helps explain the prodigious growth of London, as a finance and service centre, as a major location of manufacturing, and as a great port, which meant it had become the greatest city in Europe (and perhaps in the world) by the time the Act of Union was passed. Since the third quarter of the eighteenth century, the pace of economic change had begun to quicken, with the accelerated enclosure of common lands, increase in the area of cultivation, and improvement in agricultural output and productivity; with the construction of new roads and the creation of a national network of canals; with the development of improved techniques and capacities in coalmining; with a range of technological innovations in the textile and iron industries; and with the expansion in the financial infrastructure as many new country banks were established. None of this should be exaggerated. Like the simultaneous growth in population, these were evolutionary processes rather than revolutionary events. To be sure, agricultural change was widespread in England, and Scotland would soon have its own more disruptive and controversial variant in the form of the Highland clearances. But there was noticeably less change in Wales and Ireland, and large parts of Britain and virtually the whole of Ireland were scarcely touched by the developments in industry and mechanized production that were taking place elsewhere. Moreover, by the 1790s, the economic prospects seemed far from alluring: cyclical fluctuations were becoming more pronounced, unemployment was high, food prices were soaring, some thought that population growth was getting out of control, and there was another costly war to fight against the French.

So when Great Britain and Ireland were inadequately united and incompletely conjoined in 1800, no one could have foreseen that for much of the next century Britain would become the world's most advanced and successful industrial economy (albeit with exceptions), whereas Ireland would become one of Europe's least successful and most stagnant economies (albeit with some exceptions). In Britain, but scarcely anywhere in Ireland, the gradual shift from reliance on organic forms of power (wind and horses) to inorganic forms of power (fossil fuel and above all coal) would constitute a fundamental and seemingly irreversible transformation; while the early changes in ways of producing iron and manufacturing cotton were the harbingers of a mechanized and workshop-and-factory-based economy that would develop, albeit hesitatingly, uncertainly, and only in particular places, by the middle of the nineteenth century. Such would be the extraordinary transformation of the economy of parts, but by no means all, of the British Isles that would be accomplished by the 1850s, the beginnings of which were already under way by the 1780s and 1790s. Although contemporaries did not know it at the time, this would give the United Kingdom a decisive advantage in the short term in the latest wars against France, and in the longer term would make it the pre-eminent economic power in the world. It would put Britain far ahead of its European neighbours and competitors, and also of those Asian powers such as China which, until the third quarter of the eighteenth century, may well have been as successful and productive as the most advanced economies in Europe. Indeed, it is no exaggeration to say that it was Britain's success in becoming the first industrial nation, combined with its corresponding success in becoming the world's pre-eminent financial nation, which propelled it to such global hegemony as had been achieved by the middle of the nineteenth century.

Yet despite these broad vistas of power and prosperity that would soon be opening up, life for most people in the recently created United Kingdom of Great Britain and Ireland at the beginning of the nineteenth century was a harsh struggle. Most people died before they were forty, and many mothers did so in childbirth; there were no antiseptics, antibiotics or anaesthetics; and for those who survived into old age, there were no pensions or any form of social security. Many people could neither read nor write, their formal education would have been confined to a few years at the local church school, and their knowledge of the vibrant high culture of their times would have been minimal. Most marriages did not last long, because one of the partners died young, the unintended consequence of which may have been that there was less need for easy divorce than there would later be when both partners were living longer. Even the grandest and best-equipped houses were cold and draughty, the only artificial light was provided by candles, and hot water was rarely to be had; while at the bottom of the social scale, overcrowded housing was cold, dark and damp, and the winter months must have been almost unendurable. Towns and cities were polluted by smoke, open sewers and animal excrement, which meant that stench and squalor were an integral part of urban living, even for the rich and privileged, but especially for those who were not. This was a world in which the domestic reach of the state (beyond the generally unreliable informers employed by the Younger Pitt and his successors in the hope of preventing subversion) was still very limited, where there was no police force to maintain public order, and where there was no obligation to deploy the resources of government to help those many people who could not help themselves. Whatever its real but mistaken nostalgic appeal in certain quarters, the United Kingdom of the early nineteenth century was a world that we should be glad has been left behind.

TRADE, EMPIRE AND WAR

The recently consolidated United Kingdom, with all its problems and uncertainties, was also closely connected with much of the greater world that lay beyond its borders, and those links took many different forms. To begin with, it was closely linked to the continent of Europe by virtue of the fact that George III, like his two forebears, was not only His Britannic Majesty, but also Elector of Hanover. As such, he was both a British and a European monarch, and like his two predecessors, this meant he also enjoyed the right to elect the Holy Roman Emperor (until Napoleon abolished that position in 1806). The fact that from 1714 to 1837 the United Kingdom formed part (but the preponderant part) of a dual British-German monarchy was of considerable significance for the men in London who had to conduct the foreign policy of the United Kingdom. For when it came to European affairs, which often in practice in the eighteenth century meant European wars, they had to consider not only British interests but those of Hanover as well, and they were not necessarily the same. Yet this was not the only way in which the British and German worlds were connected across the North Sea, for the links of culture were at least as great as those royal connections provided by the House of Hanover. This had obviously been so in the case of music, when George Frideric Handel had accompanied George I from Hanover to Britain, and those links long survived the composer's death in 1759, as exemplified by the fact that Joseph Haydn composed his last London symphony while staying in the capital in 1795. There was a similar two-way relationship in philosophy, for much of Immanuel Kant's work had been in response to the writings of David Hume; while in a later generation, Goethe's writings on aesthetics and colour would profoundly influence

Turner, while his work on science would be equally influential for Charles Darwin.

Such links between the United Kingdom and the north German princely states were further cemented by a shared Protestant faith, which would be of especial significance for the marriages made by the British royal house during the 1830s and 1850s. But British ties to the European continent also transcended religion, politics and international rivalries, for in some ways, and notwithstanding 1789 and its bloody aftermath, the closest cultural connection was between Britain and France. In some ways Protestant Britons (but not Catholic Irish) defined themselves over and against the Catholic French, but in other ways, relations between the two countries and their two cultures were much closer than they were confrontational. In geographical terms France was Britain's closest continental neighbour, and in peacetime (and sometimes even in wartime) Paris was the European capital that Britons most frequently and appreciatively visited. Any seriously educated Briton spoke French, whether for diplomatic or other purposes, while the British constitution was much admired by such French *philosophes* as Montesquieu and Voltaire. For upper-class Britons seeking to acquaint themselves with European culture, France was in every sense the first destination; but it was also the point of departure for a second Catholic region of the continent with which patrician Britons also shared a strong cultural affinity, and that was Italy. During the eighteenth century, young upper-class British men would go on the grand tour to the Italian kingdoms and city states, where they might learn about the classical culture of ancient Rome, and the late medieval and early modern wonders of the Renaissance. As a result, many British country houses were adorned with ancient (or counterfeit) Roman sculptures, while Gibbon had been inspired to write his epic history of the fall of Rome by contemplating the ruins of the Forum. Throughout the nineteenth century these cultural links between the United Kingdom and Germany, France and Italy would remain strong, whatever the diplomatic stresses and tensions might be: indeed, to a well-educated person such as Palmerston or Gladstone, this 'European sense' would be an essential part of their mental makeup.

There was another part of the world with which the United Kingdom's cultural ties also remained strong, despite the recent loss of the thirteen colonies, and that was in North America. Official relations with the newly constituted United States would continue cool and occasionally conflicted across the next hundred years, but there were many other ways in which connections between the two nations remained very close. The American constitution was constructed as a purified version of the British, with its president as a surrogate monarch, and with the separation of powers between legislature, executive and judiciary, and it was an imitation and adaptation that was the sincerest form of flattery. The United Kingdom and the United States also enjoyed a shared language and thus a cultural heritage that the colonists' erstwhile allies in the War of Independence, the French and the Spanish, could not claim. And once peace had been made in 1783, correct and increasingly cordial diplomatic relations were established between the United Kingdom and the United States, so much so that by the early 1800s, George III was becoming almost as much appreciated in America as in Britain. He was no longer regarded as the tyrant who had sought to subvert the colonists' liberties, but increasingly admired as a decent, honourable and Christian gentleman. This in turn meant that although the two nations had separated in what had initially been a bloody and acrimonious divorce, the United States soon re-established itself as the most appealing destination and hospitable place for emigrants wanting to leave Britain, especially from Scotland and Ireland. By the 1790s and early 1800s they were again heading across the Atlantic, as their predecessors had done before 1776, in their many thousands. Meanwhile, Canada, which had been won by General Wolfe from the French during the Seven Years War, remained loyal to the British throne and government even as the American colonists rebelled. Thereafter, it also became a favoured destination for Britons (and especially Scots) heading west for a new life in the New World.

These connections, by turns cultural and personal, between the United Kingdom and Europe, and between the United Kingdom and North America, were significant in their own right (and would become more so as the nineteenth century drew on); but there were other important links that bound these northern hemispheric nations and regions together. For the seas that surrounded the United Kingdom served not only to secure it against foreign invasion, but also to connect it with trading partners across the English Channel, the North Sea, and the Atlantic Ocean. By the late eighteenth century, Britain had become the foremost trading nation of the world, and not only in peacetime but sometimes in wartime, too. Despite their centuries-old animosity, which had recently been intensified, the British bought luxury items from *ancien régime* France, and exported cheaper products in return; trade with Germany was as much in goods as in culture and ideas; while Britain imported raw materials (especially cotton) from the United States and increasingly exported manufactured goods to it. This, in turn, helps explain why London became such a great eighteenth-century port and, increasingly, the centre of finance for much of the world's trade. But then, as for most of the nineteenth century, the focus remained in the northern hemisphere. By comparison, there was scarcely any trade with the Spanish and Portuguese empires in Latin America, or with the Ottoman Empire, the Russian Empire or the Chinese Empire, none of which welcomed British merchants or British goods. When Lord Macartney led Britain's ill-fated expedition to Peking in 1793, the Chinese Emperor had been distinctly unimpressed by the industrial goods with which he had been presented, and although the United Kingdom was in the process of surpassing China as an economic power it would be several more decades before British traders effectively penetrated the Far Eastern market.

But although much of Britain's overseas trade was with the continent of Europe and North America, there were two other parts of the world with which it was closely and commercially involved at the end of the eighteenth century. One was the West Indies, and the other was India, and in the aftermath of the loss of the thirteen American colonies, they became much more important – in the case of the Caribbean only briefly, in the case of South Asia more lastingly. British trade with the West Indies was primarily in what were deemed by contemporaries to be two commodities, namely sugar and slaves, and the two were closely connected. Slaves were captured and acquired on the west coast of Africa, and then transported in conditions of exceptional brutality and appalling cruelty to the Caribbean, where they worked on the sugar plantations; the same ships subsequently returned to Britain carrying the sugar for which there was a growing domestic demand, before heading off, generally empty, to West Africa to begin their triangular journeys all over again. Other Western nations, with empires in the Caribbean and Latin America, such as France, Spain and Portugal, were also involved in the slave trade; but the British transported slaves in greater numbers, amounting to more than three million, and they held more West Indian islands, among them Jamaica, Barbados, Trinidad, Nevis, St Kitts and St Lucia, which for much of the eighteenth century were the most valuable parts of their overseas empire. Some Britons settled in the Caribbean and made their fortunes as planters, while many people and places in the United Kingdom also became rich as a result. Mercantile families such as the Gladstones of Liverpool did well out of the slave trade, as did landowners such as the Lascelles of Harewood House in Yorkshire, while the westward-facing ports of Bristol, Liverpool and Glasgow, which were deeply involved in the triangular trade, grew fabulously rich as a result. With such connections in high places, the slave owners and slave traders formed a powerful parliamentary lobby, as the slave trade was at its most lucrative and the British West Indies at the peak of its sugar-producing prosperity.

Although contemporaries could not have known it, the West Indies had declined from its economic zenith by the time the Act of Union with Ireland was passed, and for much of the nineteenth century the Caribbean colonies would be among the most depressed and neglected parts of the British Empire. Moreover, by the end of the eighteenth century, a very different, and ultimately more lucrative trading-cum-imperial venture was already beckoning. This was the subcontinent of India, which in the aftermath of the loss of the American colonies would become the United Kingdom's greatest imperial preoccupation – and concern – from the 1800s to the 1850s. As had originally been the case with the Caribbean, the British connection with South Asia had begun as a commercial venture, with the granting of a royal charter to the East India Company during the last years of the reign of Queen Elizabeth. By the middle of the eighteenth century, the British had established major trading posts at Calcutta, Bombay and Madras, and the Company enjoyed a monopoly in exporting manufactured goods from Britain in exchange for large supplies of raw cotton. But the East India Company was not the only such European enterprise trading with South Asia: the French were also a significant presence, and trade rivalry soon morphed into military rivalry, climaxing during the Seven Years War, when Robert Clive vanquished the Nawab of Bengal, the chief local ally of the French, at the Battle of Plassey in 1757. Thereafter, the East India Company was increasingly drawn into Indian political affairs, especially in Bengal, where the ostensibly paramount Mughal Emperor no longer carried authority, and where the British became involved, not only as traders and soldiers, but also as tax collectors and administrators. Yet these Company men were not trained to govern, and many of them were also deeply corrupt, amassing large fortunes by illicit private trading.

This was an increasingly unsatisfactory situation, and the British government carried two major pieces of legislation aiming to bring the East India Company under some degree of official control. The first was Lord North's regulating act, passed in 1773. This established the Governor of Bengal as the pre-eminent authority, and de facto Governor General of British India, with authority over the governors of Madras and Bombay. It also outlawed private trading and established a supreme court in India that would be staffed by British judges. Eleven years later, Pitt passed an additional measure, with the aim of further strengthening government oversight. It established a board of control, the president of which was effectively responsible for the affairs of British India. The post frequently carried cabinet rank, and among its holders would be Henry Dundas, Lord Castlereagh and George Canning. In 1786 Pitt sent Cornwallis to India, where he implemented important land taxation reforms in Bengal known as the Permanent Settlement, and also defeated the French-supported Tipu Sultan in the third Anglo-Mysore War. Cornwallis would eventually be succeeded in 1798 by the Earl of Mornington (subsequently Marquis Wellesley), who strongly believed that Britain's mission in India was proconsular and imperial rather than commercial and financial. To that end he established Fort William College to train young Britons to govern India, and he spent a fortune constructing a monumental Government House in Calcutta, on the grounds that the British Empire in India should be ruled from a palace and not from commercial premises.

Together, British North America, the Caribbean islands and a growing involvement in India constituted what remained of the empire after the American colonies had successfully rebelled. To these must be added Gibraltar, some slave trade forts on the West African coast, the island of St Helena in the middle of the south Atlantic Ocean (where Napoleon would eventually be exiled), and a small colony on the American isthmus. But there were two additional colonies, recently established and of marginal importance at the time, that were also signs and portents for Britain's imperial future. One of them was the settlement that had been established at Botany Bay in Australia in

1788, when what later became known as the 'First Fleet' of British convicts, under the command of Captain Arthur Phillip, arrived. Australia had been 'discovered' by Captain James Cook in 1770, and with the loss of the American colonies it seemed an appropriate and alternative destination for transported convicts. Although there was an indigenous population of Aborigines, the British authorities declared that Australia was *terra nullius*, a land belonging to no one, which meant that the new settlers were free to take it over. Having established a bridgehead in what would become New South Wales, the British later began a second colony in 1803 in what was then called Van Diemen's Land, an island to the south of the main Australian landmass, which would subsequently be known as Tasmania. Four years after the landing at Botany Bay, approximately 1,700 freed black former slaves, who had fled the United States and settled in Nova Scotia in Canada, set sail for Sierra Leone on the west coast of Africa, where they hoped to make new lives for themselves in a British colony that had witnessed several previous failed attempts at settlement. For much of the 1790s, Sierra Leone was plagued by tensions between the Canadian arrivals, the local British officials, and another group of free blacks who subsequently came from Jamaica. There was much ensuing bloodshed, especially between some of the black colonists and the authorities, but the colony survived.

As the case of Sierra Leone illustrates in an exaggerated way, all of the British colonies, whether in Canada, the Caribbean, Africa or South Asia, were ruled from London in an authoritarian manner, as were all other colonies of European empires at the same time. In each case the colonial governor, who had been sent out from Britain, was in full and complete command, and no one, whether white or black, immigrant or indigene, had a vote. As such, the late eighteenth-century British Empire was certainly more authoritarian than the government in Britain. Yet three significant caveats must be entered. In the first place many colonists who supported the British cause at the time of the American Revolution had left for Canada, where they believed they would be more free than if they had stayed in an independent United States: for these so-called 'Empire Loyalists' the British Empire was a place of liberty, whereas it was the newly created United States that was the place of authoritarian intolerance. There was also a growing demand that slavery should be abolished throughout the British Empire, on the grounds that it was an institution contrary to the laws of God, which deemed every man equal in the sight of the creator. In 1787 the London Committee for Effecting the Abolition of the Slave Trade was established, led by Thomas Clarkson and William Wilberforce, and two years later Wilberforce introduced his first bill seeking to abolish the slave trade. At the same time, efforts by successive London governments to regulate the affairs of the East India Company were motivated by concern for the interests of native peoples, and by a determination that Britons in India should not exploit and abuse them. Hence the impeachment of Warren Hastings, the first British Governor General, which began in 1788, for alleged 'high crimes and misdemeanours' against the indigenous population, and for his unacceptably authoritarian behaviour. After a trial lasting twelve years, Hastings was eventually acquitted, but his prosecution, led by Edmund Burke, was eloquent evidence of a developing sense of imperial responsibility and what would later be termed trusteeship.

Such was the British Empire as it existed, regrouped, consolidated, evolved and expanded across the twenty-odd years after the loss of the American colonies. It was an empire that to Gibbon-inspired pessimists seemed to be running down, but which to optimists held out the prospects of further expansion in Canada and in India and perhaps also in the Antipodes. It was an empire in which authoritarian modes of government and control co-existed uneasily and paradoxically with ideas of liberty, concern for native peoples, and a growing momentum for reform. It was an empire founded on a belief in the Protestant religion, but which constantly rubbed up against people who espoused an alternative form of Christianity (such as the Catholic French Canadians), or no form of Christianity at all (such as Hindus and Muslims in South Asia). It was an empire about which successive governments in London knew even less in detail than they did about their home country, and where the official mindset was one of analogical thinking, which, for example, likened the rulers of Indian princely states to great aristocrats and territorial magnates at home. It was also very much a *British* rather than just an English imperium. Scots emigrants were especially prominent among the Britons working on the Indian subcontinent, both as rulers and as soldiers; and the Younger Pitt's colleague, Henry Dundas, who himself came from north of the border, was admired for (or accused of) having ruthlessly promoted what was termed 'the Scottishization of

India' during his years as President of the Board of Control (1793–1801). It was also the case that from the time of the Act of Union onwards, the Irish would not only bridle at their semi-colonial status of subordination, but would also make a disproportionate contribution to the British Empire as proconsuls in Canada and India, as emigrants (especially to Australia), and as officers and foot soldiers in what was misleadingly termed the 'British' army.

The late eighteenth-century United Kingdom was thus a complex and convulsed country, with an unsystematically acquired and settled empire, which found itself involved in yet another global contest, as Revolutionary France declared war on Britain and its empire in February 1793. This conflict would be fought out on the European continent, on the high seas, and also much further afield, and it would last for more than twenty long years, with only the briefest of interludes between 1802 and 1803. There had been scarcely ten years since the American War of Independence had been humiliatingly concluded, and although that had been enough time for the Younger Pitt to repair some of the damage to the national morale and public finances, this was not a war which was greeted with any enthusiasm among the ruling circles in Britain; not least because it carried with it subversive and revolutionary ideologies even more disturbing to the established order than those that had been proclaimed by the leaders of the thirteen colonies in the Declaration of Independence in 1776. Pitt hoped to win this latest Anglo-French conflict with the same policy that his father had so successfully prosecuted during the Seven Years War: Britain would finance its European allies to contain France on the continent, while the Royal Navy would defeat France on the high seas and mop up its colonies. Hence the succession of coalitions, amounting eventually to seven in all, that British governments would repeatedly and laboriously put together between 1793 and 1814. But as their number suggests, this was not a policy that met with success during Pitt's lifetime, and nor would it do so for several years thereafter. The early alliances failed to contain – let alone defeat – France on the continent, and although the Royal Navy successfully prevented an enemy invasion, mopping up the French colonies proved more difficult this time. Moreover, Pitt was forced by his cabinet colleagues to countenance some British military involvement on the continent, but this was a fatal strategic compromise that did not work.

Yet at the beginning, Pitt was confident that the war against France would be short, as in mid-1792 he formed – and subsidized – the First Coalition, consisting primarily of Austria and Prussia. Initially there seemed cause for optimism, as the French attack on the Netherlands failed, there was a royalist insurrection in the Vendée and the defection of France's greatest naval base at Toulon. But the allies failed to capitalize on this promising military and political situation; the French began to organize their 'nation in arms' into what would become the best fighting force the world had yet seen; and in Napoleon they would produce a general of extraordinary genius, who for a time seemed both infallible and invincible (he would win his first twelve battles). The result was that by the spring of 1797, the Dutch, the Prussians, the Austrians had all been defeated, the First Coalition was at an end, Britain had no allies left on the continent and had been expelled from the Mediterranean. To be sure, there had been naval victories against the French at Ushant on the 'Glorious First of June' 1794 and at Cape St Vincent in February 1797; Britain had occupied Corsica, and had acquired Cape Colony, Malacca and Ceylon from the Dutch, once the Netherlands had been overrun by the French. But these were minor consolations: for eighteen months from early 1797, France was triumphant in Europe, and Britain seemed on the verge of defeat. There was domestic distress and subversion, Ireland was on the brink of insurrection, the French were planning to invade Britain, there were naval mutinies, and there was a major financial crisis, which forced the Bank of England to suspend cash payments. Bereft of allies, facing rebellion in Ireland and potentially revolutionary unrest at home, with the navy not to be relied on, and with what seemed the imminent prospect of national bankruptcy, Pitt had no alternative in 1798 but to sue for peace on terms that would have been an acknowledgement of defeat. But the belligerent party in France, flushed with triumph, regarded his proposed concessions as inadequate, and so the war went on.

The next phase of the conflict was in many ways a rerun of what had gone before, both at sea and on land. In 1798 Pitt laboriously put together a Second Coalition, consisting of Prussia, Austria and Russia, whose armies would be largely financed by his new income tax, to pursue the continental battle against France which, increasingly, meant against Napoleon. Meanwhile, the British re-entered the Mediterranean and scored further naval victories against the Dutch at Camperdown

(October 1797) and against the French at the Nile (August 1798); in India, urged on by Wellesley, they defeated and killed Tipu Sultan in the fourth Anglo-Mysore War (May 1799), and they also captured Malta (September 1800). But these maritime and colonial successes, which were a vivid reminder that this was a war of empires at least as much as it was a war of nations, could not counter the succession of defeats suffered by the Second Coalition on land. George III's son, the Duke of York, failed to deliver victory in the Netherlands, and other efforts by the British to make an impact on the continent were equally unsuccessful. Meanwhile, Napoleon, having won on land in Egypt, but lost at sea, returned to Paris in August 1799, proclaimed himself First Consul in December, and took complete control of the government. Secure in power, he moved quickly and decisively to destroy the armies of the Second Coalition, which he did with spectacular success at the Battle of Marengo (June 1800). Austria thereupon withdrew from the Second Coalition, as did Russia, and Britain was once again left with no allies on the continent. To make matters worse, in December 1800, the Scandinavian powers formed the so-called League of Armed Neutrality, to prevent the search by the British of neutral ships for contraband. More ominously, from Britain's point of view, this meant the broken French fleet would be reinforced by the strong Danish fleet, and that the Baltic, with its valuable maritime supplies of pitch, hemp and pine, would be closed to the Royal Navy.

A TEMPORARY AND TARNISHED PEACE

Accordingly, when the Younger Pitt resigned in February 1801, just one month after the Act of Union with Ireland came into effect, and on account of his disagreement with the king over Catholic Emancipation, his reputation as a war leader was significantly lower than his earlier reputation had been as a peacetime premier. He had woefully underestimated the length and cost of the war that had begun in 1793, and also the resilience and resourcefulness of the French; he had overestimated the capacity of the Russian, Prussian, Austrian and Dutch armies to confront and conquer the massed troops of the French nation in arms; and although the British had done well by sea and taken French and Dutch colonies, this hardly weighed in the balance. Pitt would go down to posterity as 'the pilot who weathered the storm'; but for all his stoicism and fortitude, he had little idea where he was going, or how to get there. So when Henry Addington succeeded him as prime minister in March 1801, he was confronted by the unresolved issues arising from the Irish Union and by the military challenges resulting from an unsuccessfully prosecuted war. Several of Pitt's cabinet, recognizing the domestic and international crises that the nation faced, stayed on in the new administration, and Pitt himself gave Addington his support from outside. Pitt had rejected Napoleon's suggestion of a negotiated peace in 1799, but Addington was more willing to talk terms, and he instructed his Foreign Secretary, Lord Hawkesbury, to begin negotiations. Despite its naval and colonial victories, Britain's bargaining position was weak, even allowing for Horatio Nelson's audacious triumph at the Battle of Copenhagen, in April 1801, when he destroyed the Danish fleet, which brought the League of Armed Neutrality to a close and ensured the Baltic would remain open to the Royal Navy.

By September 1801, a preliminary agreement was reached, and soon after, Addington dispatched Cornwallis, by now free of his Irish responsibilities, to conclude the negotiations with Napoleon's brother, Joseph Bonaparte, and his Foreign Minister, Talleyrand. Cornwallis was under strong pressure from the new British government to come to terms, and at the end of March 1802 he signed a treaty at Amiens. France came out of it decidedly better than the United Kingdom, which gave back most of its colonial conquests. The Cape of Good Hope and its former West Indian islands were restored to the Dutch; the island of Minorca was returned to Spain; the captured Indian and African stations were to be given back to France; so were the West Indian colonies of St Lucia, Tobago, Martinique, St Pierre and Miquelon; and the British government also agreed to withdraw its troops from Malta and Egypt. All that the United Kingdom retained of its earlier conquests were Trinidad, taken from Spain, and Ceylon, captured from the Dutch. But after nearly a decade of inconclusive war and protracted domestic turbulence, the peace was wildly popular in Britain, and only twenty MPs voted against it when it was debated in the Commons. However, among Pitt's former colleagues who were now out of government, there was a strong feeling that too many concessions had been made, especially as in its very later stages the war had seemed to be turning Britain's way. Lords Grenville and Spencer denounced the terms as humiliating, Henry Dundas spoke against them in private, while William Windham (Pitt's former Secretary of War) was said to be 'absolutely raving' about what he called the 'death blow' to the country. It was not 1783 all over again; but nor was Amiens a victorious or triumphant peace. Cornwallis shared these doubts, and had been far from happy with the agreement he had been bounced by the Addington government into accepting. But in the end he obeyed his instructions, for he had also worried about 'the ruinous consequences of . . . renewing a bloody and hopeless war' – even though it was widely recognized that 'renewal' would have to come.

2
Britannia Resurgent, 1802–15

The latest instalment in the protracted Anglo-French contest had ended in a stalemate that was by turns welcome yet disappointing, and no one contemplating what could only be a temporary truce would have predicted that within little more than ten years the United Kingdom would triumph absolutely over France, and that the global nineteenth century would turn out to 'belong' to Britain more than to any other nation. The Treaty of Amiens had optimistically declared that 'peace, friendship and good understanding' existed between the two countries, and during the later months of 1802 British visitors flocked to France, and French tourists swarmed across the Channel in the opposite direction, many of them eager to resume the political friendships and cultural contacts that had been broken since the early 1790s. Among the first Britons to reach Paris was Charles James Fox, who wanted to find out what kind of France the Revolution had created: but his talks with Napoleon were unrewarding, and this resulted in a disenchantment that would have significant political consequences. Other visitors included the scientist William Herschel, who conferred with colleagues at the Observatoire, the artist Turner, who filled a sketchbook with drawings of the *Exposition des produits de l'industrie française* held at the Louvre, and the philosopher, painter and critic William Hazlitt, who spent three months studying and copying Old Masters in the same museum. Meanwhile, William Wordsworth and his sister Dorothy had crossed over to Calais, where they met up with Annette Vallon, with whom the poet had fathered a daughter, Caroline, during their affair nine years earlier. Wordsworth now met Caroline for the first time, and recalling a seaside walk with her during his visit, he would later write the sonnet, 'It is a beauteous evening, calm and free'. Among French visitors to England at the same time were the wax artist Marie Tussaud, who created and curated an exhibition in London similar to her pioneering establishment in Paris; and the balloonist André-Jacques Garnerin, who staged displays in the capital, and also made the hot-air journey from London to Colchester in forty-five minutes.

Many of these cross-Channel travellers from Britain were relatively young: Hazlitt was twenty-four, Turner was twenty-seven, and Wordsworth was thirty-two. As such, they belonged to that eager and ardent generation who had initially been so enthused by the French Revolution, which had taken place when they had only been in their teens. In the same year Arthur Wellesley was only thirty-three, the Duke of York thirty-nine, the Prince of Wales forty-two, William Pitt forty-three, Lord Nelson forty-four, and Charles James Fox fifty-three. To be sure, George III was in his early sixties, but for the most part, the early nineteenth-century United Kingdom was a young man's country, at the heart of a young man's empire, and fighting a young man's war. This was partly because, with life expectancy averaging little more than forty, most Britons died well before reaching what should have been their allotted biblical span. Old age was unusual and uncommon. But the young man's country was also the result of the sudden expansion in the birth rate during the second half of the eighteenth century, which meant more young people were living in the United Kingdom than ever before, and during the first half of the nineteenth century over 60 per cent of the population would be less than twenty-four years old. Indeed, that first generation of baby boomers included the future Tory prime ministers Lord Liverpool (previously Lord Hawkesbury) and George Canning, who had both been born in 1770, and Sir Robert Peel born in 1788; while among Whig premiers, Lord Melbourne had been born in 1779, Lord Palmerston in 1784, and Lord John Russell in 1792; and Lord Aberdeen, who would head a coalition government, had been born in the same year as Palmerston. This also meant that during the first two-thirds of the nineteenth century, Britain and its empire would be governed by men whose early and most impressionable years had been spent in the shadow of the French Revolution and the Terror, and all of them had still been relatively young at the time of the Battle of Waterloo.

Despite the protestations at Amiens that 'peace, friendship and good understanding' now existed between the United Kingdom and post-Revolutionary France, it was soon clear that this forced respite would be of very short duration, as some of the treaty's substantive terms were never in fact honoured or implemented. The British refused to undertake demobilization, but instead maintained a 'peacetime' army of 180,000 men; they also declined to withdraw their troops from Malta, which was an essential staging post on the route from the United Kingdom to Bengal; while Lord Wellesley, having recently vanquished Tipu Sultan, made no attempt to implement the peace terms in India. Napoleon defied the treaty even more belligerently, sending troops into Switzerland and northern Italy, invading Hanover (a very serious matter in the eyes of George III), and dispatching a naval expedition across the Atlantic to regain control over revolutionary Haiti, and to reoccupy Louisiana, which Spain had ceded to France in 1801. With the two nations once again drifting towards war, the resumption of formal hostilities was only a matter of time, and for those British visitors still sojourning in France there were serious risks they would not get out before the fighting began again. The Anglo-Irish writer Maria Edgeworth spent the winter of 1802–03 in Paris, but then had to leave in a hurry, and she was lucky to return safely to Dover in March. Another author, Fanny Burney, was less fortunate. She travelled to Paris in April 1803 to be reunited with her husband, the Comte Alexandre d'Arblay; but the following month, in response to Napoleon's increasingly aggressive, predatory and belligerent behaviour, Britain again declared war on France. With hostilities resumed, it proved impossible for Burney to leave, and she was obliged to remain in France until the end of the war in 1815. A similar fate befell the Scottish peer Lord Elgin, caught en route travelling back to Britain from Greece, where he had recently obtained the marbles from the Parthenon with which his name would become indelibly associated, and which the government would eventually acquire from him in 1816 and hand over to the British Museum.

A LONG, HARD, GLOBAL CONFLICT

The Anglo-French struggle that was resumed in 1803 was the seventh encounter, and would be the culminating conflict, between the two nations since they had fallen out in the immediate aftermath of the Glorious Revolution of 1688, in which William III had confronted Louis XIV. But England and France had been hereditary enemies for much of the preceding half-millennium: briefly during the 1600s and before that in the 1620s, more protractedly from the 1510s to the 1550s, and at much greater length during the fourteenth and fifteenth centuries in what was arrestingly but mistakenly termed the Hundred Years War. That late-medieval confrontation had actually lasted from 1337 to 1453, but those 116 years would be surpassed by the 126 years that separated 1689 from 1815. There were other differences, too. The first Hundred Years War had been essentially a dynastic contest between the royal houses of England and France; the conflict of 1803–15 would be a struggle between the two nations and their respective empires. The Hundred Years War was largely contested by native English and French troops; but the British armies that fought between 1803 and 1815 also recruited Scots, Irish and continental soldiers, while the French forces included conscripts drawn from the many European territories that Napoleon had conquered and suborned. The Hundred Years War had been largely confined to the two belligerents of England and France, but the Napoleonic Wars were fought out across most of Europe, extending from Spain to Sweden, Portugal to Russia, Norway to Naples. The Hundred Years War was waged and decided almost entirely on land; but the confrontations of 1803–15 took place as much on the seas and the oceans as on *terra firma*, and would extend far beyond Europe to the Mediterranean, the Atlantic, the Caribbean, North America, southern Africa and South Asia. Indeed, in their length, intensity and expense, the Napoleonic Wars would dwarf all earlier Anglo-French conflicts – and since then, the two nations have never again gone to war against each other, but have invariably been on the same side.

Once the conflict was resumed, Addington's government tried to repeat the strategy against Napoleonic France that the Younger Pitt had unavailingly pursued against Revolutionary France: namely waging and winning the maritime war, conquering the enemy's imperial outposts, subsidizing European allies who would do most of the fighting on land, and mounting occasional British forays onto the continent. But for much of the time, the results would be no better than mixed, for as in the earlier Revolutionary Wars the British retained command of the seas but the French dominated the land, and it would be many years before that stalemate was decisively broken to the United Kingdom's advantage. Despite recent retrenchments in the fleet, the Royal Navy was in a relatively strong position when hostilities again ensued. It soon took back the Caribbean islands of France and its co-belligerents, among them St Lucia and Tobago, and it also captured Dutch Guyana. At the same time, the British government persuaded Russia and Austria to join a Third Coalition against Napoleon, paying them £1.75 million for every 100,000 men they put in the field. But the French army was brilliantly led, well trained, highly experienced, and very mobile; and in 1804, Napoleon not only declared himself to be Emperor of the French, but also determined to invade and conquer Britain, once again assembling a large army at Boulogne. But as had been the case for his predecessors in 1797 and 1798, Napoleon needed the assured command of the sea if he was to accomplish this objective. To be sure, he had built up the French navy, which was a considerable fighting force, and it was further strengthened in 1804 when Spain entered the war on Napoleon's side. Once again, fear of invasion ran high across the United Kingdom; but in October 1805, Nelson defeated the combined French and Spanish fleets at the Battle of Trafalgar, where he was himself struck down in his moment of supreme triumph. The threat of a French invasion had been averted, Britain's maritime supremacy would henceforward be unchallenged, and the Royal Navy had obtained a death-and-glory national hero, who was buried with unprecedented pomp and pageantry in St Paul's Cathedral.

But while the Royal Navy carried all before it on the seas and the oceans, the land-based war again went badly, as the Third Coalition, like its predecessors, suffered humiliating defeat. Even before Trafalgar, Napoleon had rushed his army from Boulogne to the upper Danube, where he annihilated the Austrians at the Battle of Ulm (October 1805), and he then moved his troops rapidly eastwards where they smashed a combined Austrian and Russian army at the Battle of Austerlitz (December 1805). Once again, the Habsburg Emperor was obliged to sue for peace, and this enabled the French to consolidate their hold of what had previously been his lands and dominions in northern Italy. These severe defeats again demonstrated how difficult it was to support and sustain a successful European alliance, especially when it was confronted – and conquered – by a military genius such as Napoleon. In 1806 Russia, Saxony and Sweden joined with the United Kingdom to form the Fourth Coalition, along with Prussia, whose earlier non-involvement had seriously weakened its predecessor. But in

October, Napoleon crushed the once-mighty Prussian army at the Battle of Jena, and soon after he captured Berlin. The Russian forces put up more stubborn resistance, but in June 1807 they were heavily defeated at the Battle of Friedland. The treaties of Tilsit, negotiated later in that year, signified the end of the Fourth Coalition. Prussia lost much of its lands and became a virtual French satellite, and Russia agreed to ban British trade and eventually to join in an alliance with France. Napoleon had achieved a continental mastery that neither Louis XIV nor Charlemagne had rivalled. Spain, Italy and the Low Countries were subservient, southern and much of western Germany were merged into the pro-French Confederation of the Rhine, western Poland was turned into the equally friendly grand duchy of Warsaw, the new kingdom of Westphalia was created, ruled by Napoleon's brother Jerome Bonaparte, and the Holy Roman Empire was declared to be at an end. Only Portugal and Sweden remained as the United Kingdom's European allies.

Having failed to launch a successful seaborne invasion, Napoleon resolved to defeat 'the nation of shopkeepers' by other means. In 1806 and 1807 he issued decrees banning any of the countries over which he ruled from trading with Britain. Since the United Kingdom depended heavily on European markets for its export industries, this was a serious threat. At the same time, Napoleon sought to accumulate the timber and other shipbuilding resources, supplied by the Baltic countries and Dalmatia, and to deny them to the Royal Navy, which constituted an additional challenge. Moreover, reduced British earnings from exports would starve London of the currency it needed to pay subsidies to any future coalition, and to purchase goods for its own expeditionary armies or to spend on maintaining the Royal Navy. Napoleon's aim, then, was to weaken the United Kingdom economically, financially and militarily; and once he had again built up the French fleet, he would be able to mount a renewed assault directly across the English Channel. The British government retaliated in 1807 by issuing the Orders in Council, which in turn banned Napoleonic France from trading with the United Kingdom or with its allies or with any neutral nations, and which empowered the Royal Navy to blockade all French and allied ports. But for a time it looked as though Napoleon's strategy was working, for the United Kingdom's heavy dependence on foreign commerce made it vulnerable to his trading ban.

In 1808, and again in 1811–12, the commercial warfare waged by the French and their European allies and clients took a severe toll on Britain's export trades: stocks of manufactured goods piled up in Thames-side warehouses, their producers unable to find European markets, and the London docks were also bursting with colonial produce that could not be re-exported. Meanwhile, the United Kingdom's national debt ballooned at an alarming rate, and there were fears that national bankruptcy could not be far off, and that Napoleon might indeed win the economic war.

To be sure, the Royal Navy could roam the seas and the oceans virtually unchallenged in the aftermath of Trafalgar, and this made possible a succession of further colonial conquests. They included the Cape of Good Hope (again) in 1806; Curacao and the Danish West Indies in 1807; several of the Molucca islands in 1808; Cayenne, French Guiana, San Domingo, Senegal and Martinique in 1809; Guadeloupe, Mauritius, Amboina and Banda in 1810; and Java in 1811. There were also further British successes in South Asia, where Lord Wellesley and his successor as Governor General, the Marquis of Hastings, waged war against the Marathas, annexing large swaths of territory in an attempt to create a British Empire on the subcontinent that would serve as an oriental counterweight to the vast French imperium that Napoleon was creating and consolidating in Europe. The most notable British victory was won by Lord Wellesley's younger brother, Arthur Wellesley, at the Battle of Assaye in September 1803. Part cause, part consequence of this substantial upscaling in South Asian warfare was that between 1790 and 1815 the East India Company increased its fighting force from 90,000 men to 200,000, making it one of the largest standing armies in the world. Britain might not be a military power on the landmass of Europe, but it was increasingly becoming one on the Asian subcontinent, where British officers commanded large numbers of Indian troops known as sepoys. Meanwhile, and closer to home, the Royal Navy successfully mounted a devastating pre-emptive strike against the Danish navy which, according to a (supposedly) secret clause in the Treaty of Tilsit, was to be taken over by France in order to further strengthen Napoleon's fleet. The government in London regarded such a transfer as unacceptable, but the Danes refused to surrender their ships to Britain instead, whereupon the Royal Navy bombarded Copenhagen in the late summer of 1807, killing 2,000 civilians, destroying many prominent buildings, and capturing eighteen Danish ships of the line, without ever even declaring war. Even Napoleon was allegedly impressed at such a ruthless display of British maritime might.

Yet none of these successes enabled the British to beat the French on land in Europe, and in other theatres of war the picture remained bleak. In the autumn of 1807 French troops invaded Portugal, one of the United Kingdom's two remaining allies, and a major route through which British goods could still enter Europe, despite Napoleon's continental blockade. Thereafter, the French forces turned east, and established direct control over Spain, which had long been a French ally. By the middle of 1808 they had taken Madrid, King Ferdinand VII had been deposed, and Joseph Bonaparte, another of Napoleon's brothers, was imposed as the new Spanish king. Later that year, there was a Spanish rebellion against the ousting of their monarch, but the French Emperor moved rapidly and decisively to crush it. In 1809 a Fifth Coalition was formed, as the Austrians again renewed hostilities with the French, in a misguided attempt to recoup their earlier losses. But the Habsburg Emperor's armies were heavily defeated at the Battle of Wagram in July, and once again Vienna was compelled to sue for peace, to cede yet more territory to France, and to pay a substantial indemnity. Wherever continental opposition arose to Napoleon's will and rule, it was swiftly and effectively dealt with, and the Fifth Coalition, which collapsed in the aftermath of Austria's latest defeat, was the shortest lived of any of the continental alliances put together since 1793. Even the undoubted successes of the Royal Navy were undermined by the tendency to fritter away many resources on misguided schemes that were ill-conceived and never likely to succeed in the first place. One such misadventure was the bombardment and capture of Buenos Aires in 1807–08, in the hope that this would open up the whole of Spanish America to British goods, but in the end the British troops were forced to withdraw. A second was the 40,000-strong expeditionary force sent in 1809 to occupy the island of Walcheren in the estuary of the River Scheldt, with the aim of capturing Antwerp, destroying the French fleet anchored nearby, and relieving pressure on Austria. But the military leadership was decidedly feeble, none of these objectives was achieved, and there was an ignominious retreat with a substantial loss of life.

As a result, the strategic stalemate continued: Britannia ruled the waves, while Napoleon lorded it over the land, but neither could make any effective inroads into the other's area of supremacy. Yet in retrospect the situation had already begun to change to Britain's advantage, for Napoleon's drive into Spain in 1808 was the first such thrust that failed to deliver a completely successful conquest. To be sure, he had vanquished the Spanish armies, but the local population, which had become increasingly disenchanted with the rule of his brother, Joseph Bonaparte, refused to acquiesce and instead resorted to guerrilla warfare. The French armies might be invincible on the battlefield, but they were not well equipped to deal with this very different form of fighting, which meant that a French force of more than 300,000 troops would soon be permanently but ineffectively tied down on the Iberian Peninsula. This Spanish insurrection also gave the British their first opportunity since 1793 to gain a foothold on the continent, and in July 1808 Lieutenant General Sir Arthur Wellesley, who had returned from India, become an MP, and was serving as Chief Secretary for Ireland, was dispatched with 9,000 men to Portugal to aid the rebels. After some initial delays and setbacks, where his early successes in Portugal were nullified by the Convention of Cintra, Wellesley proved himself an imaginative and resourceful leader. In 1809 he won victories at Oporto in Portugal and Talavera in Spain; the following year, he mounted a brilliant defence at Torres Vedras; in 1811 he beat the French at Fuentes de Onoro; in 1812 at Ciudad Rodrigo, Badajoz and Salamanca; and in 1813 at Burgos and Vitoria. These were remarkable successes, and Wellesley would be lavished with honours, eventually becoming the Duke of Wellington as well as the second great British hero of the war. The Iberian insurrection carried other benefits too: for in most of the United Kingdom's previous conflicts against the French, the Spanish had sided with the enemy. Now they were allies, and this meant not only that Gibraltar was secure, thereby strengthening Britain's position in the Mediterranean, but also that Spain and its Latin American colonies were open for British exports, despite Napoleon's decrees to the contrary.

The long-running Iberian campaign became known as Napoleon's 'Spanish ulcer', and it was the first indication that he might not be invincible after all. But the succession of defeats that Wellington inflicted on the French army to the southwest of the Pyrenees was not enough by itself to bring the emperor down, while he continued to control so much of Europe. For Napoleon to be defeated he would have to be vanquished on a major battlefield by a very large continental army; yet for that to happen, a successful coalition of European powers would have to be organized against him, but the precedents provided by the previous five alliances were not encouraging. In the end, it was the French Emperor himself who provoked such a winning combination, following his misguided decision to invade Russia in 1812. Napoleon led a Grand Army that was 600,000-strong in the assault on Moscow, but it turned out to be a disaster: 270,000 troops were killed, 200,000 more were captured, and the emperor also lost significant numbers of guns and horses that proved impossible to replace. The result was a major collapse of morale in the French army, and the successful creation of the Sixth Coalition, consisting of Spain, Portugal, Russia, Sweden and Britain, which was joined in 1813 by Prussia and Austria. Once again, the United Kingdom supplied the money and weapons, but this time the alliance held, in part encouraged by Wellington's triumph over Joseph Bonaparte's army at the Battle of Vitoria. Later in 1813, the combined forces of Austria, Sweden, Prussia and Russia vanquished Napoleon at Leipzig at the aptly named 'Battle of the Nations', in a much greater clash than anything that Wellington's peninsula army had experienced, as 365,000 Coalition troops overwhelmed 200,000 of Napoleon's soldiers in four days of brutal fighting. This was a crushing defeat for the French army, which ever since the Revolution had seemed all but invincible.

Napoleon's decision to invade Russia had proved fatal, and his decisive defeats on the eastern front marked the beginning of the end – in a way that Wellington's victories in Spain would never have done – as the French forces were expelled from Germany in the aftermath of the Battle of Leipzig, compelled to retreat west of the Rhine, and eventually driven back to their homeland. Napoleon conducted a brilliant tactical defence in northeastern France early in 1814. But by then his army was drained in strength, its morale was eroded, and it contained too many raw recruits; and now that the fighting was taking place on French soil, popular support for the emperor also began to melt away. His last hope was that the Sixth Coalition would dissolve, as the allies fell out among themselves over their ultimate war aims. But the novel prospect of victory kept them together, while any incipient disagreements were successfully papered over in the spring of 1814 by the British Foreign

Secretary, Castlereagh, who negotiated the Treaty of Chaumont with Russia, Prussia and Austria. It was agreed that the Sixth Coalition would stay in being until Napoleon was finally defeated, and the United Kingdom undertook to provide another £5 million in subsidies. Napoleon was now fighting a war on two fronts, and on French soil: as the allied armies pushed westwards, and as Wellington's troops crossed the Pyrenees and advanced from the south. Napoleon's generals reluctantly accepted that the war was lost. In April 1814 the emperor abdicated, and the man who had dominated most of Europe for the best part of a decade was exiled to the island of Elba in the Mediterranean. 'My power,' he had rightly observed in his heyday, 'depends on my glory, and my glories on the victories I have won.' Once Napoleon stopped winning, the glory was gone and the power evaporated, even in France itself. Or so, for the time being, it seemed.

In the end, then, the policies that the Younger Pitt and his successors had pursued so disappointingly for so long had borne fruit and come good, as the Sixth Coalition held together and Napoleon was vanquished. But it had been a long, hard struggle, and until its final stages the allied road to victory was neither assured nor clear-cut. This was especially true for the United Kingdom which, as a by-product of the war with France, became engaged in another conflict, on the opposite side of the world, with the recently independent United States. The separation between the American colonists and the British Empire in 1783 had been neither amicable nor complete. During and after the war Americans who remained loyal to Britain poured into Canada in their thousands, eager to remain within the British Empire, which was one reason why a renewed Anglo-American war seemed a distinct possibility. Meanwhile, the fledgling republic was beginning to entertain imperial ambitions of its own. In 1803 President Thomas Jefferson purchased Louisiana from the French, who had recently regained it from Spain, for $11 million, and British Canada seemed to some Americans the obvious next area for annexation. There were also several issues that had not been settled in 1783, concerning such matters as boundaries and fishing rights. Relations between the United States and the United Kingdom had been difficult from the very beginning, and as the Napoleonic Wars dragged on the Americans became increasingly resentful of what they regarded as Britain's high-handed behaviour in the Atlantic Ocean. The Orders in Council interfered with the free movement of American shipping, the

British claimed the right to search all neutral vessels, and thousands of American merchant seamen were impressed into the Royal Navy. The Americans were also aware that the British were making alliances with the indigenous tribes that lived along the border with Canada, with the aim of winning back territory which they had controlled before the Revolution, but which was now governed from Washington.

These tensions came to a head in June 1812, the very same month that Napoleon invaded Russia, as the United States declared war on the United Kingdom. The population of the American republic significantly outnumbered that of British Canada, but despite their military commitments in Europe the British were able to field more troops, many of whom had already seen active service, and they also had sufficient spare naval capacity to blockade the American coast. Moreover, the republic's forces were poorly organized and indifferently led, and their early incursions into Canada were easily rebuffed, as the British successfully waged a defensive war. But by the end of 1813, the United States had improved its military capability, and had gained control of Lake Erie. In the following year, the British retaliated by invading the United States, defeating American forces at Bladensburg in Maryland, occupying Washington DC for a time, and even setting fire to the White House and the Capitol. By then, it was becoming increasingly apparent that neither side could win, and in June 1814 peace negotiations were begun at Ghent. By the end of the year, the two nations had agreed that the border between America and Canada would remain in its pre-war position, and the conflict was formally concluded. But it took time for news of the treaty to reach the western hemisphere, and in January 1815 the Americans successfully repelled a British attack on New Orleans, while the British defeated the Americans at Fort Bowyer. In some ways this war was a sideshow, compared to the stirring events that were taking place in Spain and eastern Europe. Many people in Britain did not even know their country was at war with the United States; many New Englanders, who were heavily dependent on the transatlantic Anglo-American trade, remained lukewarm throughout the conflict; and the result could best be described as a draw. But just as Napoleonic France had failed to prevail against Tsarist Russia, so the United Kingdom had been unable to defeat the United States. Here was an early indication of the limits to Britain's nineteenth-century global hegemony, and they would become more pronounced as the century continued.

These seismic shifts in geopolitics lay far in the future: from April 1814 onwards the immediate task that confronted the United Kingdom, along with Austria, Prussia and Russia, was to try to make a lasting European peace. Following Napoleon's abdication and deportation to Elba, the Bourbon dynasty was restored in the person of Louis XVIII, the younger brother of the previous, guillotined monarch. The Treaty of Paris, signed in May, declared that France's borders would be returned to the boundaries that had existed in 1792, before the Revolutionary Wars had begun; and it also unified the Dutch and the Belgians within a newly created kingdom of the Netherlands, which would act as a buffer against any further French attempts at territorial expansion. In September the representatives of the great powers moved on to Vienna, where they planned to settle all remaining matters, and to consider how the hard-won peace might be maintained in post-Napoleonic Europe. These gatherings were attended by emperors and kings, aristocrats and grandees, who entertained each other extravagantly by night, even as they negotiated hard with each other by day, and the most significant British figures attending were Wellington and Castlereagh. Their aim was not only to bring peace to a continent that had been racked and ravaged by war for a generation, but also to restore the hierarchies and values which both the French Revolution and the usurper Napoleon had been misguidedly determined to overthrow, namely monarchy, legitimacy, order and tradition, the very structures and sentiments that George III had come to represent in his apotheosized old age. The United Kingdom was the only nation that had not been subjected to invasion and occupation, its unwritten constitution and parliamentary government had survived intact, it had successfully vanquished subversion and revolution at home, and if Wellington and Castlereagh had their way, those threats to stability would now be banished from the continent as well. This was a very different world, and so was Britain's place within it, from that which had existed in 1789 or 1802.

THE POLITICS OF PATRIOTISM

With one conspicuous exception, every major modern war in which the United Kingdom and the British Empire have been involved across the last two and a half centuries has been associated with significant (though not always successful) political leadership: the Elder Pitt in the Seven Years War, Lord North in the American War of Independence, the Younger Pitt in the French Revolutionary Wars, Lord Palmerston in the Crimean War, Lord Salisbury in the Boer War, Asquith and Lloyd George in the First World War, and Winston Churchill in the Second World War. These were the men who took the credit for military triumphs and bore the blame for military disasters. But the one large-scale conflict that did not conform to this general rule was the Napoleonic Wars of 1803–14. For while, in the persons of Nelson and Wellington, they threw up two military leaders who were undoubted national heroes, there was no dominant, long-serving prime minister: instead, there were six different administrations that held office from the beginning to the end of hostilities, each of them averaging less than two years in power. Here was a protracted period of high political instability the like of which had not been seen since the first decade of the reign of George III. But then, the nation had generally been at peace; whereas for most of the time between 1802 and 1814 the United Kingdom was at war, and it was – and is – widely believed that changing leaders during a major conflict was not a good idea. Yet during the years when Napoleon was battled and beaten, no prime minister succeeded in establishing a firm hold on power at home, or masterminded military operations overseas: Addington was derided as a mediocrity; when the Younger Pitt briefly returned to power in 1804, he was but a shadow of his former self; his two immediate successors, Lord Grenville and the Duke of Portland, were stop-gap Whigs; Spencer Perceval, who came next, was dismissed as 'Little P'; and no one would have believed that when Lord Liverpool (as Lord Hawkesbury had become) formed his administration in 1812, he would still be in office fifteen years later.

But while the politics of the Napoleonic Wars were unusual in that no prime minister became the supreme generalissimo and commanding strategist, the picture needs to be modified, in part because the political instability of the 1800s bore only a superficial resemblance to that of the 1760s. At the beginning of his reign George III had interfered constantly, and his views had often been decisive in determining the fate of ministers and ministries; but by the 1800s he was far less influential or involved, and apart from his determined stance against Catholic Emancipation, which had led to the Younger Pitt's resignation, the king was not personally responsible for the rapid turnover of administrations in the way he had been half a century before. Three of these governments were short lived because their leaders either died or were incapacitated while in office: the Younger Pitt was the first prime minister in modern times to expire in harness, the Duke of Portland suffered a stroke in the autumn of 1809 and had to resign, and Spencer Perceval was assassinated. Moreover, these rapid comings and goings at the very top concealed significant continuities. For much of Addington's premiership the Younger Pitt was willing to support him; when Pitt returned to power, he soon persuaded Addington to join his cabinet (as Viscount Sidmouth); and Sidmouth continued in office under Pitt's successor, Lord Grenville. Even more important was the almost constant presence in government of the Anglo-Irish patrician Castlereagh, who wielded significant influence in successive cabinets, and by the end of the Napoleonic Wars would be regarded as the one British politician of major stature in Europe. He had been the Younger Pitt's Chief Secretary for Ireland when the Act of Union was passed; he entered Addington's cabinet as President of the Board of Control, overseeing British India; he held office in all six of the wartime administrations with the exception of Lord Grenville's; and from the spring of 1812 he would be a very influential Foreign Secretary.

The first of the cabinets in which Castlereagh would hold office had been formed by Addington early in 1801, when he replaced the Younger Pitt, who had not only offended his sovereign on the subject of Catholic Emancipation, but was also worn out by the strain of almost twenty continuous years as premier. Addington was looked down on for his humble origins, and for being a less commanding figure than his predecessor; 'Pitt is to Addington,' ran the disparaging rhyme, 'as London is to Paddington'. Yet this was an unfair judgement: Addington pushed forward the peace negotiations leading to the Treaty of Amiens, which bought time and a much-needed breathing space, and once war with the French was resumed he brought a new clarity and purpose to domestic, military and foreign policy. He revised and improved Pitt's innovative income-tax scheme, by deducting the revenue at source, which simplified the system and reduced evasion, thereby laying the solid financial foundation for the subsidies that the United Kingdom would pay to successive continental coalitions until Napoleon was beaten. Addington supported the First Lord of the Admiralty, Lord St Vincent, in reforming the administration and improving the discipline of the Royal Navy, which meant there would soon be enough ships available to defend home waters and to dispatch a substantial force to the West Indies. He also demonstrated resolution and determination in maintaining public order and in suppressing potential revolutionary threats, especially in the case of Colonel Edward Despard, an Irishman and former army officer, who for reasons of personal grievance devised an improbable plot to assassinate the monarch and his ministers. But Despard was arrested and executed in 1803, and thereafter the rumours of conspiracy and oath-taking, and the reports of clandestine night-time meetings and secret conferences in the back rooms of London taverns, all died away. There may have been greater concern about a possible French invasion in 1803–05 than there had been in 1797–98, but the fear of domestic subversion was considerably less.

This was not only on account of the successful government crack-down on potential revolutionaries. It was also because, in the realm of popular politics, protest was increasingly being superseded by patriotism. Early enthusiasm for the French Revolution had largely evaporated, as Napoleon was increasingly seen, not so much as the embodiment of liberty, equality and fraternity, but rather as a usurper, tyrant and megalomaniac determined to suborn the whole of the continent to his will, the United Kingdom included. The principal indication of this changed public mood was Addington's remarkable success in strengthening home defences, and in rousing the nation to action and (literally) to arms. By early 1804 he had raised 85,000 men for the militia and over 400,000 volunteers for the home guard, and in the vulnerable and exposed southern counties approximately half of all men aged between seventeen and fifty-five joined up. There are no precise figures available, but it seems possible that 800,000 men, or one in five of the male population of military age, were in armed service at that time, which was the greatest popular mobilization to take place in Hanoverian Britain, and it was mobilization on the side of order not subversion. From the authorities' point of view this was just as well, for never before, and never since, have so many Britons been given weapons. Meanwhile, such erstwhile supporters of the French Revolution as the poets Wordsworth, Coleridge and Southey had long since abandoned their earlier radical beliefs, and so, more recently, had such previously dissident Whigs as Charles James Fox, disenchanted by his Parisian encounter with Napoleon, and the playwright and Whig politician Richard

Brinsley Sheridan. As the invasion scare intensified, so did popular patriotic fervour. George III was increasingly seen as the embodiment of decency, probity and legitimacy, compared to the cross-Channel tyrant, upstart and usurper, and the king was cheered and acclaimed whenever he reviewed and inspected corps of Volunteers.

These were not the only ways in which popular patriotism expressed itself. The London stage was monopolized in the early 1800s by invasion drama and anti-French propaganda, including Shakespeare's *Henry V*, Roger Boyle's Restoration play *Edward the Black Prince*, the song 'Britons Strike Home', and many other rousing patriotic productions. Caricatures were published of Britannia, weighing in her scales the fate and future of Europe, in which a well-endowed John Bull easily overwhelmed a short and skinny Napoleon. In 1799 the sculptor John Flaxman had proposed erecting a statue, 230 feet high, atop Greenwich Hill, celebrating 'Britannia, By Divine Providence Triumphant', in recognition of the recent British naval victories. In the end it was never built, but as its extended title implied, British patriotism during these years was as much a matter of religion as it was of politics. To be sure, the established Church of England was failing to keep pace with the growth and shifting distribution patterns in the population: in 1800 there were seven hundred parishes in rural and agricultural Norfolk but only seventy in industrializing and urbanizing Lancashire. But Protestantism remained a binding and unifying force, as exemplified in the oft-proposed toast: 'Church and king'. In the public mind, the Evangelical movement may have been most closely associated with its campaigns against the slave trade and slavery; but Wilberforce and his friends not only demonstrated that there was still energy and commitment to be found among members of the established church, but also remained strongly in support of the whole established order. So did the Methodist movement, even though it had separated from the Church of England following John Wesley's death in 1791. But Methodism retained the Tory attitudes and authoritarian instincts with which he had imbued it. In the words of Jabez Bunting, one of Wesley's most influential successors, 'Methodism hates democracy as it hates sin.' In the final wars against Britain's traditional enemy, popular Protestantism was an energizing and potent force.

Yet Addington failed to benefit from or to personify the very mood of popular patriotic defiance that he had done so much to help promote.

The Younger Pitt had initially supported him, but came to resent his successor. As a result, Addington's position became untenable, and in May 1804 Pitt returned to 10 Downing Street. But in his final, brief administration he was no more successful a war leader than he had been between 1793 and 1801. The Additional Force Act was an attempt to raise more recruits than Addington had done, but it conspicuously failed to do so. The triumph of Trafalgar in 1805 meant the second French invasion scare was over, but Napoleon's extraordinary victory at Austerlitz the same year effectively brought to an end the third and last of Pitt's abortive European coalitions. Yet more damaging were the charges of corruption brought against Pitt's closest political colleague, Henry Dundas (recently given a peerage as Viscount Melville), who had returned to office with him as First Lord of the Admiralty. Early in 1805 Melville was accused in the tenth report of the Public Accounts Commissioners of misappropriating public funds; his formal censure in the Commons reduced Pitt to tears, and Melville was compelled to resign in May. 'We can get over Austerlitz,' Pitt later complained, 'but we can never get over the tenth report.' In fact, Pitt could not get over either of these defeats: mortally stricken with a grave intestinal disorder that may have been cancer of the bowel, his command of the war effort, his authority over the Commons, and his grip on government business were all gradually weakening. Had he not died in January 1806, he would almost certainly have been brought down in the Commons later that year. One version of his last words, 'Oh, my country! How I leave my country!', suggests that he, too, recognized his deficiencies (another version was 'I could do with one of Bellamy's meat pies'). Pitt's funeral was a further grand affair; but it took place in Westminster Abbey rather than St Paul's Cathedral, and it could not compare with Nelson's magnificent obsequies.

Nelson had been killed at the age of forty-seven. The Younger Pitt had died at forty-six; and on his demise, George III turned to Lord Grenville. He had been Pitt's Foreign Secretary between 1791 and 1801, but since then had moved towards the Foxite Whigs, and he now put together a cabinet known optimistically but misleadingly as the 'Ministry of All the Talents'. There was a greater change in governing personnel than in 1801 or in 1804, and Pitt's closest followers were conspicuously absent from this new administration. But Sidmouth remained in office, and Fox returned for an all too brief and too late spell of power (he had last been in government in 1783) as Foreign Secretary and Leader of the Commons. George III, who had never liked him, had grudgingly agreed to his appointment, perhaps mollified by what seemed to have been Fox's changed attitude to Napoleonic (as distinct from Revolutionary) France. But the administration was more shambolic than successful: Grenville was not a commanding figure, an increase in the property tax was unpopular, and attempts to switch men and resources from the militia and the Volunteers to the regular army were ill-judged. Meanwhile, and contrary to the king's hopes, Fox made peace overtures to the French on what seemed like abject terms. But Napoleon, who was then rampaging across continental Europe, was in no mood to accept them; and in any case, Fox died of drink and dissipation in September 1806. This meant the government had lost its only commanding figure in the Commons, it would only be a matter of time before it fell, and it was the vexed issue of Catholic Emancipation that was once again the reason. The Foxite Whigs had always been in favour of such a measure, but George III remained implacably opposed, and all the Grenville government could do was to introduce a lesser measure that would have allowed Catholics to take up commissions in the armed forces. Even this was too much for the king, who demanded that his ministers never again raise the Catholic issue. They were unwilling to comply, and Grenville and his cabinet resigned in March 1807.

But shortly before Grenville's ministry fell, it succeeded in passing one significant piece of legislation, namely the abolition of the slave trade throughout the British Empire. This was the culmination of the twenty-year-long campaign, initiated and driven forward by Evangelicals led by William Wilberforce, which dated from the establishment of the London Committee for Effecting the Abolition of the Slave Trade in 1787. Thereafter, Wilberforce regularly introduced bills to outlaw this appalling trafficking in human life, and in 1792 the Commons had agreed to abolish the trade in stages. But the West Indian planters were a well-organized lobby, and they were much more powerful in the Lords, which was far less sympathetic to abolition, and refused to pass the bill sent up from the lower house. The result was a growing public campaign against the trade, and petitions poured into parliament, supported not only by middle-class Evangelicals but also by many members of the working class. Yet during the troubled 1790s neither the government nor parliament was minded to give in to popular pressure on any subject, while a succession of slave rebellions during the first half of the decade only strengthened the planters' determination not to give way. But by the mid-1800s the slave trade was becoming less important, as the fall in death rates in the Caribbean meant the slave population was becoming virtually self-sustaining; and Britain's West Indian colonies were shifting their emphasis away from slave-based sugar production to the re-exporting of British manufactured goods to Latin America. The return of the Foxite Whigs to power, and the death of Fox himself, who had been an ardent abolitionist, concentrated the minds of his colleagues on passing a piece of legislation to which he had been committed for so long. The planters were no longer so opposed as they had previously been, and early in 1807 the measure passed both the Commons and the Lords by substantial majorities. Slavery itself remained permitted throughout the British Empire, but it would only be a matter of time – albeit a long time – before it, too, was eventually outlawed.

On the resignation of the Grenvillites, George III turned to the Duke of Portland, who was another Whig, and had been the titular head of the Fox-North coalition in 1783. But he, too, had gone over to Pitt in 1794, and unlike Grenville, he had stayed loyal to his leader thereafter, serving in Pitt's last administration as Lord President of the Council. Portland's cabinet included three future prime ministers: Spencer Perceval as Chancellor of the Exchequer, Lord Hawkesbury (who became Lord Liverpool in 1808) as Home Secretary, and George Canning as Foreign Secretary, while Castlereagh also returned to office as Secretary of War and for the Colonies. But Portland was old and infirm, he was often too ill to attend his own cabinet meetings, and he failed to provide adequate leadership or keep his colleagues in order. Despite the Royal Navy's maritime supremacy and the shelling of Copenhagen, the war against Napoleon did not go well: the Fourth and Fifth Coalitions both collapsed, the Walcheren expedition was a disaster, the cabinet was deeply divided over the Orders in Council, and the Whig opposition was vehemently opposed to them. There was also a succession of damaging scandals. Sir Arthur Wellesley's campaign on the Iberian Peninsula had got off to a bad start with the Convention of Cintra, under which 26,000 French troops were repatriated to France by the Royal Navy; and the result was a huge public outcry in the course of which Wordsworth denounced the government's action as 'abject, treacherous, and pernicious'. Then the Duke of York, one of George III's delinquent sons, but who had been commander-in-chief of the British army since 1798, found himself embroiled in a public scandal brought on by revelations made by a former mistress, Mary Anne Clarke. Confronted with both a public inquiry and a parliamentary motion accusing him of corruption, the duke had no alternative but to resign. Finally, in September 1809, two of Portland's cabinet colleagues, Castlereagh and Canning, became such bitter rivals that they fought a duel on Putney Heath. Canning was slightly wounded, and both men promptly resigned. The ministry disintegrated early in October, and Portland himself was dead before the end of the month.

The king now looked to Spencer Perceval, a fervent Evangelical and determined opponent of Catholic Emancipation; but since the Whigs held aloof, and neither Canning nor Castlereagh could yet be decently brought back to public office, Perceval found it difficult to form a government. Of the major figures from the previous administration, only Lord Liverpool stayed on, while Arthur Wellesley's elder brother, the former Governor General of British India and by now the Marquis Wellesley, became Foreign Secretary. For much of the time Perceval and his colleagues were preoccupied with royal affairs. In October 1809, the same month in which he formed his administration, George III entered the fiftieth year of his reign, and his Golden Jubilee occasioned an unprecedented outpouring of popular adulation and state-sponsored benevolence, for which the momentum had built in the months before Perceval became prime minister. Behind all the pageantry and partying lay a growing recognition that, since the 1790s, the monarch had come to embody, in a way that no transient, controversial politician ever could, the nation's determined resolve to fight against the French; and there was an increased and widespread awareness that George III had come to stand for order, hierarchy and legitimacy against the chaos, terror and usurpation of the Revolutionary and Napoleonic regimes that had taken hold in Paris. The young sovereign who had earlier been criticized for interfering too much in politics, and for the disastrous loss of the American colonies, had morphed and evolved in old age into a venerable and venerated figure, revered and acclaimed as 'the Father of his People'. Across the length and breadth of Britain and its empire, there were spontaneous celebrations and church services in villages, towns and cities; landowners put on feasts and festivities for their tenants; the king and his family attended a thanksgiving service at Windsor; and in London public buildings were decorated and illuminated. At the same time debtors were discharged from prison, non-French prisoners of war were allowed to go home, and an amnesty was declared for all deserters willing to rejoin their ships or regiments.

But his Golden Jubilee would turn out to be George III's last hurrah. He was increasingly blind and infirm (hence unable to attend the public service of thanksgiving in his honour held in London), and he had already been subjected to three prolonged attacks of what seemed to have been madness, in 1788, 1801 and again in 1804. Each time he had recovered, but in October 1810 the death of his youngest and favourite daughter, Princess Amelia, precipitated a final and permanent mental collapse. This meant the king's eldest son, the Prince of Wales, would now be installed as Regent, as he had hoped would happen in 1788. All his life the prince had been a friend of the Whigs, in opposition to his father's support of the Tories, and he had been particularly close to Charles James Fox. In 1788 the prince had intended to wield full royal powers as Regent, and to use them to put the Whigs in government. In the end George III had recovered, and the prince was disappointed. But mindful of that earlier episode, and still hoping that the king might get better, Perceval's government passed legislation in February 1811 preventing the Regent from taking any irreversible actions, to which on recovery the king might object, for the next twelve months. During that time the Prince Regent reflected on his relations with the Whigs and with the government, and also on the course of the war on the eastern front and in Spain. At such a critical juncture, he concluded, his duty was simply to support those in power, and he accordingly 'ditched' the Whigs. This decisive act of royal approbation emboldened Perceval to strengthen his administration in the spring of 1812: on Wellesley's resignation from the Foreign Office, he brought Castlereagh back; at the same time Sidmouth returned as Lord Privy Seal; and the second Viscount Melville, son of the Younger Pitt's disgraced Scottish henchmen Henry Dundas, became First Lord of the Admiralty, as his father had been before him. These were the men who would see the war through to its successful conclusion; but Perceval himself would not, for he was shot dead by a deranged bankrupt in the lobby of the House of

Commons on 11 May 1812, thereby becoming the only British prime minister to be assassinated.

It was still unclear, at the time of Perceval's death, that Napoleon could be beaten, despite Wellington's continued advances on the Iberian Peninsula. The economic war was going particularly badly, there was a growing campaign among the commercial and mercantile sectors of the population for the repeal of the Orders in Council, and tensions were mounting alarmingly between the United Kingdom and the United States. Yet such was the apparent disconnect between the global war that Britain was waging beyond its borders and the narrowly enclosed world of Westminster politics that it took *eleven weeks* of political manoeuvring and infighting before Lord Liverpool eventually emerged as the next prime minister in July 1812. But despite such lengthy and protracted negotiations, the new administration closely resembled that which it replaced. Like Perceval and Portland before him, Liverpool was a Pittite, and he had been almost continually in government since 1799, except for the brief period when the 'Ministry of All the Talents' held office. He was also hard working, efficient, untainted by scandal, and exceptionally lucky, for he became prime minister at just the point when the war finally turned in Britain's favour. To be sure, the Orders in Council were repealed too late to prevent conflict with the United States, but Napoleon's failed invasion of Russia suddenly transformed the military situation to Britain's advantage, and as one of the major architects of the Sixth Coalition, Castlereagh emerged as a European statesman of the first rank. Doubting that Liverpool's administration would last long, Canning had refused to serve under him, and quixotically absented himself from British politics by accepting the Lisbon embassy. This would prove to be a terrible error of judgement on Canning's part, for it enabled Castlereagh to keep the Foreign Office; it was he more than anyone who kept the European allies together until the spring of 1814, and once the French Emperor had abdicated, it was Castlereagh and Wellington who together represented the United Kingdom in the great-power negotiations that followed.

Even as the Liverpool ministry was establishing itself, and overseeing what seemed to be the final phase of the war against Bonaparte, it was also concerned about domestic matters, especially the growing pressure from the commercial and dissenting middle classes in favour of policies more sympathetic to their interests. The repeal of the Orders in Council was one response; the admission of Unitarians to the benefits of the Toleration Act of 1689 was another; and the reduction of the commercial privileges of the East India Company when its charter came up for renewal in 1813 was a third. The Company was allowed to continue in business, but the terms under which it could conduct its future operations were significantly altered. The Company's monopoly of the China trade was confirmed, but its London-based monopoly of commerce with India and southeast Asia was abolished: in future, all Britons were allowed to trade with the subcontinent, and they could henceforth do so through many regional ports as well as through London. These changes were a triumph for private enterprise and for overseas merchants based in the provinces, and they represented an irreversible diminution in the Company's powers. At the same time, the importation of finished cotton goods from India to Britain was prohibited, a catastrophic measure for the subcontinent that would effectively destroy its indigenous textile industry. In future and instead, India would become merely a major supplier of cheap raw cotton, gradually superseding the United States, and these low-priced imports would be the key to the rapid expansion of the Lancashire cotton industry. Once again, it was provincial business that would benefit in Britain, while the Indian economy would suffer, not only through the loss of its domestic cotton industry but also as the price of its raw cotton exports was inexorably forced down. A further condition of the renewal of the Company's charter was that the previous ban on Christian missionaries from entering India would also be repealed. This provision was largely in response to an avalanche of petitions reminiscent of those that had earlier been sent in against the slave trade, and it was a high point in interdenominational co-operation.

By the time these measures were passed, the war against Napoleonic France was entering its final phases, and in displays of patriotic fervour reminiscent of the Jubilee celebrations of 1809 the inhabitants of the United Kingdom gave themselves over to the peace festivities that began in early June 1814 and extended throughout the summer. En route from their Paris to their Vienna meetings, the sovereigns and generals of the victorious allies – Austria, Prussia, Russia, Sweden and some of the German states – paused to visit the United Kingdom, at the invitation of the British government and the Prince Regent, and they arrived in London on 7 June. Among them were Tsar Alexander of Russia, King Frederick William III of Prussia and Prince Metternich of Austria. For the next two weeks there was a non-stop succession of entertainments and ceremonies: court levees, the Ascot races, an honorary degree ceremony at Oxford University, a City of London banquet, and a naval review at Portsmouth. But these celebrations were generally confined to the illustrious personages for whom they were intended, and ordinary people could do little more than watch. In contrast, the 'Grand Jubilee' held on 1 August was a much more popular affair: not only was it the culmination of the peace celebrations, it also commemorated the centennial of Hanoverian rule and the sixteenth anniversary of Nelson's victory at the Battle of the Nile. Hyde Park, Green Park and St James's Park were packed with crowds, and an array of entertainments and festivities celebrated what seemed to be both a triumphant victory and a glorious peace. Of these, the most impressive took place in Green Park where, during a spectacular firework display, a 'Gothic fortress of considerable extent' was miraculously transformed into a 'Temple of Concord'. On its walls were a series of allegorical images: strife and tyranny were banished by peace and victory, and beneath them processed a triumphant Britannia, preceded by 'Prudence, Temperance, Justice and Fortitude' and followed by 'the Arts, Commerce, Industry and the Domestic Virtues'.

THE SINEWS OF WAR AND THE PRICE OF VICTORY

The (highly selective) list of the activities and attributes that allegorically accompanied Britannia on the walls of the 'Temple of Concord' was something of a mixed bag, especially as it related to the United Kingdom's recent triumphant war effort. Among the nation's leaders, 'Temperance' had not been much on display, least of all in the case of the Younger Pitt, whose daily consumption of port wine was excessive even by the standards of the time, and may well have hastened his death; in a rare display of common purpose, albeit recreational rather than political, Charles James Fox had also been a renowned tippler. Nor, in regard to the management of the country's finances, had 'Prudence' been seriously practised, as the cost of fighting the Revolutionary and Napoleonic Wars had been unprecedentedly high, in terms both of financing the United Kingdom's own war effort and of subsidizing a succession of European coalitions. As a result, Pitt's earlier efforts to restore the national finances in the aftermath of the loss of the American colonies had largely been undone. Total expenditure from 1793 onwards was not much short of £700 million, which was almost six times greater than the amount spent on the war against the American colonists. To be sure, much of this dramatically increased outlay was financed by raising taxes, which took 14 per cent of the national income in 1802, rising to 19 per cent by 1806. But the waging of war by land and sea around the globe also required the raising of more loans and bonds: between 1793 and 1816 the national debt quadrupled, reaching the unprecedented level of £900 million by the time hostilities finally ended. This was approximately three times the gross national income, the interest payments alone taking up a quarter of all government expenditure, and there were widespread fears that the risk of national bankruptcy was even greater than it had been in 1797.

But while this did not look like 'Prudence', as the British government raised and spent more money than ever before, and ran up unprecedented debts, it was, from another perspective, an impressive expression of 'Fortitude'. The ruling classes had held their nerve during a war that was not only long and hard, but ever more expensive, and the British state had become much more efficient than the French, both at raising money and thus at providing the necessary resources for waging a world war. The revenues that accrued by both indirect and direct taxes kept going up: customs and excise receipts increased over threefold from £13.5 million in 1793 to £44.8 million in 1815, while the yield from the new income and property taxes rose almost tenfold from £1.67 million in 1799 to £14.6 million in the final year of the war. This enabled successive governments to maintain Britain's domestic and international creditworthiness, which meant they were able to raise much larger loans than Napoleon was willing to contemplate or able to obtain. During the final critical years of the war, the British government was borrowing more than £25 million annually, and the increases in revenue enabled it both to service this growing debt and to pay for the costs of the war. One banker who was especially well placed to take advantage of these opportunities was the Jewish financier Nathan Rothschild, who had arrived in Britain from Frankfurt in 1804. Thanks to his four brothers, his continental connections were unrivalled, and he raised money to fund the British war effort in Germany and Holland – and even in France itself – much of which was directly conveyed to Portugal for the use of Arthur Wellesley and his army. This was, Rothschild claimed, 'the best business I ever did'. By 1814 he had loaned nearly £1.2 million to the British government and established himself as one of Europe's leading bankers; indeed, he was revealingly described as the 'Napoleon of finance'.

At the same time, the civil and armed services of the British state became more efficient at procuring the products that were needed, and at making the necessary arrangements, for the effective waging of war. The conflict was eventually won by the admirals and sailors on the seas and the generals and soldiers on the battlefields, but it was the Whitehall bureaucrats, the 'silent men of business', who oversaw the building of the ships, the provision of weapons, and the essential supplies that made possible Britain's victories on land and water. During the 1800s men such as John Herries, the Commissary-in-Chief of the Army, Henry Bunbury, the Deputy Quarter-Master General and later Undersecretary of State for War, and William Marsden and John Barrow, who were the First and Second Secretaries at the Admiralty, transformed the nature and working of the British military state, and greatly enhanced its logistical efficiency and capability: corrupt practices were abolished, hours of work were increased, recruitment was improved, financial management was tightened up, interdepartmental communication was encouraged, and the number of clerks was doubled. As a result, the Victualling Board, which was responsible for providing the rations that ensured the British army was fed, became one of the most efficient agencies of government; and it needed to be, because when Wellington's peninsula army reached its peak, it was consuming forty-four tons of biscuit a day. In the same way, the regular refitting of ships was raised to new levels of efficiency by the Admiralty, which meant the Royal Navy could fight longer, harder and more lethally than the enemy ships. Small wonder that on their triumphant tour in the late summer of 1814, the allied sovereigns visited Portsmouth to see the huge mills where the blocks used in the rigging of warships had been manufactured, and indeed, mass produced.

The 1800s not only witnessed a revolution in the performance of the government of the United Kingdom as a war-making machine, with a new generation of young men committed to hard work and disinterested public service rather than apathy, self-interest and corruption. It also witnessed an unprecedented collaboration between state sponsorship and business, as many of the essential materials of war were provided and manufactured by private enterprise: between 1803 and 1815, for instance, more than three-quarters of the navy's ships were built by individual contractors rather than in royal dockyards. As this example suggests, these years in which British government made unprecedented demands on the British economy also witnessed a significant advance in industrial output, in part (but only in part) as a direct result of the increased demand for iron, coal and timber, and for the weapons and the ships with which the war was fought. The production of pig iron went up almost fourfold from 68,000 tons in 1788 to 244,000 tons in 1806, and rose further to 325,000 tons by 1811; textile manufacturing had been insignificant at the time of the American War of Independence, but imports of raw cotton from India and Egypt almost doubled from £32.5 million in 1789 to £62 million in 1804 and had more than doubled again to £132.4 million by 1810, and cotton textiles were the nation's largest export by 1815. These years also witnessed the zenith of the 'canal mania', with substantial investment in inland waterways in England (the Grand Union and Grand Junction Canals), in Wales (the Llangollen Canal, with its spectacular Pontcysyllte aqueduct) and in Scotland (with the initial work on the Caledonian Canal); and the beginning of the construction of a new road linking London to Holyhead in the aftermath of the Act of Union. Such industrial advances and infrastructural provisions help explain why one marvelling contemporary observed in 1814 that it was 'impossible to contemplate the progress of manufactures in Great Britain within the last thirty years without wonder and astonishment. Its rapidity, particularly since the commencement of the French revolutionary war, exceeds all credibility.'

'Fortitude' and 'Industry' were both much in evidence in Britain during the first fifteen years of the nineteenth century, and so in turn was 'Commerce'. Despite Napoleon's best efforts to the contrary, the growing industrial output meant that the United Kingdom's overseas trade continued to increase, although for a time many of the most lucrative European markets were officially closed, as were those in the United States between 1812 and 1814. This was partly because neither the European nor thc North American markets were in practice ever completely shut off. Throughout the whole period from 1807 to 1814 large quantities of British manufactures and colonial re-exports were directly shipped to Spain and Portugal, or were smuggled to other continental destinations with the connivance of bribed local officials in such places as Heligoland and Salonika; and in the same way, British goods were also still exported to New England indirectly via Canada. Overseas trade also increased because the British export economy was increasingly sustained by the rise in business with regions untouched by Napoleon's Continental System or the United States' 'non-intercourse policy', namely Asia, Africa, the West Indies and the Middle East (and of these, the markets in the Caribbean and India were of particular importance). And finally it also increased because, as a result of diplomatic and military developments at opposite ends of the European continent, Napoleon's attempt at an all-pervasive blockade failed as completely as did his attempted conquest of Russia. When Spain changed sides in 1808, this opened up not only a large part of the continent to British export goods, but also large areas of Latin America, which became for the first time a significant (if sometimes overestimated) overseas market. Four years later, Napoleon's invasion of Russia allowed British goods to pour into the Baltic and northern Europe, which more than compensated for the (only temporary and never complete) loss of the American market.

Behind these developments in finance, industry and trade was further evidence of a nation undergoing unprecedented transformation. The population of the United Kingdom continued on its rapidly upward path, increasing from roughly sixteen million in 1801 to almost twenty-one million by 1821. Many towns, especially in the north and the Midlands, grew more rapidly during these two decades than before or since: Manchester from 75,000 to 126,000, Birmingham from 71,000 to 102,000, Liverpool from 82,000 to 138,000, Glasgow from 77,000 to 147,000. But the great metropolis of London remained pre-eminent, becoming during this time the nation's (and Europe's) only million-person city: 959,000 in 1801, 1,139,00 in 1811, 1,379,000 in 1821. Not surprisingly, the 'great wen' of London overwhelmed the rest of urban Britain: it was the location of the court, the government and parliament; the focus of culture and intellectual life; a major manufacturing centre; and the greatest port in Europe, if not the world.

Indeed, the first decade of the nineteenth century saw a massive increase in its docks, as new facilities were built downstream from the old Pool of London, among them the West India Docks (1802), the East India Docks (1803), the London Docks (1805) and the Surrey Commercial Docks (1807). Underpinning this unprecedented expansion in population, industrial production, overseas trade and commerce was the simultaneous expansion in agricultural output and productivity, as there were more Britons to feed than ever before. As early as 1783, Great Britain had become a net importer of wheat, but it was much more difficult to do so in wartime, and this further added to the pressure to increase output. As a result, the rate of enclosure of common land dramatically increased during the 1790s and 1800s, and so did the acreage that was cultivated, extending higher up the hillsides than ever before – or, indeed, since. Here, then, was a nation in which all the indicators were that every aspect of its economy, throughout the years when Napoleon was being fought, was booming and thriving as never before.

This in turn may help explain why 'the Arts' were also flourishing during the first fifteen years of the nineteenth century, albeit with varying degrees of engagement with the military and political events of the time. The Romantic poets who had previously welcomed the French Revolution now increasingly turned towards conservatism and in on themselves: Coleridge took to opium, Wordsworth settled in the Lake District, while Southey wrote a life of Nelson in 1813 and was appointed Poet Laureate in the same year. Sir John Soane, an architect of exceptional originality and creativity, was at work on the Bank of England, the Royal Hospital at Chelsea, and the Dulwich Picture Gallery, buildings respectively devoted to finance, the military and art. The architect James Wyatt created some of his most elaborate Gothic extravaganzas at Fonthill Abbey, Belvoir Castle and Ashridge, constituting an embattled rebuke to the revived classicism of Revolutionary France. Once established as Prince Regent, and as the war turned in Britain's favour, George III's eldest son commissioned John Nash to replan much of central London. Between 1811 and 1815 Jane Austen published *Sense and Sensibility*, *Pride and Prejudice*, *Mansfield Park* and *Emma*; while *Northanger Abbey* and *Persuasion* would appear posthumously in late 1817 and early 1818 respectively. Although Austen's brothers fought in the Napoleonic Wars, her novels referred to the conflicts only very occasionally, preferring to dwell on the lives and loves of those men and women at the intersection of the middle and upper classes. By contrast, Turner's historical painting of *Hannibal Crossing the Alps* (1812) drew parallels between the Punic Wars and the Napoleonic Wars, and was a deliberate response to Jacques-Louis David's portrait of *Napoleon Crossing the Alps*, which he had seen on his visit to Paris in 1802. Two years after Turner painted *Hannibal*, Sir Walter Scott published (anonymously) his first novel, *Waverley*, which was set in Scotland during the time of Bonnie Prince Charlie. The eponymous hero was initially captivated by the Jacobite cause, but later transferred his allegiance to George II – an appropriate fictional outcome as *Waverley* was published in the centennial year of the Hanoverians' accession to the British throne.

Perhaps surprisingly, Britannia, although attended by 'the Arts' in the triumphant pageant of August 1814, received no such support from 'the Sciences', even though they, too, had been flourishing throughout the period of the Revolutionary and Napoleonic Wars. The Royal Institution of Great Britain had been founded in 1799 just before the Act of Union by (among others) Henry Cavendish, Sir Joseph Banks and Sir Benjamin Thompson, with the aim of promoting a better public understanding of 'the application of science to the common purposes of life'. The Institution was soon accommodated in impressive buildings in Albemarle Street in Mayfair. Its first lecturer and laboratory-based researcher was the lowly born scientist Humphry Davy, who immediately established himself as a metropolitan celebrity and charismatic performer at the podium. His 'Discourse Introductory to a Course of Lectures on Chemistry', delivered at the Institution in 1802, was one of the earliest proclamations in Britain of the belief in the possibilities and benefits of scientific progress. Later in the decade, Davy would discover the chemical elements of sodium, potassium, calcium and magnesium, and he would employ the young Michael Faraday as his laboratory assistant. In 1815 Davy invented the safety lamp, which dramatically reduced the number of casualties among coalminers, and four years later he would be granted a baronetcy. A generation earlier, a similar honour had been bestowed on the much better-born naturalist and botanist Sir Joseph Banks, who had accompanied Captain Cook on his voyage to the Pacific in the *Endeavour* and who served as President of the Royal Society from 1778 to 1820. Banks advised George III on his botanical collection at Kew, sought to keep scientific contacts open with France during the Napoleonic Wars, and sponsored later expeditions to the Pacific by, among others, George Vancouver and (the ill-fated) Captain Bligh. Along with Davy, Banks was the pre-eminent British scientist of his time, and both combined great intellectual credibility with wide public acclaim and significant official recognition.

As for the 'Domestic Virtues' that had been acclaimed and proclaimed at the Grand Jubilee of 1814, they had certainly been exemplified by George III and his wife, who had lived lives of decency and dutifulness, as captured and conveyed in the conversation pieces painted by the German-born artist Johannes Zoffany. But the same had not been true of their sons, who had without exception led rebellious existences characterized by both financial irresponsibility and sexual promiscuity, with the Prince Regent being merely the most notorious example. The conjugal decency of 'Farmer George', as the king was known, no doubt contributed to his late-life apotheosis; but by the 1810s the fashion among some members of the upper classes was tending towards a greater degree of self-indulgence, with which the Regency of the future George IV became synonymous. But this was not the whole story, as men of government such as Lord Liverpool and Robert Peel, perhaps influenced by the Evangelical movement and its 'call to seriousness', combined private virtue with public duty. Among the middle classes, both in London and the provinces, there was a growing separation of workplace and home, which provided greater opportunity for cultivating a loving and companionate domesticity that increasingly centred on the wife and mother as 'the angel in the house'; but it is difficult to know whether these attitudes and arrangements filtered down to the lower classes and orders beneath. As for 'Justice', the United Kingdom had certainly avoided the spate of judicial killings and excesses of the guillotine associated with the reign of terror in France, and having been immune to invasion its inhabitants had avoided the rape and pillage so often characteristic of occupying armies. But in the dark days of the 1790s there was no doubt that British liberties had been compromised and constrained, both by the legislature and the judiciary, and this would continue until the war against Napoleon was finally won.

Yet, as so often, the political picture was more complex than that. For while the ruling regimes of these years were in some ways undoubtedly repressive, the metropolitan culture of late-Georgian London was in other ways remarkably undeferential to established authority. The first two decades of the nineteenth century were among the most creative and critical years for political caricature and satirical cartoons, as embodied in the brilliant and often scatological works of Thomas Rowlandson, James Gillray and the young George Cruickshank. These images reached a wide audience because, following the precedent established by William Hogarth, they were often engraved and made available in cheap editions. Rowlandson and Gillray had both been born in 1756: Rowlandson studied at the Royal Academy and in Paris; Gillray, after spells as a soldier and strolling player, also studied at the Academy. Both had turned to caricatures during the early 1780s, both were dismayed by the French Revolution, and in John Bull they created an iconic symbol of British national identity and embattled patriotic pride. But they were also deeply hostile to what they deplored as the dull and parsimonious monarchy of George III, and the wayward self-indulgence of his sons, especially the Prince of Wales; and they were equally scornful of most of the major politicians of their time, including the Younger Pitt and Addington. Their cartoons and caricatures were often irreverent and lewd in the extreme; but they went uncensored, and this gave them a freedom of expression that did not exist in the empires of Napoleon or the Russian Tsar. Gillray died in 1815 and Rowlandson in 1827. Their successor was George Cruickshank, at least during the early stages of his career. He, too, delighted in depicting John Bull and in deploring Britain's enemies; but he also took great pleasure in attacking and caricaturing the royal family – so much so that in 1820 he was (ineffectually) bribed 'not to attack His Majesty [George IV] in any immoral situation'. Cruickshank, who was born in 1792 and lived on until 1878, would later find a different form of fame as the illustrator of many books, including some of the early novels of Charles Dickens.

As Cruickshank's change in career trajectory suggests, the British tradition of satirical cartoons and scatological caricature, when all authority figures were legitimate targets, would largely die out by the time Queen Victoria acceded to the throne, and 'respectability' increasingly became the watchword. But another aspect of early nineteenth-century libertarian London would survive and endure for the whole of the queen's reign, namely the welcome and shelter it gave to those fleeing revolution or persecution in other countries, both in Europe and the western hemisphere. Among the thousands of Americans loyal to the British throne who returned to the mother country when the thirteen colonies declared their independence was Sir Benjamin Thompson. Knighted by George III in 1784, Thompson was a prolific inventor and one of the moving forces behind the establishment of the Royal Institution. From 1789 the victims of the first American Revolution had been joined by many of the losers in the French Revolution, fleeing to London to avoid the Terror and the guillotine. So great was the influx that in 1793 the government had passed an Aliens Act, which required that these recent arrivals must be registered with the local Justices of the Peace. But the measure did not prohibit immigration, and it would be revoked in 1815. By then, at least ten thousand French men and women, among them aristocrats and Catholic priests, were living in Britain, the majority in London, where they settled in Soho and Fitzrovia, respectively to the south and north of Oxford Street. Soon after, these victims of revolution were joined by would-be perpetrators of revolution from Latin America, among them such long-term residents as Francisco de Miranda, Andrés Bello and José María Blanco White who, during the 1800s and 1810s, worked to promote rebellion in Spanish America. So, too, did Simón Bolívar, who briefly resided in London during 1810 before returning, like Miranda before him, to take up arms against the Spanish authorities. Some of these immigrants were temporary, others stayed much longer; some were hostile to revolution, others in favour. There would be many more such arrivals in London during the decades that followed.

Early nineteenth-century Britain may have been an attractive place and appealing destination for those fleeing or plotting revolution, but many of the exiles and emigrés endured hardship and poverty, loneliness and depression. And in any case, for much of the time, it was far from clear that the United Kingdom would prevail in the struggle against France. For just as the naval victories of 1797–1805 were followed by a long period of strategic and military stalemate, and a very late turning of the tide against Napoleon in 1812, so in the short run the British economy and the British people often appeared on the brink of buckling under the protracted and unrelenting strain of a seemingly interminable and unwinnable war. The price of foodstuffs rose to unprecedented levels; there were major banking panics and financial crises; industrial production and output fluctuated alarmingly in the short term; export markets were also often unreliable; and as late as 1812, facing both the continental blockade of Napoleon and the Americans' 'non-intercourse' policy, the economic outlook seemed at best uncertain, at worst bleak. This in turn meant that, although the Revolutionary disturbances of the 1790s seem genuinely to have abated, there was still serious – and wholly justifiable – popular discontent. The soaring price of bread meant that there were regular food riots; the conditions in the rapidly expanding industrial towns of the north were often deeply squalid; there was high unemployment at times of economic downturn; and there were protests against new machinery that seemed to be robbing traditional workers of their time-hallowed means of livelihood. These different forms of alienation, distress and protest never coalesced into a coherent movement with a political agenda that required the overthrow of the established order. But they certainly present a very different image from the patriotic culture and armed volunteers with which they uneasily co-existed, and they were a vivid reminder that life for most people during these years was indeed hard.

At every stage during the Napoleonic Wars there were protests from those below and anxieties registered by those above as to whether the economy could be kept on track or the social fabric would hold. The resumption of war in 1803 resulted in new strains on the British economy, and a sudden upsurge in food prices. There had been strikes by shipwrights in government dockyards in 1801 and in civilian shipyards on the Thames in the following year; by textile workers in the south-west and in the northern counties in 1802 and in 1808; and by Tyne keelmen in 1803. Even more ominous was the growth of a movement called 'Luddism', named after its mythical founder, 'Ned Ludd'. Most of the Luddites were highly skilled textile and hosiery operatives whose job security and high wages were being threatened by the introduction of new machinery, and they operated in well-defined areas of the country, especially the old-craft woollen textile districts of Wiltshire and Gloucestershire, the East Midlands and the West Riding of Yorkshire. Their attacks were carefully selected and their activities were supported by a community that had developed a strong sense of craft-identity now threatened by the new technology. Accordingly, in 1802, the attack on cloth-finishing machinery by Wiltshire craftsmen was accompanied by calls that such machinery should be declared illegal and by petitions to parliament to uphold craft traditions and practices. Thereafter, matters remained uncertain, and 1808 was an especially bad year in the aftermath of Napoleon's decision to blockade Britain. Exports that in 1806 had been worth £40.8 million dropped within two years to less than £35.2 million; Liverpool's raw-cotton imports fell from 143,000 sacks in 1807 to 23,000 the following year, while grain imports fell to one-twentieth of what they had been in 1807. Slackening demand produced low wages and short-time working, while the dislocation to the grain trade meant the price of corn skyrocketed from 66s a quarter in 1807 to 94s in 1808.

Not surprisingly, there were frequent protests and violent strikes, of which those in Manchester were the most ominously noteworthy. At the same time there were growing demands among the few radicals in parliament for some measure of parliamentary reform. So it was scarcely surprising that the Perceval ministry supported the ill-fated British attempt to capture Buenos Aires, in the hope it might open up more markets and relieve pressure on the economy. This was overly optimistic, but Napoleon's Continental System did begin to relax (or retreat) during the next few years. Nevertheless, 1811 turned out to be even worse for the British economy than 1808 had been, as a crisis in overseas trade coincided with major challenges to government expenditure and also a financial panic. British exports to northern Europe fell to only 20 per cent of what they had been in 1810, while those to the United States declined to 76 per cent, and trade with South America was barely two-thirds of the level it had been during the previous year. Indeed, total British exports, which had recovered to almost £61 million in 1810, now suddenly plummeted to a mere £39.5 million. Such a contraction in trade, of which the collapse of the South American market with the bursting of a speculative boom was the most noteworthy, naturally reduced government revenues, and at a time when its outgoings were higher than ever before: payments to the successive coalitions went up from £6 million in 1808 to £14 million in 1810, the war in Spain was ever more costly, as was the purchase of increased quantities of grain after the bad harvests of 1809 and 1810. To this dismal picture was added a major financial crisis, sparked off by the bankruptcy of five of Manchester's biggest firms in the late summer of 1810, which led to a wave of failures and stoppages across the country. By May 1811, Lancashire labourers were working a three-day week, if they were working at all, insolvencies had reached record levels, and there was profound and widespread gloom about the country's economic prospects.

The late 1800s and early 1810s were a period of bad harvests, heavy taxes, rising prices, unemployment and fluctuating export markets, and they witnessed constant and growing unrest. In 1809 there were 'price riots' in Covent Garden and disturbances in Nottingham; but it was the years 1811–12 that saw the greatest discontent, especially in the hosiery industry in the East Midlands. Indeed, it has been estimated that one thousand stocking frames were destroyed in the year from March 1811 to February 1812. In Yorkshire skilled workers protested against the introduction of shearing frames, machines were broken and some larger woollen mills set on fire, and at least one mill owner was killed. The government tried to compensate for the inefficiency of local control by making the penalties for such behaviour more ferocious, and it became a capital office to smash frames, damage property, or take Luddite oaths. Troops were moved in to the affected districts, and in January 1813 seventeen Luddites were executed and six more were transported to Australia. But while this cowed the Luddites, it did not crush them: they persisted in greater secrecy than before, and they continued their activities until the end of the war. Some of them were also political radicals, calling for parliamentary reform, and others may have gone further, in part provoked by government reprisals, and sought the overthrow of the whole system of parliamentary government. There were certainly rumours of secret meetings, drilling with live weapons in remote parts of the country, and plans for violent revolution. These may not have amounted to all that much, but there were certainly celebrations in Nottingham and Leicester when Spencer Perceval was assassinated, there was a riot at the Manchester Exchange in April 1812 after the Prince Regent's decision to support the incumbent administration rather than turn to the Whigs, and protests against both the government and the war continued until hostilities ceased.

By 1812 popular protest was both industrial and political in its causes and motives, and it merged with more conventional radical opposition by those few MPs in Westminster who sought to build connections with the discontented world beyond. However efficient the government and civil service might have become at logistics and at the provision of men and weaponry and money to fight the war, the Melville scandal, the Duke of York's misdemeanours, and the unpopularity and indebtedness of the Prince of Wales led radicals in parliament to demand reform as a way of purging what seemed to them a thoroughly rotten and corrupt system. Writers like William Cobbett and John Cartwright fanned the flames of such discontent, and the rich radical Sir Francis Burdett, having been incarcerated in the Tower of London for defying the Speaker of the House of Commons, became chairman of the so-called Hampden Clubs, which were intended to continue the agitation for radical reform that had been initiated by the corresponding societies in the 1790s. Here was another challenge to government authority. But as with the Luddites, the threat may have been less than it seemed. The membership of the Hampden Clubs was restricted to people who could afford it, and Burdett, Cobbett, Hunt and Cartwright were more interested in reforming the system than in overthrowing it. But there were other areas of popular protest, too. Many middle-class people agitated against the Orders in Council, on the grounds that they were damaging the economy, and they were duly repealed in June 1812. The following year an attempt was made to introduce legislation that would keep the price of corn at the high levels it had recently reached. Manufacturers, shopkeepers and working men joined in meetings and in riots across the United Kingdom, which continued throughout 1814, even as the war ended and the peace was being celebrated. And for the time being they did so successfully, for no legislation of importance concerning corn was passed.

From 1802 to 1814 the United Kingdom was a nation that combined disaffection and popular protest with patriotic zeal, which witnessed severe short-term fluctuations in the economy along with longer-term growth, which suffered financial crises even as its capacity to raise money and to spend it was unprecedented, and which for much of the time seemed to be losing the war with Napoleonic France even as it eventually emerged victorious and triumphant. There was also the matter of Ireland, recently but incompletely incorporated into the Union. To be sure, there were no repetitions of the French invasions and domestic insurrections of the late 1790s, and many Irishmen served in the British armies that fought Napoleon. But the 1800s also witnessed greater coercion in Ireland than in Great Britain, widespread regret that the intransigence of George III made Catholic Emancipation impossible, and a growing recognition that the Irish remained a different and largely inferior race, and that the Union was not working. In 1814 the two pre-eminent British figures in Europe, the one military, the other political, were Wellington and Castlereagh. Both were descended from Irish landed families, but Wellington 'preferred England because my friends and relations reside there', while Castlereagh was thought to be 'so very unlike an Irishman'. On the tenth anniversary of the implementation of the Act of Union, Henry Grattan, who was undeniably an Irishman, and had sat in the Dublin parliament before being compelled to relocate to Westminster, inquired: 'Where is the consolidation? Where is the common interest? Where is the heart that should animate the whole?' In the same year, 1812, Edward Wakefield produced his two-volume *Account of Ireland, Statistical and Political*, in which he declared that the country was being ruled as 'a distant province' under a Union which was at best 'half effected'. If Spain was Napoleon's ulcer, then Ireland would be Britain's running sore, and it would fester for much longer.

But in the short term it would be Scotland where there would be the greatest signs of discontent and disaffection, resulting from what were known as the 'Highland Clearances'. From one perspective they were the most controversial and brutal version of the agricultural enclosure and improvement that were still taking place in England and Wales. From another they were an attempt to deal with depressed and marginalized rural regions characterized by poverty, limited opportunity and an expanding population that simply could not be supported by the resources available. From yet a third viewpoint, they were rational efforts by large-scale landowners to improve and transform their under-developed and under-capitalized estates, by investing extensively in transport improvements and harbour construction, and by leasing their lands more profitably to commercial sheep farmers. But the thousands of tenants who, as a result, were forcibly removed from their smallholdings, saw themselves as the victims of heartless and unrelenting landlord rapacity and brutality. The ensuing evictions were often violent, many houses and outbuildings were burnt to prevent those expelled from attempting to resettle, and those who were dispossessed were often compelled to emigrate, especially to British Canada and, subsequently, Australia. The tempo of clearances had notably accelerated during the 1790s, but the most notorious phase began during the late 1800s and would last until 1820. It was especially associated with the evictions carried out on the vast, million-acre estate of the Countess of Sutherland in the very north of Scotland. They were overseen by her factor, Patrick Sellar, who used brutal methods that were so widely condemned that the Countess eventually dismissed him. Elsewhere in the Highlands the clearances continued down to the 1840s, and they would leave a lasting legacy of bitterness and resentment. It was not only in Ireland that there was disenchantment and resentment on 'the Celtic fringe', and as in Ireland, Scottish protests at the iniquities of the landlord system would again become especially pronounced during the 1880s.

A FINAL BUT TAINTED TRIUMPH

Such was the varied and uncertain condition of the United Kingdom of Great Britain and Ireland, by turns traumatized and triumphant, divided and united, disenchanted and loyalist, as the statesmen of Europe sought to resettle the continent in the aftermath of Napoleon's defeat and abdication in the spring of 1814. But while the Congress of Vienna was sitting and deliberating, Bonaparte escaped from exile on the island of Elba late in February 1815, and thus began the final act of the Napoleonic Wars known as the 'Hundred Days'. The restored Bourbons had not exactly won over the hearts or minds of their French subjects. Taxes remained high to pay the crippling costs of the long wars that had been waged since the early 1790s, and the proclamation of a militant Catholicism as the revived national religion had alienated many among the educated and free-thinking population, who for an entire generation had embraced the anti-clericalism of the Revolution. By the time Napoleon returned to France the following month, his old comrades hastened to join their fabulous and fabled leader; and on 20 March he triumphantly entered Paris, where the populace rose en masse to acclaim him. But the earlier alliance between Austria, Prussia, Russia and the United Kingdom still held, and faced with a possible return to more years of war, each nation agreed to keep 75,000 men under arms until Napoleon was definitively defeated and a final peace treaty signed. Thus was established the Seventh and last Coalition, to which the British government pledged another £9 million. On 18 June

Napoleon and his 68,000 troops confronted an allied army of 72,000 British, Dutch and German troops, led by Wellington at Waterloo, eight miles south of Brussels. The battle was long, hard and for much of the time indecisive, and it was only the arrival late in the day of the Prussian forces under Field Marshal Blücher, that gave the allies the edge. By early evening the French army's resistance had crumbled. Napoleon returned to Paris where he abdicated a second time, and he was exiled to the remote British island of St Helena in the South Atlantic, from which there would be no escape and where he died on 5 May 1821 aged fifty-one.

Wellington later described the Battle of Waterloo as 'the nearest run thing you ever saw in your life'. This was certainly true, and in terms of the nationalities participating in that final confrontation it was at least as much a victory for the Seventh Coalition as a whole as it was for the British in particular. But the odds were always against a successful Napoleonic comeback, for even if the Iron Duke and Blücher had lost on that fateful day, fresh Russian and Austrian armies were rapidly approaching from the east, and sooner or later they would have delivered a knock-out blow to what would have been a seriously weakened French army. As so often in protracted and geographically extended wars between great powers, it was not just a matter of tactics and strategy and generalship, important though those were and are, but of the relative distribution of resources between the combatants. And once the British, Austrians, Prussians and Russians had finally (and belatedly) agreed to unite in a concerted campaign against the French usurper, as they had successfully done since 1813, their overwhelming superiority in men and money and materiel compared to what the French nation in arms could muster and mobilize, meant that Napoleon must sooner or later be beaten, notwithstanding his undoubted military brilliance. So ended more than twenty years of virtually continuous European and global warfare, in which an estimated 5 million people had died, and in which the British had been involved from almost the beginning to the very end – and not just in Europe, but in every theatre of conflict around the world. In the aftermath of Waterloo, the military leaders, the politicians and the statesmen returned with relief to Vienna to complete their work of redrawing the map of Europe, and of reconstructing its polities and its politics. Despite his undoubted indebtedness to Blücher, Wellington was more than ever the hero of the hour, and Britain's prestige in Europe stood higher than it had ever been before, or than it would ever be again, even in 1945.

Along with Russia, the United Kingdom was the chief beneficiary of the Napoleonic Wars, and it was no coincidence that both of these two nations (and empires) were at the extremities of continental Europe. For it had been Russia's overwhelming resources in manpower, allied with Britain's growing economic advantage and undeniable naval supremacy, that had in the end felled the Corsican tyrant and thwarted his project of continental domination. As late as the 1780s France's economic performance (like that of China) had not been significantly inferior to that of Britain. But whereas the Revolutionary and Napoleonic Wars had both stimulated the British economy and had thus enabled the United Kingdom to mount an unprecedented war effort, they had been a disaster for France, which had failed to match the United Kingdom in terms of population growth, which had failed to industrialize, and which had failed to raise sufficient funds to prosecute the war to a successful conclusion. The eventual peace settlement was generous in that France retained its pre-1793 borders and was given back most of its Caribbean colonies. But the Napoleonic dreams of a global empire encompassing Louisiana, Egypt and South Asia had long since been abandoned. Nor was France the only nation whose imperial aspirations had been thwarted, for it would only be a few years before the increasingly ramshackle empires of Spain and Portugal in Latin America would also be overthrown by colonial rebellions, many of them led by men who had once lived in London, that would become as successful as the revolt of the American colonists in 1776. Meanwhile, Britain had acquired the Ionian Islands, Malta, Heligoland, Trinidad, Tobago, St Lucia, Guyana, Ceylon, Mauritius and the Cape Colony, as well as trading posts in the Gambia in West Africa. It also possessed many of the best harbours in the world, including Gibraltar, Malta, Sydney, Cape Town, Halifax in Nova Scotia and Kingston in Jamaica.

By the end of 1815, then, the United Kingdom and the British Empire had emerged victorious and triumphant as the strongest, richest and most powerful country in the world. Many of its poorer inhabitants, eking out their short lives and hard living at the lower end of the social scale, might not have seen it that way; but Britain's ruling classes could celebrate an astonishing recovery of their country's fortunes that had seemed so ruined in 1783, so threatened in 1793, so low in 1797, so uncertain in 1802, and so desperate in 1812. For just as Edward Gibbon's *Decline and Fall of the Roman Empire* had caught that earlier national mood of pessimism, anxiety and disenchantment, so now in the year of Waterloo, Patrick Colquhoun's *Treatise on the Wealth, Power and Resources of the British Empire* celebrated the very different feelings of providential deliverance and imperial resurgence that were widespread by 1815. His figures, like most contemporary efforts at political arithmetic and national quantification, were often suspect; but their orders of magnitude are suggestive, they can often be updated and augmented by more modern statistics, and the overall position was clear. In 1750, Colquhoun calculated, the population of Britain's dominions, including the American colonies, had been 12.5 million; by 1815, even after the loss of America, it was 61 million. Both figures need revising upwards, and by 1815 it seems likely that the total number of people living in the United Kingdom and the British Empire was approaching 200 million, or one-fifth of the inhabitants of the globe. This was a very big number, which meant that in terms of aggregate population alone, Britain was indisputably a great world power. Colquhoun also calculated that the combined military and naval strength of the British Empire, including the militias at home and the East India Company forces in South Asia, amounted to a million men. This unprecedented mobilization represented a trebling of Britain's armies and navy since 1793, and for the first time, it put British military forces on something approximating to equal footing with the continental armies and navies of Austria (600,000), Russia (1.5 million) and France (1 million in 1811, but only 200,000 after 1815).

Once again, these were very big numbers; but the global geographical distribution of these British forces was equally significant. The Royal Navy had consolidated its dominance in the northern waters of the Baltic, the North Sea and the Atlantic Ocean; while in the Mediterranean it had replaced French, Spanish and Venetian paramountcies in the west, and it would soon destroy Ottoman supremacy in the east. In the Persian Gulf, across the Indian Ocean and up and down the Red Sea, British maritime power had superseded the Dutch and the French as the dominant force, and in the Pacific Ocean, recently explored by Captain James Cook, British ships were unchallenged by any competing Western naval power. For the first time, that proudly boastful claim, made in 1745, that Britannia ruled the waves, had serious and global substance to it, and in the death-and-glory of Nelson the Royal Navy possessed an exhortatory myth of leadership and genius, duty and sacrifice that would resonate down the decades. At the same time, the British army had reasserted itself on the continent, invading France over the Pyrenees, and in Wellington it could boast a general of genius and a national hero who had delivered a succession of European victories the like of which had not been seen since the days of the Duke of Marlborough. Of even greater long-term significance was the growth of British power in India, which meant the United Kingdom could now challenge the Russian Empire on land in Asia, as it had for so long done on the sea in the Mediterranean. These enlarged military resources and cumulative successes had secured correspondingly massive gains. Colquhoun estimated the total private and public property of the British crown at £4 trillion, reckoning that one-quarter of it had been captured during the recent wars, in India, Africa and the Caribbean. In aggregate wealth this meant that Britain, when coupled with its empire, was richer than either France or Russia, and in addition it was the most advanced economy and the greatest international trading nation in the world.

In all these ways, it seemed as though the United Kingdom of Great Britain and Ireland was the exceptional and the indispensable nation, as European peace was finally made in 1815. Yet for the next century, its direct and sustained military involvement in continental affairs would be less than at any time since before 1688, while its engagement with the wider world of imperial rule and international trade, investment and migration would be greater than ever before. Only at the beginning of the twentieth century would Britain again commit itself wholeheartedly to the continent. In the aftermath of Waterloo it would be widely believed that the United Kingdom's political culture was innately more 'liberal', to use a word that was coming into fashion at just this time, borrowed from those in Spain who had opposed Napoleon, than that of any of the European powers: the monarchy was more constrained, the press was freer, there was a greater degree of religious tolerance, and public opinion was more influential, than was the case in France or Russia or Austria or Prussia. And as industrialization gathered force, and the condition of the people gradually improved, Britain would enjoy a long period of social peace and political stability that would be the envy of the world, while also making possible the gradual, ordered transition from the narrow politics of oligarchy to the broader politics of a mass electorate. On the continent there would be many revolutions and much bloodshed in the decades ahead, but thanks to the abilities of its people, and the blessings of a divine providence, the United Kingdom would avoid them. It would successfully combine liberty and order, prosperity and progress, in ways that no other nation could match. Such were the stories that the British would come to tell about themselves, and increasingly to believe about themselves, during their century of world dominance established in 1815. And as such, they would be enjoying, and acting out, the most uplifting and self-regarding national narrative of any country on earth.

Up to a point, this quintessentially Whiggish account was true; but only up to a point. For like all the myths that countries devise and develop about themselves, it was a resonant amalgam of fact and fiction. Despite the economic, constitutional, imperial and global pre-eminence that it eventually came to enjoy, there had been nothing inevitable or preordained about the victory of the United Kingdom over Revolutionary and Napoleonic France; even as late as the closing months of 1812, the outcome had still been in doubt. Britain's political culture might in some ways be uniquely 'liberal', but during the first half of the nineteenth century this would not always be obvious, and its separateness from, and superiority to, developments on the continent was often exaggerated. The decades from the 1800s to the 1840s were scarcely a time of social peace or political stability, while during the second half of the nineteenth century many European nations would come to embrace democracy more enthusiastically than Britain's leaders were by then willing to do. Like many European nations, the United Kingdom was a composite monarchy, containing many nationalities, and the Act of Union was no guarantee that the Irish would remain contented and quiescent. And while there might be liberty in the United Kingdom and the colonies of settlement, the rest of the British Empire would be governed on much more authoritarian lines. Nor was Britain's global pre-eminence, though undeniable, providentially preordained or of indefinite duration. Much of it depended on luck, and on the lack of rivals, both in Europe and beyond; and that absence of continental and global competitors would not last. Only in retrospect does Britain's success during the nineteenth century seem so complete, so certain and so self-assured. But it rarely seemed like that at the time, and it certainly did not do so in 1815 – as the men of power contemplated an uncertain future, both at home and abroad, as the entrepreneurial and business classes rightly feared another economic downturn, and as the majority of the people, once they turned aside from the euphoria of victory, resumed the toil and insecurity of their everyday lives.

3
Great Power, Great Vicissitudes, 1815–29

In the aftermath of Waterloo, Wellington rejoined Castlereagh at the Congress of Vienna, where Britain and the other victorious allied powers returned to their task of restoring to Europe the traditional order of monarchies and aristocracies that had seemed mortally threatened by the subversive ideologies unleashed in 1789. Despite the best efforts at Vienna of such *ancien régime* admirers as Talleyrand and Metternich, respectively supporting the restored French throne and the victorious Austrian monarchy, that right royal restitution would never be complete, but until 1848 the belief in monarchical legitimacy would remain the animating impulse of many of the kings and their counsellors who ruled in Madrid and Paris, Naples and Vienna, and Berlin and St Petersburg. In the United Kingdom, by contrast, there was no need for such a deliberate and contrived counter-revolution, since the established order of throne and altar, government and parliament, supported by widespread popular patriotic sentiment, had held firm during a quarter century of global warfare. Subversion in Britain had been overcome and rebellion in Ireland suppressed. The United Kingdom had prevailed on the oceans of the world and (eventually) on land as well, and it had lavishly bankrolled a succession of continental coalitions which, after many false starts, had brought Napoleon down. As the victory bells rang out in the aftermath of France's final defeat, it did indeed seem as though Britain's wartime successes were the fitting result of the bravery and courage, the duty and discipline, the heroism and professionalism of its soldiers and sailors, its statesmen and state servants. The late eighteenth- and early nineteenth-century fiscal military state, underpinned by an expanding industrial economy, able to raise unprecedented amounts of money, and governed by hard leaders and administered by able men, had eventually delivered successful results and triumphant outcomes, which were as deserved as they were providentially ordained.

But from another perspective the picture seemed very different, for in 1815 the United Kingdom was an exhausted nation on the brink of what would soon become a severe post-war recession, and it also seemed to many to be mired in profligacy and sunk in corruption, as it was led by an increasingly distanced, selfish, rapacious and self-interested elite. As peacetime premier, the Younger Pitt had sought to make the British government more rational, responsible and efficient, and to put its finances on a more stable footing in the aftermath of the costly conflict with the American colonists. But the Revolutionary and Napoleonic Wars had spelled the end of those attempts, and the subsequent quadrupling of the national debt, along with the corresponding increase on the service charges, were not so much signs of the United Kingdom's underlying economic strength and fiscal resilience, but rather a portent of national bankruptcy and financial ruin. Moreover, according to writers and critics such as William Cobbett and John Wade, the public finances of the United Kingdom were made worse by the existence of what they described and denounced as 'Old Corruption', as hundreds of thousands of pounds were paid out each year to families such as the Grenvilles and the Wellesleys in sinecure fees and pension charges. Far from being acts of heroic endeavour and disinterested patriotic duty, Cobbett and Wade contended that military and government service amounted to the selfish, self-seeking endeavours of unscrupulous men on the make, with their hands in the till and their snouts in the trough. The royal family was even worse, as the Prince Regent and the Duke of York remained heavily indebted, embroiled in marital and sexual scandal, and as a result continued deeply unpopular and widely disliked. George III might have won a late-life apotheosis as the 'father of his people', but he had been a failure as the father of his own sons.

This picture of a governing oligarchy that seemed more self-interested than patriotic, and that had been irresponsible rather than prudent in its stewardship of the national finances, was further corroborated by two events that took place in the immediate aftermath of Waterloo. The first concerned the income tax that had been introduced by Pitt in his budget of December 1798 as a temporary measure and wartime expedient, and then re-established by Addington when hostilities with France had recommenced in 1803. The tax had been essential to the successful financing of Britain's military effort. Faced with an unprecedented amount of public debt and with interest payments that would average £32 million a year between 1816 and 1820, the highest level across the whole of the nineteenth century, Lord Liverpool's government sought to retain the 'temporary' income tax in peacetime. But the Commons refused to ratify such a proposal: many MPs disapproved of high government spending once the war was over, and there were some who thought the revenue generated by the income tax had merely financed the costs and payments of 'Old Corruption'. The result was that Liverpool's government was left with no coherent policy to deal with the servicing of the debt, and the Chancellor of the Exchequer, Nicholas Vansittart, was forced to resort to a variety of short-term expedients, including further heavy borrowing from the City of London and the Bank of England. The second episode, which would turn out to be even more controversial, and engage the passions of the politicians and the people into the 1840s and even beyond, was the passing of the Corn Law, which banned the import of any foreign wheat until the domestic price reached 80s a quarter. Earlier attempts to carry such a measure had failed; but the government now insisted such legislation would provide a strong incentive to domestic producers and would ensure a steady supply of wheat at a stable price. But to members of the middle and working classes such arguments were humbug, for they regarded the Corn Law as the self-interested act of a landowner-dominated legislature, which cared nothing for the well-being of the population as a whole.

While the Corn Bill was debated in parliament, protest meetings were held across the country, and there was also an extensive petitioning campaign. Crowds gathered outside the Palace of Westminster and had to be dispersed by troops, while the London homes of known supporters of the measure were attacked. This agitation seems to have been genuinely spontaneous, the outraged behaviour of poor working people who were being compelled to pay more for their bread, so as to preserve the rentals and comforts of those who were landed, leisured and rich. But it was all in vain, and this provided further evidence in support of those radicals and would-be reformers who were calling for change in the structure of the legislature to make it less self-serving and more representative of the people and the nation as a whole. These demands were further driven by the sudden economic downturn beginning late in 1815, as peace brought neither prosperity nor plenty to a war-weary people. On the contrary, immediate cuts in public spending, combined with the cancellation of orders for additional military supplies, and the return of thousands of demobilized soldiers often vainly seeking work, brought increasing hardship to many. Conditions may not have been as bad as elsewhere in Europe, where famine, disease and high unemployment were widespread, resulting in the most violent and widespread grain riots since the French Revolution. But a depressed and disordered continent was not a good market for the products of British industry, and between 1814 and 1816 the volume of the United Kingdom's overseas trade fell by one-third. As the economy went into recession (there were more than 20,000 unemployed weavers in the London district of Spitalfields alone), it seemed as though the febrile and subversive 1790s were returning, and far from certain that victory abroad would guarantee political or social stability at home. For Lord Liverpool and his colleagues the challenges of peace seemed as great as the problems of war, and in that, at least, they had something in common with the people whom they governed.

EUROPE, THE EMPIRE AND THE WORLD

Yet despite these many domestic uncertainties, Lord Liverpool's administration would last until he became incapacitated early in 1827. His fifteen continuous years in office would be the longest since the Younger Pitt's, and more extended than any of his nineteenth-century successors. A further indication of this high-political continuity was that the United Kingdom's overseas relations during this time were conducted by only three people: by Castlereagh and Canning as successive Foreign Secretaries and also as Leaders of the House of Commons, and by the third Earl Bathurst, who for the entire period combined the offices of Secretary for War and for the Colonies. Castlereagh was not only the scion of a great Anglo-Irish aristocratic dynasty, and heir to the Marquis of Londonderry, but he was also a protégé of the Younger Pitt's. He had been continually in public life since the mid-1790s, playing a major part in passing the Act of Union. Having then been a crucial figure in negotiating the settlements at the end of the Napoleonic Wars, his aim thereafter was the preservation of European peace by maintaining the balance of power between defeated France and the victorious allied nations, thereby enabling Britain to pursue its imperial interests and maritime avocations instead of being further distracted by and drawn into renewed continental conflicts. Castlereagh's long-time political rival and eventual successor was Canning; he was also a Pittite, and had already served as Foreign Secretary under the Duke of Portland, but his lowly background was very unusual for a nineteenth-century Foreign Secretary. In addition to having an actress for a mother, Canning's father was a lawyer in modest circumstances, and for ten years he himself was MP for Liverpool (along with Gladstone's father), which meant he was in close touch with commercial and mercantile opinion, and his foreign policy would be widely regarded as more assertive in those interests than Castlereagh's. Bathurst was another patrician landowner, a close friend of Wellington's, and he would preside over the consolidation and expansion of the Colonial Office and, correspondingly, over the significant expansion of the British Empire.

What resources did Castlereagh, Canning and Bathurst have at their disposal to project and maintain British power overseas? Behind them lay the all-conquering Royal Navy and the (eventually) victorious British army, and the linkage between power and policy was clear as Bathurst combined War and the Colonies in a single ministerial portfolio. At the end of the hostilities against France, there was massive pressure inside and outside government to reduce military expenditure, so the British army and the Royal Navy were both appropriately scaled back; but the Indian army, which was financed by the subcontinent not by the British taxpayer, was kept at its wartime levels, while the Royal Navy still remained greater in numbers than the fleets of the next three maritime powers combined. But the administrative apparatus that Castlereagh, Canning and Bathurst had at their disposal, both in London and overseas, was exiguous in the extreme. In 1821 the entire staff of the Foreign Office consisted of twenty-eight men, including two under-secretaries and a Turkish interpreter; while the annual cost of the whole British diplomatic service abroad was less than £300,000. The Colonial Office had been responsible for twenty-six colonies in 1792, but it was in charge of forty-three by 1816, yet it remained smaller than the Foreign Office in terms of the personnel it employed in Whitehall. In neither department was there a rational or meritocratic system of recruitment, as appointments were made on the basis of ministerial patronage and family connection. Castlereagh's ambassadors included Earl Cathcart (a soldier, to St Petersburg), Lord Stewart (his half-brother, to Vienna) and Henry Wellesley (younger brother of Wellington's, to Madrid). None were particularly noteworthy. At the Colonial Office the Duke of Kent, another of George III's sons, had been appointed Governor of Gibraltar in 1802, but had soon been recalled because his harsh discipline had provoked a mutiny, although he continued to hold the job nominally until his death in 1820. The Duke of Manchester had been sent out as Governor of Jamaica in 1808, and would remain in post until 1827. Both men were professional soldiers, neither was distinguished, and Kent's only lasting claim to fame was that he fathered the future Queen Victoria a year before his death.

Castlereagh's post-war objective was to ensure the continent was not engulfed in another long, costly and destructive conflict. To this end he sought to establish what would become known as 'the Congress System', or the 'Concert of Europe', which would put the meetings recently held in Vienna on a semi-permanent footing, as representatives of the four great powers of the United Kingdom, Austria, Russia and Prussia agreed to meet regularly in future, to maintain international co-operation, talk over questions of common interest and settle any differences between them. (A further advantage, although it greatly increased his workload, was that Castlereagh would be conducting these negotiations himself, rather than leaving them to his not very impressive ambassadors.) In the aftermath of Waterloo, Castlereagh worked with the other three victorious powers to bring about this arrangement, and they were as eager as he was to build secure collective defences, one indication of which was that the four nations agreed they would immediately declare war on France if any member of the Bonaparte family should return to power during the next twenty years. The first of these congresses was held in 1818 at Aix-la-Chapelle, and was attended by the Emperors of Russia and Austria, by the King of Prussia, and by Castlereagh and Wellington as the British representatives. The chief aim was to conciliate the French: the amount of compensation demanded by the allies for the destruction caused by the Revolutionary and Napoleonic Wars was reduced, their occupation that had lasted since 1814 was brought to an end, and France was admitted to the Congress of Europe. But thereafter, Britain's interests began to diverge from those of Russia, Prussia and Austria, who had already formed their own separate 'Holy Alliance' in defence of absolutist monarchy and firm government, and to ensure further revolutions would be suppressed. This was the compact that Castlereagh memorably dismissed as 'a piece of sublime mysticism and nonsense'.

To be sure, Castlereagh was no friend to revolution, and the Duke of Wellington was even less so: both had been appalled by the events of 1789 in France and those of 1798 in Ireland. But they were also answerable to parliament in ways that absolute monarchs and their ministers were not: Castlereagh to the Commons as Foreign Secretary, Wellington to the Lords, having joined Liverpool's government as Master-General of the Ordnance late in 1818. Neither Castlereagh nor Wellington liked the idea of 'public opinion', and did not feel personally bound by it or pay much heed to it, but they could not be wholly indifferent to it, both in their own country and in other parts of Europe, too. These divergences of view between the British and the absolutist members of the Concert of Europe emerged into the open in 1820, a year that witnessed uprisings and revolutions in Spain, Portugal, Naples and Piedmont. In Spain the Bourbons had been restored in 1813 following the abdication of Napoleon's usurping brother, Joseph Bonaparte, but the returning legitimate sovereign, Ferdinand VII, rejected the liberal constitution that had been passed in 1812, and sought to restore absolute monarchy at home and to defeat the Latin American rebels who were at the same time seeking independence from Madrid. But by 1820 the Spanish government was virtually bankrupt, taxes had been raised to unacceptable levels, there was widespread discontent in the army, and public dissatisfaction boiled over. Ferdinand was forced to recognize the 1812 constitution, to summon a legislative assembly (or Cortes), and to make way for a liberal government. The crisis in Spain was compounded by similar events in Italy, where King Ferdinand of Naples and the Two Sicilies was also compelled to adopt the Spanish constitution of 1812, and King Victor Emmanuel I of Piedmont-Sardinia was forced to abdicate in 1821. And in Portugal there was another liberal uprising, inspired and encouraged by that in neighbouring Spain, demanding the return of the exiled royal house from Brazil but also the establishment of a legitimate constitutional monarchy.

The view of the Holy Alliance was that they should intervene against such revolutionary threats. By contrast Castlereagh, while eager to uphold the balance of power, and instinctively on the side of the established order, thought it wrong to interfere in the internal affairs of other states 'in order to enforce obedience to the governing authority', and he insisted that the Neapolitan, Spanish and Portuguese revolutions were merely 'domestic upsets' on the periphery of the continent, which did not threaten the peace and security of Europe as a whole. Having distanced Britain from Austria, Russia and Prussia, he refused to attend the congresses held at Troppau (Opava) in 1820 or at Laibach (Ljubljana) in the following year, which were intended to formulate robust responses to the uprisings on the Iberian Peninsula and in Italy. Instead Castlereagh sent his half-brother, Lord Stewart, but as an observer rather than as a participant, and he restated the British position that intervention in the domestic affairs of other European nations was contrary to international law and to the terms of their original alliance, and in any case it would do no good. But in 1821 the two pro-Russian principalities of Moldavia and Wallachia revolted against Ottoman rule, and early in 1822 a self-styled Greek National Assembly issued its own declaration of independence from 'the cruel yoke of Ottoman power'. Public opinion in the United Kingdom was overwhelmingly on the side of the Greeks, but this time Castlereagh took a different view of the relative merits of rebellion and suppression. He worried that the Hellenic revolt would seriously weaken the already enfeebled Ottoman Empire, which he was eager to support as a counterweight to Russia for, like many subsequent Foreign Secretaries, he feared that the Tsar harboured predatory designs on Britain's Indian Empire and dominant position in the Mediterranean. More immediately, he was afraid that Russia might support the Greek rebels in the hope of inflicting serious and lasting damage on the Ottomans. Accordingly, Castlereagh resolved to attend the next congress, scheduled to meet at Verona in the autumn of 1822 so he could discuss these issues face to face with the Russian Emperor. But overwork had taken its toll: during the summer Castlereagh suffered a serious mental collapse, and in August he took his own life.

By then, he had been Foreign Secretary continuously for a decade, a span of office that would not be equalled again until the time of Sir Edward Grey in the early twentieth century. During the last year of his life Castlereagh had inherited the family estates and titles, and was thus briefly translated to the House of Lords as the Marquis of Londonderry (though it is as Castlereagh that posterity has always identified him). For most of his time in power he had been preoccupied with European issues, although towards the end of his life he had become increasingly sceptical of the Congress system, and was urging a policy of British non-involvement in European affairs. Indeed, at the Verona meeting of October 1822, which took place two months after Castlereagh's death, and where Wellington represented the government, the United Kingdom permanently detached itself from the Concert of Europe and the Congress System. Castlereagh had remained a firm believer in monarchy, aristocracy and hierarchy, and in the need to maintain peace in Europe; but towards the end of his life the rise of Russia as a great power had obliged him to concern himself with Britain's Asiatic affairs, while the persistent and widespread rebellions against imperial rule in Spain's Latin American colonies had also begun to demand his attention. He was succeeded at the Foreign Office by Canning, who had recently served as Lord Liverpool's President of the Board of Control, responsible for overseeing the affairs of the East India Company. Canning's rhetorical talents and presentational skills were greatly superior to those of his predecessor, he was the first Foreign Secretary to publish large quantities of diplomatic correspondence to justify his actions, and he enjoyed much better relations with the press than Castlereagh had ever done – or, indeed, had ever wanted to do. Canning's priorities were also more global and extra-European than continental, although it was events in Spain and Portugal that served to bring Latin American affairs to the centre of his attention.

By the mid-1820s the earlier hopes that both Spain and Portugal might accept liberal constitutions with limited monarchies had been disappointed. In 1823 the French invaded Spain with the aim of restoring Ferdinand VII, who had recently been deposed by the Cortes because of his determined efforts to thwart the liberal constitutionalists. Adhering to Castlereagh's policy of non-intervention in the affairs of another European state, and further constrained by recent cuts in defence spending, Canning resolutely refused to involve Britain. Ferdinand was duly restored and would reign for another ten years, during which time he consolidated his despotic monarchy and took revenge on the liberals who had earlier opposed him. Meanwhile, in Portugal, John VI had returned from his Brazilian exile in 1821 and initially accepted a liberal constitution that limited his own powers, according to the recent Spanish model. But the French intervention in neighbouring Spain encouraged the Portuguese military to stage a coup, which overthrew the existing parliament and constitution and gave increased powers to the king. John VI reigned until 1826, and he would eventually be succeeded by his son, Miguel I, who as a friend and admirer of Prince Metternich's would prove to be even more authoritarian. On this occasion Canning decided that he would intervene and dispatched a military expedition in the hope of preserving Portugal's liberties. But there was no chance it would succeed, and Miguel would reign as a domineering monarch until 1834 when he would be overthrown and forced into exile. For all Canning's unrivalled skills in defending it in parliament and the press, this scarcely amounted to a coherent or successful foreign policy, and the eventual outcome in both Spain and Portugal was the strengthening of despotic royal regimes, which were more amenable to the monarchs of the Holy Alliance than to the British government, and which were enthusiastically backed by the equally reactionary Charles X of France.

Canning's response, as he put it in his most oft-quoted remark, was to turn away from these regrettably revived European despotisms on the western extremity of the continent and to 'call in the new world to redress the balance of the old'. Since the French invasion of the Iberian Peninsula, the Spanish and Portuguese colonies in Latin America had been seeking to emancipate themselves from their imperial overlords in Madrid and Lisbon, be they Napoleonic usurpers or the legitimate but authoritarian sovereigns who had subsequently been restored, and by the early 1820s they had largely succeeded. Having failed to prevent the re-establishment of these despotic regimes in Spain and Portugal, Canning resolved to recognize the independence of the recently established republics of Buenos Aires (subsequently Argentina), Mexico and Columbia, and he did so early in 1825. Soon after, he negotiated a settlement with John VI of Portugal, who belatedly recognized the independence of Brazil, where his own son, Dom Pedro, had proclaimed himself as an independent, constitutional sovereign (indeed, an emperor) three years earlier. In recognizing the independence of these new nations, Canning sought to show his preference for liberal internationalism rather than continental despotism, but he also had an eye for business, for he strongly (perhaps excessively) believed that the open markets of Latin America that were now established would soon become major outlets for British goods and investment. That was what he meant when he triumphantly declared in the Commons that 'Spanish America is free, and if we do not badly mismanage our affairs, she is English!' Canning's prime interest in Latin America was the potential for trade rather than the desire for further dominion. He also sought to align British policy with the Monroe Doctrine recently adumbrated by the President of the United States, which had not only recognized the independence of the former Spanish colonies, but had also strongly discouraged any future political or military intervention by the European powers in the nations and affairs of the western hemisphere.

Canning's Latin American policy was highly popular with the press, public opinion and the commercial and mercantile sections of the population, and it served to distract them from his less successful efforts on the Iberian Peninsula, although it was not so well received elsewhere. George IV (as the Prince Regent had become on the death of his father in 1820), the Duke of Wellington and other High Tories disliked his support of rebels, whom they believed were misguided and mischievously inflamed by 'French principles' against their legitimate sovereigns. But Canning took the view that the Spanish and Portuguese Empires were in terminal decay, that he was backing the winners against the losers, and that this would also be to Britain's long-term commercial advantage. Across the other side of the world, the Ottoman Empire was another once-great imperium in decline. Yet towards the Greek revolt, which had taken place near the very end of Castlereagh's time in office, Canning adopted a rather different policy. To ardent philhellenes such as the poet Lord Byron the Greeks were like the Latin Americans in that they were also fighting for freedom against despotism. But Canning shared Castlereagh's view that it was important to prop up the Ottoman Empire as an essential counterweight to Russia's expansionist ambitions in Asia and the Mediterranean, and the fact that Russia was supporting the Greek rebels only strengthened him in that belief. He was willing to recognize the Greeks as belligerents, and he did so early in 1823; but he insisted that the United Kingdom would remain neutral, and he refused to join a continental congress, for fear the result would be the coercion of the Ottomans and a triumph for Russia. But the death of the assertive Tsar Alexander I in early December 1825 opened up an opportunity, and Wellington was dispatched to negotiate with his successor, Tsar Nicholas I. The result was the Treaty of London of July 1827, under which Britain, Russia and France agreed to confirm Greece's sovereignty over its internal affairs, but the Ottoman Sultan retained overall control.

Canning's support for liberals and revolutionaries was thus constrained by broader geopolitical considerations. He would back them if they were rebelling against governments of which Britain disapproved, successfully in Latin America, although to no avail on the Iberian Peninsula; but he was at best lukewarm if they were in revolt against regimes that the United Kingdom wished to sustain, especially the Ottoman Empire. He was undoubtedly a more flamboyant and media-smart Foreign Secretary than Castlereagh, but he worked in circumstances so different that debates about how much continuity and discontinuity there was between the two men seem rather beside the point. There were similar continuities and discontinuities during Lord Bathurst's long tenure at the Colonial Office, where he would carry on the United Kingdom's relations with many parts of the world, and where he would be more in sympathy with Castlereagh than he would later be with Canning. Bathurst was a close friend of Castlereagh's and Wellington's, he shared their view that the prime task of government was to help preserve the established order both at home and overseas, and he had briefly served as Foreign Secretary in Spencer Perceval's government. Bathurst also loathed the French Revolution, and one of his most agreeable tasks as Colonial Secretary had been to select the remote island of St Helena as an appropriate place for Napoleon's final exile. He enjoyed sinecure incomes as Teller of the Exchequer and Clerk of the Crown, and as such was another beneficiary of the 'Old Corruption' that radicals found so abhorrent. He was a High Churchman and a High Tory, a significant Gloucestershire landowner, a lifelong supporter of the agricultural interest, and he had voted in favour of the Corn Law of 1815. As such, Bathurst was a loyal and committed upholder of what might be termed the United Kingdom's ruling 'military-agrarian complex'; and as Colonial Secretary he would be hostile both to the establishment of representative institutions, and to the development of a free press in Britain's overseas dominions.

Taking up office during the last phase of the Napoleonic Wars, Bathurst had been more concerned between 1812 and 1815 with the military side of his portfolio than with his colonial and imperial responsibilities: he was in charge of the successful provisioning of Wellington's peninsular armies and subsequently of his troops at Waterloo (hence their later close friendship) as well as for organizing the United Kingdom's other war against the United States (hence his later concerns about British Canada's security). But it was also during these wartime years that the Colonial Office first became a recognizable and significant department of state, as it took on the tasks of overseeing the additional colonies that Britain had acquired during the Napoleonic Wars. This was a substantial increase in burdens and business, but the Colonial Office remained small and under-resourced, and it did not grow in size as the British Empire did. Indeed, Bathurst's total Whitehall staff was never more than twenty, and they were in charge of what was becoming the largest, the most rapidly expanding and the most geographically dispersed empire in the world. But at a time of high taxes and soaring national debt, expenditure on the Whitehall bureaucracy had to be kept to a minimum, and to this necessity the administration and government of the British Empire were no exception. Bathurst himself was an efficient and conscientious bureaucrat, but he ran the Colonial Office on traditional, paternalistic lines. As at the Foreign Office, domestic appointments were still made on the basis of patronage, and overseas postings also owed much to status and to family contacts. In 1814 Bathurst appointed Lord Charles Somerset, a son of the Duke of Beaufort, to be Governor of Cape Colony in South Africa; four years later, when his own brother-in-law, the Duke of Richmond, was in financial straits, Bathurst sent him out to be Governor General of Canada; and in 1826 he would appoint his son-in-law, Major General Sir Frederic Ponsonby, to be Governor of Malta.

As befitted someone who had previously been President of the Board of Trade, and who combined the War and Colonial portfolios, Bathurst's conception of the British Empire was both commercial and militaristic. As a supporter of the Corn Law, he was in favour of trade within the empire, not least because the revenues derived from it might help defray the expense of colonial administration. He was also a firm believer in what has been termed 'authoritarian agrarianism', an imperial ideology and practice that stressed the importance of the Anglican Church, the traditional social hierarchy, arable farming, aristocratic appointments abroad, military preparedness and firm civil authority. Hence the development of elaborately ceremonial proconsular regimes, often centring on titled and aristocratic governors, in Cape Colony, the

Ionian Islands and Malta. Hence the migration of agrarian experts from Scotland and Ireland to the empire, where they preached agricultural reform. And hence a further phase of imperial assertiveness and acquisitiveness in South Asia, as the British conquered the interior of Ceylon between 1814 and 1815, including the lands of the King of Kandy, and began to turn their attention further east. Java had been conquered in the aftermath of Napoleon's invasion of the Netherlands; but it had been restored to the Dutch in 1815. One of the young men who had been involved in that fighting was Stamford Raffles, and between 1819 and 1823 he acquired the island of Singapore for the empire, intending to make it the great entrepot port for British trade between India and China. It was an early example of what would soon become a recognizable phenomenon: a local figure, exceeding his authority, and peremptorily extending the empire, to the unavailing dismay of the authorities in London.

Bathurst also sought to implement a series of reforms, assisted by his under-secretary, Henry Goulburn, and by James Stephen, whom he had appointed as the Colonial Office counsel. Some were unsuccessful. His most ambitious scheme, to unite Upper and Lower Canada, was meant to strengthen Britain's position in North America in relation to the United States, but the inhabitants of Ontario and Quebec wished to retain their separate colonial identities, and it was withdrawn. Although he owned a plantation in Jamaica, Bathurst was a friend of William Wilberforce, and was eager to try to improve the living conditions of Caribbean slaves. Supported by Stephen, he pressed his colonial governors hard; but the local legislatures were unwilling to adopt such programmes or to consider gradual emancipation, and this intransigence in turn provoked slave rebellions in Demerara in 1823. Bathurst was more successful in other areas of imperial governance and administration. In 1821 he brought together the coastal settlements and former slave-trading ports that were scattered in the Gambia, the Gold Coast and Sierra Leone into a single colony of British West Africa; but on the Gold Coast there would be a protracted war between the British and the Ashanti people, the former resented the ending of the slave trade from which they had profited, and which would last from 1824 to 1831. Meanwhile, in Whitehall, Bathurst enlarged and restructured the Colonial Office: the clerks were graded and new recruits were appointed 'with a view to obtain more effective service', and the post of second under-secretary, abolished in 1816, was restored. He also assigned the handling of different parts of the empire to particular senior officials. One under-secretary dealt with the western hemisphere: the Caribbean, Guiana and Honduras, the North American colonies and Newfoundland. The other was responsible for Gibraltar, Malta, the Ionian Islands, British West Africa, Cape Colony, New South Wales and Van Diemen's Land, Ceylon, Singapore and Mauritius, and also for relations with Morocco, Algiers, Tunis and Tripoli.

But as this list of the varied and multifarious colonial territories makes plain, their acquisition and administration did not amount to a coherent or systematic imperial 'project'. There were naval bases, settlement colonies, penal colonies, tropical colonies, and former slaving ports on the coast of West Africa; and the Colonial Office was also conducting relations with some areas of the world that were in no formal sense part of the British Empire at all. Moreover, whereas Ceylon came under the jurisdiction of the Colonial Office, Britain's expanding empire in India, which was by some margin its most important overseas territory, did not. In the aftermath of the charter renewal in 1813, it was still essentially the commercial fiefdom of the East India Company, overseen by a Governor General with very considerable de facto powers, and by the Board of Control in London, which in practice provided little by way of sustained oversight. For ten years, beginning in 1813, the Governor General was Lord Hastings, who was an imperial expansionist in the mould of Lord Wellesley. Between 1817 and 1819 Hastings waged Britain's final war in the Indian interior against the Marathas, who this time were conclusively defeated. As a result the British were established as the paramount power on the subcontinent, with an Asiatic victory to match the earlier triumph of Waterloo in Europe, and there would be more to come. 'We must look,' observed one proconsul, 'to an increase of territory by conquest over our enemies in the interior of India', and he had no doubt 'that opportunities will arise for effecting such conquests'. So, indeed, they did, and there was a further phase of expansion under Hastings's successor as Governor General, Lord Amherst. In 1821 the Burmese Kings of Ava annexed Assam, and two years later invaded British territory. This was a serious provocation. The resulting Burmese War cost over £13 million, but the British captured Rangoon, the King of Ava was eventually obliged to sue for peace, and the British Empire in India was greatly augmented eastwards. By 1825 it extended all the way along the Burmese coast from the Irrawaddy northwards without a break to Chittagong; and in the following year the kingdom of Assam was also annexed.

Even though the East India Company was still ostensibly a trading enterprise, albeit with a very large army attached, the inexorable expansion of the territory it controlled meant the company's employees, and the British officials who oversaw them, increasingly came to see themselves as rulers rather than traders. Sir Thomas Munro, who governed Madras between 1820 and 1827, believed 'we should look upon India, not as a temporary possession, but as one which is to be maintained permanently, until the natives shall in some future age have abandoned most of their superstitions and prejudices, and become sufficiently enlightened, to frame a regular government for themselves'. More particularly, such men believed that it was by promoting Western education that the people of India could best be fitted for future self-government. As Mountstuart Elphinstone, the Governor of Bombay from 1819 to 1824, explained: 'If there be a wish to contribute to the abolition of the horrors of self-immolation and of infanticide, and ultimately to the destruction of superstition in India, it is scarcely necessary now to prove that the only means of success lie in the diffusion of knowledge.' These views were held with especial conviction and determination by Lord William Bentinck, who would succeed Amherst as Governor General in 1828, and who would drive through many determined policies of modernization and Westernization, including the reorganization of the judicial system, the restructuring of taxation in the northwest provinces, the abolition of suttee (the burning of widows on their husbands' funeral pyres), and the suppression of thuggee (a fraternity of professional assassins). Bentinck's tenure of office would be the most sustained and systematic assertion yet of Western proconsular power, with the aim of advancing what he regarded as primitive South Asian society to the high level of civilization epitomized by the nineteenth-century British.

Such were the official connections, in the aftermath of Waterloo, between the United Kingdom and Europe, the British Empire, Latin America and the wider world. But the nation's engagements with the peoples and the politics of the globe took place on many different levels, beyond and below those of government, and they were carried on in many different ways. Between 1815 and 1830 more than 150,000

Britons emigrated to the United States, many of them Scottish or Scots-Irish, disillusioned by high taxes in war and by the agricultural depression and the industrial slump that followed. Even more Britons emigrated to what was called British North America, but its vast territory contained fewer than a million inhabitants, relations between the English-speakers and the French were tense, imports and exports were minimal, and the colonies were a serious drain on the British exchequer. During the same period fewer than 10,000 people left the United Kingdom for Australia. But under the governorship of Lachlan Macquirie, New South Wales began to transition from penal to settlement colony: he constructed roads and public buildings, instituted a police force, and Sydney became a substantial settlement of 10,000 inhabitants, with cottages, shops, churches and schools. A second colony had been begun on Van Diemen's Land in 1803, and in 1829 a third was established in Western Australia, with an initial settlement at Perth. As the British settlers began to move inland, they encountered further opposition from the indigenous Aborigines, whose lands they coveted. There was also some minimal emigration to the recently acquired Cape Colony, but relations between the new British arrivals and the Boers, descended from the original seventeenth-century Dutch settlers, were not good: their languages, cultures and religions were different, as were their attitudes to the indigenous inhabitants. In 1828 the British government declared that black and white men were equal before the law, a decision that horrified and alienated the Boers.

Emigration from the United Kingdom would grow significantly during subsequent decades, and it would soon increase the latent tensions between the British settlers and the French in Canada and the Boers in South Africa, and it would also lead to brutal clashes with the indigenous peoples of Australia, New Zealand and South Africa. But this largely lay in the future. Meanwhile, one further consequence of the Revolutionary and Napoleonic Wars was that London had replaced Amsterdam as the trading and financial capital of Europe, and thus of the whole world. Once peace was restored in 1815, British trade resumed with continental Europe, although the post-war recession was not an easy time; cultural contacts were also re-established, as exemplified by Beethoven's decision in 1824 to dedicate his Ninth Symphony to the Philharmonic Society of London, which had commissioned the work seven years earlier. The United Kingdom also benefited from the rapid development of the American economy after 1815, and became the main entrepot for the new world's burgeoning trade with the old, with British investment the prime financial driver of the westward expansion of the United States. Trade with South Asia was also increasing, in part because of the ending of the monopoly of the East India Company, which meant more businesses could participate, but also because the importing of finished cotton goods was replaced by the purchasing of raw cotton, which was good for the Lancashire mill owners but effectively destroyed the indigenous Indian textile industry. Further east, the Company retained its monopoly on the China trade, for which Singapore soon became a major staging post. In exchange for silks, porcelain and (especially) tea, the British traded opium, even though its importation had been banned by the emperor in the late eighteenth century. But by 1830 there were millions of addicts, and the Company sold 18,000 chests of the ostensibly prohibited substance to Chinese middlemen. On the other side of the world, Latin American opportunities also beckoned as independence from Spain and Portugal meant a lifting of the high imperial tariffs that previously had been in force; and during the early 1820s Argentina and Brazil became especially attractive markets for British traders and investors, just as Canning had hoped.

IDEAS, BELIEFS AND DOUBTS

By the middle of the 1820s the United Kingdom was more fully connected with more parts of the world, as a trading and investing nation, as a maritime force and imperial power, than any other country on earth, or than any other country had ever been. Yet it was still governed on what was in many ways a recognizably eighteenth-century basis, of family connection and a division between Whigs and Tories. As so often, the party identities were far from clear. The Younger Pitt had always called himself an 'independent Whig', and when the Prince Regent had asked Lord Liverpool to form an administration in 1812, the incoming premier told the prince that he would undertake it on 'Whig principles'. Yet the Pittite tradition, which ran directly in terms of policies and personnel from the Younger Pitt via Spencer Perceval to Liverpool, and would be reinforced and perpetuated until the

1840s by Wellington and Peel, was essentially and undeniably Tory. Family connections were also weakening, although the Whig cousinhood remained strong, with Spencers, Devonshires, Bedfords, Grenvilles intermarried many times over; and this, combined with the growing cult of Charles James Fox, no doubt helped to keep them going during their long and virtually unbroken years of opposition from 1793 to 1830. On the whole, the Tories were more committed to the established order, firm government and the Church of England, whereas the Whigs were more willing to entertain 'liberal' ideas such as Catholic Emancipation, and enjoyed disproportionate support among the 'middling orders' and nonconformists. As a result they would become increasingly sympathetic to parliamentary reform, which Lord Grey in the upper house and Lord John Russell in the Commons would embrace during these years, albeit intermittently and rather erratically. Even in times of subversion and distress, they could afford to take a relatively relaxed attitude to the affairs of the day: for the Whig dynasties were generally greater, grander, richer and more self-regarding than their Tory equivalents.

Whatever their party allegiances might have been, many public figures, such as Liverpool or Peel or Russell, were men with serious cerebral interests, and they were deeply engaged with the developing ideas and public controversies of the day, or with what was known to contemporaries as 'the march of intellect'. Much of this discussion and debate took place in the recently founded and highly partisan periodicals, beginning with the *Edinburgh Review* (1802 and generally Whig) and the *Quarterly Review* (1809 and usually Tory). By the 1820s these divisions were further sharpened by the founding of *Blackwood's Edinburgh Magazine* (1817 and extreme or 'Ultra' Tory), and the *Westminster Review* (1824 and avowedly radical). Whatever the specific subjects of consideration and particular areas of disagreement, all these periodicals served a similar purpose in that they provided a forum in which the impact of the great public events, and the developing ideas of the time, could be assimilated and assessed. The influence of the Scottish Enlightenment persisted, via Adam Smith, in the writings of the political economist David Ricardo who, following his mentor, was a staunch advocate of Free Trade and opposed the Corn Law of 1815. Yet Enlightenment rationality had also been behind the American and French Revolutions, which had not only been subversive of the established imperial (British) and domestic (French) orders, but had also been godless revolutions no less subversive of the established religious order: the United States of America enforced a strict separation of church and state, while the French Revolution was as atheist as it was anarchic, symbolized by the conversion of the Pantheon in Paris from a sacred church into a secular mausoleum. These broader concerns took concrete form in an issue that preoccupied British writers and intellectuals of the 1810s and 1820s even more than politicians: was the church-and-state constitution of the United Kingdom fit for purpose, or not? Was it Britannia triumphant, to be celebrated and safeguarded, or was it mired in 'Old Corruption', to be reformed or even overthrown?

Behind these questions, and their alternative and antagonistic answers, lay the shades of Edmund Burke and Tom Paine, who seemed almost reincarnated, during the 1810s and 1820s, in the persons of Samuel Taylor Coleridge and Jeremy Bentham; together they were for their generation the embodiment of what by this time had become the competing creeds of romanticism and rationalism. Coleridge, who had long since recanted of his earlier radical opinions, was indeed the heir to Burke, convinced that society was a web or an organism, which had developed slowly and traditionally over time, as a result of the divine handiwork of God. Its guardians and directors, he insisted, should be those whom he called the 'clerisy': an intellectual and cultural elite who, via politicians and administrators, should exercise tutelage over the lower orders, to protect, think and (if needs be) coerce them in a paternalistic sort of way. Bentham, by contrast, argued that there was no such thing as a Coleridgean animating 'spirit of the state'; instead, he insisted that the key to human progress was individual self-interest combined with a broader notion of utility. From these very different perspectives Bentham argued in a pamphlet published in 1818 that the reform of parliament was essential, for only if elected on the basis of universal adult male suffrage could the legislature carry through the sweeping programme of change that was urgently needed. By then, Bentham was gathering around him some able followers, many of them based in Bloomsbury and who would become the animating force behind the *Westminster Review*. They also founded a non-sectarian university college in 1826, which would provide cheaper higher education than Oxford and Cambridge and for a broader social range, which would challenge clerical narrowness and monopoly in higher education (to which Anglicans would respond by founding King's College three years later). For Bentham and his followers, known as utilitarians, society was not a Burkean organism, evolving slowly according to a grand providential design. On the contrary, it was a machine that developed in a logical and rational way, so as to bring the greatest amount of happiness to the largest number of people. 'Every Englishman,' the utilitarian philosopher John Stuart Mill wrote in 1840, 'of the present day is by implication either a Benthamite or a Coleridgian.'

For all their divergent and discrepant visions of British society, Bentham and Coleridge both thought and wrote, like most of their contemporaries, within an essentially Christian and religious framework. For utilitarians and Coleridgeans alike needed a God, and they took His existence for granted, even if they disagreed as to His identity and purpose. In the one case, He was envisioned as a mechanic, who wound the clocks and kept them going, and in the latter case, He was more of a weaver, who held the fabric of society together. Either way, He was essential, and this shared assumption pervaded and contained the many other religious disagreements of the time. Within the Church of England, High Churchmen believed that Scripture was the repository of divine knowledge; they stressed the necessity of church attendance, and the sacramental and sacerdotal aspects of religion, and believed that the purpose of worship was to proclaim respectability and to maintain the established social structure. By contrast, the Evangelicals were more concerned with conversion, faith and enthusiasm, and with the issue of slavery, although they, too, shared a commitment to law and order. The abolition of the slave trade in 1807 had not led to an improvement in the condition of those who were already slaves, and in the spring of 1823 such prominent middle-class Evangelicals as William Wilberforce, Thomas Clarkson and Zacharay Macaulay established the Anti-Slavery Society, the aim of which was the immediate improvement of the condition of slaves and the ultimate abolition of slavery itself. A second Evangelical enthusiasm, fuelled by the popular religiosity and patriotic feeling generated by the wars against the French, was for conversion work overseas. The 1790s and 1800s had witnessed the establishment of many missionary societies, and in 1813 the reform of the charter of the East India Company meant that for the first time missionaries could work on the subcontinent. By 1821 the collective income of the British missionary societies amounted to over £250,000 a year, and during that decade the so-called 'missionary frontier', in South Asia, on the west and southern coasts of Africa, and in the Antipodes, was as dynamic and expanding as the imperial and mercantile ones.

Despite these undoubted signs of religious energy and engagement, drawing on the revived popular Protestantism stimulated by the French wars, there were many indications that the established church was failing to respond to the challenges of the time. By the 1810s there were widespread and well-justified fears that the Church of England was no longer engaging with the growing population in the rapidly expanding towns and cities of the industrializing north. In mushrooming communities, such as early nineteenth-century Birmingham, the provision of church space for the majority of the population was lamentably inadequate; in 1821 that town boasted a mere five parishes to accommodate a population growing by 30 per cent per decade. Indeed, in many of the newly expanding urban areas, there was church accommodation for scarcely one-tenth of the population. Among High Churchmen and Evangelicals alike this became a serious worry in the aftermath of Waterloo, and once the French had been finally beaten, there was a growing feeling that more churches should be built, not only as an urgent religious necessity but also as a national thank-offering for victory. In 1815 the Reverend Richard Yates published a pamphlet entitled *The Church in Danger*, pointing out the deficiency in the provision of places of public worship. But where was the money to come from to deal with this problem? In the countryside benevolent local landowners might provide adequate accommodation, but what was to be done in the industrial towns and cities, where such figures rarely existed, and where local government had neither the resources nor the authority to intervene? At the end of 1815 the High Churchman Joshua Watson and the Evangelical John Bowdler combined forces to send a memorial to Lord Liverpool urging the case that more churches should be constructed, and at the state's expense. At a time of high anxiety about the unprecedented national debt this seemed an audacious and rather desperate suggestion.

But as these would-be ecclesiastical reformers feared, both popular and elite anti-clericalism were increasingly pronounced in the years after Waterloo. William Cobbett complained that the Church of England was a fundamentally misconceived enterprise, because there should never have been a Reformation, and the monasteries should not have been dissolved. Jeremy Bentham, in *Church of Englandism Examined* (1818), deplored it as part of the discredited political, legal and ecclesiastical establishment, criticized its role in misguidedly educating the young, and opposed the efforts to widen its appeal to the new industrial working classes as merely another way of perpetuating its abuses. Soon after, John Wade's *Black Book* (1820) depicted the Church of England as an essential element in 'Old Corruption', mired in waste and extravagance in its higher echelons, and wielding sinister and unaccountable political influence. The Archbishop of Canterbury and the Bishop of Durham each enjoyed incomes of £19,000 a year, in addition to what were, literally, palatial places of residence, while the eighth Earl of Bridgewater was from 1780 to 1829 a 'prebendary' of Durham Cathedral, had two family benefices in Shropshire and lived in Paris, surrounded by dogs and cats dressed as humans. At the same time, the radical MP Joseph Hume argued that, in times of post-war economic hardship, the state should not be spending so much money on the established church; instead, its own resources should be redistributed away from the rich cathedrals and well-supported rural livings and towards the heathen towns instead. As such, the Church of England was not only increasingly out of touch with the growing numbers of town and city dwellers: it was also unacceptably privileged, wealthy and costly. Indeed, one pamphleteer calculated that while post-Revolutionary France spent a mere £35,000 annually on its church per million of inhabitants, and 'priest-ridden' Italy only £40,000, the Church of England cost an astonishing £3.25 million per year, making it the most expensive ecclesiastical provision in western Europe.

These figures were undoubtedly exaggerated, but in drawing attention to the Church of England's privileges and perquisites, limitations and shortcomings, they must have motivated both reformers and critics alike. At the same time, various strains and strands of Nonconformity were offering a more engaged and plausible religion compared to that provided by what seemed the corrupt and discredited Anglican establishment. Many of the leading business and entrepreneurial families were Quakers or Unitarians or Baptists, such as the Darbys of Coalbrookdale or the Whitbreads in brewing. Among ordinary people, the most powerful appeal of Nonconformity came in the form of Methodism, which had blossomed during the years of the Revolutionary and

Napoleonic Wars. Wesley himself had died in 1791, having denounced the godless republicanism of the French Revolution, and having asserted his firm and abiding loyalty to the British crown; and he also consistently maintained that he had no wish to break away from the established Church of England. But the majority of the early recruits to Methodism had been workers, and at a time of falling real wages, they had been attracted and converted by what has been termed 'apocalyptic anxiety', namely the consoling and reassuring prospect of a paradise in the next world they would never know in this. It was new converts such as these who forced the drive to separation from the Church of England that took place soon after Wesley's death. The next upsurge in Methodist conversion and recruitment came in the early 1820s, but this was a very different time, and those who joined were from correspondingly different backgrounds: more middle class than proletarian. By then, many of the working-class Methodists had splintered and seceded, as Primitive Methodists, Bible Methodists, Reformed Methodists, Independent Methodists, and those who remained in Wesley's original church were increasingly middle class in their backgrounds.

Dissenting faiths in general, and Methodism in particular, expanded considerably in England during the early decades of the nineteenth century, as the established Anglican Church seemed increasingly unfit for purpose – so much so that Jeremy Bentham thought it should be ended by 'euthanasia'. Elsewhere in the United Kingdom, the position was different again. Methodism was increasingly appealing in both rural and industrializing Wales, where differences of languages served to reinforce what was becoming the growing divide between church and chapel. In Scotland the Presbyterian Church was less associated with the ruling establishment than the Church of England was south of the border, and was less hierarchical in its organization; but it, too, was in difficulties when it came to appealing to the growing urban and industrial populations of Glasgow and the Forth-Clyde valley. In Ireland the picture was different again, for industrialization was limited in its impact, and the major religious issue was not so much a decline in numbers as a growth in antagonisms that were increasingly politically inflected. The established Church of Ireland appealed to the British-oriented, Protestant, landowning elite, and members of the professional classes in Dublin. But in Ulster, where many of its inhabitants were of Scots-Irish ancestry, popular Protestantism was generally Presbyterian, whereas elsewhere in Ireland the overwhelming majority were Catholic, and hostile both to the Anglo-Irish Ascendancy families and to the Union with Britain. These were very different religious problems from those confronting the Church of England in the industrializing towns and cities; but they were undeniably significant, and they would become more so as the century wore on. Moreover, even the fact that the British government subsidized the Catholic seminary at Maynooth was viewed as unacceptable by many Tory Protestant MPs; and the Evangelical attempt to bring a 'New Reformation' to Catholic Ireland during the 1810s and 1820s failed to produce mass conversions, but had hardened religious divisions instead.

It was impossible for contemporaries to know whether the United Kingdom was a more Christian nation in the 1820s than it had been in the 1780s; nor could they know that those decades also witnessed the highest rates of population growth in the nation's history, peaking in the years from the early 1800s to the mid-1820s. Between 1801 and 1821 the number of people living in the United Kingdom would increase from almost sixteen million to nearly twenty-one million, and by 1831 it would reach twenty-four million. By the standards of the time, this was spectacular growth; it was as much in evidence in Ireland as in Great Britain, and it was primarily the result of a rise in the birth rate that in turn was explained by the falling age of marriage. Although the majority of the population still lived in the countryside and in small towns, it was in the new industrial centres that the increases in numbers were most striking. Between 1801 and 1831 the populations of Birmingham and Liverpool more than doubled, those of Glasgow, Manchester and Merthyr Tydfil almost tripled. To be sure, the United Kingdom was in no sense a predominantly industrial society by 1830, but it was in that sector that the most significant growth and transformation occurred, aptly symbolized by the opening of the first commercial passenger railway, from Stockton to Darlington, in 1825. In textile manufacturing, imports of raw cotton trebled between 1815 and 1830, and the number of cotton mills in Manchester grew from fewer than sixty to more than ninety. Over the same period, coal production rose from approximately sixteen million tons to just under thirty million tons, emphatic proof that the transformation of the economy from wind and water and animal power to coal and steam power was continuing and consolidating. As a result, industry became more efficient and more competitive, the prices of its products fell, and British manufactured goods became markedly cheaper than those produced on the continent. But as their prices declined, increased sales more than made up and increased revenue, as exports grew from £35 million in 1815 to £47 million in 1830.

As the poet and journalist Robert Southey rightly noted, the advent of such machines as 'the steam engine and the spinning engines' meant there would be 'increased activity, enterprise, wealth and power'; and in retrospect this was the direction in which the British economy was headed. But he also recognized that in the short run they would bring 'wretchedness, disaffection, and political insecurity', as short-term fluctuations were more in evidence and made a greater impact than long-term trends. As a result of the post-Waterloo recession, the years from 1815 to 1820 witnessed a protracted period of distress and unrest greater than anything that had taken place during the preceding quarter century. It was more widespread geographically, and it made a much greater impact in Britain's rapidly expanding industrial towns and cities, especially in Lancashire and Yorkshire, and also in London itself. In some parts of the country, the trigger to popular agitation was the high price of bread; in parts of Lancashire and Nottingham there were protests against the introduction of machinery that threatened the jobs and livelihood of skilled workers; in Merthyr Tydfil in south Wales there were riots against wage reductions; in Birmingham the unemployed took to the streets, and in Glasgow serious disorder led to bloodshed. But it was in London that the most famous disturbances took place at the end of 1816, when 'Orator' Henry Hunt addressed a meeting on 15 November at Spa Fields, calling for parliamentary reform as the only way to prevent 'the people' being burdened by higher prices and higher levels of taxation. The meeting was attended by up to 10,000 people and Hunt's escort carried the tricolour flag and a revolutionary cap perched on top of a pike. There was a second meeting at the same location held early in December, but before Hunt could address it, the protest was taken over by the followers of the even more radical Thomas Spence. It was widely – perhaps wildly? – rumoured that there were revolutionary plots to capture the Tower of London and even take over the Bank of England, and they were given mild credence when, early in the new year, a stone was thrown at the Prince Regent as he made his way to parliament.

This was a very febrile atmosphere, in which a secret committee reported in February 1817 that the distress of the labouring and manufacturing classes had been exploited by dangerous radicals seeking 'a total overthrow of all existing establishments'. This may have been exaggerated; but the economy remained depressed, and some plans were laid in such areas as Yorkshire and the east Midlands for a general uprising. More particularly, in the following month, unemployed Manchester weavers set out on an abortive protest march to London, and in June 1817 there was an equally futile display of dissent from unemployed framework knitters. During the second half of the year there was a better harvest, and trade began to pick up, and 1818 witnessed less radical protest and political activity. But the following year saw renewed economic distress, short-time working and high unemployment, as well as increased numbers of protest meetings calling for an end to high taxes and 'Old Corruption' and for significant parliamentary reform. In January 1819 mass meetings were held in London, Birmingham, Leeds and Manchester, many of them addressed by Orator Hunt. In August, against a background of rising tension and amidst reports that revolution was being plotted by the industrial workers of Lancashire, Hunt was invited back to Manchester to address a meeting at St Peter's Field, attended by at least 60,000 people carrying banners with revolutionary inscriptions. There was no disorder, but the magistrates lost their nerve, and sent in the local yeomanry to break up the meeting. The result was that 600 of the crowd were injured, and between eleven and seventeen people were either killed outright or later died of their injuries. The incident was soon dubbed 'the Peterloo Massacre', in satirical homage to Wellington's battle with Napoleon, and resulted in widespread indignation. Indeed, for the rest of 1819 the radical cause became truly national for the first time, and there were riots and protests up and down the country.

Matters did not ease during the following year. A recently released prisoner, Arthur Thistlewood, who had been involved in the Spa Fields riot of 1816, and was a known associate of Thomas Spence, hatched a (wholly unrealistic) plan for the seizure of London and the assassination of the entire Liverpool cabinet, which he hoped would lead to a general and revolutionary insurrection. The so-called 'Cato Street Conspiracy' was planned to take place on 23 February 1820; but thanks to informers the details were already known to the authorities, the conspirators were duly arrested, and five of them, including Thistlewood, were executed for high treason on 1 May. The timing of the conspiracy had been partly determined by the death of George III in January 1820. Thistlewood and his friends thought the ensuing period of uncertainty, with a new and unpopular monarch not yet established, was the ideal time to strike. They were wrong to think their own cause would prosper; but they were right to think the new monarch's would not. George IV came to the throne with a well-earned reputation for sexual and financial irresponsibility; he was also determined to divorce his wife, Queen Caroline, who was as morally wayward as he was, and from whom he had been estranged from the very beginning of their marriage. But while the king urged Lord Liverpool's government to introduce a Bill of Pains and Penalties in the Lords as the preliminary to a divorce, Caroline returned to London in June 1820 from what had been her long and licentious continental exile, and implausibly yet persuasively represented herself as the people's champion against a corrupt government and monarch. She was welcomed by large and enthusiastic crowds, she was lionized in the radical press, and petitions rained down on parliament in her support. Having only passed the Lords by a narrow margin, the government withdrew the Bill, rather than risk defeat in the Commons, and the streets of London were illuminated as radicals celebrated their great triumph. But it soon turned to ashes: the Queen was no real friend of the people or a champion of liberty, she accepted a state pension, made a laughing stock of herself at the King's coronation, and in August 1821 she died.

The end of the 'Queen Caroline Affair' brought to a conclusion half a decade of recession, distress and protest, along with the anxiety, insecurity and repression that were the pervasive response in official circles. Perhaps this helps explain why so many artists and writers put so much of their effort into nature, escapism and fantasy during these years. The vast, apocalyptic paintings of John Martin, such as *The Fall of Babylon* (1819), *Belshazzar's Feast* (1820) and *The Seventh Plague of Egypt* (1823), depicted human beings as tiny, ineffectual creatures, overwhelmed by natural catastrophes expressing the wrath of God. Although Turner and Constable engaged with the new industrial technology, both were more at ease in expressing the Romantics' engagement with landscape – in Turner's case wild and grand panoramas, in Constable's peaceful and more intimate Suffolk scenes. Mary Shelley, the daughter of William Godwin and Mary Wollstonecraft, took refuge in science fiction, exemplified by the Gothic-Romantic horror of *Frankenstein* (1818), in which a young scientist created a monster he could not control, and in the apocalyptic futurism of *The Last Man* (1826), which depicted a world ravaged by plague. Sir Walter Scott, by this time well into his stride as a novelist, produced a succession of be-tartened historical romances, among them *Rob Roy* (1817), *The Heart of Midlothian* (1818), and *The Bride of Lammermoor* (1819); English novels extending from the twelfth to the sixteenth centuries, such as *Ivanhoe* (1819), *Kenilworth* (1821) and *Peveril of the Peak* (1822); while at Abbotsford in the border country he created the prototypical Scottish baronial mansion, festooned with turrets, gables, suits of armour and coats of arms. Meanwhile, Byron, Shelley and Keats represented a new generation of poets who offered a more intense and traumatized version of the Romantic genius: brilliant, unstable, anti-establishment, often in ill-health, invariably in and out of love, and dying young and overseas (Keats in Italy, 1821; Shelley in Italy, 1822; Byron in Greece, 1824) in those hauntingly beautiful Mediterranean lands to which so many Britons were drawn.

By the early 1820s, the British economy was significantly improving, and it would continue to do so until late 1825, fuelled by a major export boom to the newly independent republics of South America, by a significant rise in share prices on the London stock exchange, and by easy loans that were being made available, with the encouragement of the Bank of England, from country banks. As employment opportunities picked up, popular protest died down, radicalism lost some of its earlier appeal, and cultural angst was superseded by cultural confidence. The great new building to house the rapidly expanding British Museum was begun to the designs of Sir Robert Smirke in 1823. The following year the National Gallery was established, thanks to the government's purchase of thirty-eight paintings from the heirs of John Julius Angerstein, and to a donation of sixteen further canvases from Sir John Beaumont. Lord Brougham, who earlier in the decade had been the darling of the London crowds for the part he played in parliament in defending Queen Caroline, turned his attention to advocating the education of working men and in 1826 founded the Society for the Diffusion of Useful Knowledge, which aimed to 'leave nothing undone until knowledge has become as plentiful and as universally diffused as the air we breathe'. But by then, such optimism and confidence again seemed seriously misplaced, for in December 1825 there had been a severe banking crisis, as the South American export bubble burst, leading to a sudden collapse in the price of shares and the failure of many under-capitalized companies. The resulting crash was unprecedented in its depth and in the damage that it caused. Eighty country banks failed in the early months of 1826, the Bank of England itself was only saved by an influx of gold bullion from the continent, and 500 of the 624 companies formed during the bubble of 1824–25 had collapsed by 1827. At the time of George IV's coronation in 1821, *Blackwood's Magazine* had predicted that 'quiet times' had come. But they did not last long, and the political consequences of their ending would be momentous.

LIBERAL TORYISM

In terms of its influence on global affairs the United Kingdom was widely recognized in the years after 1815 to be the greatest power in the world, and that influence was firmly and carefully wielded by Castlereagh and Wellington, and more flamboyantly by Canning. But domestically, the position was far less settled or secure, for it was the economic vicissitudes, the ensuing social upheavals and the resulting political discontents that provided the uncertain background against which Lord Liverpool and his colleagues sought to maintain order and restore the public finances. The 'present discontents' of the early 1790s might have been temporarily damped down and allayed during the years of strain and conflict, but they now re-emerged even more strongly during the years of victory and peace. As a result, the two ministers concerned with domestic affairs in the immediate post-war period would be associated with a policy of consistent repression, which showed scant sympathy for the impoverished circumstances and political aspirations of ordinary people. One of them was the former prime minister Henry Addington, who had been created Viscount Sidmouth in 1805, had held minor office under Spencer Perceval, and served as Lord Liverpool's Home Secretary from 1812 to 1822. The second was the Earl of Eldon, a self-made lawyer who had made a fortune at the bar, and was Lord Chancellor almost continually from 1801 to 1827.

Eldon was loathed by most radicals and Whigs for being (according to Lord John Russell) 'the incarnation of prejudice and intolerance', and he embodied just the sort of judicial conservatism, legal delay and lawyerly procrastination that Dickens would later pillory in *Bleak House* (1852–53). When Byron turned his attention to Britain's domestic affairs, he denounced Sidmouth and Eldon for their unyielding conservatism, he put Castlereagh in the same category, and he ridiculed Southey as the High Tory apologist of Liverpool's regime in *The Vision of Judgement* (1822).

But it was not only Whigs and radicals who took against Lord Liverpool and his administration. In his novel *Coningsby* (1844) the future Conservative leader Benjamin Disraeli would scorn his predecessor as 'the Arch-Mediocrity, who presided, rather than ruled, over a Cabinet of Mediocrities'. Yet while Liverpool was (and is) the least well known of all the United Kingdom's long-serving prime ministers, this was not the whole truth of things. He had been Addington's Foreign Secretary, Portland's Home Secretary, and Bathurst's predecessor as Secretary of State for War and the Colonies under Spencer Perceval, where he had been responsible for provisioning Wellington's peninsular army until he handed the job on to Bathurst. Liverpool's own administration contained six other men who had been or would be prime minister: namely Sidmouth, Canning, Goderich, Wellington, Peel and Palmerston (who was Secretary at War from 1812 to 1827, but outside the cabinet). To be sure, Sidmouth and Goderich were no political titans, but none of these men were mediocrities, and Castlereagh and Eldon were also formidable figures. In skilfully keeping his government together for so long, and in managing these many outsize personalities, Liverpool was no 'Arch-Mediocrity'. He was tactful, experienced and highly intelligent; he could keep elder statesmen in check while also bringing on new talent; and he saw himself as upholding the best of the Pittite tradition in politics and public life. Although it merely confirmed the view of critics that the system of representation was rotten to the core, Liverpool won four successive general elections in 1812, 1818, 1820 and 1826, which suggests he must have enjoyed at least a modicum of support in some parts of the country. He showed determination and ruthlessness in suppressing disorder and discontent, especially during the years from Waterloo to Peterloo, when the memories of 1789 and 1798 remained vivid and cautionary. (Liverpool himself had witnessed the storming of the Bastille when staying in Paris as a nineteen-year-old.) But he also supported some limited domestic reforms, and increasingly so during the early 1820s when economic improvement brought a less harsh social climate.

The government's shifting and uncertain course between repression and reform soon became apparent in the immediate post-Waterloo years. One unresolved issue, too long delayed by the exigencies of war-time, concerned the nation's currency. The Younger Pitt had suspended cash payments during the financial crisis of 1797, which meant bank-notes could no longer be exchanged for gold. The resulting over-issue of paper money by the Bank of England was one reason why domestic prices rose so dramatically during the 1800s: that, at least, was the conclusion of a Commons inquiry, known as the Bullion Committee, which in 1810 had recommended a return to gold. But no action was taken in the midst of war, and it was not until 1819 that the question was addressed again by a committee chaired by Robert Peel, who had recently relinquished the post of Chief Secretary for Ireland but who remained a rising figure among the Tories. Peel's committee also urged a return to gold, and after a long and tense debate in the cabinet and in the Commons it was decided that the Bank of England should resume cash payments. This led to an immediate improvement in the United Kingdom's position on the foreign exchanges, as gold began to flow in from overseas. But prices fell again, and even more sharply than they had after 1814, while the circulation of notes of less than £5 in value virtually collapsed. There was widespread criticism from landowners and industrialists that their interests were being subordinated to those of the City of London. Indeed, Thomas Attwood, the Birmingham ironmaster and banker, believed that what he called 'Peel's Act' had created 'more misery, more poverty, more discord, more of everything that was calamitous to the nation, except death, than Attila caused in the Roman Empire'. On the other hand, the leaders of the cotton industry, the very sector of the economy where the Peel family had made its fortune, were strongly in favour, because they believed the return to gold would help stabilize international trade and lead to a growth in exports.

The blatant class bias of the 1815 Corn Law, combined with the pursuit of a financial policy that in the short term yielded at best mixed results, along with the severe economic downturn, did not ease the transition to peace for the majority of the population. In response to the huge meeting at Spa Fields in December 1816 and the subsequent disturbances, Addington piloted through parliament what were known and deplored as the 'gag acts', with the aim of curbing civil liberties in the name of stamping out sedition. Once again, habeas corpus was suspended, the law of 1798 was extended making seditious meetings illegal, the 'seduction' of soldiers and sailors from their allegiance to the crown was outlawed, and all the safeguards against treasonable activities granted to the king were extended to the Prince Regent. Two years later, in the aftermath of the protests against the 'Peterloo Massacre', the government rushed through yet more repressive measures known as the 'Six Acts'. They made illegal 'seditious assemblies' of more than fifty attendees; they prevented meetings for the training of people in the use of arms; they increased the penalties for blasphemous and seditious libels; they imposed stamp duties on all periodical publications in the hope of making the radical press too expensive for working (or unemployed) people to buy; and they gave powers to local magistrates to search any private property for weapons. As had been the case during the 1790s and 1800s, the Home Office organized a national network of spies and agents provocateurs to infiltrate radical groups, and the implementation of justice undoubtedly became more severe: in England and Wales between 1811 and 1819, committals and convictions increased threefold, the number of death sentences given out rose from 359 per annum to 1,206, and actual executions went up annually from 45 to 108. In all these measures the heavy hands of Sidmouth and Eldon may be discerned, but in the short run they intensified still further the very discontent they were designed to stamp out.

Yet even as these repressive steps were being taken, the government planned to withdraw these new Acts, or at any rate to ignore them, when popular discontent quietened down (the Seditious Meetings Prevention Act was repealed in 1824, and two of the Six Acts survive today in modified form). Moreover, the policy of repression co-existed with some early attempts at social amelioration. Expenditure on poor relief, which was largely paid for by landowners, increased from £4.1 million in 1803 to £5.7 million in 1815–16, peaked at £7.9 million in 1817–18, and remained high thereafter. In 1817 the Chancellor of the Exchequer, Nicholas Vansittart, steered through parliament the Poor Employment Act, which made available £1.75 million in the form of exchequer bills to individuals or corporations willing to invest in employment-generating public works schemes; the money was expended on cutting canals, making roads and draining marshes to bring more land under cultivation. These schemes marked a notable extension in state-sponsored activity. In 1819 the government carried the Cotton Factory Act, which was a significant piece of social legislation because it forbade the employment of children under the age of nine, and limited the hours of work for those aged between nine and sixteen to twelve hours a day. Three years later, the first measure was passed regulating the treatment of animals, which made it a crime to treat cattle, oxen, horses and sheep cruelly or to inflict unnecessary suffering on them, and the first prosecution was brought soon after. This legislation represented a significant change in sensibility (and it would be reinforced by the founding of the Society for the Prevention of Cruelty to Animals in 1824). These early interventions by the state into the workings of the free market were only of the most tentative kind, and throughout the nineteenth century there would be strong parliamentary opposition to any further such intrusions, as Gladstone would later discover when trying to deal with the vexed problem of Irish land. But even in an age where the prevailing ideology was one of laissez-faire, such interventions indicated that government regulation of the market and of private property would become an increasingly important and contentious issue.

At the same time, Lord Liverpool's government sought to respond to the growing concerns, already described, that the Church of England was no longer engaging pastorally with the inhabitants of the nation's industrializing towns and expanding cities, and was thus becoming increasingly vulnerable to radical and Nonconformist criticism. In 1817 a committee was appointed to form a society for 'promoting public worship by obtaining additional church-room for the middle and lower classes', while in the following year parliament granted £1 million for the building of new churches and set up a Commission to oversee the disbursement of this money and the construction of the buildings. Liverpool regarded this as the 'most important' measure he had ever introduced. In some cases the Commission provided the entire sum to fund the costs of construction, whereas in others the difference was met by private donations and public subscription; in no case did the Commissioners make a grant in excess of £20,000. By February 1821 eighty-five churches had been constructed, with seating for almost 150,000 worshippers; but almost all the money had been spent, so applications for twenty-five more churches were postponed because of lack of funds. Three years later, a further £500,000 was made available by a second parliamentary grant, which was given out in smaller sums and spread more widely, helping to fund the construction of an additional 500 churches. Here was another unprecedented state intervention, to promote the state religion in the aftermath of war, revolution, population growth and massive urbanization; but the construction of these 'Waterloo Churches' or 'Commissioners' Churches', as they were known, did not fully solve the problem of providing sufficient Anglican places of worship in the rapidly expanding towns and cities of the British Isles. On the other hand, the church may have been more successful when it came to the provision of elementary education; by 1820 the majority of children attended a school for some part of their childhood, and most of the British population was basically literate, though rates were lower than in France, Prussia or Scandinavia.

By the early 1820s the worst years of post-war deprivation and unrest seemed to be abating, as the economy began to pick up; and these welcome improvements coincided with two significant political changes. The first was the accession of the Prince Regent as George IV at the very beginning of the decade. He had not been a popular heir, and he would not be a popular monarch either. He was a great supporter and patron of the arts, but this involved him in spending more money and running up more debts. His treatment of Queen Caroline did him further considerable damage in the public eye, and his coronation was appropriately overblown and extravagant as grandeur degenerated into farce. But the death of the queen gave him a brief breathing space, during which he made two innovative visits to the more distant parts of his kingdom, touring Ireland later that year and Scotland the following year. To his critics, the accompanying ceremonials were as much a sham as his earlier coronation had been. But to his admirers, among them Sir Walter Scott, who had planned the Edinburgh extravaganza complete with tartans and kilts, they were a welcome and reassuring indication that the monarch was reaching out to his people as never before. Unlike his father, George IV was more interested in an ornamental than in an interventionist monarchy – more eager for show and display, less eager to engage actively with his ministers in the laborious business of government. To be sure, he caused trouble for Lord Liverpool by insisting on the introduction of the Bill of Pains and Penalties, and he would later create many difficulties over the issue of Catholic Emancipation; but by the end of his reign, the monarchy's capacity to intervene decisively would be significantly less than it had been when he had assumed the Regency twenty years before. On the king's death in 1830, Wellington would describe George IV as 'a medley of the most opposite of qualities, with a great preponderance of good'. *The Times* was less forgiving: 'There never was,' it observed, 'an individual less regretted by his fellow creatures than this dead king.'

The second and greater political change was Liverpool's reconstruction of his cabinet, in part (but only in part) made necessary by Castlereagh's suicide in August 1822. But several years before that, the prime minister had resolved to bring in new talent and to strengthen his administration in the House of Commons, where it had suffered too many reverses (as in the case of the voting down of the income tax in 1815). He had begun to make changes in early 1818, bringing in Frederick John Robinson as President of the Board of Trade (he had been Vice President since 1812) to succeed the unimpressive Earl of Clancarty. In January 1822 Liverpool appointed Peel, who had recently chaired the committee advocating the return to gold, to replace Sidmouth as Home Secretary. In September that year he brought back Canning, who had been Portland's Foreign Secretary, and President of the Board of Control from 1816 to 1821, to replace Castlereagh at the Foreign Office and also as Leader of the House of Commons. In January 1823 Robinson would succeed the accident-prone Vansittart as Chancellor of the Exchequer, and in October that year William Huskisson entered the cabinet, having earlier replaced Robinson as President of the Board of Trade. To be sure, Wellington and Eldon remained as representatives of the more 'Ultra' form of Toryism, the shift in the conduct of foreign policy from Castlereagh to Canning was more one of style than of substance, and the reforms that would soon be implemented by Peel at the Home Office and Huskisson at the Board of Trade were already being discussed before they took charge. But the change in personnel, drawing in more figures from mercantile and business backgrounds, combined with the easier times brought about by the economic upturn, meant the second phase of Liverpool's government was more 'liberal' than the first, and he was happy to allow

Huskisson, Robinson, Canning and Peel to stamp their marks indelibly on their respective departments.

But when in 1823 George IV, anticipating by more than twenty years Disraeli's later strictures, informed Lord Liverpool that his administration was 'a government of departments', he did not mean it appreciatively. On the contrary, he thought his prime minister failed to provide adequate let alone vigorous leadership, and he regarded him as little better and little more than 'a sort of maître d'hotel'. This was scarcely fair, in that Liverpool's co-ordinating leadership was very skilful, albeit unobtrusively so; but there were undeniably tensions between some of the ministers, which were the result both of their outsize personalities and controversial policies. Huskisson, for example, resented the fact that Liverpool had not immediately brought him into the cabinet when appointing him President of the Board of Trade, and was constantly prey to paranoia and self-pity. Wellington had enjoyed close and cordial relations with Castlereagh, and had been in harmony with him over foreign policy from the Congress of Vienna onwards; but he never took to Canning, disliking both his policy of recognizing the independence of the Latin American republics and the histrionic flamboyance with which he presented it in the Commons and the country. There was also a broad division in Liverpool's cabinet between the High or 'Ultra' Tories and those of more liberal leanings. Wellington, Eldon and Sidmouth could neither forget nor forgive 1789. They believed in firm government, in opposing dangerous and irresponsible measures of reform, in the landed interest's God-given right to rule, and in the importance of the established church. But men such as Canning and Peel, who came from different social backgrounds, and who recognized the growing importance of trade and industry, were less convinced that landed aristocrats were born to rule, and believed that government should work for the benefit of society as a whole, 'animating industry, encouraging production, rewarding toil, correcting what is irregular, purifying what is stagnant and corrupt'.

Such views certainly animated the reforms Huskisson and Peel now set about implementing. At the Board of Trade, Huskisson sought to maximize the advantage of Britain's early industrial lead and the booming economy of the early 1820s. In 1823 he carried the Reciprocity of Duties Act, breaking the principle of protection embodied in the Navigation Acts, which had mandated that only British ships could carry goods being imported into Britain. Henceforward, they could be brought into the United Kingdom in vessels from any European country that extended the same right to Britain itself. During the next two years Huskisson reduced tariffs on a wide variety of imported goods including raw materials, repealed more than one thousand Customs Acts and rationalized the remaining duties at a much lower level – in the case of imported manufactured goods establishing a general rule that the duty should be 20 per cent rather than 50 per cent. He also lifted the restrictions preventing Britain's colonies from trading directly with foreign countries, and removed the requirement that all such commerce should be carried on via London. Such were the 'liberal principles' that Huskisson hoped to 'establish', and there seemed to be a clear connection between the freeing up of trade and the growth in exports. As the economy prospered, and as political agitation melted away, this also seemed the right time to repeal the Combination Acts passed in 1799 and 1800, which had outlawed trades unions and strike action. This was duly done in 1824, but the legalized unions that sprang up promptly organized strikes in favour of higher wages and better conditions, and some of them turned violent. In 1825, under pressure from the manufacturers, Huskisson decided the new legislation must be severely modified, and a second measure was passed declaring illegal any conspiracies or combinations deemed to be 'in restraint of trade'. To be sure, trades unions could still legally exist, but the limits on their actions were very tightly prescribed.

Meanwhile, Peel was implementing a series of reforming and rationalizing measures at the Home Office. His motives were not so much humanitarian as technocratic, for he wanted the law to work better (and as a deeply religious person he also believed in punishing sin). Lacking proper codification, the English law (Scotland and Ireland were separate jurisdictions) was complex and chaotic, while judges could hand out savage sentences to those convicted of relatively minor offences. As a result, some of the most draconian punishments were imposed for crimes against property rather than against people, while juries often acquitted if there was likely to be a heavy sentence, regardless of whether the accused was innocent or guilty. Peel first attempted to tackle these problems by introducing five statutes in 1823, significantly reducing the number of crimes for which judges could impose the death penalty, in the belief that greater respect would be shown towards laws that no longer prescribed execution for relatively trivial crimes. In the same year he carried the Gaols Act, which was a pioneering piece of reform, marking the beginnings of a national policy on prisons: each county and large town was required to maintain its own gaol out of local rates, gaolers were to be properly paid, and there was also provision for regular prison inspections. Women prisoners were to be supervised by women gaolers, all prisoners should receive some education and medical care, and the use of irons and manacles was banned. In 1825 Peel carried another piece of consolidating legislation, the Jury Act, which rationalized more than eighty existing statutes and clarified the process of jury selection. This measure also paved the way for two further consolidating statutes: the first was aimed at improving the administration of central justice; the second replaced ninety-two confusing and repetitive statutes on theft with just five. Between them, these two Acts covered more than three-quarters of the most common offences. By then, Peel had also abolished the use of spies and agents provocateurs, who had been brought back during the late 1810s but whose reports were often inaccurate and exaggerated.

These were significant domestic accomplishments on the part of Lord Liverpool's government; but there were several other matters, all of them potentially more important and also more divisive, that sooner or later were going to demand attention. The first was the abolition of slavery. In 1823, the same year that the Anti-Slavery Society had been established, Thomas Fowell Buxton moved the Commons resolutions against slavery, which prompted Lord Bathurst's unsuccessful efforts to promote reform in the British Caribbean; and the following year Wilberforce and Buxton both spoke in further parliamentary debates, vainly calling for abolition. This was also a cause that most Whigs supported, in part out of loyalty to the memory of Charles James Fox. For the time being there was no further progress, because the planters' lobby remained strong in parliament. But if the legislature was reformed, then the chances of abolition would greatly increase; and parliamentary reform was also a Whig cause. The young Lord John Russell had declared himself in favour of disenfranchising rotten boroughs in 1819; two years later he succeeded in getting the notoriously corrupt constituency of Grampound in Cornwall disenfranchised, and its two MPs reallocated to Yorkshire (although not to the industrial city of Leeds, as he had hoped). In 1822 Russell declared that 'the House of Commons does not possess the esteem and reverence of the people', but his attempt to take away 100 seats from small boroughs and redistribute sixty of them to the counties and the remaining forty to large towns was defeated by 269 votes to 164, and a similar motion would be turned down by a bigger majority in 1826. Yet while the causes of abolition and parliamentary reform languished during the better years of the 1820s, there was always the hope (or fear) that their prospects might dramatically improve if the economy turned down again, and popular discontent and radical agitation once more took centre stage, as they had in the years between Waterloo and Peterloo.

The revival of those two issues lay in the (not very far off) future; but of greater immediate concern to Lord Liverpool was the condition of Ireland and the still unresolved issue of Catholic Emancipation. It would soon be a quarter of a century since the Act of Union had been passed, but the promised companion measure, allowing Catholics to sit in parliament and hold high public office, had still to be made law. As father and son, George III and George IV had little in common; but the one political view they both shared was a stubborn hostility to supporting emancipation, which they regarded as inconsistent with their coronation oath to uphold the Church of England and the Protestant religion by law established. Not until Queen Victoria took umbrage at the prospect of Home Rule would a British monarch again be so damagingly intransigent on an Irish issue. Like his mentor the Younger Pitt, Lord Liverpool believed he should defer to George IV's hostile views, which had been neither modified nor moderated by his successful visit to Ireland in 1821; and he accepted that it was not a matter that he or any of his cabinet colleagues should formally raise with the king. But Liverpool had a further reason for trying to keep Catholic Emancipation off the political agenda, because his ministers were deeply divided on the issue, and in unexpected ways. Among the more liberal Tories, Peel was vehemently against, whereas Huskisson was in favour. Despite their common background as Irish patricians, and their closeness in foreign policy, Wellington and Castlereagh had long since differed on this matter, the duke being as intransigent as Peel in his opposition, whereas Castlereagh had been in favour ever since he had helped carry the Irish Union. And whereas Castlereagh and Canning had been rivals and enemies, and conducted – or presented – their foreign policies very differently, they both supported the policy of bringing Catholics fully into the public life of the country.

Lord Liverpool's response to these divided counsels was to obtain from all of his ministerial colleagues an undertaking that they would refrain from discussing Catholic Emancipation with the king, and never raise the issue in cabinet. But while he could effectively prevent his colleagues from bringing up the subject, he could not stop others from doing so, and least of all in Ireland itself. One such figure was Henry Grattan, who had vehemently opposed the Act of Union as an Irish MP; having got himself elected to the British parliament in 1805, he immediately began campaigning for emancipation as the first essential step towards repealing the Union. Some of his proposals passed the Commons, only to be thrown out by the Lords; but his last vain attempt, in 1819, had been supported by both Castlereagh and Canning. Two years later, William Plunkett, the MP for Dublin University, carried a measure through the Commons, but again the upper house rejected it. By then, a Dublin-based lawyer named Daniel O'Connell had concluded that the only way to achieve emancipation would be by bringing the powerful pressure of Irish public opinion to bear on the Westminster parliament; and to that end he formed the Catholic Association in 1823, which enjoyed unprecedentedly widespread support among the Irish people and the Catholic clergy, whom O'Connell sought to mobilize against Liverpool's administration in London by delivering eloquent and inspiring speeches. But the government regarded him as a demagogic and inflammatory agitator, and they dissolved the Catholic Association in 1825, the same year that another emancipation measure was passed by the Commons but rejected by the Lords. O'Connell responded by founding the New Catholic Association. Like its predecessor, it collected funds from ordinary Irish people, often with the help of the Catholic priesthood. The Association supported pro-Emancipation candidates at the general election of 1826, and many were returned, defeating the candidates put up by the Protestant landlords. Despite Liverpool's wishes to the contrary (and also those of the king), it was only a matter of time before the emancipation issue became live again.

By early 1827 Lord Liverpool had been in office continually for fifteen years, holding together a broad coalition of Tories, and regularly refreshing and renewing his cabinet, with an impressive combination of managerial skill and physical stamina. But he found the unrelenting demands of office and the succession of crises a constant strain, and for much of the 1820s he suffered from what we would now term 'stress-related disorders'. It was becoming increasingly difficult for him to keep his cabinet colleagues together, and there were unresolved issues that could not be indefinitely papered over or postponed. Despite the evidence produced by the opponents of 'Old Corruption', the number of 'placemen', pensioners and sinecurists on which governments could rely for parliamentary support had been almost continually diminishing since the days of the Younger Pitt, which meant that Liverpool's Commons majorities were constantly at risk. Although they had seemed in abeyance by the mid-1820s, the causes of abolition and parliamentary reform might again unite popular opinion and the parliamentary Whigs if and when the economy turned down again. In 1825 the issue of Catholic Emancipation had driven Liverpool and Peel to the brink of resignation, and Daniel O'Connell was bound to keep pressing for it in the aftermath of the election results the following year. And in December 1825 the economy had indeed crashed again, leading to increased unemployment and popular protest during the next twelve months. All this was too much for a wearied and weakened prime minister. In February 1827 Lord Liverpool collapsed, having suffered a stroke; two months later he retired from public life, and in December 1828 he died.

THE POLITICAL CRISIS BEGINS

At his death, Pitt had been just forty-six, and Liverpool was only fifty-eight. These were greater ages than the average life expectancy for males in the United Kingdom at that time, but still an indication that long spells in the highest office took their toll, especially at a time when the demands of war and peace were equally challenging, albeit in different ways. But whereas Liverpool had been able to perpetuate and revive what had become known as Pitt's Tory regime, there was no prospect that anyone who succeeded him would be able to repeat that achievement. Liverpool's policy during the 1820s had been to support established institutions, and allow some moderate measures of reform, while keeping contentious issues off the agenda. But henceforward it would no longer be possible to maintain such a Tory coalition on that basis, in part because opinions were becoming increasingly polarized, and in part because the pressure of events would further intensify that polarization. Ultras like Wellington and Eldon did not want any more change, and least of all in terms of Catholic Emancipation or parliamentary reform. Peel wanted more penal reforms, but he, too, shared Wellington's views against Catholics and parliamentary reform. But there were others, such as Canning, Huskisson, Robinson (now ennobled as Lord Goderich) and Palmerston, who were, with varying degrees of enthusiasm, willing to contemplate broader measures of reform, and who increasingly felt they had more in common with the Whigs in opposition than with the more reactionary members of their own party with whom they were in government. From this perspective, Liverpool's departure portended a major political realignment of the parties, as would occur again after 1846, 1859 and 1885.

This impending reconfiguration was made plain when, in April 1827, George IV turned to Canning to form a government, having resisted lobbying from the Tory right urging him to appoint Wellington instead. Canning had been the most talented man in Liverpool's cabinet, and he was opposed to parliamentary reform, on the grounds that the legislature was already effectively answerable to public opinion and the press. But his humble origins, his theatricality and flamboyance, eloquence inside and outside parliament, 'liberal' foreign policy and support for Catholic Emancipation, were anathema to those who now called themselves 'Ultra' Tories; and as a result, Wellington, Eldon and Bathurst refused to serve under him, as did Peel who, although no Ultra, still held firm in his opposition to Catholic Emancipation. Canning also became Chancellor of the Exchequer, and he was able to carry over into his own administration liberal Tories such as Goderich, Huskisson and Palmerston. To fill the remaining gaps he brought in such Whigs as the Duke of Portland, the Marquis of Lansdowne and the Earl of Carlisle, who were the least committed to historic notions of party loyalty. This greatly annoyed the Whig leadership, namely Lord John Russell in the Commons and Lord Grey in the upper house, who by contrast regarded all Tories as anathema; but it was a sign that the party allegiances that had generally held throughout the 1810s and 1820s were becoming seriously unstuck. Canning hoped to carry a bill, which had been agreed upon at the very end of Liverpool's administration, to modify the Corn Law of 1815; but Wellington insisted on a wrecking amendment in the Lords, and Canning was forced to withdraw the measure. More positively, Canning's government hardened the protocol Wellington had signed in St Petersburg in April 1826, which had established a measure of Greek independence from the Ottoman Empire; the Treaty of London, which guaranteed these provisions, was signed in July 1827 by Britain, France and Russia, and the three powers sent fleets to the eastern Mediterranean to enforce these terms if the Ottomans refused to co-operate.

There were also hopes (or fears) that Canning would enact Catholic Emancipation; but he died in early August 1827 after barely four months as prime minister. He was only fifty-seven, another premier worn out by the burdens of office. The king thereupon turned to Lord Goderich, who, as Liverpool's Chancellor of the Exchequer, had earned the nickname 'Prosperity Robinson' during the economic boom of the mid-1820s. His aim was to continue the alliance that Canning had recently constructed between liberal Tories and conservative Whigs, and he successfully carried over many of his predecessor's appointees into his own administration. But Goderich lacked toughness and determination (he was described by one contemporary as being 'as firm as a bullrush'), and he was unable to hold his cabinet together, in large part because he allowed George IV to bully him into appointing a high Tory, John Charles Herries, to be Chancellor of the Exchequer. But Herries had been a strong opponent of Huskisson's liberalizing financial policies during Liverpool's administration, and it soon became apparent that the two men could not work together under Goderich, in whose administration Huskisson was Secretary for War and the Colonies. Herries and Huskisson duly fell out, over whether the Whig, Lord Althorp, the son and heir of Earl Spencer, should be appointed to the chairmanship of a finance committee. Huskisson was strongly in favour, Herries implacably against. Goderich felt he could not afford to lose either of them, but was almost bound to do so. He wrote to the king in December 1827, explaining his feelings of inadequacy, and George IV chose to regard this epistle as a letter of resignation. Further intrigues ensued, and the following month Goderich was informed that the king regarded his administration as being at an end. It had lasted barely five months, and had never met parliament, which for the whole of that time had been in abeyance. 'Never surely,' Huskisson wrote to a friend, 'was there a man at the head of affairs so weak, undecided and utterly helpless.'

Since the king would not countenance the Whigs, the only man left to whom he could turn was Wellington. Although he had joined Liverpool's cabinet in 1818, the victor of Waterloo was by experience and temperament more a military man than a politician. He lacked the common touch and was utterly indifferent to public opinion. His parliamentary speeches were often unpremeditated to the point of bluntness, and he was ill-fitted for the negotiations, compromises and accommodations that were an integral part of political life. Yet despite these significant shortcomings, Wellington showed considerable political skill in constructing his ministry. He kept the liberal Tories, among them Huskisson and Palmerston, and he brought Peel back as Home Secretary and Leader of the Commons. He refused to give places to such 'Ultra' Tories as Eldon and Westmoreland, and he left out the Whigs who had served under Canning and Goderich. The result was a more middle-of-the-road administration than might have been expected. Indeed, having given 'due consideration' to 'talent, family connections and the circumstances of the Empire', Peel's father concluded that the new government would provide 'general satisfaction'. But neither the king nor his prime minister were themselves middle-of-the-road men. On commissioning Wellington to form a government, George IV had stipulated that 'the Roman Catholic Question was not to be made a cabinet question', and that Catholics be debarred from prominent positions in the Irish administration. As for the duke, it was his belief that no changes were needed to put parliamentary representation 'on a footing more satisfactory to the people of the country than it now is'. But as would very soon become clear, these were positions, on both Catholicism and the constitution, that would no longer be compatible with 'general satisfaction'.

4
The Iconoclastic Years, 1829–41

The Duke of Wellington was never able to control his cabinet or the legislature in the authoritarian way that he had commanded his troops in the earlier, military phase of his life; and since that was the only style of leadership he knew, this meant his administration was in difficulties almost from the very beginning. In February 1828 Lord John Russell, the Whig leader in the Commons, introduced another bill to repeal the Test and Corporation Acts, in response to a campaign by Protestant Dissenters that had once more revived after a lapse of nearly forty years. As measures that discriminated against Nonconformists, and which upheld the supremacy of the established Church of England, they had long been Tory articles of faith. But Russell's proposal passed the Commons by forty-four votes, and despite mutterings from the 'Ultra' Tory right, the Lords also let it through. In practice, the repeal of these two Acts made little difference to the lives of dissenters, for both measures had long since fallen into desuetude. On the other hand, such a measure, chipping away as it did at the legal and religious basis of the established order, might be seen as a welcome or ominous precedent for further constitutional and ecclesiastical reform, especially regarding the vexed issue of Catholic Emancipation. This represented an immediate challenge to Wellington and Peel, both of whom had in recent years privately (but not publicly) abandoned their earlier intransigence, and had reluctantly come to recognize that such a measure might indeed have to be carried. To this end they would need the support of the more liberal members of the government; but in the spring of 1828, soon after the repeal of the Test and Corporation Acts, Huskisson, Palmerston and two other 'liberal' Tories resigned, ostensibly over an amendment to the Corn Laws of 1815, which was carried, but in practice because Wellington had mishandled his relations with them.

In attempting to reconstruct his government in the aftermath of the departure of its most liberal members, Wellington inadvertently triggered a major political crisis by appointing an Irish country gentleman named Vesey Fitzgerald, who was well thought of by Peel and was personally in favour of Catholic Emancipation, to be President of the Board of Trade. By the conventions of the time (and they would last for the remainder of the century), MPs accepting ministerial office were obliged to seek re-election; and although this was normally a formality, such by-elections could on occasion give rise to surprising and portentous upsets. Fitzgerald's attempt to secure re-election was just such a case, for he represented the Irish constituency of County Clare, and he now found himself opposed by none other than Daniel O'Connell, who was buoyed up by the rising support he was receiving across much of Ireland from the New Catholic Association, which he had founded three years earlier. As a landowner who was in favour of emancipation, Fitzgerald expected to win again; but the electors followed the advice of their priests, and O'Connell duly won the contest by more than a thousand votes. However, as a Catholic, he was legally barred from taking his seat in the House of Commons, and given the agitated state of Irish popular opinion the government feared that enforcing this prohibition would not only disregard the clear decision of the County Clare electorate but might also provoke a broad-based Irish insurrection. Such a rebellion would put in jeopardy the Anglo-Irish Union that was barely a quarter of a century old, but which was increasingly regarded as being non-negotiable in British governing circles. With no relish, but with determined resignation, Wellington and Peel concluded that if the choice lay between denying emancipation and putting the Union at risk, or conceding emancipation and preserving the Union, they must adopt the second course of action.

This was a brave and bold decision, as Wellington and Peel now sought to carry a measure they had spent much of the early 1820s opposing, but it had serious and disruptive political consequences. To many 'Ultra' Tory MPs and peers, who believed their party existed to uphold the established church, this amounted to a treacherous act of 'betrayal' by their own leadership; the more sympathetic 'liberal' Tories, who might have supported Wellington and Peel in cabinet, had recently resigned; and on this matter George IV remained as intransigent as his father had been. This meant that in the immediate aftermath of the County Clare by-election, which had taken place in July 1828, neither Wellington nor Peel could go public with their decision, as they needed time to work on the 'Ultras' and the king. The result was that tensions mounted on both sides of the issue, and on both sides of the Irish Sea. There were mass Catholic demonstrations in Ireland in support of O'Connell and emancipation, and these were countered, especially in Ulster, but also in England, by the creation of 'Brunswick Clubs', pledging their loyalty to the Union and the Protestant religion. It also took time for Wellington to persuade the angry and resentful monarch that emancipation was unavoidable, which in the end he grudgingly conceded. But the result was that it was not until February 1829 that Wellington and Peel were publicly able to announce their decision to carry emancipation, by which time it appeared more a forced concession to Irish mob violence rather than a statesmanlike recognition of the will of the County Clare electors. Nor were the Catholic Irish fully satisfied with the measure when it finally received royal assent in April: some senior government posts still remained restricted to Protestants; many political societies, including O'Connell's New Catholic Association, were explicitly banned; and the Irish franchise was significantly narrowed, as the voting threshold for smallholders was raised from owning land worth 40s. a year to land with an annual value of £10.

Amidst so much political uncertainty, Peel managed to pass one piece of domestic legislation that was the culmination of his career as a reforming Home Secretary, at least in retrospect. In 1822, worried by the threats to public order recently manifest during the Queen Caroline affair, he had set up a parliamentary committee to explore the possibility of establishing a police force in London, which would be 'consistent with the character of a free country'. Several Scottish towns, including Edinburgh and Glasgow, had already formed such forces by private Acts of Parliament; Dublin had been given its own police in 1808; and Peel himself had created a national Irish constabulary six years later, though that was as much concerned to root out political subversion as to deal with criminal activity. The committee refused to come up with the recommendation Peel wanted, but he was more successful the second time around, and in July 1828 a second committee recommended in favour of instituting 'an efficient system of Police in this great

Metropolis, for the effective protection of property, and for the prevention and detection of crime'. Peel thereupon introduced legislation, which reached the statute book by June 1829, creating a constabulary, under the direction of the Home Office, which would police the whole of the metropolis with the exception of the City of London itself. Critics denounced the new force as being 'unconstitutional', and there were fears that it would evolve into a secret police on the continental (or Irish?) model, which were hardly allayed when it emerged that the first two Commissioners appointed to oversee the force came from Irish families, and that most of the senior officers and many of the constables had military backgrounds. Although metropolitan crime dropped significantly after 1829, it took many years to establish the force on a satisfactory footing, and it would be a quarter of a century before the London police model was generally imitated across the cities and counties of England and Wales.

THE GREAT REFORMING ACT

But during 1829 it was Catholic Emancipation that was the burning political issue. To be sure, its passing had finally complemented the Act of Union with Ireland, as the Younger Pitt had originally intended nearly thirty years before. Yet despite Peel and Wellington's best intentions, its consequences were more disruptive than ameliorative. For the majority of the Irish, emancipation was too little and came too late to win them over to the Union or to government from London; and it also precipitated a major high-political crisis, which in turn would usher in a decade of unprecedentedly broad-ranging reform. How did this come about? From one perspective the dominant issues of 1828 and 1829, both in parliament and in the press, had been primarily religious and ecclesiastical, concerned with lessening legal discriminations against dissenters and Catholics. But these also had far-reaching political consequences. The repeal of the Test and Corporation Acts, combined with the passing of Catholic Emancipation, had abruptly and fundamentally undermined the confessional nature of the British state, which had previously restricted full participation in its public affairs to those who were fully compliant and communicant members of the Church of England; and this was not just a religious matter but a constitutional one as well. Many Tory Anglicans had felt betrayed by Wellington and Peel: to them, their leaders were not brave statesmen, responsibly and sensibly changing their minds in accordance with dramatically changed circumstances, but cowards and apostates, who had, in the words of one contemporary, 'murdered the constitution'. As a result, Wellington forfeited the backing not only of the liberal Tories, who had left his government early in 1828, but also by the following year of many 'Ultra' Tories as well.

As the duke's government lost its way and credibility with large numbers of its own erstwhile supporters in the Commons and the Lords, the issue of parliamentary reform began to surface again, and it did so from several directions, both within the Palace of Westminster and outside. One group who now took up the cause of reform were those men who, paradoxically, had hitherto been most opposed to any constitutional change, namely the Tory 'Ultras'. They rightly noted that Catholic Emancipation had been deeply unpopular among the people of predominantly Protestant Britain, which meant that a more representative House of Commons would never have passed such a measure. They accordingly embraced reform in the hope of preventing any such similar calamity in the future. At the same time the disaffected liberal Tories and the Whigs began to press more determinedly for parliamentary reform, partly because they needed a new cause now that the Catholic question had been settled, and partly because there were signs that the issue was gaining traction more widely for the first time since the late 1810s. It resurfaced at county meetings, and was also taken up by many new extra-parliamentary organizations that were known as political unions. The first of them was founded in Birmingham in January 1830, and by the spring of 1832 there would be similar associations in being in more than one hundred towns across the United Kingdom, including Leeds, Manchester, Bristol and Blackburn. As the term 'political unions' suggests, they often encompassed a wide range of social backgrounds and an equally broad spectrum of political opinion, from 'Ultra' Tory bankers such as Thomas Attwood in Birmingham to working-class radicals. It was these combined developments, both inside parliament and outside, that help explain Russell's proposal, in February 1830, to transfer parliamentary seats to Manchester, Leeds and Birmingham, and Grey's later denunciation of Wellington and his colleagues as being unfit to govern.

These popular protests and parliamentary proposals were rendered the more significant and ominous by the growing economic uncertainty and social distress that coincidentally characterized these years. The effects of the banking crisis of 1825–26 lingered, delaying the return of investor confidence and stunting economic growth. Between 1827 and 1832 food prices rose in the aftermath of a succession of bad harvests, while the movement of industrial prices, profits, wages and investment was markedly downwards. In the north, rising unemployment provoked strikes in Lancashire and Yorkshire, while in the south, there were outbreaks of rural arson and machine-breaking, carried out by local radicals and agricultural labourers, which were known as the 'Captain Swing' riots and named after their imaginary leader. The intensifying mood of popular discontent and public disenchantment may have been further strengthened by the inauguration of Andrew Jackson to the United States presidency in March 1829, which was seen as marking a new and stronger era in American democracy; and also by another revolution in France in July 1830, when the would-be absolutist Charles X was overthrown and replaced by a more constitutionally minded sovereign in the person of Louis-Philippe. In the midst of this mounting parliamentary crisis, growing public dissatisfaction, and international uncertainty, the death of George IV in June 1830 necessitated a general election that could hardly have come at a worse time for Wellington's government. Voting took place during July and August, and although only a limited number of constituencies were contested, fifty seats changed hands, which was more than at any recent election. This was a significant vote of no confidence, undoubtedly so in the case of Wellington's government, which was further weakened, and perhaps in the parliamentary system as a whole.

The early autumn of 1830 was thus a jittery and febrile time, and two more events intensified this climate of anxiety and unease. The first occurred on 15 September, soon after the election was over and just before the new parliament was due to meet: William Huskisson was run over by a train, pulled by George Stephenson's *Rocket*, at the opening of the Liverpool and Manchester Railway, which was also attended by Peel and Wellington. Huskisson was the first man to be killed in a railway accident. Here was another shocking and violent version of the future, by turns fast and dangerous, mechanized and technological, industrialized and urban, every bit as ominous and unsettling as that embodied in the new political unions. The second event took place in early November, as public pressure for parliamentary reform was building, when Wellington declared in the House of Lords that he was implacably opposed to any such measure, since it was his firm conviction that the current representational arrangements were beyond improvement: 'Britain,' he insisted, 'possessed a Legislature which answered all the good purposes of legislation, and this to a greater degree than any legislature ever had answered in any country whatever.' Wellington's aim was to reassure his supporters that, having earlier changed his mind over Catholic Emancipation, he would not do the same over parliamentary reform. But he had failed to consult his cabinet colleagues beforehand, several of whom were beginning to recognize that some measure was probably unavoidable; while the duke's ill-chosen words merely confirmed the widespread view that he was wholly out of touch with the popular mood. Two weeks later his government was defeated on a finance bill, and Wellington resigned. As the victor of Waterloo, he had been the undisputed national hero since 1815; but as the unyielding enemy of parliamentary reform, he would be the most unpopular man in the country for the next two years.

The new monarch, William IV, was the younger brother of George IV, and he was no more in favour of parliamentary reform than his predecessor had been well disposed to Catholic Emancipation; but his opposition would likewise prove ultimately ineffectual. On Wellington's resignation, the king turned reluctantly to Earl Grey, who had been Whig leader for almost a decade. It was an abrupt change in his party's fortunes. From the late 1710s to the 1760s the Whigs had been the dominant force in British politics; but thereafter they had only briefly been in office in 1783, and again in 1806–07 (when Grey had served as Foreign Secretary in the Ministry of All the Talents), and they now returned to power thanks to an extraordinary combination of causes, crises and coincidences. The long-term rise in population, the gradual transformation of parts of the economy, and the growth in size of towns and cities had led to widespread dissatisfaction with a representational system that seemed increasingly anomalous, anachronistic and corrupt. The abrupt downturn in the economy in the late 1820s provoked popular discontent with parliament that was more intense and focused than in the late 1810s. The fracturing of the Tory Party in the aftermath of Catholic Emancipation led to a high-political crisis, and the death of George IV in 1830 necessitated a general election which, in however limited a way, linked popular politics and high politics together. It propelled the Whigs back into power, where their challenge was to pass a measure of reform that both parliament and the public would accept. But Grey was no radical. On the contrary, he was a broad-acred and well-connected Northumberland landowner whose essentially moderate views on parliamentary reform had been settled since the 1790s. In November 1831 he would tell the House of Lords: 'There is no one more decided against annual parliaments, universal suffrage and the ballot, than I am.'

Throughout the ensuing months of crisis, Grey's aim was 'not to favour, but to put an end to such hopes and projects', and in that essentially conservative endeavour he would eventually succeed. The composition of his cabinet gave further abundant evidence that revolution was not on the new government's agenda. It included one 'Ultra' Tory grandee, the Duke of Richmond, who was Paymaster General; he supported parliamentary reform so that measures such as Catholic Emancipation could never be passed again. There were three former 'liberal' or Canningite Tories, Lords Goderich, Palmerston and Melbourne, respectively at the Colonial, Foreign and Home Offices: they, like Grey, had no intention of capitulating to the demands of the mob. The majority of the cabinet was made up of very grand Whigs. The broad-acred third Marquis of Lansdowne was Lord President; he had been Chancellor of the Exchequer in 1806–07 and Foreign Secretary in Goderich's ill-starred administration in 1827–28. The Lord Privy Seal was the Earl of Durham, who was Grey's son-in-law and owned rich coal-bearing acres in his titular county, and who would draft the Reform Bill. The Chancellor of the Exchequer was Viscount Althorp, heir to Earl Spencer, which placed him at the very heart of the Whig cousinhood. Lord Stanley was Chief Secretary for Ireland, and heir to the Earl of Derby, whose Lancashire acres were even more valuable than those of Lord Durham across the Pennines. Stanley did not enter the cabinet until June 1831, when he joined with the Paymaster General, Lord John Russell, who was a son of the Duke of Bedford, one of the richest Whigs of them all, in part thanks to Thomas Cubitt's recent development of his London estate in Bloomsbury. In terms of its wealth, lineage and status, Grey's was a far grander administration than Wellington's had been; indeed, he boasted that his colleagues owned more acres than any previous cabinet.

To the king's evident displeasure and dismay, Grey had made it a condition of his appointment that he would introduce a measure of parliamentary reform; he spelt this out in his first speech in the Lords from the government benches in November 1830, and the next eighteen months would be a time of high drama and tension, both inside the Palace of Westminster and across the United Kingdom as a whole. Wellington believed that 'the country was in a state of insanity about reform', and when Russell introduced the government's more radical than expected reform bill in the Commons on 1 March 1831, there were many in parliament who felt that Grey and his colleagues had also succumbed to what they regarded as the prevailing revolutionary madness. Greater representation would be given to the counties, eleven large towns including Manchester, Birmingham and Leeds would be apportioned two MPs each, and sixty English rotten boroughs would be completely disenfranchised – this final proposal being greeted with incredulous laughter by Tory MPs. But the Whig case for such a wide-ranging (but far from revolutionary) measure was that nothing less would satisfy public opinion in its current agitated state, and if public opinion was not successfully appeased, then recent disturbing events in France offered a cautionary tale as to what might happen instead. The bill eventually passed its second reading in the Commons, but only by one vote, which meant it was sure to be destroyed at the committee stage. Soon after, the government was defeated on a motion for the grant of supplies, and Grey persuaded the king to dissolve parliament, even though the previous election had been held less than twelve months before. The ensuing contest, which took place between April and June 1831, was characterized by widespread demands for 'the bill, the whole bill, and nothing but the bill'; and public opinion against its opponents was so inflamed in the capital that the windows of Wellington's London mansion, Apsley House at Hyde Park Corner, were smashed by angry protesters.

The election result was a landslide victory for Grey, as the Whigs registered more than thirty gains in the English counties and double that in the boroughs. With the government being triumphantly vindicated, Russell introduced a second reform bill, virtually identical to its predecessor, which passed its second Commons reading in early July 1831 by 136 votes. It was then held up in committee, as the Tory opposition tried to delay it and to dilute its provisions (by, for example, extending the right to vote to so-called 'tenants at will', whom it was thought would do their landlords' bidding), but it eventually passed its third reading by a clear majority in September. The action then shifted to the upper house, where in October their lordships rejected the measure by forty-one votes. This blatant defiance of the lower house and of the will of the people led to a renewed outburst of popular indignation, which was the most sustained, angry and violent of the whole Reform Bill crisis. There were serious disturbances in London (more grandees' windows were broken), Derby (where the rioting led to fatalities), Nottingham (where the Duke of Newcastle's residence was burned to the ground) and Bristol (where the bishop's palace and property belonging to the Tory town corporation were targeted). As Home Secretary, Lord Melbourne acted firmly, but even he was 'frightened to death' by events in Bristol, where it took three days for soldiers to regain control and restore order; while Princess Lieven, the wife of the Russian ambassador, feared Britain was 'on the brink of a revolution'. This was probably an exaggeration, but it certainly concentrated the minds of the politicians. Moderate Tory peers such as Lords Wharncliffe and Harrowby began negotiations with ministers in the hope of agreeing upon a compromise bill that they would support in the upper house; while Grey believed he had extracted a grudging promise from the king that, if the upper house remained obstructive, he would create sufficient new peers to ensure the measure passed the Lords.

The action then shifted back to the Commons, where a third version of the Reform Bill was introduced, with some of its provisions slightly watered down, and it eventually passed its third reading in March 1832, by the still-commanding majority of 355 votes to 239. And so for the second time, the measure was sent up to the Lords. Armed with what he believed was his assurance from the king, Grey was able to persuade some Tory 'Waverers', and also several of the bishops, to change their votes, and on 14 April the reform bill passed its second reading in the upper house by the narrow but sufficient majority of 184 to 175. Three weeks later, however, the government was defeated on a resolution moved by Lord Lyndhurst, the former Tory Lord Chancellor in Wellington's government, postponing consideration of several key clauses relating to the disenfranchisement of the rotten boroughs. Notwithstanding his earlier promises, the king refused to honour his pledge to create sufficient peers to ensure the bill passed; this put Grey and his colleagues in an impossible position, and so on 8 May they resigned. The king now turned once again to Wellington, hoping he would lead an administration that would come up with a moderate measure of reform acceptable to the Lords. Despite his previous hostility to any tampering with the constitution, the duke dutifully accepted his sovereign's commission, convinced that, in the febrile atmosphere of 'the days of May', the king's government must nevertheless be carried on. But popular opinion would not countenance another Wellington administration, and the radical leader Francis Place sought to intensify this hostile and uncertain mood by urging depositors to withdraw their money from the Bank of England, thereby intensifying the already deep economic crisis. To this end he coined the ominously resonant slogan: 'To stop the duke, go for gold.'

This campaign might not have deterred Wellington from doing his patriotic duty; but his efforts proved unavailing when Peel made plain he would not change his mind on this second great issue, and his refusal to serve alongside his former chief meant it was effectively impossible for the duke to form a government. But Wellington (and, perhaps, Peel too?) now realized that some sort of parliamentary reform was unavoidable, and that only a restored Whig government, enjoying the full support of the sovereign, could carry it. Thus advised by the duke, the king sent for Grey once more, and this time gave him definite and unequivocal assurances that he was prepared to create the necessary number of new peers to ensure the bill would pass in the Lords. At the same time Wellington urged his followers in the upper house to abstain when the Reform Bill was reintroduced, thereby conceding the measure, but at least fending off a massive influx of new peers. The result was that the bill finally passed its third reading in the Lords on 4 June 1832, by a vote of 106 to 22: most Tory peers did indeed abstain, as Wellington had urged, and only a diehard minority pursued their opposition to the bitter and ineffectual end. Three days later, amidst widespread scenes of popular relief and rejoicing, the Great Reform Act received the royal assent (and soon after, separate but similarly intentioned measures were carried for Scotland and Ireland). For Grey, this was a remarkable personal triumph: he had kept his cabinet together through twenty difficult and testing months; he had held his nerve in the face of what seemed like revolutionary threats to the established order; and he had faced down a hostile and uncooperative monarch. With the new representational structure established, he insisted on another general election; and in December 1832 Grey was rewarded with an even greater victory than in the previous year, winning an overwhelming Commons majority of 300 seats.

The Whigs' commitment to constitutional reform was as genuine as their wish to stave off revolution, and they achieved both objectives by carrying what seemed to be a carefully judged and brilliantly calculated measure, which was both the most that parliament would grudgingly tolerate and the least that the people would be prepared to accept. All along the Whigs' basic aims had been clear, simple and straightforward. They sought to strengthen the traditional 'landed interest', to which so many of them belonged, as the natural leaders of the nation, by enfranchising the sober, upright, independent, property-owning middle classes who, as Grey insisted during the debates, 'form the real and efficient mass of public opinion, and without whom the power of the gentry is nothing'. Once again, but this time domestically (instead of internationally) and under Whig (instead of Tory) leadership, a new world was being called in to redress the balance of the old. Yet both of these assumptions were psephologically dubious and sociologically ill-grounded. Despite being out of power for so long, the Whig grandees had tended to assume that the landed interest was monolithic and that they stood at the head of it and therefore represented it; but it had always been divided between Whigs and Tories; and although the Whigs carried the majority of rural constituencies at the general elections of 1831 and 1832, many of them would return, later in the decade, to what seemed to be their natural Tory allegiance, much to the Whigs' disappointment. As for the middle class, in whose probity and decency Grey and his colleagues reposed their hopes for sustaining the landed interest: in reality there was little solid information about who these people were, what they did, where they lived, or how many of them there were, and they formed a more diverse and varied group than these highly generalized characterizations suggested.

But for all their limitations, inaccuracies and vagueness, these basic assumptions served the Whigs well when it came to drawing up the detailed terms of their reforming measure. One way of safeguarding the landed interest was to render it less vulnerable to charges of being corrupt and self-serving: hence the complete disenfranchisement of fifty-six rotten boroughs in England, containing fewer than 2,000 inhabitants; and the removal of one of the MPs from another thirty with populations of fewer than 4,000. A second way to buttress the traditional rural hierarchy was by giving greater representation to the larger rural constituencies, and in England and Wales the number of county seats was increased from 92 to 159, while the franchise continued to be based (with some additions and modifications) on the traditional 'forty-shilling freehold'. At the same time, the provincial urban middle classes were also recognized: twenty-two large towns were enfranchised, and each given two MPs, among them Manchester, Leeds, Birmingham, Bradford and Sheffield; and another nineteen would receive one MP each, including Huddersfield, Rochdale and Salford. Moreover, the franchise was standardized on the basis of householders owning property worth £10 annually, or holding a long-term lease on which they paid a similar sum in rent. In Wales, county representation was slightly augmented, while Swansea and Merthyr Tydfil were also assigned MPs. In Scotland, similar county constituencies were effectively established for the first time, and Glasgow was given two MPs. Irish representation at Westminster was slightly enhanced by the addition of five new borough MPs. As for the electorate: in England and Wales it grew (significantly) from 350,000 to 650,000; in Scotland (spectacularly) from 5,000 to 65,000; but in Ireland (miserably) from 75,000 to only 90,000.

These were indeed important changes, but as the Whigs had intended, they were more designed to preserve the old system than to yield to radical demands for change. The landed interest remained pre-eminent, albeit still divided between Whigs and Tories, and it would dominate the Commons, the Lords and cabinets well on towards the end of the nineteenth century. Although the most corrupt of the rotten boroughs had been abolished, many constituencies remained effectively in the gift of great landowners: among them Newark, owned by the Duke of Newcastle, where the young Mr Gladstone was returned as a Tory MP in 1832. Despite the considerable extension of the franchise, the continuing vagaries and anomalies of the system meant that in some boroughs the numbers of voters actually went down in the decades following 1832. Across the whole of the United Kingdom, one in seven of adult males could now vote; but this meant that the overwhelming majority remained unenfranchised; and the Reform Act was the first legislation of its kind which formally declared that only men, but not women, could vote. In future, less than 10 per cent of the adult population would be involved in selecting members of the House of Commons, while the House of Lords remained almost exclusively hereditary. There had been some slight redistribution of parliamentary seats away from England and towards Wales, Scotland and Ireland, but the representational system remained overwhelmingly Anglo-centric in its bias. The proportion of male voters also varied significantly across the British Isles: it was one in five in England and Wales, one in eight in Scotland, but only one in twenty in Ireland, where the franchise remained deliberately more restricted according to the qualification imposed at the time of Catholic Emancipation. This was further evidence, for the disaffected Irish, that their country had not done well out of the Act of Union.

Even after the passing of the Reform Act, then, the representational system of the United Kingdom remained in many ways anomalous, inconsistent, restricted and oligarchic. But that was what Grey and his colleagues had always intended: there had been no capitulation to radical demands for annual parliaments, universal suffrage, or the ballot, and this would leave many of those who had campaigned and protested between 1830 and 1832 feeling at best disappointed, at worst betrayed. Yet when Wellington later described (and lamented) the accomplishment of such limited constitutional reform as a 'revolution by due process of law', he was not wholly mistaken. A House of Commons that had been widely derided as unrepresentative had in the end bowed to the views of the British people, as expressed, however imperfectly, via two general elections and countless riots, protests and petitions; and in the end, neither the peers nor the monarch could stand against the public and the lower house combined. Russell may have oversimplified when he declared it was 'impossible that the whisper of a faction should prevail against the will of a nation', but the national will did indeed triumph over the factious whisper. Perhaps that was what the future radical MP John Bright meant when he observed that the Reform Act 'was not a good bill', but it was 'a great bill when it passed'. But in addition to demonstrating determination and command, the Whigs had also been very lucky: in part because William IV in the end proved malleable, Wellington was a maladroit opponent, and Peel kept his head down; and in part because the British economy picked up soon after the Act reached the statute book, which meant that much of the popular protest and political discontent, which the Whigs had both dreaded but needed to get their measure through, died away – at least for a time.

The passing of the Great Reform Act was not only a domestic success story, but took on an even greater significance when seen in the broader context of the widespread disorder and revolution that convulsed much of Europe during the years from 1830 to 1832. In addition to the July 1830 revolution in France, there were uprisings in the Netherlands, parts of Italy and Germany, on the Iberian Peninsula, and in Russia, and it looked for a time as though the restored order of royal legitimacy that had been put in place in 1815 was in serious jeopardy. By comparison, Britain seemed a much more stable state, and a much more robust polity, and in the aftermath of the Reform Act it had some claims to be regarded as the freest and the most liberal nation in Europe. The franchise might still be very narrow, but by continental standards the new electorate of the United Kingdom was very large indeed. A greater proportion of men could now vote than in France or Spain, while Austria, Denmark, Russia and Greece had no elected national assemblies at all. Indeed, in the Europe of 1832, only in parts of Scandinavia were the boundaries of active citizenship set wider. But among the major powers of Europe the United Kingdom was unique in simultaneously avoiding revolution, extending the franchise, and maintaining governmental stability. The result was that the British parliament acquired a new legitimacy, and the British constitution a new respect, not only at home but also abroad. Some thirty years later Charles Dickens would memorably lampoon such chauvinistic complacency in *Our Mutual Friend* (1864–65), in which Mr Podsnap explains the position to a somewhat bemused 'foreign gentleman': 'Our Constitution, Sir. We Englishmen are Very Proud of Our Constitution, Sir. It Was Bestowed Upon Us By Providence. No Other Country is so Favoured as This Country.'

Yet in addition to fulfilling many of the Whig government's aims, and while making the United Kingdom's polity and constitution the envy of much of Europe, the passing of the Great Reform Act also resulted in many unintended consequences, which did bring about some significant changes in the nature and practice of politics. This was especially the case concerning the balance of power between government (including the monarch), parliament (especially the Commons), the electorate and the public, which did begin to shift; and in general terms this would be away from government, which henceforward would have to take more notice of the views of the legislature, the voters and the people. The significant reduction in the number of rotten boroughs meant that the Commons became more independent of the executive, as an increasing number of MPs felt they must try to be responsive to constituency pressures concerning particular issues, and were also obliged to recognize the importance of public opinion more generally. This in turn meant that the old world of authoritarian Toryism, embodied in the Younger Pitt, Castlereagh, Liverpool and even Wellington himself, was gone for ever, at least in regard to the government of the United Kingdom, though this would be less true of the (non-settlement) colonies of the empire, and also of India, where authoritarian attitudes would remain strong among many members of the proconsular elite. It also meant that on occasions MPs responding on some constituency issues could unseat governments at non-election times, as in 1855 when the Aberdeen coalition would be brought down by an adverse vote in the Commons. But the most portentous results of the Reform Act derived from the fact that in future all voters had to be registered, more constituencies would be contested, and the increase in the number of seats where local influence no longer dominated would give greater opportunities for intervention by the party leadership.

As a result, the two main parties were increasingly obliged to organize for elections on a national basis, and from the 1840s onwards the verdict of the electorate could be decisive in unseating governments in ways that had not been true before 1830. Hence the establishment of the Carlton (Tory) and the (Whig) Reform Clubs on Pall Mall in London, as the permanent headquarters of new, centralized party organizations, which oversaw the getting and disbursement of funding, liaised with many of the constituencies, supported the work of registering voters, and provided accommodation for party activists from the provinces. As such, the Carlton and the Reform Clubs represented a deliberate and determined accommodation to the new electoral landscape, superseding the traditional aristocratic, political-cum-gambling clubs of Whites (Tory) or Brooks's (Whig) on St James Street near Piccadilly, which had previously served as the two parties' unofficial foci, but which seemed too social, amateur and exclusive in the new age of reform. More broadly still, the momentum of public support that had built up behind the Reform Bill, and which had been essential to its passing, had also aroused additional popular expectations that the Whig governments and a reformed parliament would engage with many pressing contemporary issues that had been too long neglected under the old Tory regime. And if parliament could reform itself once, so as to re-establish its legitimacy, then why should it not do so again, if further changes in the economic and social circumstances of the country warranted? Soon after its passing, Lord John Russell declared the Great Reform Act to be a 'final' constitutional settlement; but he would later come to regret that observation, and subsequent events, in which he himself would play a part, would render his prediction invalid.

WHIGS IN CLOVER?

After many decades in futile and seemingly permanent opposition, the Whigs would remain continuously in power for all but a few months from 1830 until the summer of 1841. In a way that had no precedent and no equivalent across the whole of the nineteenth century, this meant the 1830s would be a continually reforming decade, for which the epochal legislation of 1832 was both the necessary precondition and the essential precursor. The Whig leadership had more than proved itself in carrying the Reform Act; this reinforced its self-image as a rich and privileged coterie who, nevertheless, stood for liberalism and liberty, progress and improvement; and for the rest of the 1830s the Whigs believed they had a mandate to try to identify and address the major economic, social, religious and political problems by which the whole of the United Kingdom seemed beset. They were further emboldened by the fact that compared to Wellington's fissured and faltering Tory government, Grey's cabinet seemed exceptionally talented, including as it did one previous prime minister, Lord Goderich, and four future prime ministers: Lords Melbourne, John Russell, Palmerston and Stanley. (A less admiring view was that it was a government of 'prima donnas' which Grey found hard to hold together.) The result was that the Whig decade witnessed a more extensive legislative programme, longer parliamentary sessions and a more professionalized government than had ever been seen before. The era of the fiscal military state, epitomized by the Duke of Wellington, where the purpose of government was primarily to raise money and wage war, was passing; the age of improvement, which meant legislative engagement with contemporary issues, was beginning.

An early indication of these changed priorities and of a more interventionist approach concerned the issue of child labour in factories, although the precedent had in fact been established by Lord Liverpool's government in 1819. The initial impetus to reform came not from the Whig administration but from two Tories, the MP Michael Sadler and Lord Ashley (later the seventh Earl of Shaftesbury), whose genuine concern for the welfare of children was reinforced by a lofty disdain for manufacturers and laissez-faire capitalism. In the aftermath of the Reform Act, they secured a parliamentary committee on child labour in the textile industry, which produced extensive evidence of deplorable exploitation and resulted in the Factory Act of 1833. This was carried through the Commons by an alliance of landowners and radicals, against the protests of manufacturers and laissez-faire ideologues. The measure applied to most textile factories, and mandated that those aged between nine and thirteen could only work forty-eight hours a week, and must attend school for at least two hours a day, and that no one younger than eighteen could work for more than twelve hours a day. But in some ways the most innovative and influential provision of the Factory Act was the creation of an inspectorate to police the enforcement of these regulations. This established the precedent and became the model for government intervention in other areas of business, production and manufacturing, which had previously been entirely free from state surveillance or supervision. After Sadler and Ashley, the most influential figure in drafting the Factory Act was the young Edwin Chadwick, a devoted and energetic disciple of Jeremy Bentham, who for the next twenty years would be the most influential civil servant involved in matters of public health and social reform.

In the case of Ireland, the Whig government sought to break from the Tory policy that had been continuously implemented since the Act of Union, by putting less emphasis on coercion, which presupposed the Irish were incorrigibly different, backward and hostile, and making greater efforts at conciliation, by treating them as fellow Britons, who should be governed and legislated for on the same principles as those that applied on the mainland. This was a generous and high-minded position, but it was difficult to maintain given the manifest failure of Catholic Emancipation to reconcile Daniel O'Connell and his followers to the Union. On the contrary, it emboldened him and them, urged on by the Catholic priesthood, to campaign and agitate for its repeal, and after the election of 1832 O'Connell was supported in the United Kingdom parliament by nearly forty like-minded Irish MPs. In the short run, this continued Irish discontent meant Grey's cabinet was compelled to pass another Coercion Act, giving the Lord Lieutenant of Ireland powers to declare curfews and suppress subversive meetings; but it was implemented sparingly, and three years later it was replaced by a much milder measure that was never put into operation and was completely repealed in 1840. By then, in its first attempt at conciliation, the government had carried the Irish Church Temporalities Act, which sought to reform the (minority Protestant) Church of Ireland so as to make it less unacceptable to the (majority Catholic) population, by abolishing two archbishoprics and ten bishoprics, consolidating Anglican dioceses in under-populated areas, and relieving Irish ratepayers of the obligation to contribute to the upkeep of what most of them regarded as alien Anglican churches.

But like so much nineteenth-century Irish legislation, this did not work – on either side of the Irish Sea. Because it made no provision to apply any of the money saved from reforming the Church of Ireland to supporting the Catholic Church in Ireland, O'Connell and his followers both inside and outside parliament were further offended. At the same time diehard Tory opponents of reform deplored this measure for the very opposite reason, namely that it was a further unjustified and sacrilegious assault on the established church and the Anglican hierarchy. In the end, the bill was only passed with the support of Peel and Wellington; but soon after, in May 1834, four of Grey's colleagues resigned. Moreover, three of them – Lord Stanley (who as Chief Secretary for Ireland was less conciliatory than most of his cabinet colleagues), Lord Ripon (as Goderich had by this time become) and Sir James Graham (a north-country landed baronet, formerly First Lord of the Admiralty) – would eventually join Peel's government in 1841. Grey duly set about reconstructing his cabinet, but his efforts foundered amidst further disagreements over the renewal of the Irish Coercion Act. As a result Lord Althorp, the popular leader of the House of Commons as well as an influential Chancellor of the Exchequer, also resigned. Grey did not think he could go on without Althorp, and so in early July he, too, quit. Once again, it was affairs in Ireland that had brought an administration to an end. Grey would live for another eleven years in retirement on his Howick estate in Northumberland, but he never held office again. He would be fittingly commemorated in a majestic street and a statue atop a lofty column in the centre of nearby Newcastle-upon-Tyne.

On Grey's resignation the king turned to Lord Melbourne, who was Grey's heir apparent, and his cabinet closely resembled its predecessor. The new premier's first task was to implement the recommendations of a Royal Commission that the Whig government had set up in 1832 to inquire into the workings of the Poor Law, on which spending had escalated from £2 million annually in the mid-1790s to £7 million that year. Heavily influenced by Benthamite ideology, and reinforced by the ubiquitous presence of Edwin Chadwick, the commission concluded that too much money was being spent on subsidizing the poor: the parish-based structure of 'outdoor relief', financed by locally levied rates, was intrinsically wasteful and inefficient, while those eligible for poor relief were paid too generously, which meant they had no incentive to seek employment. The commissioners had little understanding of the complex and varied causes of poverty, they gave more attention to rural than to urban areas, and they were determined to produce a more rational and uniform system, which would also save the ratepayers money. In 1834 their proposals were embodied in the Poor Law Amendment Act, the key principle of which was that those unwilling or unable to work should not be so generously subsidized that they would be better off than those who were in employment. 'Outdoor relief' was effectively abolished, and in future the relief of poverty would primarily be carried on via the workhouse where, on the basis of the principle of 'less eligibility', conditions would be worse than outside, so as to discourage people from using workhouses except in the last resort. Their administration would be overseen by locally elected Poor Law guardians, who would have a strong incentive to keep costs down since they depended on ratepayers' votes for re-election; and a new central body would be established to supervise the whole system.

From the outset the Poor Law Amendment Act was controversial. It certainly cost the government less, at least in the short term: by 1837 spending on poor relief in England and Wales had fallen to £4 million annually. Yet while this may have pleased the economy-minded ratepayers, the poor would come to detest the forbidding workhouses, with their prison-like supervision and lack of privacy, as the embodiment of a hated and heartless welfare regime that would last until 1929. But having passed this measure, Melbourne's administration was effectively dismissed by William IV in November 1834, because he had become increasingly alarmed that the Whigs might be contemplating a further undermining (as he saw it) of the established church in Ireland. At this point the monarch turned to Sir Robert Peel (he had inherited his father's baronetcy in 1830), who formed a minority government, since the Whigs were still numerically dominant in the Commons; he would be the last prime minister to be installed by the sovereign against the wishes of the majority of MPs. The king granted Peel's government a dissolution, and for the general election held early in 1835 he produced his 'Tamworth Manifesto', in which he urged that the Conservative Party (the name by which the Tories were increasingly being known) would incline to liberal rather than 'Ultra' policies, notably by accepting the Great Reform Act as 'a final and irrevocable settlement'. The Tories made substantial gains, but the Whigs remained the majority party, and after a succession of defeats in the Commons, Peel resigned in April 1835, whereupon the king was compelled to recall Melbourne. This was a significant defeat for the king, as the electorate had refused to endorse his dismissal of the Whigs: royal will was no longer enough to sustain a government against the wishes of the voters. (During this period of high-political uncertainty two select committees were appointed to inquire into domestic and colonial sinecures, and their recommendations would lead to the effective end of 'Old Corruption'.)

Having reformed parliament and the Poor Law, the Whigs now turned their attention to local government, and towards the end of 1835 they passed another major piece of legislation, the Municipal Corporations Act, which was the community counterpart to the Great Reform Act and implemented the recommendations of another Royal Commission that had been set up in 1833. The main aims of the measure were twofold: first, to break open the closed, oligarchical (and often Tory) corporations that existed in many boroughs, where power was concentrated in the hands of a narrow and self-perpetuating group; and second, to enable large industrial towns, which had recently gained parliamentary representation, but which possessed only the most rudimentary form of local government, to establish their own authorities. Accordingly, the Act of

1. Warehouses on the Isle of Dogs, 1802: the maritime and commercial strength of the United Kingdom vividly displayed.

2. Nelson's funeral procession arrives at St Paul's Cathedral, 9 January 1806.

3. 'Britannia weighing the fate of Europe, or John Bull too heavy for Buonaparte': anonymous caricature of the British fear of French invasion, December 1803.

4. The 'Free-born Englishman, envy of nations', by George Cruickshank, 1819, protesting against the repressive post-war measures of Lord Liverpool's government.

5. Wellington and Blucher meet at Waterloo: a print of 1819, celebrating their decisive victory four years earlier.

6. Rioters being driven back from an ironworker's residence at Merthyr Tydfil in 1816.

7. The arrest of the Cato Street conspirators, 23 February 1820.

8. Lord John Russell sweeping away parliamentary abuses: caricature by George Cruickshank, March 1831.

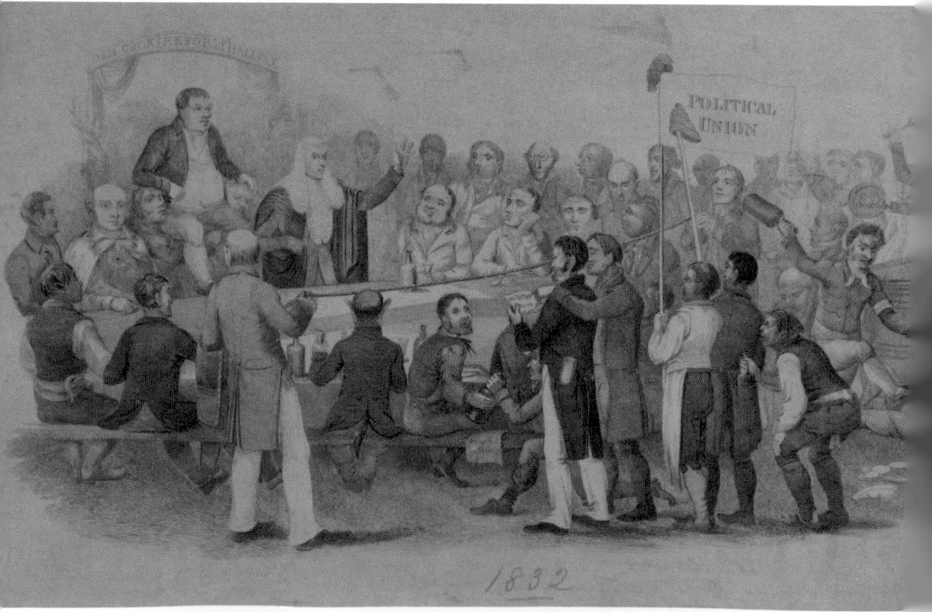

9. Cartoon of a public meeting in favour of the Reform Bill, 1832. Note the banner proclaiming 'political union'.

10. (*above*) A British traveller in New Zealand, *c.* 1827, with Maori bearers carrying his belongings.

11. The British army entering the Bolan Pass en route to Kabul at the beginning of the First Anglo-Afghan War, March 1839. Scarcely any of them would return.

12. Starving Irish queuing for a meal of imported corn at the time of the Great Famine, 1847.

13. Crowds assembled at Kennington Common in London, 10 April 1848. This was Chartism's last hurrah.

14. The Menai Straits, *c.* 1860: the bridges by Thomas Telford and Robert Stephenson linked London to Holyhead by road and rail, and passengers could then travel on to Ireland by sea.

15. Crystal Palace, 1851, viewed from Kensington gardens, by Augustus Butler. The juxtaposition of a pioneeringly modern building with large crowds of visitors is especially noteworthy.

16. St George's Hall, Liverpool, 1852. One of the great civic buildings of the nineteenth century, proclaiming the wealth, power and cultural aspirations of towns and cities beyond London.

17. Policemen in Manchester, *c.* 1850: only later in the century did the constabulary become generally accepted as the agents of law enforcement.

1835 replaced 178 closed boroughs with groups of councillors elected by the ratepayers, who held office for three years, along with nominated aldermen who served for six years, and a mayor who was elected annually by the whole corporation. At the same time it made provision for unincorporated towns to petition to obtain similar municipal provision, and by 1838 both Birmingham and Manchester had done so. Although the franchise was restricted to ratepayers who had been resident for three years, the Municipal Corporations Act was one of the major reforms of the nineteenth century, for it fundamentally restructured local government, and empowered the new corporations to levy rates to deal with such pressing issues as water supply, paving and lighting; and it also required them to establish police forces, which would be under local control rather than, as in the case of London, responsible to the Home Office. As such, the measure reflected the Whig government's genuine concern for strengthening local communities and fostering responsible citizenship, along with a more partisan hostility to traditional strongholds of Tory power and influence.

In reforming parliament, factories, the Poor Law and local government, the Whigs had intervened in the nation's affairs more energetically and widely than any previous administration: setting up commissions and committees of inquiry, and carrying measures of improvement and intervention that were high minded yet also politically partisan. These were sometimes (as in the case of the Factory Act) hostile to laissez-faire and the free market, while on other occasions (as in the case of the New Poor Law) more sympathetic to them. Either way, there was a clear recognition that in a nation undergoing unprecedented disruption and change, the government had to find out what was happening and intervene more. It was no longer sufficient to claim that its sole purposes were the raising of revenue and the safeguarding of the realm. This was certainly the view of Lord John Russell, who accelerated the pace of reform at the Home Office, where he had succeeded Melbourne, that Peel had initiated in the previous decade. He obtained powers to appoint inspectors of local prisons, steered through a series of acts that abolished the death penalty for most cases of burglary, robbery, forgery and arson, and for all non-violent crimes except treason, and also empowered Justices of the Peace to establish police forces in the counties. Bear-baiting and other cruel sports were abolished, and so was the pillory. In 1836 the civil registration of births, marriages and deaths was made compulsory, ending the old system that had depended on Anglican clergymen providing such a service in each parish. At the same time marriages conducted by non-Anglican clergymen were made fully legal, including those conducted by dissenting ministers in Nonconformist chapels – a further undermining, as its most committed supporters saw it, of the established position of the Church of England by the Whig government.

In June 1837 King William died, and he was succeeded by his niece, the young, unmarried Queen Victoria, daughter of the Duke of Kent and the oldest legitimate grandchild of George III. She was forbidden from becoming Elector of Hanover owing to her gender, and that title passed to her uncle, the Duke of Cumberland, who was George III's last surviving son. But a change of sovereign once again required a general election in the United Kingdom. Peel's Conservatives made further large gains, but overall the Whigs were still thirty MPs ahead, although this lead would be eroded over the coming years. Meanwhile, the Whigs continued with their reforms, including the Rural Constabularies Act obliging the county authorities to create their own police forces, and an Education Act that provided government grants for the construction and maintenance of schools. But the government's efforts were mainly focused on religious issues and Irish matters. In the case of the Church of England, their aim was not to undermine it further, as their Tory critics alleged, but rather to make it better equipped to face the challenges of an urbanizing and industrializing society. Already, in 1836, the Whigs had abolished payment in kind by the Tithe Commutation Act, substituting compulsory rent charges; they had standardized episcopal salaries at £4,000 a year and established a permanent Ecclesiastical Commission to redraw the boundaries and redistribute the income of dioceses. Between 1838 and 1840 they passed legislation prohibiting clergymen from holding two church livings simultaneously if more than ten miles apart, and abolishing many non-residential cathedral appointments. Such economies were a further contribution to the ending of 'Old Corruption', and they also made possible a renewed emphasis on church building in towns, in the hope this would enable the ecclesiastical establishment (in the words of the Archbishop of Canterbury) to 'resist the attacks of its enemies'.

During the second half of the decade, the Whigs were more successful than they had been earlier in seeking to replace coercion and hostility in Ireland with a policy of conciliation and equivalence. It was important, Russell declared, that Britain should stop looking at Ireland 'through a police telescope', and that the two nations should be treated in the same way, with 'the closest possible assimilation of the laws and institutions of both countries'. This had already started in 1831 with the creation of an Irish Board of Works, on the British model, with the aim of bringing Ireland up to British standards regarding communications and economic infrastructure. Moreover, by the later 1830s, relations between the Whig government and Daniel O'Connell and his parliamentary followers were greatly improved, so much so that O'Connell was prepared to abandon his opposition to the Union provided the Whigs were willing to bring about 'an identity of laws, an identity of institutions, and an identity of liberties' between the two parts of the United Kingdom. To this end, and as they became increasingly dependent on the support of Irish MPs as their Commons majority fell away, the Whigs passed a series of measures applying to Ireland the legislation recently enacted for Britain. Hence the establishment in 1836 of an Irish constabulary, along the lines of that which Peel had set up for London in 1829, and which was meant to be less militaristic than the earlier organization he had created in Ireland in 1814. Hence, two years later, the extension of the New Poor Law to Ireland and the settling of the tithe question along the lines established in Britain two years earlier. And hence the reform of Irish local government in 1840, intended to be 'upon the same principles as those of the acts which have already been passed for England and Scotland', but which was much modified by the Tory (or Conservative) opposition, who feared an almost complete transfer of power from their own allies to O'Connell and his followers.

ECONOMIC VICISSITUDES AND POPULAR DISCONTENTS

This unprecedented glut of reforms, following on the quasi-revolutionary legislation of 1832, meant the 1830s were the pivotal decade in the history of the nineteenth-century United Kingdom; and they also witnessed the beginnings of more formal party organization among both the Whigs and the Tories. The reforms were sponsored by government (rather than backbenchers) to a degree that had not been true before; they depended on the novel collection (and sometimes deliberate manipulation) of large amounts of data and supporting evidence; and they gave opportunities to a new group of civil servants, inspectors, experts, professionals and zealots who were in a righteous rage to get parliament and government to engage with the pressing social problems of the day. But although Grey and Melbourne were more politically adroit than Wellington had been, neither of them could keep their cabinet colleagues together all of the time. Moreover, across the 1830s, the support for the government in the Commons gradually ebbed away; on several occasions Grey and Melbourne could only get their legislation through with the (much-advertised) support of Peel and of O'Connell and his Irish followers; in the Lords they were even more vulnerable because of the Tories' inbuilt majority; and William IV was no friend to the Whigs. In addition, Grey and Melbourne carried their reforms against an ominous background of economic transformation, cyclical disruption and popular agitation that gave the Whigs extra leverage over their opponents in the legislature, but which also meant there would be many erstwhile supporters who would be dissatisfied and feel betrayed that the government had not done enough.

When William Huskisson was run over by George Stephenson's *Rocket* in 1830, no one could have guessed that within a decade the opening of the Liverpool and Manchester Railway would portend a revolution in transport (and in much else besides) that would transform the economy and society of Britain and, eventually, much of the world. By then, and following the earlier precedent of the Stockton to Darlington, many railways had been planned, linking the major towns and cities of Britain, and most of them would be constructed during the 1830s. There was no central co-ordination, as was famously demonstrated in the difference in the railway gauges. Most companies, in deference to Stephenson's original northern routes, settled on four feet eight and a half inches; but Isambard Kingdom Brunel, the flamboyant chief engineer to the Great Western Railway, preferred a broader gauge of seven feet, and some of these lines would last until late in the nineteenth century. In 1835 there were only 338 miles of railway open in Britain; but by 1841 the figure had risen to 1,775 miles, and most of the chief trunk routes either had been completed or were being built. Within fifteen years of the opening of the Stockton to Darlington line in 1825, tracks had been laid from London to Southampton, Portsmouth, Brighton and Dover; to Bristol (with an extension to Exeter under construction); to Birmingham (and continuing on to Liverpool, Manchester and the towns of the northwest); and to Hull, York and Newcastle (and the towns of the northeast). By this time, too, many cross-country routes were also in operation, from Carlisle to Newcastle, Liverpool to Leeds, and Derby to Bristol. Not surprisingly, Wellington was as ill-disposed to railways as he had been to parliamentary reform; he feared they 'encouraged the lower classes to travel about'.

The sudden expansion in the railways during this decade (their construction peaked in 1836) was as revolutionary a development as the passing of the Great Reform Act; but it was only one indication that the economy of the United Kingdom was being fundamentally transformed during these years, as the output of coal, iron, cotton and textiles increased dramatically, and as together they formed an ever larger share of Britain's exports. Here are some indications of the extent of the changes during this decade. In 1831 the population of England, Wales and Scotland was a little above 16.2 million; ten years later it stood at more than 18.5 million. Meanwhile, the population of Ireland had gone up, across the same decade, from 7.7 to 8.1 million. These were significant increases in absolute numbers, even as they represented a slowing down (especially in Ireland) from the higher rates of growth that had characterized the 1800s, 1810s and early 1820s. But the cumulative impact was extraordinary: by 1841 the population of the British Isles was more than twice what it had been in 1783, which meant it had grown more rapidly than anywhere on the European continent. Only because of the expansion of the industrial economy was it possible for such an increase to be sustained. In addition to the construction of the railways, there was abundant evidence for that expansion. Between 1835 and 1840 consumption of raw cotton rose from £318 million to £459 million, while the number of cotton factory hands increased from 220,000 to 264,000. To a greater extent than before, the industrializing regions of the Midlands, the northwest and the northeast, along with Glasgow and Belfast, were seen as the harbingers of a new industrial world – and by the late 1830s Manchester had acquired a uniquely global reputation as 'Cottonopolis' and as the 'shock city' of the time.

But several qualifications must be added to this picture, for it is only in retrospect that the ever onward march of factory-based industry, often powered by fossil fuel, and using large-scale methods of mass production seems the leitmotif of this decade. The first qualification is that technological change was localized and rarely involved large-scale factory production at this stage. Only twenty-five of Lancashire's one thousand spinning and weaving firms employed more than a thousand workers in 1841, and the average number was less than two hundred, while most woollen factories in the West Riding of Yorkshire employed fewer than fifty workers. Aside from textiles, more than three-quarters of British industry remained small-scale, cottage-based and powered by water, humans or animals rather than by steam; in towns such as Birmingham, increased production was the result of the multiplication in the number of small workshops rather than their replacement by large factories; and as the greatest manufacturing centre of all, London continued to produce footwear, clothes, soap, paper, beer and other goods on a small scale and by traditional and essentially pre-industrial methods. Nevertheless, the growth in output necessarily meant that the aggregate number of people involved in production increased, which meant that the populations living in towns such as Birmingham and Manchester expanded greatly during this decade; and (the second qualification) the pollution, stench and squalor also increased correspondingly, since before the passing of the Municipal Corporations Act many of the largest industrial towns possessed only the most rudimentary forms of local self-government. As a result, they were inadequately paved, policed and lit, and many of their inhabitants lived and worked in circumstances of extreme squalor and degradation, which were given particular attention in many of the reports and inquiries that were conducted in the 1830s and on into the next decade.

A further qualification must be that whereas some areas of the British Isles underwent industrial transformation, there were other parts, and also certain traditional occupational groups, that were the victims rather than the beneficiaries of such undoubted but uneven progress. In the 'old' wool-producing counties of Wiltshire, Norfolk and Suffolk there were spectacular examples of de-industrialization to balance changes elsewhere in the opposite direction, as production failed to keep up with the new methods being pioneered and practised in Yorkshire. Across the Pennines in Lancashire the handloom weavers were unable to compete with the new powered weaving, and both their wages and their numbers declined dramatically: between 1835 and

1845 they would lose two-thirds of their numbers. In Ireland the textile industry went into a severe decline during the 1830s, leading to substantial levels of emigration to England and North America. The result was that except for Dublin, Belfast and parts of Ulster, the Irish economy remained essentially agricultural, while too many of its population were dependent on a single crop – the potato. The final qualification is that while contemporaries recognized that fundamental changes and undoubted transformations were taking place, they were as much exercised by the evidence that the economy was blighted by serious fluctuations, especially when bad harvests coincided with downturns in investment and industrial production. The 1830s began and ended with just such periods of intense and acute depression: the agitation over parliamentary reform, extending from 1829 to 1832, coincided with a severe economic downturn and high unemployment; and the same was true after 1836, when the recent series of good harvests came to an end, and a serious depression in business, trade and exports meant unemployment increased dramatically, while railway construction came to a standstill.

This was the turbulent background against which the Whigs tried to govern, reform and maintain order during the 1830s. It was, Thomas Attwood later claimed, 'the distress of the country which primarily led to the agitation of [parliamentary] reform', and by the time Grey took office late in 1830, popular agitation was already in full flow. As the new prime minister told the new king: 'We did not cause the excitement about reform. We found it in full vigour when we came into office.' But while the Whigs were eager to use the evidence of widespread and growing popular discontent as a lever over the king, the Lords and the Commons, they were as much on the side of law and order as Wellington himself. As Home Secretary, Melbourne was particularly hostile to the 'Swing Riots' that swept the agricultural districts in 1829 and 1830. The destruction of threshing machines, the burning of hayricks and corns tacks, the protests against tithes, the demand for living wages, and anger against the working of the Poor Laws were all elements in a spontaneous movement, otherwise known as the 'village labourers' revolt', which spread in the south of England from Kent to Dorset. In response, Melbourne issued a circular to the magistracy, asking it to act with determination, promptitude, vigour and decision, and he went on to repress the outbreaks with ruthlessness and without mercy. In

December 1830 special commissioners were sent to put the rioters on trial, as a result of which nineteen labourers were hanged, more than four hundred were transported, and as many again were imprisoned. Refusing to countenance positive measures to alleviate distress, Melbourne insisted that, even as they sought to pass a reform bill, the Whigs were determined to maintain the constitution of civil society and the rights of property.

The Whig government's ambivalence towards popular pressure and protest continued as long as the Reform Bill crisis lasted. On the one hand, and in the aftermath of the rejection of the bill by the Lords in October 1831, the Whigs more than ever needed public opinion, and even public agitation, in their support, as they sought to side with the people against the forces of obscurantism and reaction. But many of those popular voices were not only too radical for William IV and Wellington, they were also too strident and subversive for the prime minister and his cabinet. Indeed, it was Grey's hope (and also his strategy) that once parliamentary reform was settled, 'all the sound part of the community would not only be separated from, but placed in direct opposition to, [radical] associations whose permanent existence every reasonable man must feel to be incompatible with the safety of the country'. That, for Grey, was his essentially conservative objective; but it was also why, once the Reform Act was passed, many of those who had agitated for it felt let down, disappointed and disillusioned as they came to understand the details and consequences of the measure. As the *Poor Man's Guardian* put it in October 1832, echoing Grey's own remarks, but drawing very different conclusions: 'The promoters of the Reform Bill projected it, not with a view to subvert, or even re-model, our aristocratic institutions, but to consolidate them by a reinforcement of sub-aristocracy from the middle classes.' In the same journal Henry Hetherington and his friends of the so-called National Union of the Working Classes dismissed the Great Reform Act as a 'delusive, time-serving, specious and partial measure', which only gave votes to a 'small and particular portion of the community'.

Had the economy remained depressed in 1833 and 1834, and popular agitation continued, the passing of the Great Reform Act might have seemed less of a triumph of Whig statesmanship (and calculation) than in retrospect it did. The government was lucky, for between 1833 and 1836, and in part stimulated by the first railway 'boom', the economy picked up, there were good harvests (the price of wheat fell from 63s a quarter in 1832 to 36s by late 1835), and there was something approaching full employment for the first time in almost a decade. The result was that radical agitation for more drastic political reform fell away, and attention shifted to trying to improve the wages and conditions of those in work by promoting and organizing trades unions. These efforts met with varying success. In October 1833 the Friendly Society of Agricultural Labourers was founded at Tolpuddle in Dorset by George Loveless, a farm worker and Methodist lay preacher, with the aim of securing for all its members, by fair means, a just remuneration for their labour. As such, it was merely one example of a widespread movement to establish trades unions for unskilled rural labourers, whose pay was being inexorably reduced as a result of increased mechanization. But Melbourne, remembering the disturbances caused by the 'Swing Riots' scarcely a year before, took a dim view of these developments, and lent his support to a local magistrate, who had Loveless and five of his associates arrested for violating an act passed in the harsh year of 1797, prohibiting unlawful oaths. There was a hurried and rigged trial, presided over by Melbourne's brother-in-law, at which the six men were sentenced to be transported to Australia for seven years. They immediately became known as the 'Tolpuddle Martyrs', and duly took their place with the Levellers and the Luddites in what would eventually become the labour pantheon. There was also a public outcry at the time, led by prominent radicals, and in 1836 Melbourne's successor at the Home Office, Lord John Russell, granted the martyrs a free pardon.

Among those heading the campaign to free the Tolpuddle Martyrs were the leaders of the impressively named Grand National Consolidated Trades Union. It had been founded in February 1834 and drew its inspiration from the ideas of Robert Owen, a cotton manufacturer and paternalist employer, who had formed model communities at New Lanark, near Manchester, and across the Atlantic at New Harmony in Indiana, dedicated to workers' welfare, co-operative ownership and utopian socialism. Following these precepts, the aim of the GNCTU was to promote the unity of the working classes, so that by peaceful means they could gain control of their industries and reorganize them into co-operative ventures, thereby ushering in the Owenite millennium. As such, the GNCTU was the biggest working-class organization created during these years, with an elaborate national structure; but it was far from being the only one, as separate unions were also established for builders, cotton spinners, potters and clothiers. The result was the widespread proliferation of industrial unrest, and a series of increasingly militant strikes, pitting combinations of workmen against counter-combinations of employers. But such protests achieved little. The GNCTU's claim to speak on behalf of all workers everywhere was belied by the many other unions that came into being, both in industry and agriculture. It also failed to keep control of its members in particular trades and localities, especially London, and for all its high-sounding and pacific rhetoric, some of the strikes in which it became involved were violent and ugly. Moreover, there was never the remotest likelihood that the socialist utopia would be ushered in. By August 1834, scarcely six months after it had come into being, the GNCTU had effectively collapsed, and most of the other unions had also been defeated – to the Whig government's great relief.

Between the failure of the GNCTU in the autumn of 1834 and the early months of 1837, working-class political activity was largely confined to small minority groups, of which the most important would turn out to be the London Working Men's Association, founded in June 1836 by the veteran radicals Henry Hetherington, Francis Place and William Lovett. It was pledged to promote independent action on the part of all labourers and working men, not only to secure better pay and conditions but also to investigate the deeper 'causes of the evils that oppress them', and this seemed to portend a return to further political campaigning against the limited and inadequate Whig reforms of the decade. Soon after the LWMA was founded, such a change of strategy became essential as the economy once again lurched downwards: the sequence of good harvests ended, driving wheat prices back up; there was a financial crisis sparked off in the United States; and the recent trade boom imploded. By the summer of 1837 there were complaints from Birmingham and Liverpool of the 'alarming consequences' of the 'intense distress', and in Manchester there were 50,000 workers who were either unemployed or on short time. Although there were some signs of a qualified revival in 1838 and 1839, there continued to be high unemployment in particular areas, such as Lancashire and the Midlands, and there was serious popular unrest as a result. There was a slight improvement in 1840, but in the following year recession plunged into depression. Prolonged business difficulties, a collapse in domestic investment and overseas trade, and four years of harvest dearth made the United Kingdom 'unhappy and afraid, a country of conflict and despair'. What was the use of being 'the workshop of the world', if so many of the workshops and the workers were idle?

By definition, the unemployed could not negotiate for shorter hours, better conditions or more pay, while the Whig government's enactment of the New Poor Law and its harsh imprisonments had generated cumulative resentment and growing hostility among many members of the working class (especially those who were no longer working). Meanwhile, attempts by such radical MPs as George Grote, Joseph Hume and J. A. Roebuck to secure annually elected parliaments and household suffrage failed to gain any traction among their fellow MPs; and soon after it was established the LWMA published a scathing indictment, entitled *The Rotten House of Commons* (1836), drawing attention to the continued dominance of landowners, now augmented by businessmen and capitalists who were no friends to their labouring employees. The lower house, the authors concluded, ought to be 'the People's House', and 'there our opinions should be stated, there our rights ought to be advocated, there we ought to be represented, or we are serfs'. It was as a result of the combined efforts of thwarted radical MPs and the LWMA that a Charter was drawn up in 1837 containing six demands, at least three of which Earl Grey had deliberately ensured would not be part of any package of parliamentary reform for which he would be responsible: universal adult male suffrage, the secret ballot, parliamentary constituencies of equal size, the abolition of property qualifications for MPs, the payment of MPs and annual parliaments. At a time of renewed depression, the Charter constituted a dramatic return to the radical demands that had been articulated between 1829 and 1832, another period witnessing an economic downturn; and the aim was to bring about a representational system much more progressive and democratic than that which the Whigs had created in 1832. The final draft of the Charter was settled in May 1838, and soon after, working-class groups agreed to merge their particular aspirations and agitations in one common struggle.

Hence Chartism, which has rightly been described as the first organized mass movement of British radicalism in the modern sense, and which would be the most significant form of popular protest across the

United Kingdom for the next ten years. Indeed, in some ways it was the most important popular protest of the nineteenth century, even though it achieved none of its six aims in the short term (and one of them, annual parliaments, remains unrealized to this day). At the outset, support for Chartism was widely based among members of the working classes in London and the new big industrial cities such as Manchester, Leeds, Birmingham and Glasgow; but also among occupational groups and regions that were in decline, such as the hand-loom weavers and areas like the southwest that had been bypassed by the new industrial capitalism. As such, it became the focus for discontents that were many and varied, and its appeal was greatly increased because of the energetic and aggressive involvement of Fergus O'Connor, an Irish Protestant lawyer, who became closely involved with the London Democratic Association, which was a more radical rival to the LWMA. Towards the end of 1838 he began publishing the Leeds-based newspaper *The Northern Star*, which soon became the main journalistic organ of Chartism, and which O'Connor used to advocate a general strike of all workers, and even hinted at more revolutionary activity. During the course of 1838, enormous rallies and huge meetings were held in the great towns and cities across the nation, the ostensible purpose of which was to elect representatives to a National Chartist Convention that would meet in London and prepare a petition to parliament. The Convention duly met in February 1839; a petition advocating the Six Points, to which were appended an estimated one million signatures, was presented to parliament; but in July the Commons peremptorily rejected it by 234 votes to 46.

At the Convention that had drawn up the petition there had been a clear division between those who advocated 'peace, law and order' and those who preferred violent, even revolutionary, tactics. Parliament's refusal to take Chartism seriously intensified the split in the movement; the moderates became increasingly alienated from the Convention; and after long and acrimonious debates, it was dissolved in September 1839. During the second half of that year, there was widespread rioting in many of the large towns and cities, and in Newport, Monmouthshire, John Frost, a radical draper, former mayor and ex-magistrate, led an armed rebellion against the government of the county. Once again, the Whig government came down heavily on the side of law and order, calling in the military to suppress the protesters: many Chartist leaders were arrested, including Fergus O'Connor, and sent to gaol; fourteen of the Newport insurrectionists were killed by troops; and Frost was sentenced to death, later commuted to penal transportation. By June 1840 more than five hundred Chartist leaders had been imprisoned, and for the next two years it seemed as though the movement's impetus was indeed spent. But General Sir Charles Napier, a distinguished and unusually intelligent soldier, who had been appointed to command the Northern District of the army, and who dealt with the unrest more sympathetically (and more reluctantly) than many of his colleagues, was not so sure. He was appalled by what he saw of the squalid and polluted state of industrial Manchester: 'the chimney of the world' was his most understated description of it; or, yet more graphically, 'the entrance to hell realized'. Small wonder, he thought, that people there and elsewhere were protesting. 'Would that I had gone to Australia,' he noted in his journal, 'and thus been saved this work, produced by Tory injustice and Whig imbecility; the doctrine of slowly reforming when men are starving is of all things the most silly: famishing men cannot wait.'

GLOBAL AGENDAS

Napier's involvement with Chartism was a brief domestic interlude in a military career spent largely overseas, for he had previously served in the Iberian Peninsula with Wellington and in the Mediterranean, and he would later end his days of command in India. Nor was this the full extent of his overseas interests, for during the 1830s he had been offered (but refused) employment in Canada, and had been involved with schemes to set up a new colony in South Australia (hence the Antipodean allusions in his words quoted in the previous paragraph). As such, Napier's career is an instructive reminder that these years of Whig ascendancy, and of unprecedented change and disruption at home, were also a time when the British people were becoming ever more engaged with the wider worlds beyond their own shores. One indication of this greater global involvement was that Lord Palmerston served as Foreign Secretary throughout the administrations of Lords Grey and Melbourne from 1830 to 1841 (apart from the few months in 1834–35 when the Tories vainly tried to form a government). He was the most influential figure in the making of the United Kingdom's foreign policy, and he would effectively maintain that position until his death in 1865. There was no such comparable continuity at the Colonial Office, in the way that Bathurst had provided during the years of Lord Liverpool's administration. On the contrary, between 1830 and 1841 there were no fewer than six Secretaries for War and for the Colonies, which suggests that imperial matters ranked significantly lower than foreign affairs in terms of the United Kingdom's overseas engagement. But the impulses to empire were many and varied, they were becoming more significant during these years, and they were more significant than the Colonial Secretaries themselves.

Beginning in the 1830s Palmerston would cultivate a reputation, plausible and deserved up to a point, for supporting 'liberal' regimes in Europe as 'the champion of justice and right', and for vigorously asserting British power and interests in other parts of the world by what was termed 'gunboat diplomacy'. As a result he became the embodiment of a particular kind of robust patriotism, which claimed to be both nationally beneficial and also globally enlightened, and which was well summed up in his own phrase that Britain had 'no eternal allies and no permanent enemies. Our interests are eternal, and those interests it is our duty to follow.' More particularly, this meant that one of Palmerston's main aims, as had previously been the case with Castlereagh and Canning, was to maintain the balance of power in Europe, which in the 1830s meant keeping post-Napoleonic France in check, and also curbing Russia's expansionist ambitions in the eastern Mediterranean and in Asia vis-à-vis India. Ideally, Palmerston hoped to constrain France and Russia by diplomatic rather than military means, thereby avoiding direct continental involvement, and this was also his preferred way of supporting those emerging liberal nationalists who were struggling to win independence from autocratic regimes such as the Russian and Ottoman Empires. His wish was that these new nations would be established, in emulation of Britain, on the basis of liberty and the rule of law: 'Constitutional states,' Palmerston told the Commons, using this essentially shorthand term, 'I consider to be the natural allies of this country.' More globally, Palmerston aimed to protect and extend the opportunities for British traders and investors overseas, thereby consolidating and expanding the United Kingdom's recently established position as the world's pre-eminent fiscal, industrial and trading power.

This was the international agenda that Palmerston sought to implement as Foreign Secretary, regardless of the domestic disputes and disturbances which disrupted the high and low politics of the 1830s. Yet it would be misleading to over-personalize Britain's increasing overseas involvement during this period, by focusing exclusively on Palmerston. For there were broader developments, both in Britain and abroad, of which he was simultaneously the agent and beneficiary, but which extended far beyond the making of official policy. Just as the 1830s were the hinge decade in domestic politics, so they also witnessed the beginning of a new era internationally, as most of the favourable global conditions began to emerge and converge that would form the basis of British world power during the remainder of the nineteenth century. Among these were a gradually dawning recognition that there would be no repeat of the Napoleonic threat to Britain, and that France could be effectively contained and constrained; a developing realization that the balance of power in Europe could be maintained, which meant there would be no reiteration of the gruelling wartime years from 1793 to 1815; a growing appreciation of the fragility of the Ottoman, Persian and Chinese Empires, which would mean new opportunities for British trade and investment; the complete detachment of the Latin American republics from their Iberian masters, which held out similar prospects; and a massively increased capacity on the part of Britain to export goods, people and capital overseas. The result was a significantly enhanced British engagement with many parts of the world, involving many people and interests, which was never the result of a single master plan drawn up by the Foreign Office (or the Colonial Office), but was rather the expression of the 'chaotic pluralism' of the many means whereby the United Kingdom increasingly engaged with more regions of the world, and in more ways, than it had ever done before.

But on taking over the Foreign Office in 1830, Palmerston's immediate problems were essentially European, as he was faced with four successive revolutions that for a time seemed to place the settlement reached in 1815 in jeopardy: initially in France, where Charles X was ejected in favour of Louis-Philippe; then in the Low Countries, where the Belgians sought independence from the Kingdom of the Netherlands; next in Italy, where there were rebellions against the papal and Austrian states; and finally in Poland, where there was a revolt against

Russian overlordship. Palmerston considered these revolutions to be 'decisive of the ascendancy of Liberal principles throughout Europe', but they also threatened the balance of power and in practice his policies were less assertively 'liberal' than his rhetoric suggested. In France, where Palmerston had no choice but to let events take their course, the new 'July monarchy' did indeed turn out to be more liberal than its predecessor. The Belgian revolution was of particular concern, since Britain had no wish for the opposite coastline of the English Channel to come under the influence of a potentially hostile foreign power (as had happened in 1793). Accordingly, Palmerston convened a conference in London, and persuaded the other major continental powers to recognize Belgian independence, while Queen Victoria's uncle, Leopold of Saxe-Coburg-Gotha, was installed upon the throne (arrangements that were finally confirmed in 1839). Elsewhere, Palmerston was much less influential. The Russians put down the Polish rebellion in 1831, which had attracted the support of liberals and intellectuals throughout Europe; but there was nothing that Britain could actually do. Likewise in Italy, Palmerston refused to intervene, and the revolutionaries were duly suppressed. So when he claimed, in 1832, that 'there never was a period when England was more respected than at present in her foreign relations, in consequence of her good faith, moderation and firmness', he was giving a very one-sided picture of what he had been able to achieve.

Such had been Palmerston's continental diplomacy while the Great Reform Bill was making its way through parliament: less liberal than he claimed, but successful in avoiding military involvement. Once that measure had been passed, the Whigs could turn to a broader reform agenda, and the empire was naturally a part of it. The last time they had been in power, in 1807, they had abolished the slave trade; after 1832 it was almost universally assumed that one of the earliest acts of a more representative parliament would be the abolition of slavery itself. During the intervening years there had been a sustained campaign for abolition, led by such ardent and persistent evangelicals as William Wilberforce, Thomas Clarkson and Elizabeth Fry. In the aftermath of the Reform Act, the West Indian planters' lobby was no longer as strong in parliament as it had previously been, and by this time both slavery and the sugar trade were of declining importance. Moreover, the years from 1830 to 1832 had witnessed a greater degree of popular petitioning in favour of abolition than for parliamentary reform itself. Altogether the legislature received more than five thousand petitions signed by one and a half million people, and there was a special, additional ladies' petition, which stretched for half a mile and bore 187,000 signatures. In 1833 Lord Stanley, who had recently become Colonial Secretary, successfully carried the abolition of slavery throughout the British Empire, and Wilberforce lived just long enough to learn of its passing. The planters were generously compensated to the tune of £20 million, and all slaves received their freedom within a year, but with mandatory apprenticeship to their former masters until 1838. Thereafter, the Royal Navy would continue to patrol the coasts of Africa, the Caribbean and South America, seeking to capture slave ships flying under other flags, and to free their slaves; as Foreign Secretary, Palmerston vigorously supported such actions, and received widespread public approval.

Across the next half a century other Western powers would follow Britain's pioneering example; but in practice emancipation did little to improve the condition of former slaves, and the British West Indian islands would languish as a colonial backwater, while imperial action and priorities shifted increasingly to the east. One indication of these changed agendas was that in the same year that slavery was abolished, the Whig government rescinded the last remaining trading monopolies of the East India Company, which were with China, and the Company was further instructed 'with all convenient speed to close their commercial business', thereby opening up a new overseas market to private enterprise and initiative. In fact, the Company did no such thing, but continued to trade in one infamous commodity in which, after 1833, it was joined by many new British dealers – namely opium. Between 1821 and 1837 sales of opium increased fivefold, and by 1838 the British were selling 1,400 tons of the narcotic annually to China. The following year the emperor determined to eradicate the trade in such a 'harmful drug', which was doing so much damage to so many of his subjects, by confiscating and destroying the cargoes carried in British ships. This dispute escalated into what became known as the First Opium War, and eventually the Royal Navy bombarded the port of Canton and its environs in January 1841 and again in May. Palmerston justified this action, not so much in terms of the opium trade, but rather because of the need to open up Chinese markets more generally. But the government survived a Commons' censure motion by only nine votes, and Gladstone denounced Palmerston for defending 'the infamous contraband traffic', which he deemed 'a national iniquity towards China'.

China was one of the weakening empires that seemed to offer new and exciting trading and investment opportunities, but of more pressing concern was the decaying Ottoman Empire. It straddled the land routes to India, which were becoming of ever-increasing concern to the British. If it broke up, there was every likelihood that Russia would be the prime beneficiary in terms of territory, which the British did not want. And ever since Napoleon had invaded Egypt, the French also cherished acquisitive ambitions in the region. In 1831 the Egyptian Pasha, Mehmed Ali, renounced his allegiance to the Ottomans, captured Syria, and threatened to march on Constantinople, the Sultan's capital. Palmerston disliked Mehmed Ali as an 'ignorant barbarian', and was determined to support the Ottoman Sultan against him. But the cabinet refused to back him when he proposed sending military aid, and as a result it was Russia that gained enhanced influence in Constantinople. In 1839 Mehmed Ali and the Sultan went to war again, but this time Palmerston brokered a deal between France, Russia, Austria and Prussia that guaranteed the integrity and independence of the Ottoman Empire. But the French government, which preferred to side with Mehmed Ali than support the Sultan, was unhappy with this agreement. Palmerston thereupon negotiated another deal, from which the French were excluded. Eventually, Mehmed Ali renounced his claims to Syria, and was confirmed in his hereditary possession of Egypt on condition of paying an annual tribute to the Sultan. The Ottoman Empire had been preserved, and Russia had been successfully thwarted in its predatory ambitions. So, too, had France; but its relations with Britain were seriously soured as a result, and there was talk in both Paris and London of an impending 'Eastern Crisis'.

Nevertheless, Palmerston had shown great diplomatic skill and persistence in dealing with the Ottoman Empire. He hoped it would now enjoy 'ten years of peace', during which it might be reformed and reorganized and 'become again a respectable power'. Moreover, in 1838 he had secured an agreement that opened up the Sultan's dominions to foreign merchants and traders, which he hoped might bring new and lucrative markets to British business, while also increasing the likelihood of domestic reform in Constantinople itself. But although

Palmerston had blocked Russia's expansionist ambitions in the eastern Mediterranean, he feared – perhaps exaggeratedly? – that British India remained vulnerable to potential Russian advances, to which the invasion of Persia or Afghanistan might serve as the prelude. As he told Charles Ellis, Britain's ambassador to Teheran, in 1835: 'The interests and policy of Russia, as regards Persia, are in almost all things not only different, but opposite, to those of Great Britain.' Palmerston hoped to establish Persia as a buffer state between Russia and British India, but by 1838 it was widely believed that Russia had obtained an unassailable hegemony there, because Palmerston had made insufficient efforts to conciliate it or to warn the Russians off. Lord Auckland, the Governor General of India since 1836, now concluded that if Persia was unrealizable and unsound, an alternative buffer state must be created in neighbouring Afghanistan. His plan was to send British forces there to depose the pro-Persian (and thus pro-Russian) Emir, Dost Mohammad, and replace him with the pro-British Sha Shuja, who had been expelled from the kingship in 1809 and lived in exile in British India. In 1839 British forces invaded Afghanistan, occupying Kandahar and Kabul, Dost Mohammad fled, and Sha Shuja was duly installed as the new Emir – at least for the time being.

For a time it seemed as though this policy was working, and that Russia would be as effectively contained vis-à-vis British India as it had already been successfully checked in the eastern Mediterranean. This was all the more essential, because in recent years British India had been the setting for a major instalment of reform and reconstruction, part evangelical, part utilitarian, that had been pushed through by Lord William Bentinck, Auckland's predecessor as Governor General of India, from 1828 to 1835. Like Napier, Bentinck had previously seen military service in the Peninsular Wars and the Mediterranean, and had previously been Governor of Madras. He was a Whig grandee and zealous reformer, determined to improve the efficiency of British administration and promote what he termed 'the moral regeneration of India': hence his suppression of suttee and thuggee. Although he did not wish to interfere with the native princes of India, Bentinck deposed the Raja of Coorg for cruelty, annexed Cachar at the request of the inhabitants, and assumed the entire administration of Mysore owing to the misgovernment of the Raja. But Bentinck's most influential – and controversial – decision concerned education, where he declared that the United Kingdom's 'great objective ought to be the promotion of European literature and science among the natives of India'. Supported by Charles Trevelyan (a zealous Indian civil servant) and his brother-in-law Thomas Babington Macaulay (an MP between 1830 and 1833, joining the Supreme Council of India in 1834), Bentinck prescribed English as the proper language of instruction and official administration in the most uncompromising terms.

By such means Bentinck hoped India might be regenerated, just as Palmerston hoped that the Ottoman and Chinese Empires might be brought closer to the higher standards of Western civilization as a result of being opened up to British trade and commerce. Such concerns and anxieties as the Whigs may have felt about their capacity to govern and reform the United Kingdom were conspicuously absent when it came to the confidence and certainty they displayed in governing and reforming India. Elsewhere, the British Empire was administered and augmented in largely haphazard and unforeseen ways, with the initiative often lying with individuals and private enterprise rather than the imperial authorities in London. Two acquisitions were of undeniable strategic importance: the Falkland Islands (1833) was a secure base on the route from the South Atlantic to the Pacific, while Aden (1839) was a major harbour on the route from the Red Sea to India and was, indeed, administered from Calcutta. But much of the interest in the empire during this decade came from radicals who were eager to promote, by private enterprise, the cause of emigration. Pre-eminent among them was Edward Gibbon Wakefield, who established the joint-stock South Australian Association in 1836 to settle a new colony in the British Antipodes. But the company was not well run, the government was obliged to intervene to suspend its charter, and soon after South Australia became a crown colony. By then, Wakefield had shifted his attention to another area of the Antipodes, having founded the New Zealand Company in 1839. Colonists under his scheme first landed there in the following year, when they negotiated the Treaty of Waitangi with the native Maori chiefs, and soon after New Zealand was reluctantly acquired as another British crown colony. This was partly to ensure Wakefield's company adhered to the terms of its charter, partly to protect the rights of the indigenous inhabitants vis-à-vis the British settlers, and partly to pre-empt a possible French annexation.

In both Australia and New Zealand, British officialdom grudgingly followed where private enterprise settlement schemes had led; but in South Africa, by contrast, the impulse to imperial expansion came from the Governor of Cape Colony, Sir Benjamin D'Urban, who between 1834 and 1837 sought to annex the neighbouring frontier territories occupied by the native Kaffir tribes, in the hope of soothing the relations between the indigenous inhabitants and the British and Boer settlers. But his local initiative was doubly counter-productive. The British government refused to accept the annexation, since it was convinced that the violence of the settlers had goaded the Kaffirs into attacking them; while the Boer farmers, annoyed at the abolition of slavery and with what they saw as misplaced official concern for the natives, repudiated British rule by embarking on the 'Great Trek' across the Orange River to establish a new nation where they hoped to live undisturbed. By 1837 the Great Trek was over, but in the same year the British faced another form of imperial rejection, with rebellions occurring simultaneously in Upper Canada (overwhelmingly British, based around the Great Lakes, and focused on Toronto) and Lower Canada (predominantly French, centred on the St Lawrence River and Quebec): in Upper Canada because of disagreements between the legislative council and the assembly over money and taxation; in Lower Canada because of French hostility to what they saw as increasing British encroachment. The rebellions were put down, but Lord Melbourne sent out Lord Durham to investigate. His report, published in 1839, recommended the union of Upper and Lower Canada, with a single parliament and limited powers of self-government, proposals that were embodied in the Canada Act of Union passed in 1840.

At the official level Britain's broadening engagement with the wider world during the 1830s was carried on by diplomats, proconsuls, civil servants and soldiers; most were drawn from aristocratic or agrarian backgrounds; and as a result there were many paradoxes and inconsistencies. At the Foreign Office, Palmerston's policies were more cautious and contradictory than his belligerent rhetoric suggested: he supported revolutionaries in Belgium but not in Poland; he preferred to keep the Ottoman Empire intact rather than break it up; and like many Foreign Secretaries, he probably exaggerated the threat that Russia presented to British interests. Some British administrators in India were confident they could refashion this 'backward world in their own enlightened image'. According to Macaulay, Lord William Bentinck had 'infused into Oriental despotism the spirit of British freedom' and 'never forgot that the end of government is the happiness of the governed'. But others thought the British regime was too military and authoritarian, and had failed to establish any close identity with native peoples. The picture was different again in the settlement colonies. In Australia and New Zealand the Colonial Office reluctantly followed private initiative, but tended to take a more favourable view of indigenous peoples than the new arrivals from Britain. In Canada, Lord Durham's report and recommendations patched things up between the British and the French, the colonists and the mother country. But in South Africa relations remained vexed between the British, the Boers and the indigenous peoples.

In any case, and notwithstanding the undeniable constraints and contradictions of British official policy, this was only one of the many ways in which an increasing number of Britons were interacting with the wider world. Indeed, according to Lord Palmerston, the Whig administration saw itself as serving rather than dominating some of these connections, as when he told parliament in 1839 that 'the great object of the government in every quarter of the world was to extend the commerce of the country'. Some of these expectations, especially regarding Latin America, the Levant and China, would at least in the short run turn out to be exaggerated. Britain's overseas trade during the 1830s remained overwhelmingly focused on Europe and North America. But there can be no doubt that as the British economy grew, so its overseas trade expanded correspondingly, and this decade additionally witnessed the beginning of significant overseas investment, running at more than £4 million a year. The 1830s were also the first years when the number of migrants from Britain became significant: in 1832 more than 100,000 departed for destinations beyond Europe, and by 1840 over 700,000 Britons had emigrated, which was almost twice the number that had left between 1815 and 1830. More than 40 per cent went to the United States, but the majority went to British North America and the Antipodes. Here was the beginning of that greater Britannic world that would loom so large by the end of the nineteenth century. Finally, there were missionaries, in Africa, in the Antipodes and now also in China, where they became unprecedentedly active during this decade; and their view of empire did not necessarily align with that of traders, settlers or officialdom. Their aim was to convert the natives to Christianity and urge the British government to protect them from settler incursion and itinerant traders selling guns and alcohol.

In all these varied, different, pluralistic and uncoordinated ways the engagement of the peoples of the United Kingdom with the rest of the world significantly expanded and increased during those years of Whig government, as evidenced by a parliamentary committee of 1837, which noted that 'the situation of Great Britain brings her beyond any other power into communication with the uncivilized nations of the earth'; and, for good measure, it might have added, with 'civilized nations' as well. But official involvement, though undoubtedly increasing, diplomatically, militarily and imperially, and driven vigorously (if selectively) forward by Palmerston, was as ever only a part of it, as many different lobbies and interest groups in Britain, and ambitiously acquisitive men on the periphery, also pressed for annexation and forced the pace of imperial expansion. Either way, the result was that internationally as well as domestically the 1830s were the pivotal decade of the British nineteenth century. Thus regarded, they offered ample evidence of a vigorous and expanding society, whose national resources were increasing and whose global horizons were broadening; but that was far from being the whole picture. The fact that people were leaving the United Kingdom in unprecedented numbers – either to settle as migrants overseas or to sign up for military service within the empire – suggests that for many Britons conditions at home gave little cause for optimism or hope; moreover, the transportation of convicts to New South Wales and Van Diemen's Land would continue until 1853. The industrial strength of the United Kingdom may have been increasing, and its place in the world may have been extended and enhanced, but there was little sign these developments were yet improving the quality of life for ordinary people.

CULTURES OF CRISIS

Despite the economic vicissitudes, social unease, popular protests, cultural angst and religious anxieties that characterized the 1830s, there were also positive developments exemplified by the new construction taking place in central London near the southern end of Regent Street.

John Nash had put up a neoclassical building at the corner of Pall Mall and Waterloo Place for the United Services Club, which had been established in 1815, for senior officers of the army and navy. On Waterloo Place itself a column and statue were erected between 1832 and 1834 as a monument to the Duke of York, his tarnished earlier reputation having recently been forgotten; and in 1840 the laying out of Trafalgar Square was begun, to plans by Charles Barry, along with the installation of Nelson's Column to designs by William Railton. These building works were vivid and lasting reminders of the Revolutionary and Napoleonic Wars, but there were other developments in the same neighbourhood suggesting different cultural sensibilities and national priorities. In 1824 the Athenaeum had been founded as a meeting place 'for literary and scientific men and followers of the fine arts', and in 1830 its clubhouse, designed by the young Decimus Burton, was opened on the opposite side of Waterloo Place and Pall Mall from the United Services Club, in matching classical style. Between 1832 and 1838 the National Gallery was constructed on what would become the north side of Trafalgar Square, to designs by William Wilkie; it was located equidistant between the east and west ends of the great metropolis, in the hope that it would attract members of the working class as well as those who were better off. The consolidation of a London-based intelligentsia, combined with the provision of opportunities for the improvement and edification of ordinary people, seemed a world away from the agitation over the Reform Bill at the beginning of the 1830s or the Chartist demonstrations at the end.

Although they were most vividly in evidence in London, such developments were not confined to the capital. In the rapidly expanding industrial towns and cities across the United Kingdom, environmental degradation was offset by the construction of new buildings devoted to culture and enjoyment. In Birmingham a Town Hall was put up in the classical style to designs by Joseph Hansom, and was opened in 1834. Despite its name, the building's prime purpose was to serve as the home for the Birmingham Triennial Music Festival, which had been established in 1784, and also as a venue for political meetings (John Bright and Joseph Chamberlain would often speak there). Not to be outdone, Liverpool retaliated later in the decade by commissioning St George's Hall, which would be an even grander neoclassical edifice, housing a concert hall for the city's own music festivals, and also providing ample accommodation for the local assize courts. These years also witnessed the heyday of many local literary and philosophical societies, which had been founded in places such as Leeds, Liverpool, Manchester, Hull and Leicester, across the half-century from the 1780s, and which often housed libraries and hosted lectures and discussions on politics, sciences and the arts. These buildings and organizations were based in the centre of provincial towns. But as their environments became more degraded and polluted, many members of the middle class moved to settle in the new suburbs that were being developed by this time; and there, too, they established more voluntary societies, ranging from asylums for the care of the deaf, the dumb and the blind, to botanical gardens, full of rare and exotic species. The middle-class provincials who came to maturity in the quarter-century after Waterloo, and in such unprecedented numbers, were both genealogically and culturally the direct descendants of their late eighteenth-century Enlightenment forebears. They were also the people to whom the Whigs had been most eager (although not always able) to give the vote. They were often Evangelicals or Dissenters, concerned with issues such as slavery, and they would be an increasing cultural as well as political presence in the decades ahead.

The 1830s also witnessed new ways of thinking, about the earth, about humankind, and about scientific approaches to understanding the material world, much of it conveyed in the form of lectures to the 'Lit and Phil' societies. Humphry Davy's last work, *Consolations in Travel*, published posthumously in 1830, contemplated questions of permanence, immortality and change across vast reaches of time and space, as if anticipating the disruptive decade that would follow. In the same year the mathematician Charles Babbage produced *Reflections on the Decline of Science*, which denounced the Royal Society as an outwork of 'Old Corruption', urged the need for improved scientific education in Britain, and argued that the governing classes must embrace more rational modes of thought. By contrast, John Herschel's *Preliminary Discourse on the Study of Natural Philosophy*, which also appeared in 1830, took a very different viewpoint, outlining how the sciences could be used to define good character and appropriate ways of behaviour, and offering a quietly utopian vision of science and its public purposes. Four years later Mary Somerville published *On the Connexion of the Physical Sciences*, addressed to a broad general audience, which urged that the many recent advances in astronomy, experimental physics and chemistry should not result in the creation of separate disciplines, because they could be unified and synthesized mathematically. Between 1830 and 1833 Charles Lyell produced his *Principles of Geology* in three volumes, arguing that it was a sophisticated scientific subject which should be pursued without any reference to the Scriptures. And in 1835 the Edinburgh lawyer and lecturer George Combe published a cheap edition of *The Constitution of Man*, which sought to reformulate the very basis of scientific knowledge, by arguing that the mind depended for its working on the physical qualities of the brain.

These scientific writings resonated powerfully with the political perturbations of the 1830s, and some of them also contained potentially subversive implications for the biblical authority of Christianity. At the same time a new breed of self-appointed social investigators, such as James Kay, Peter Gaskell and Andrew Ure, produced a series of reports, often based on evidence drawn from Manchester, which drew attention to the increasing plight of the urban and industrial working classes. But the most significant and enduring critic of these deteriorated contemporary social conditions was the young Thomas Carlyle. In 1829 he had published 'Signs of the Times' in the *Edinburgh Review*, denouncing the new forces of industrialization and mechanization, and their accompanying ideologies of utilitarianism and laissez-faire, for together destroying human individuality and repressing the human spirit. Ten years later he published *Chartism*, where he broadened his assault on the 'mechanical society' that was coming into being. The new form of economic activity known as 'industrialism' might hold out the hope of improved general welfare, but this could scarcely be reconciled with the dramatic degradation in the circumstances of the urban poor that had recently taken place and was still going on. This in turn meant that what Carlyle termed 'the condition of England' was far from being healthy or hopeful, and the Chartist disturbances were but a symptom of a much deeper and more dangerous disease by which the country was increasingly afflicted. If the government did not act, he insisted, to recognize and ameliorate the plight of the working classes, there might well be a revolution. What was most urgently needed, he concluded, was a new and 'real' aristocracy (which bore some resemblance to Coleridge's clerisy), to lead the people and the nation through these disruptive changes and wrenching vicissitudes and to a better, more settled and more humane future.

It was not only the social investigators and cultural critics who were anxious and pessimistic during these years. Many Tories remained enraged that their own leadership had betrayed them over Catholic Emancipation, and they were equally annoyed by the measures passed by the Whigs which had further undermined the privileged and exclusive position of the Church of England. They regarded the cholera epidemic, which reached London from India via Hamburg in February 1832, and in which thousands of people perished, as God's wrathful judgement on Catholic Emancipation. And they viewed the destruction of most of the old Houses of Parliament by fire in 1834 (which was vividly depicted by Turner) as another act of divine vengeance for the passing of the Great Reform Act. Two young, ambitious politicians, who would become lifelong rivals, were at this time agreed in their loathing of what the Whigs had wrought. Late in 1835 the would-be MP Benjamin Disraeli published *A Vindication of the English Constitution*, in which he denounced the Whig government, liberal utilitarians and all reforms based on mistaken abstract principles, expressed his disdain for ideas of popular sovereignty and the notion of universal suffrage, and reasserted the importance of tradition, precedent and aristocracy, and the timeless Englishness of organic community. Three years later the young William Ewart Gladstone, already acclaimed as 'the rising hope of those stern and unbending Tories', published his first book, *The State in its Relations to the Church*, in which he urged, in vain defiance of a decade's legislation enacted by both parties, that the British State and the Anglican Church ought to be co-equal and co-extensive, the Church hallowing the State, and the State supporting the Church.

This reassertion of traditional constitutional arrangements and ecclesiastical structures, in protest against the secular and industrializing trends of the times, was widespread during the 1830s. On Bastille Day 1833, John Keble preached a sermon in the University Church in Oxford on 'National Apostasy', protesting against the 'treason' of the Whig government in passing the Irish Church Temporalities Act; this marked the beginning of what would become known as the Oxford Movement. Keble's views were shared by John Henry Newman, the Vicar of the University Church, and E. B. Pusey, the Regius Professor of Hebrew.

In a series of pamphlets entitled *Tracts for the Times* and published between 1833 and 1841, Keble, Newman and Pusey denounced the current idea that the Church of England was a mere theological convenience, insisting instead that it was a divinely inspired repository of doctrine, wisdom and tradition, with which governments interfered at their peril. A second and more secular version of the same alienated impulse was the 'Young England Movement', led by aristocrats such as George Smythe and Lord John Manners, who were much indebted to the novels of Sir Walter Scott and Sir Kenelm Digby's *The Broadstone of Honour* (1822). They, too, opposed radicals, manufacturers, industry and utilitarianism, and thought the best hope for the future lay in a revived paternal aristocracy allying itself with the workers in defence of traditional, deferential, organic communities. Their medieval revivalism reached its climax in the Tournament held at Eglinton Castle, Ayrshire, in 1839, which re-enacted the chivalric deeds of Sir Walter Scott's romances. Ten thousand people attended, to watch knights in armour jousting; but it rained torrentially, and there was widespread criticism of such an expensive, self-indulgent fantasy at a time of economic recession, high unemployment and popular discontent.

This hostile reaction to the modernizing tendencies of the age found its most creative and enduring form in the Gothic Revival. In 1831 Thomas Love Peacock published a novel, *Crotchet Castle*, whose eponymous, pseudo-medieval dwelling was inhabited by a Mr Chainmail, who believed that everything had been going downhill since the twelfth century. Mr Chainmail's fictional views were shared in the real world by a young architect of almost demented genius named Augustus Welby Pugin, who converted to Roman Catholicism in 1834 and was a passionate believer in Gothic art, because it embodied a more wholesome view of society than the conventional wisdoms of these iconoclastic times. Two years later he published *Contrasts*, the architectural manifesto for medieval revivalism, in which he praised the 'Noble Edifices of the Middle Ages' as being far superior to the 'Buildings of the Present Day'. Medieval towns, he insisted, were characterized by church spires, hospitals making benevolent provision for the poor, and water running pure and free from Gothic fountains; whereas the new industrial towns were blighted by chimneys, factories, pollution and the cruel and heartless workhouse. Pugin's great opportunity to put these theories into practice came with the decision to construct a new legislature in the Gothic style, partly to harmonize with Westminster Abbey and Westminster Hall, and partly because it was deemed to be the English style par excellence. The plan of the new Palace of Westminster was drawn up by Charles Barry, but Pugin was responsible for the decoration and ornament. The result was a faux-medieval fantasy, embodying a deeply conservative vision of the constitution and the social order: a plain and unadorned House of Commons, a splendidly ornamented House of Lords, and spectacularly ornate royal apartments. This was 'Young England' in masonry and mortar – and not quite the world that the proponents of the Great Reform Act had intended to bring into being.

5
The 'Hungry' Forties, 1841–48

By the late 1830s the momentum of political reform was slowing down, running out of steam, coming off the rails, and in danger of hitting the buffers (all novel but apt metaphors, given the beginnings of what would soon become the railway age and the many accidents and fatalities in these early years). The Whigs had been unable to maintain the large Commons majority they had won at the general election at the end of 1832, and the subsequent resignations of Lord Stanley, Lord Ripon, Sir James Graham and Viscount Althorp had severely weakened the cabinet. Moreover, Lord Melbourne possessed fewer leadership skills and qualities than Grey, and despite the recent growth in constituency and central organization, it became increasingly difficult for the Whigs to maintain party discipline in parliament. As a result, the political initiative gradually but inexorably began to pass from the government to the opposition, and Peel's 'Tamworth Manifesto', produced for the general election of 1835, was of undoubted and innovative significance. It put forward a programme for the whole electorate, and recognized that politics was becoming increasingly national rather than merely local, and that it would in future be as much concerned with legislation as with governing. The manifesto also signalled a significant shift in attitudes by the party leadership, away from the days of Wellington's constitutional intransigence, by recognizing the strength of the national feeling for change, and by promising that when the next Tory government gained power, it would embark on 'a careful review of institutions, civil and ecclesiastical, undertaken in a friendly temper, combining, with the firm maintenance of established rights, the correction of proved abuses and the redress of grievances'. As such, the leadership of what was increasingly being called the Conservative

Party declared itself in favour of preserving established structures of government and authority, while also recognizing that there was a need for sensible yet limited reform.

By such means, Peel set out to reposition the Conservatives, rejecting the obscurantism and disenchantment of Young England, the Oxford Movement and the Gothic Revival, and accepting that the spirit of the age was that of improvement and reform, to which the party must adjust and adapt. To be sure, his brief, minority administration of early 1835 had owed its existence to little more than a royal whim, and had not lasted. But although the Whigs returned to power, the Tories had made gains of approximately one hundred seats at the ensuing general election, which vindicated Peel's espousal of moderate, responsible policies. For the remainder of the decade he presented the Conservatives as a disinterested, patriotic party, willing to support the Whig government when it undertook sensible, moderate reforms, and it was only with Tory support, especially in the Lords, that Melbourne and his colleagues were able to carry their Irish legislation. After the election of 1837 the Whigs remained the largest party, but the Conservatives had made significant additional gains: the government's majority was reduced to less than fifty, and for the first time since 1832 the Tories had won a preponderance of MPs sitting for English constituencies. Thereafter, Melbourne's parliamentary position weakened still further, and in May 1839 he survived a Commons motion of no confidence by only four votes. He resigned, to the dismay and distress of the young Queen Victoria, who had formed an infatuated attachment to her first prime minister. She reluctantly sent for Peel, but refused to accept his suggestion that she should replace her ladies of the bedchamber with women who were more sympathetic to the Tory point of view. Rebuffed by his sovereign, Peel abandoned his attempt to form a government and Melbourne returned to power. But in June 1841 he was defeated by one vote in the Commons, and in the general election that followed the Conservatives were returned with a seemingly secure majority of over seventy seats.

In the short run at least, Peel's huge victory at the polls was a triumph for his strategy of moderate Conservatism and discerning support for the Whigs' reforms: to many, indeed, he had appeared more altruistic than self-interested while in opposition, and more statesmanlike than partisan in his political conduct. Having established himself as a prime minister in waiting, Peel was now the prime minister in power, and by some criteria at least, this was a propitious time to have won the keys to 10 Downing Street so convincingly. Across the 1840s as a whole the economy of the United Kingdom, and in particular its expanding manufacturing and mercantile sector, performed very well. The number of jobs in the cotton industry increased from 260,000 in 1838 to 330,000 in 1850, which was by almost one-third. During the same period registered British shipping rose from 2.8 million tons to 3.6 million. The 'railway mania' reached its zenith in the mid-1840s, and between 1841 and 1851 annual gross national product grew from £452 million to £523 million. No other nation on the globe could compete or compare with such an extraordinarily vigorous and innovative economic performance. Yet these figures must be treated with considerable caution, for the 1840s look much better to historians in retrospect than they did to contemporaries in prospect. Indeed, it can be argued that they were the most disturbed decade of the nineteenth century, economically and socially, politically and ideologically, domestically and internationally. Across the whole of the United Kingdom distress and despair, anxiety and anger, were the mood and the modes of the times. Ireland endured a catastrophic famine, which was the greatest natural demographic disaster ever to befall the United Kingdom, while the Chartist revival and the well-orchestrated agitation against the Corn Laws presented the greatest threat to the established order in Britain since the riots over the Reform Bill. From this perspective the 1841 general election was not so much a good one for the Tories to win, but an even better one for the Whigs to lose.

THE CONDITION OF THE COUNTRY

The years from 1837 to 1843 were a period of severe and unrelenting depression that was greater and deeper than the earlier economic downturns that had taken place between Waterloo and Peterloo, or during the crisis that lasted from the carrying of Catholic Emancipation to the passing of the Great Reform Act. Indeed, poverty and unemployment were so pervasive that there were many who believed the industrializing and urbanizing United Kingdom was not so much pioneering a hopeful and optimistic path to the future, built around progress and improvement, but instead was rapidly heading in the wrong direction, and that it might even be on a downward path to complete national self-destruction. Thomas Carlyle had repeatedly put forward this view during the 1830s, and there were good reasons why he continued to adhere to it during the decade that followed. In 1843 he published his latest thoughts on the systemic social and economic crisis by which it seemed to him the country had been blighted and battered, in a book entitled *Past and Present*, where he again denounced the unfettered working of free-market capitalism, the mechanization and degradation of labour, the exploitation of the powerless by the powerful, and the low wages and pervasive poverty. Small wonder, he thought, that there was unprecedented protest and widespread popular discontent. 'The Condition of England,' he observed (and he could, with equal plausibility, have referred to the British Isles as a whole), 'on which many pamphlets are now in preparation, and many thoughts unpublished are going on in every reflective head, is justly regarded as one of the most ominous and withal, one of the strangest, ever seen in this world.'

The adverse effects of industrialization and urbanization, combined with rapid population growth, were more negative in their impact on the standard of living and quality of life of ordinary people during the 1840s than they had been before. Between 1841 and 1851 the population of England, Wales and Scotland rose from 18.5 million to 20.8 million, but even the impressive overall growth of the British economy across that decade was insufficient to keep these extra mouths at an even meagre level of economic comfort. In 1840 the per capita consumption of sugar was only half what it had been forty years previously, while the intake of tea and bread probably fell as well. Indeed, the 1840s may have been the worst decade for life expectancy since the Black Death in those parishes that were undergoing urbanization and industrialization, and where those in search of work were living in squalid and overcrowded conditions, being easy prey to such infectious diseases as cholera, typhus, tuberculosis and dysentery. So whereas in 1841 the national average life expectancy was forty years, it was only twenty-seven in Manchester and in Liverpool just twenty-eight. The Liverpool averages also masked significant class differentials: the life expectancy of members of the professional classes was thirty-five, but for tradesmen it was only twenty-two, while for mechanics, servants and labourers it was even less. These are difficult numbers to grasp. All large industrial towns had brothels, gin shops, alehouses, thieves' dens, and filthy courts, rookeries, communal privies, cesspools, middens, dung heaps and ill-paved streets crawling with wild dogs, feral cats and rats. Most of them were noisy, smelly, smoky and polluted; they were dark at night-time, freezing cold in winter, and flea- and lice-ridden at all times.

Thus regarded, the condition of the United Kingdom was far from sound or good during the 1840s. The expansion of the economy may have meant the standard of living of the growing numbers of people declined less than it would have done had there been no such industrial development and transformation; but it declined even so, which was why Carlyle was far from being alone in concluding that industrialization and urbanization seemed more like a curse than a blessing. To these adverse trends were added the dramatic worsening of living conditions, as the years from 1837 to 1843 witnessed a decline in manufacturing output, a falling off in exports, and a succession of bad harvests, when the price of wheat stayed consistently at near or above 77s a quarter. In addition, there was another severe financial crisis in 1839 that further deepened the depression. Railway construction slowed significantly, few new lines were promoted between 1839 and 1843, and across Britain the year 1842 was the most gloomy and distressed in the whole of the nineteenth century. Factory workers in the Lancashire and Cheshire cotton industries were either on short time with lowered wages or were totally unemployed. In Stockport more than half the master manufacturers were said to have failed. In Leeds the number of families on poor relief rose from approximately 2,000 at the end of 1838 to over 4,000 at the beginning of 1842. 'Four years of unrelieved depression,' one local lamented, 'have brought this borough into a state more lamentable than the oldest inhabitants can remember.' And from Paisley in Scotland a memorial was sent to the government early in 1842 urging that 17,000 citizens were 'enduring a gradual starvation with exhausted resources, and manifestly impaired health and strength on the part of the people, and with failing funds on the side of the relief committee'.

Under these circumstances it was scarcely surprising that many of the unemployed, impoverished and discontented turned to Chartism again. In 1840, and following the example of the Universal Suffrage Central Committee that had been established in Scotland, there were significant signs of renewed activity in the Chartist heartlands across England. The aim was to re-establish the movement on a sounder organizational basis, and to this end a great conference was held in Manchester in July, at which a National Charter Association was established. After the failed attempt to receive a hearing from parliament in 1839, the Chartists, while retaining their six demands, adopted a new strategy of peaceful persuasion, which involved lectures, pamphlets, fund-raising and the co-ordination of many local branches and bodies. In the summer of 1841 the Irish Protestant lawyer and *Northern Star* publisher Fergus O'Connor was released from gaol, whereupon he headed the effort to produce a second and more widely supported petition, drafted by the National Charter Association, which was eventually presented to parliament in May 1842, with the appended signatures, it was claimed, of more than three million people. London was at the top of the list (with 200,000); then came such major industrial centres as Manchester (99,000), Newcastle (92,000), Glasgow (78,000), Bradford (45,000), Birmingham (43,000) and Leeds (41,000); and there was even support from such smaller towns as Brighton (12,000), Cheltenham (11,000) and Worcester (10,000). But again, it was to no avail, as the petition was rejected by 287 votes to 59 in the Commons, an even poorer result than in July 1839. Macaulay might have supported the Great Reform Act, but he deplored the demand for universal suffrage as being 'fatal to the purposes for which government exists' and 'utterly incompatible with the existence of civilization'. Even the radical John Arthur Roebuck, first elected in 1832, denounced O'Connor as 'a malignant and cowardly demagogue'.

In retrospect, the rejection by the Commons of the second Chartist petition was the beginning of the end for the movement; but that was not how it appeared at the time, for economic conditions continued to worsen during the second half of 1841, resulting in the intensification of unrest and disorder, in which many Chartist leaders were involved but for which they were rarely responsible. By the early summer of 1842 the combination of high unemployment, low wages and widespread distress had brought the industrial districts of England and Scotland to a state bordering on desperation. In Lancashire and Yorkshire gangs of the unemployed, often armed with sticks and iron bars, demanded relief, attacked shops and clashed with police. In Staffordshire attempted reductions in wages by colliery owners led to strikes and lockouts, and in Leicestershire there were walkouts by colliers, stocking-makers and glove-makers. The spread of unrest and discontent was such that by August 1842 the whole central industrial area of England, bounded by Leeds and Preston in the north, and Leicester and Birmingham in the south, was in a state of upheaval, and there were also strikes and disturbances on Tyneside, in south Wales, and in the mining and industrial districts of Scotland. Across the length and breadth of Britain, prisons, town halls, police stations, law courts, railway stations, workhouses, shops and bakeries were attacked and ransacked, along with the private dwellings of magistrates and clergy. In November the diarist Charles Greville summed up the issues and the problems: there was, he recorded, 'an immense and continually increasing population', but also 'deep distress and privation, no adequate demand for labour [and] no demand for anything', and the result was 'universal alarm, disquietude and discontent'. Indeed, he had 'never seen in the course of my life so serious a state of things as that which now stares us in the face'. What a humiliation, he concluded, for a nation that 'according to our own ideas' is 'not only the most free and powerful, but the most moral and [with] the wisest people in the world'.

In many ways Greville's anxieties echoed those of Thomas Carlyle: the United Kingdom might, by various criteria, be the most successful, the most innovative and the richest nation on the globe, but the widespread existence of so much misery, poverty and discontent suggested that it was, indeed, heading along the wrong tracks – perhaps, even, along the wrong railway tracks. Indeed, a year before Carlyle had published *Past and Present*, Edwin Chadwick had produced his *Report on the Sanitary Condition of the Labouring Population of Great Britain*, in which he compiled a grim record of undrained streets, impure water and crowded, unventilated dwellings, all of which he saw as being closely related and intimately connected to the widespread prevalence of crime, disease and immorality. Two years later the young Friedrich Engels depicted Manchester, where he himself superintended the family-owned cotton mills, as a place so deeply sundered and so antagonistically divided between the profit-obsessed factory owners and the downtrodden and exploited workers that the outcome must eventually be some sort of proletarian revolution. In 1845 Benjamin Disraeli published *Sybil, or The Two Nations*, which was avowedly a 'condition of England' novel, in which he described the whole of Britain as being deeply sundered, like Engels' Manchester, between 'the rich' and 'the poor', who were 'as ignorant of each other's habits, thoughts, and feelings as if they were dwellers in different zones, or inhabitants of different planets'. Nearly a decade later, but harking back to these earlier years of misery and disenchantment, Charles Dickens would produce *Hard Times* (1854), its title making plain his hostile attitude to industrialization and his sympathy for the suffering workers, compelled to labour – or to be unemployed – in places such as 'Coketown', a fictional creation, modelled on Manchester and Preston, 'of unnatural red and black, like the face of a painted savage'.

In some senses all these works misled or exaggerated. Chadwick was so eager to make the case for sanitary reform that he deliberately withheld evidence that might have modified his picture of unrelieved gloom and suffering, by focusing on conditions in the years of economic downturn. Engels was in error in supposing the social structure of Manchester was so deeply divided, and also in thinking that the rest of Britain was in the process of becoming like Manchester. (Moreover, his book would not be published in English for forty years, so it could scarcely have influenced contemporary perceptions or discussions.) Disraeli likewise erred in depicting Britain as being 'two nations', between whom 'there is no intercourse and no sympathy'. For in reality the social structure was much more complex and varied, even in the new industrial towns and cities, being an elaborately graded hierarchy, extending from the extremely rich at the very top to the desperately poor at the very bottom, but with many different groups and occupations and levels of income in between. And in suggesting that all the new industrial communities and cities were the same, by creating an amalgam of them in 'Coketown', Dickens also misled; for towns such as Manchester and Birmingham were very different in their economic and social structures as, again, were the spinning and weaving towns of Lancashire, the wool towns of Yorkshire, the port towns of Liverpool and Glasgow, and the mining villages scattered across the industrial north. But while these authors and many other contemporary commentators did not fully understand what was happening to their country during the 1840s, or appreciate what the more positive longer-term implications might be, they were not entirely wrong. Indeed, in their recognition that there was widespread misery, suffering, deprivation, poverty and distress, and on an unprecedented scale, they were undoubtedly correct.

But it was not only among the labouring (or the unemployed) classes of the United Kingdom that disenchantment and alienation reached such high levels during the early 1840s. Many members of the business and manufacturing classes were equally disappointed and disillusioned, in their case because (and as its authors had indeed intended) the Great Reform Act seemed to have entrenched, rather than undermined, the continued influence of the aristocracy in and on government. And the most flagrant symbol and provocative sustainer of that entrenched and continued influence were the Corn Laws, which had been passed in 1815 and modified thirteen years later. The purpose of that legislation had been to offer privileged treatment to the landed classes and the agricultural interest; and it was scarcely coincidence that these were the very groups who were dominant in both houses of parliament. In essence the Corn Laws protected British agriculture from foreign competition, by banning imports of wheat until the domestic price was so high as to threaten famine at home, as came close to happening during the succession of bad harvests between 1837 and 1842. Thus regarded, the Corn Laws became the symbol of what was deplored as 'aristocratic misrule': a piece of blatantly self-interested legislation, keeping agricultural prices and rents at artificially high levels. This was at odds with the alternative and more altruistic doctrine of aristocratic paternalism, and it also ran counter to those who believed that the United Kingdom's future lay in exporting its manufactured goods and importing its raw materials and foodstuffs from overseas. For businessmen and entrepreneurs, the repeal of the Corn Laws held out the prospect that their workers would have cheaper food, and so would be less likely to demand higher wages. And lower prices and a sure supply of imported wheat would mean an end to famine, cheaper food and more regular employment for the labouring classes.

But the ensuing campaign against the Corn Laws was based on more than just a powerful sense of material self-interest. For free trade, which it was believed the repeal of the Corn Laws would usher in, was not just an economic doctrine: it was also a quasi-religious article of faith. If, so this argument ran, all the nations of the world abolished their tariffs, following the example pioneered by the United Kingdom, the result would be not only unhindered worldwide commerce but also international amity, global fellowship and universal peace. All that was standing in the way of bringing about this New Jerusalem was the ignorant self-interest of the landlords: 'the bread-taxing oligarchy, unprincipled, unfeeling, rapacious and plundering', as one contemporary pamphleteer described them. To promote these great causes, the Anti-Corn Law League had been established in Manchester in 1839, and it soon became the most well-organized and well-financed pressure group yet to appear in nineteenth-century Britain. In Manchester the League's supporters constructed the Free Trade Hall as their headquarters and as a great venue for public meetings. By 1845 there were 225 affiliated societies in England (the majority in London, Lancashire and Yorkshire), and another thirty-five in Scotland (mainly along the industrialized and urbanized Forth-Clyde axis). Three newspapers were founded to bombard and engage a national readership, *The League*, the *Anti-Bread-Tax Circular* and the *Anti-Corn-Law Circular*, and publishers took innovative advantage of the new penny post: in a single week during February 1843 more than nine million tracts were mailed out. From one perspective the League pioneered a new politics of theatre, with its money-raising bazaars, fancy-dress balls, tea parties and banquets. But it was also a religious crusade, supported by Anglicans and Nonconformists alike, as exemplified by a meeting held in August 1841, when seven hundred ministers proclaimed the cause of repeal to be 'the politics of the gospel'.

The two most prominent and influential leaders of the Anti-Corn Law League were the radical MPs Richard Cobden and John Bright, who were elected to the House of Commons respectively in 1841 and 1843. (Such men also preferred the increasingly used appellation 'Liberal' to distinguish them from the patrician Whigs.) Cobden was a Manchester merchant and calico-printer and an Anglican, Bright a Quaker mill owner from nearby Rochdale. Both men were deeply religious, inspiringly eloquent in the Commons and at public meetings, opposed to what they condemned as the aristocratic 'monopoly', and passionately committed to the cause of world peace. Under their leadership, the Anti-Corn Law League represented a much more serious threat to the established order than Chartism: the League was well financed, superbly organized, a unified national movement and a quasi-religious crusade, and it focused relentlessly on the single issue on which landowners and the agricultural interest were particularly vulnerable. As such, the League represented the biggest political threat to the established order since the agitation over the Reform Bill a decade before;

and although attempts to bring the League and the Chartists together made little progress, Peel and his colleagues became increasingly concerned that the League's itinerant orators were gaining support for their cause among the working (and unemployed) classes. So alarmed was the government at the prospect of such a broad-based campaign that in December 1842 two members of Peel's cabinet published (and with his approval) an article in the *Quarterly Review* entitled 'Anti-Corn Law Agitation', in which they denounced the League as subversive, immoral, irresponsible and unconstitutional. This was a tacit admission that it represented a much greater threat than Chartism, not least because of what seemed to be its malign influence on the labouring classes, who were already disaffected enough and needed no additional goading into protest, agitation or subversion.

To make matters worse, the hardships endured by the industrial workers in England and parts of Scotland and Wales would be as nothing compared to the travails suffered later in the decade by the non-industrial workers of Ireland. As in Britain, the Irish population had grown with unprecedented rapidity since the last quarter of the eighteenth century, rising from approximately five million in 1800 to almost seven million in 1821, and to more than eight million twenty years later. Even though downgraded after the Act of Union, Dublin remained one of the major cities of Europe, Belfast and Cork were also substantial towns, and there was a significant textile industry, largely organized on the domestic system of production. But after 1815, textiles had gone into decline, in the face of factory-based competition from Scotland and the north of England, and as arable prices fell, Irish agriculture shifted from the production of food crops to the rearing of cattle and sheep for export. Over one million people left Ireland between 1815 and 1845, and at least one-third of those who stayed were wholly dependent on the single crop that was still grown in large quantities, namely the potato. This meant that the lives of many Irish men, women and children were constantly at risk. Potatoes could not be stored for more than twelve months, so the surplus of one crop in one year could barely make up for the scarcity in another year; and if bad weather or disease meant the crop failed, there would be no alternative source of food for many millions. By 1830 localized failures of the potato harvest in Ireland had become dangerously frequent; but between 1845 and 1848 potato blight, caused by a fungus for which the scientists of the day could find no remedy, ruined almost the entire crop.

The resulting imbalance between numbers and resources, too many people and insufficient food, was a terrible Malthusian catastrophe: 'a famine of the thirteenth century acting upon a population of the nineteenth', in the words of Lord John Russell.

The consequences of the Great Famine were even more long-lasting than were those of the deprivations that had been endured by the British working classes earlier in the same decade, and they were also geographically much more widespread and dispersed. Almost one million people, or one-eighth of the entire Irish population, died directly or indirectly of starvation and malnourishment: the young were struck down by dysentery, while the elderly and the infirm were carried off by typhus. As Russell's apocalyptic words suggested, this was a demographic disaster on a scale that was unparalleled anywhere else in contemporary Europe. And for those who were lucky enough to survive, Ireland seemed to be a place where there was no future, and which was better left. During the ten years following 1845, at least one and a half million Irish sought refuge abroad. Many fled to Britain, where jobs building the railways awaited; while others emigrated to the new world, especially to the eastern seaboard of the United States, where they hoped a better and brighter future awaited. But they did not forget what they had suffered and endured: Boston would, later in the nineteenth century, become a bastion of support for the Fenians and those who favoured Irish Home Rule; and on into the twentieth entry, hostility to Britain would remain strong. As for those who remained in Ireland: for many the Great Famine provided further evidence that the Union with Britain had been a terrible mistake, and when it came to celebrating Queen Victoria's Diamond Jubilee in 1897, they would arrange counter-demonstrations with the slogan: 'Sixty Years: Starved to Death'. If the experiences of impoverished and unemployed workers on one side of the Irish Sea are added to those of the malnourished and dying and emigrating peasants on the other, then it is easy to see why their dismal and depressed decade would later go down in posterity as the 'Hungry Forties'.

PEEL IN POWER

From the autumn of 1841 until the summer of 1846 it was the responsibility of Sir Robert Peel and his Conservative colleagues to try to deal with the disaffected and impoverished working classes, the organized moral outrage of the Anti-Corn Law League, and the beginnings of the Great Famine in Ireland. Nor were these the only challenges they had to face, for the economic downturn that had begun in 1837 meant that, on taking office, Peel's government was also faced with a major crisis in the national finances. By 1841 there had been budget deficits for three years running, Peel's government inherited an accumulated shortfall of £7.5 million, and the predicted negative balance for 1842 was another £2.4 million. These were hard times – and hard facts – indeed and, as Cobden wrote to his brother soon after Peel took office, 'The aristocracy and people are gaping at him, wondering what he is going to do.' In the aftermath of the so-called 'Bedchamber Crisis', he was further hampered, at least initially, by the hostility of the queen; but nevertheless Peel thought himself more than up to the task. He was the son of a self-made Lancashire cotton manufacturer, but he was also a landed baronet who had been educated at Harrow and Christ Church, Oxford (where he took a double first in classics and mathematics), and this meant that Peel brought to the job of prime minister a fine intellect and a Christian conscience, a broader range of outlook and connection than was common among the ruling elite, and extensive experience of government going back to 1812. He had been brave and statesmanlike enough to change his mind over Catholic Emancipation, when it became clear that the cabinet to which he belonged must carry the measure; in his discerning support for what he regarded as the Whigs' more sensible reforms during the late 1830s, he had shown himself a strong believer in executive government; and he had successfully revived his party from the electoral battering it had received in 1832.

During the course of his administration Peel would show himself to be a creative statesman, perpetually searching for constructive solutions to the social problems and political challenges that confronted him, and doing so with such resourcefulness and success that his most famous protégé, Gladstone, would later opine that he was 'the best man of business who was ever prime minister'. Yet not everyone saw Peel in such a favourable or admiring light. Perhaps because of his relatively lowly social origins, he was shy and reserved, which meant he often appeared cold and haughty. Lord Ashley, who never got on with him, once described Peel as 'an iceberg with a slight thaw on the surface'. Daniel O'Connell, who was equally unsympathetic, famously compared Peel's smile to the gleam of the silver plate on a coffin lid. When addressing the Commons, the prime minister spent too much time proclaiming the purity of his motives, the rectitude of his conscience, the integrity of his character, and the high-mindedness of his actions, and such self-righteousness did not go down well with those to whom politics was a great game for power, place and preferment. Moreover, there were still many Tories, both in the Commons and the Lords, who could not forgive Peel for passing Catholic Emancipation, which they did not see as a brave piece of responsible and disinterested statesmanship, but as an act of treachery and betrayal, which meant he could never be wholly trusted again. For his part Peel took a dim view of the Tory Party in the Commons and the Lords. He thought many of its members were intellectually negligible and excessively partisan, whereas he himself harked back to the earlier days of Pitt and Liverpool when (or so it seemed in retrospect) governing was in the interests of the nation rather than of any particular party. In short, Peel was not so much a party leader as a leader around whom a party could gather – or, alternatively, a leader from whom and against whom it could disperse and divide.

Moreover, in 1841 the Tories had won a massive majority in the English counties and the smaller boroughs, but these were the constituencies where the voters were strongly in favour of maintaining the Corn Laws, as were the MPs whom they returned to the Commons. Indeed, Peel had recognized the importance of the Tory protectionist wing by appointing one of their foremost champions, the Duke of Buckingham, to be Lord Privy Seal. The rest of his cabinet was more moderate and centrist in its composition. Wellington was an inevitable member of any Conservative administration, but this time as a minister without portfolio, lending a tone rather than running a department, and always available to help steer contentious measures through the House of Lords, if needs be. There were three renegade Whigs, who had resigned over Irish church issues in 1834 and embraced Peelite conservatism by the end of the decade: Lord Ripon at the Board of Trade, Lord Stanley at the Colonial Office, and Sir James Graham at the Home Office. Lord Lyndhurst had already been Lord Chancellor under Canning, Robinson, Wellington and (briefly) Peel, and was now restored to that post again. Lord Aberdeen was Foreign Secretary, having previously held office under Wellington. Henry Goulburn, who had also been in Wellington's cabinet, was given the Exchequer (in which Peel would much interest himself). Benjamin Disraeli was not offered a post, and nor, given that he was a backbench MP of only four years' standing, was there any reason why he should have been; but in May 1843 Gladstone would succeed Ripon at the Board of Trade. Peel's cabinet thus contained not only two former prime ministers, namely Ripon and Wellington, but also, in Aberdeen, Stanley and Gladstone, three future premiers; and among many of his colleagues, especially the younger ones, notably Gladstone, Peel inspired a fervent, quasi-religious devotion.

Peel's immediate problems were the state of the national economy and its finances, both of which had worsened since the downturn that had begun in 1837; and the premier himself, rather than his chancellor, introduced the government's first budget early in 1842. His most significant proposal was the reinstatement of the income tax, which had previously been imposed as a temporary wartime expedient and had been abolished in 1817. It was only levied on incomes in excess of £150 a year, at the fixed rate of seven pence in the pound, and once again it was presented as a short-term measure. To resort to such direct taxation in peacetime was a bold and controversial step, but it enabled Peel to begin plugging the hole in the public finances. It also provided him with an alternative source of income that meant he could continue the work of cutting import duties initiated by the liberal Tories during the 1820s and carried further by the Whigs in the following decade. Accordingly, Peel reduced protective tariffs for some 750 of the remaining 1,200 items on which they were still levied, arguing that such reductions in indirect taxes would assist industry, expand trade, and improve the lot of the 'labouring classes of society' by lessening the cost of consumer goods. He also reduced the duties on wheat that had been set by the modified Corn Law of 1828, in the hope that this would end the high prices then prevailing, and enable many poor and hungry people to afford the cost of a loaf, while still allowing the agricultural interest to stay solvent. But although Peel carried his budget, many farmers and landowners regretted his undermining of agricultural protection. In the Commons eighty-five Conservatives rebelled against the budget, while the Duke of Buckingham, the self-styled 'farmers' friend', resigned from the cabinet after only four months in office, assuring the prime minister he would continue to support the government on everything – except any further modifications to the Corn Laws.

Peel regarded his first budget as merely the beginnings of his efforts to engage with the current and unsatisfactory condition, not just of England, but of the United Kingdom as a whole. By increasing the supply of cheaper wheat, he hoped to improve the standard of living of the labouring classes while at the same time enhancing their employment opportunities by cutting costs to manufacturers. Yet there was little else the government could realistically do to alleviate poverty and unemployment, and most of the running in terms of additional factory legislation was made by Lord Ashley, who was by this time well established as a philanthropist and social reformer, but was also a protectionist and no particular admirer of Peel's. Nevertheless, the government was sympathetic to imposing further legal restrictions on the hours that women and children should work, and in 1842 it passed the Mines Act, which prohibited girls and women, and also boys under the age of ten, from being forced to labour underground in collieries. Two years later a Factory Act was passed, further limiting the hours of work for children (aged between nine and thirteen) and women in industry; and to discourage evasion, it was further decreed that all employees should begin work at the same time each day, the first occasion when the hours of work of adult males were regulated by parliament. As the example of Lord Ashley suggests, on issues such as this the government was more likely to secure the support of agriculturalists and protectionists. Many members of the landed interest disliked businessmen and industrialists, with their strident demands for Corn Law repeal, and they were happy to retaliate by denouncing the money-grubbing capitalists as being in thrall to the doctrine of laissez-faire and selfishly indifferent to the sufferings of their employees, and also by supporting legislation that circumscribed their freedom of action as employers.

Between the passing of these two pieces of ameliorative legislation, Peel's government went through a difficult period, which in retrospect did not bode well for the future. In 1843 it was far from clear that the brave and innovative budget of the previous year would begin to restore the nation's finances, as there was yet another deficit of £2 million at the Exchequer, instead of the modest surplus of £500,000 that had been hoped for. In January Peel's secretary was killed by an assassin who mistook him for the prime minister, while in the Commons, Cobden blamed Peel personally for the disturbed state of the country, and he retaliated by denouncing the menacing tactics of the Anti-Corn Law

League. Later that year the League moved its headquarters from Manchester to London, with the aim of exerting greater influence on parliament and courting and cultivating voters in advance of the next general election. Meanwhile, the landowners and farmers, who had been alarmed by Peel's reduction in the duties levied on corn in his first budget, established their own organizations: the 'Anti-League' in the case of the former, and the Central Agricultural Protection Society in the case of the latter, headed respectively by the Duke of Richmond and the Duke of Buckingham. The result was an increasingly poisoned and antagonistic political climate, inside parliament and across the nation, as the tensions mounted between businessmen and manufacturers who wanted free trade, and landowners and farmers who wanted to retain protection. Many Tory MPs felt increasingly alienated from their leader, albeit for diametrically opposite reasons, either because he was insufficiently committed to free trade or because he was too much in favour of it. This parliamentary disenchantment was made vividly and ominously plain in the summer of 1844, when Peel's attempt to reduce the duties on imported sugar was initially defeated in the Commons by just such an alliance between free traders, who wanted them abolished, and protectionists, who did not want them reduced at all.

In the end Peel got his way, but only at the price of alienating many of his disaffected followers still further; they resented his dismissive and authoritarian attitudes, while he expected more loyalty than many displayed. Amidst these (partly self-inflicted) political travails, the encouraging news for the government was that by the mid-1840s the economy in Britain (though not in Ireland) was emerging from the long and deep recession. This was partly because trade and manufacturing revived, and exports recovered well in 1842–43, but it was also because of a second and greater boom in railway building. The late 1830s and early 1840s had witnessed scarcely any new construction; but 442 Railway Acts would be passed between 1844 and 1847, and more than 2,000 miles of track opened as a result. During these peak years railway building absorbed half the private investment of the country, and paid a £16 million wage bill for 250,000 navvies (many of them immigrants from famine-starved Ireland), while new lines were built linking the peripheries of Britain to the metropolitan centre. Robert Stephenson constructed the Britannia Bridge, crossing the Menai Strait in Wales, connecting London and Anglesey by train (and Anglesey and

Dublin by sea), and he also built the Royal Border Bridge at Berwick, which linked London with Edinburgh on the east coast main line. Joseph Locke oversaw the creation of a new northern railway network connecting Carlisle with Glasgow and Edinburgh. And Isambard Kingdom Brunel laid down an atmospheric railway in Devon and Cornwall, linking the West Country to the Great Western Railway he had earlier built between London, Bristol and Exeter (and which Turner memorably depicted in *Rain, Steam and Speed*, first exhibited at the Royal Academy in 1844). By the late 1840s it was possible to journey by rail from London to Bristol in less than four hours and from London to Manchester in just over eight, and more than forty million passengers were travelling each year on the trains.

As a result of this significant improvement in the economy, the public finances of the United Kingdom showed a healthy surplus of £4 million in 1844: both customs and income tax had yielded more than expected, while expenditure had been less than forecast. In this more buoyant climate, and with his budgetary policies vindicated, Peel sought to implement further financial measures that he hoped would go some way to preventing any recurrence of the sort of violent and damaging fluctuations which the economy had witnessed from the late 1820s to the early 1840s. The Bank Charter Act of 1844 rounded out his earlier legislation of 1819 by strengthening the Bank of England's adherence to the gold standard, by underscoring its national dominance in the issuing of currency, and by stabilizing banking credit by enforcing a stricter relationship between paper money and gold reserves. Peel's aim was to prevent any repetition of the 'reckless speculation' in railway shares that had characterized the mid-1830s, and the Companies Act, which was also passed in 1844, was a companion piece of legislation, with a similar regulatory objective. Henceforward, all companies had to be registered, and they were obliged to publish their prospectuses and balance sheets. But the railway companies had been individually sanctioned by parliament, which meant they were exempt from this legislation. Nevertheless, since 1839 they had been subject to a minimum amount of oversight from the Board of Trade, which appointed inspectors to inquire into the causes of accidents, and to recommend ways of avoiding them in future. During his brief tenure as President of the Board of Trade, Gladstone passed the Railway Regulation Act of 1844, which laid down minimum standards for third-class passenger travel, and gave parliament the power to bring railway companies under state control if they made too much money (though since they never did, this provision would never in fact be invoked).

With the economy picking up and the public finances mending, Peel turned his attention to Ireland, where the picture was much less encouraging. Despite Catholic Emancipation, which he had carried at such great personal and political cost, Peel recognized that most of the Irish remained unreconciled to the Union, and he sought ways to lessen this continued hostility. His solution was to give additional financial backing to the Catholic seminary at Maynooth, which trained the majority of Irish priests, in the hope of gaining their approbation and, through them, that of the Irish people more broadly. Accordingly, in 1845 Peel proposed an increase in the annual government grant from £9,000 to £26,000, with an additional one-off allocation of £30,000 for building works. The principle of state support for the seminary had been accepted since the late eighteenth century, and these were scarcely large sums; but anti-Catholic feeling had hardened in Britain since emancipation, and Peel's proposals were greeted with widespread outrage and dismay. An Anti-Maynooth Committee was set up to organize the agitation out of doors, and over 10,000 petitions, containing more than one and a quarter million signatures, poured into Westminster. In parliament many Tories denounced Peel for yet again betraying the Protestant cause, but he insisted the government must do the right thing, 'without reference to the past and without too much regard for what party considerations must claim from them'. Provoked rather than mollified by such words, nearly half of the Conservative MPs voted against the measure, which was only carried with Whig support. Despite his admiration for his leader, Gladstone, who still believed in the righteousness and necessity of a state church, resigned in protest from the Board of Trade.

By this time Peel was in no mood to give way to the increasingly militant Protestants and disaffected protectionists in his party, and despite the trouble they had made over the sugar duties and Maynooth, he determined to make further advances in the direction of free trade. In 1845 he again introduced the government's budget himself, proposing that the income tax be (temporarily) renewed for another three years, and that such additional revenue would make possible a further reduction in tariffs and indirect taxes on consumer goods. The duties on sugar would be lowered again, those on most of the remaining imported articles would be removed (among them cotton and other industrial raw materials), while export duties on all British goods would be completely abolished. This budget virtually finished the task of restoring the public finances. It also helped rebalance them by shifting income away from indirect towards direct taxation, and it further stimulated the economic recovery already under way. Peel duly carried his proposals, and they won him the plaudits of the merchants and the manufacturers, and of the labouring classes who would soon be enjoying cheaper imported food. But the Tory agriculturalists were far from amused, because his budget had done nothing for the farming sector. As Peel ironically observed in May 1845, addressing those who were allegedly his followers: 'We have reduced protection to agriculture, and tried to lay the foundation of peace in Ireland; and these are offences for which nothing can atone.' But despite the growing distrust and disappointment felt by many in his own party, Peel was by this time a figure of unrivalled command who enjoyed widespread popularity in the country; and under the tutelage of her husband, Prince Albert, whom she had married in 1840, even Queen Victoria had by this time become an ardent admirer.

At some point before the middle of 1845 Peel privately decided the Corn Laws must be completely repealed. Their continued existence (albeit with the duties much reduced since 1842) was the last great barrier to the wholesale achievement of free trade, and repeal would conclusively demonstrate the government's deep concern for the labouring classes (who would get cheaper bread) and for the manufacturing classes (who would face less pressure from their employees for higher wages). Peel also believed that he had a better understanding of the long-term interests of the landed and agricultural sectors than many landowners and farmers themselves. As an 'improving' landlord himself, he was convinced that repeal would not ruin British agriculture, by subjecting it to the lower-priced produce of international competitors, but would rather stimulate it to become more efficient. And as a firm believer in the need for a responsible and respected aristocracy, Peel had concluded that repeal would be an essentially preserving measure (as Grey had earlier insisted in the case of the Reform Bill). It would remove the last legitimate popular grievance against the landowners, namely that the continued existence of the Corn Laws demonstrated that they were self-interested rather than public-spirited. As Peel saw it, the repeal of the Corn Laws would be a great and imaginative act of social amelioration and national reconciliation, and his preferred plan was to carry the measure just before the next general election, which meant no later than the early months of 1848. But then events in Ireland took an unexpectedly adverse turn, as the unseasonable weather, combined with a poor harvest and the sudden appearance of the devastating blight, meant that by November 1845 it was clear the potato crop was ruined, that the same might well be true next year, and that a great social calamity and demographic disaster were about to unfold. From Ireland there came urgent pleas for government aid and assistance, and in Britain the Anti-Corn Law League, sensing this disaster might be their opportunity, demanded the opening of all ports for the free entry of wheat and grain.

It was at this point that Peel decided the Corn Laws must be repealed immediately, so as to make available additional supplies of wheat for the Irish, and thereafter he always gave the need to avert the impending famine as his reason for his decision. Yet in some ways this was more a pretext than an explanation: Peel could have suspended the Corn Laws temporarily rather than repealed them completely; in the short run, repeal did little to alleviate the increasingly dire situation in Ireland; and in any case it would be gradually phased in over three years. The Potato Famine was thus more the occasion for than the cause of Peel's decision; but there was more to it than that. Faced with such a developing and unprecedented calamity, the government had to be seen to be doing *something*, but the options were limited. Moreover, if Peel had only temporarily set aside the Corn Laws, he would have been obliged to reinstate them, which would undoubtedly have provoked even more agitation by the League, and might have meant that the next general election would be fought on that single issue alone. This would have deepened and intensified the social and political divisions in the country, which was the last thing Peel wanted. He set about persuading his cabinet that repeal must be immediately carried, but by early December 1845 Lord Stanley had resigned as Colonial Secretary, and with backbench hostility bound to be even greater than over the sugar duties or Maynooth, it seemed impossible for the government to carry the measure. Peel thereupon resigned. Meanwhile, Lord John Russell, who had replaced Melbourne as Whig leader, had publicly announced that he was a convert to the same cause, and the queen invited him to form a government, whose avowed purpose would be to implement repeal. But Russell could not carry his senior colleagues with him, and in any case, it was unlikely that a Whig administration could get such a measure through the Lords, with its inbuilt Tory majority.

After eleven days of inconclusive discussion, Russell was compelled to decline the queen's commission and Peel returned to power, accepting the 'poisoned chalice' of repeal with an almost messianic glee that was shared by many of his young acolytes. In January 1846 he duly introduced repeal in the Commons. It would be phased in, and it was linked with other measures: the abolition or reduction of duties on a large range of foodstuffs, such as cheese, butter and dried fish (which it was hoped would result in a fall in the cost of living); and a scheme of state-funded drainage loans to landlords and farmers (which were aimed at increasing agricultural productivity). As such, repeal was part of a broader programme, intended to appease both the landed interest and the labouring classes, and Peel also hoped that by carrying the measure sooner rather than later he would put an end to the campaigning and intimidation of the Anti-Corn Law League. But he faced serious personal and political difficulties. In the Commons the outrage of the Tory agriculturalists was given voice by Lord George Bentinck and Benjamin Disraeli (and they were now joined by Lord Stanley). Bentinck, the son of a duke and devotee of the turf, denounced Peel as a dishonest jockey who had cheated his backers and should be barred from the racecourse; Disraeli, seeking revenge for Peel's (wholly justifiable) decision not to offer him a cabinet post in 1841, viciously attacked the prime minister for again betraying his party and his principles, as he had earlier done over Catholic Emancipation and Maynooth. As the bill made its way through the Commons, two-thirds of the Tory MPs voted against it, and the measure was only carried with Whig support; and it only got through the Lords because Wellington put aside his own protectionist preferences out of loyalty to Peel, and secured enough proxy votes from absent peers to ensure the measure passed.

In repealing the Corn Laws, Peel was playing for much higher stakes than mere party advantage. His aim was nothing less than to relegitimate and thus preserve aristocratic government, and to heal the social divisions in the country, by promoting 'so much of happiness and contentment among the people that the voice of disaffection should no longer be heard'. These were noble aims, but they would be unrealized in Ireland, and the political price he paid for repeal was high. Had Peel been more conciliatory and less condescending to his backbenchers, he might have significantly lessened the number of defections on the crucial votes. But he seemed eager to court the martyrdom that awaited him, while the savage denunciations of Bentinck, and the brilliant sarcasms of Disraeli, were long savoured and remembered by those enraged Tories who felt themselves once again betrayed. They were also quick to take their revenge. Repeal was intended to ease Irish distress and discontent, but in the short run the troubled state of the country necessitated the simultaneous introduction of another Irish Coercion Bill, which was just the sort of measure Tory Protestants and protectionists habitually supported. But not this time, for the division on its second reading in the Commons, which took place on 26 June 1846, coincided with the final vote in the Lords that would enact repeal. Determined to bring their loathed leader down, many Tory MPs allied with the Whigs to defeat the bill, and three days later Peel resigned. His final speech as prime minister was wholly in character. It was self-righteous and self-congratulatory, as he concluded by expressing the hope he would leave 'a name sometimes remembered with expressions of good will in the abodes of those whose lot it is to labour, and to earn their daily bread by the sweat of their brow, when they shall recruit their exhausted strength with abundant and untaxed food, the sweeter because it is no longer leavened by a sense of injustice'.

THE EMPIRE AND BEYOND

As had been true of the Whig governments that had gone before him, Peel's administration was much preoccupied by the pressure and controversy of domestic affairs; it was over fiscal matters that the Tory Party split, and Peel was eventually defeated by an adverse vote on an Irish issue. This sense of priorities was also reflected in the fact that three of Peel's closest colleagues in the Tory cabinet were Goulburn at the Exchequer, Graham at the Home Office and (for some of the time) Gladstone at the Board of Trade, all holding domestic portfolios; while Peel's experience of government had never involved him directly in foreign affairs, but had been confined to Ireland, the Home Office and (very briefly) the Exchequer, the latter position having been held in tandem with his premiership during the 'Hundred Days' administration of 1834–35. Yet such was the nature of the United Kingdom's evolving industrial economy, and such were the nation's multi-layered and expanding involvements overseas, that domestic issues were often inescapably global issues, too. The cyclical ups and downs, which caused Peel and his colleagues such concern, were as much determined by international patterns in trade and investment as they were by the internal dynamics of industrial and agricultural production. In the same way, the blight that afflicted the Irish potato crops had originated in Mexico, and it reached Ireland via the eastern seaboard of the United States. Likewise, those who insisted that the repeal of the Corn Laws would result in the ruin of British agriculture assumed (mistakenly as it turned out, at least in the short term) that there would be a sudden and massive influx of wheat from western Europe and from Russia. And those who advocated repeal with religious fervour did so because they believed that the ending of the Corn Laws and the embracing of free trade would result not only in domestic quietude and reconciliation, but also in a new global order of universal peace and brotherhood. Thus regarded, British politics in the 1840s were not only local and national, but they were also inescapably global and imperial.

However fragile and unbalanced the United Kingdom economy may have seemed to concerned contemporaries for much of the 1840s, it bears repeating that all the long-term trends and performance indicators were heading onward and upward; and they explain how, during this period, the United Kingdom consolidated its position as the first and foremost industrial nation, establishing by the end of the decade a broader global spread of international engagement and influence than ever before. In 1830 the nominal value of British exports had been £38 million; yet by 1850 it had increased to £52 million, with northern Europe, the United States, Canada, Australia and British India taking the lion's share, but with clear indications that there were also expanding markets in Latin America, the Ottoman Empire and the Far East. By the late 1840s, 1,500 British mercantile houses were trading around the world, and nearly two-thirds of them were based outside the European mainland, with forty-one in Buenos Aires alone. The 1840s were also the decade when the increase in migration became yet more marked, as the numbers leaving the United Kingdom more than doubled from

700,000 in the 1830s to 1.6 million in the 1840s. The United States remained the preferred destination, but Canada, Australia and New Zealand were becoming increasingly attractive. To be sure, many of those who left were fleeing Ireland after the Great Famine, but whatever their circumstances and motivations this diaspora was a key component in the continuing creation of a greater British world overseas. At the same time missionary endeavour was also more intense and determined than ever: by the late 1840s the London-based societies had mapped out a global field of operations, extending from China to Africa, from India to the Caribbean. And all of these groups – traders and exporters, migrants and missionaries – would increasingly pressure the government in London to support them, to defend their (not necessarily compatible) interests overseas, and to annex territory.

Such was the United Kingdom's growing engagement with the wider world during the 1840s; and as before, it was the Foreign Secretary and the Colonial Secretary who were officially (but not always happily) responsible for overseeing most of it. For virtually the whole of Peel's administration only two people held these posts and, unusually, both had previous experience of running their respective departments. The Foreign Secretary was Lord Aberdeen, a lifelong Tory who had been in public life since the 1800s, and a protégé of the Younger Pitt. Aberdeen had spent most of his early career in foreign affairs and diplomacy and he had already served as Foreign Secretary under Wellington between 1828 and 1830. The Colonial Secretary was Lord Stanley, who had begun his career as a Whig, and who had been briefly in charge of the colonies under Lord Grey between 1833 and 1834, when he had carried the legislation abolishing slavery throughout the British Empire. He had fallen out with the Whigs over Irish Church reform, subsequently joined the Tories, and would remain at the Colonial Office until he resigned in protest in December 1845 against Peel's decision to repeal the Corn Laws. Both Aberdeen and Stanley were landed grandees, well-educated noblemen, well-read patricians and also well-travelled aristocrats. Aberdeen was a serious historian of architecture and antiquities; Stanley was a fine and accomplished classicist who would later translate the *Iliad*. Aberdeen had not just gone on the conventional grand tour, but had also journeyed extensively in the Levant, while Stanley had travelled widely in Italy, as well as in Canada and the United States. As such, they were well qualified to conduct Britain's official relations with the rest of the world, and their jurisdictions sometimes overlapped: when, for example, Aberdeen negotiated with the United States, this invariably had implications for Canada, which meant Stanley was involved as well.

On taking office under Peel, Aberdeen and Stanley contemplated the portfolios to which they had returned with mixed feelings; for internationally as well as domestically, the 1840s would look much better in retrospect for the United Kingdom than they did in prospect. Despite the powerful and simultaneous drivers of industrialization, emigration and religion, which across the decade would so increase Britain's advantageous engagement with the rest of the world, it seemed to Aberdeen and Stanley that they had inherited from the Whig governments a global situation every bit as unhealthy and uncertain as that which Goulburn and Graham (and Peel) had inherited at home. This was largely the result of the combination of belligerence and bluff that had characterized what they regarded as Palmerston's increasingly irresponsible foreign policy, which by 1841 had left Britain's relations with France and the United States in the western world, and with China and Afghanistan in Asia, in a wholly unsatisfactory state. France was still estranged as a result of the Eastern Crisis of 1840, while a series of border incidents had brought Britain and the United States to the brink of war. In Afghanistan the British armed forces were increasingly at risk, as the local population turned against them, while in China the Opium War seemed 'most unsatisfactory'. As Lord Aberdeen put it to Princess Lieven shortly before assuming office, 'In foreign affairs, we have enough on our hands; a war with China, a quasi war in Persia; a state of affairs in the Levant which does not promise a continuance of peace . . . In addition to all this, our relations with the United States are worse than ever.' During the next five years Aberdeen would be forced to deal with crises in every part of the world, while Stanley would be more concerned to safeguard the prosperity and security of Britain's existing overseas possessions than to acquire any further territories. As such, he meant his colonial policy to be consistent with Peel's domestic preoccupations with sound finance and social stability.

Faced with wars and rumours of wars in many parts of the world, with the Royal Navy largely confined to home and Mediterranean waters, with the army fully extended in China and Afghanistan, and with distress and unrest spreading and intensifying at home, Aberdeen's foreign policy was perforce one of conciliation and compromise rather than the bluster and gunboat diplomacy that had been the hallmarks of his predecessor. He was especially anxious to reduce the tensions between Britain and France, and between Britain and the United States: the enmity of either was a serious national concern; the enmity of both might well spell calamity, as it had done in the later stages of the American War of Independence. The 1830s had not been a good decade for Anglo-American relations: the unhappy memories of the war of 1812 still lingered; several of the American banking houses in London had defaulted on their obligations when the economy had turned down; and Southern slave owners resented British criticism that mounted after the abolition of slavery throughout the empire in 1833. There was also a long-running boundary dispute between the United States and Canada where Maine bordered on New Brunswick, over which negotiations had been dragging on inconclusively, and with increasing acrimony, since the late 1820s. There was a further problem arising from the Canadian rebellions of 1837, in the course of which an American had been killed by Canadian troops on American soil, and the United States had retaliated by arresting a Canadian, Alexander McLeod. It seemed likely that McLeod would be executed, in which case diplomatic relations between the United Kingdom and the United States would be terminated, and as a result there was widespread and warlike talk on both sides of the Atlantic. Equally aggravating was America's refusal to accept Britain's claim to 'police the seas' in order to enforce the abolition of the slave trade, which all the major powers had agreed to by the mid-1820s. The Americans thought the British were high-handed and interfering; the British thought the Americans were unofficially condoning the slave trade's continued but illegal existence.

On returning to the Foreign Office, Aberdeen was eager to try to reach a general settlement with the United States. Neither government actually wanted war, and the American Secretary of State, Daniel Webster, sought good relations with all the major European powers, Britain included. Since there would be detailed discussions of the boundary dispute, it was agreed that negotiations should take place in the United States; and to that end, Aberdeen dispatched Lord Ashburton to Washington late in 1841. He was neither a politician nor a diplomat, but he was a member of the Baring family, whose bank was the most important in London in financing America's economic expansion. As a result,

Ashburton had excellent contacts in the United States; moreover, his wife was American, and he was well acquainted with many prominent figures in the republic, including Webster himself. The ensuing negotiations were lengthy and protracted, but generally successful from a British point of view. The legal issues concerning Alexander McLeod and the slave trade were finessed in the diplomatic language of ambiguity and compromise, but the northeast boundary dispute concerning Maine and New Brunswick was not amenable to such an easy solution. Aberdeen thought it would be a mistake to 'go to war for a few miles, more or less, of a miserable pine swamp'; but Peel and Wellington thought Ashburton should adopt a more determined and intransigent position. Aberdeen feared such a posture would jeopardize the general settlement he was determined to achieve, but a compromise was eventually reached, and the Ashburton-Webster Treaty was ratified and agreed between the United States and the United Kingdom in the early autumn of 1842. Palmerston inveighed against it as a 'capitulation', but it was generally well received. As Charles Greville noted, 'Everyone was alive to the inconvenience of having this question left open, and there was a universal desire to settle our various differences with America upon such terms as would conduce to the restoration of honour and good will.'

It was Peel's considered opinion, despite earlier misgivings, that the Ashburton-Webster Treaty had secured a 'permanent and satisfactory peace' between the United Kingdom and the United States, and he was eager that Aberdeen might accomplish a similar success in his dealings with France. During the 1830s relations between the two countries had become increasingly strained, particularly in the eastern Mediterranean where the British policy remained that of supporting the Ottoman Empire as a counterweight to Russia, whereas the French backing for Mehmed Ali's rebellion had threatened to undermine the Sultan's authority. But Peel and his colleagues were only too eager to blame Palmerston for the earlier estrangement of France, while the cultivated anglophile French prime minister, François Guizot, who had been appointed by Louis-Philippe in the autumn of 1847, and was also acting as his own Foreign Minister, was likewise eager to improve relations with Britain. The affairs of the Ottoman Empire were a good starting point. During the time of Aberdeen and Guizot, Britain and France were generally able to agree on leaving the Sultan alone, and on supporting reforms that seemed likely to increase the stability of his regime. Relations between the two countries had also improved when, in the autumn of 1843, Queen Victoria and Prince Albert visited France, the first of several occasions on which the British and the French royal families met; and as a result, Louis-Philippe and Guizot were eager to proclaim that the *entente cordiale* had been re-established between the two countries. There were, to be sure, continuing tensions over French involvement in the affairs of Spain and Greece, and further anxieties about French expansionist ambitions in North Africa and the Pacific. But on the whole Peel and Aberdeen ensured that the Anglo-French *entente* remained intact for the duration of their time in office together.

As a result of Aberdeen's initial efforts, the likelihood of war between Britain and the United States, or between Britain and France, or between Britain and both powers in hostile alliance, had been significantly reduced. In the United States the headlong expansion of the 'slave south' (which was the main source of Lancashire cotton), and of the commercial northeast (with its close ties to Liverpool, London and Glasgow), meant there was a growing degree of mutual economic interdependence between the two nations, which made Aberdeen's efforts at settlement and pacification both more necessary and also more likely. And in Europe, now that the eastern crises of the 1830s had passed, it seemed unlikely that there would be another French-inspired war against Britain, which had been a constant fear in Whitehall from Yorktown to Waterloo, and thereafter. 'Nothing,' Wellington would tell Peel in 1845, 'can be settled in Europe or the Levant without war, unless by good understanding with France; nor can any question be settled in other parts of the world, excepting by the good understanding between France and this country.' On the whole that good understanding now held; and like the United States, France was in addition a major trading partner, with whom cultural ties and connections were also close. With improved transatlantic relations, better contacts with France, and a relatively stable Europe, British trade and investment not only expanded in the United States and on the continent, but also spread far beyond. This was certainly so in the case of the Ottoman Empire, where the treaty extracted by Palmerston in 1838 opened up the Sultan's lands to British exports and enterprise, and led to the formation of Anglo-Greek firms like the Rallis or Rodocanachis. It was also true in Latin America where the British navy was a constant presence on the Atlantic coast, patrolling and suppressing the slave trade, and where several large British firms held commanding positions, especially in the coffee and sugar trades in Brazil.

One of the bases for the Royal Navy's South American operations was the Falkland Islands, which had been occupied in 1833, and where a permanent settlement was established ten years later, named Port Stanley in honour of Peel's Colonial Secretary. But this recognition may have been scant consolation for, like Lord Aberdeen at the Foreign Office, Stanley had inherited from his predecessor too many problems in too many places for comfort, while unforeseen events, strategic anxieties, missionary activity and local commercial considerations were all driving forces working inexorably in favour of further annexations and continued colonial acquisitiveness. Stanley himself shared the generally cautious view of his Colonial Office staff, and was no imperial expansionist. On the contrary, he regarded colonial aggrandizement as a regrettable cause of unnecessary expense and needless conflict, and he did not believe that Britain should acquire any additional overseas territories. The existing empire provided a base upon which to build and extend commerce and trade, but it was secure profits, rather than colonial possessions, that Stanley believed were the key to Britain's international pre-eminence. His imperial views were thus pragmatic rather than doctrinaire: he was determined to preserve the British Empire that already existed, he believed that the country's naval supremacy made that possible, and he thought this was the key to Britain's global power. This in turn meant that he was strongly in favour of maintaining colonial preference, but was hostile to demands for colonial self-government. For Stanley, imperial protection and preferential colonial tariffs were the necessary fiscal basis for ensuring Britain's economic prosperity and for maintaining its close links with its colonies. Hence his opposition to radical demands for colonial constitutional reform, and to Peel's determination that colonial preference must be replaced by free trade.

All this lay in the future when Stanley assumed office with Peel in August 1841, and when two inherited conflicts, each with very different outcomes, awaited his attention. The first was the continuing war in Afghanistan, where British troops and Indian sepoys had been bottled up and beleaguered in Kabul for the best part of two years, surrounded by an increasingly hostile Afghan populace. Their leader, General Elphinstone, believed he had obtained an assurance of safe conduct to bring his forces home; but in January 1842 he and his 16,000 soldiers were massacred on the northwest frontier, on their retreat from Kabul to British India. The British agent in Kabul was also killed, as was the Emir Shah Shuja, and Stanley was consequently faced with the stark reality that British influence in the region had completely collapsed. At just this time, Lord Auckland, who had sent the British troops into Afghanistan because of his (probably exaggerated) fears of Russian influence, was replaced as Governor General by Lord Ellenborough. He immediately sent a military expedition to relieve the besieged garrisons at Kandahar and Jalalabad, and to rescue the surviving British prisoners in Kabul. The British forces then withdrew, and returned to India, allowing Dost Mohammad to resume his throne in Afghanistan. Ellenborough's success in recovering the prisoners helped to restore British standing, though he was convinced that keeping British troops in Afghanistan was an unnecessary commitment, only courting further potential disaster. Stanley agreed with Ellenborough; he was delighted that the Governor General's prompt and determined actions had got Britain 'out of a mess'; and he defended the withdrawal of troops to within the bounds of the Indus as a preferred alternative to the irresponsible Palmerstonian daydreams of universal British dominion in Asia.

In dealing so directly with Ellenborough, Stanley was involving himself in Indian affairs that were more properly overseen by the Board of Control. In fact, Ellenborough had his own imperial agenda, and despite his hostility to maintaining a British military presence in Afghanistan he was resolutely expansionist elsewhere. One area that attracted his attention was the province of Sind, which was south of the Punjab and formed part of the long border with Afghanistan. It occupied a position of obvious strategic importance, and as a result of the British military setbacks in Afghanistan, the Emirs of Sind began to question whether they should continue their traditional policy of loyalty to the East India Company and support for the British. Determined to face the Emirs down, Ellenborough dispatched Sir Charles Napier, who rapidly conquered Sind and annexed it unilaterally. ('Peccavi,' he wittily telegraphed Calcutta in Latin: 'I have sinned.') 'We have no right to seize Sind,' Napier recorded in more measured tones in his diary, 'yet we shall do so, and a very advantageous, useful, humane piece of rascality it will be.' With great reluctance, Peel and Stanley accepted this fait accompli;

but when Ellenborough went on to incorporate Gwalior as a protected state, they became increasingly alarmed at his continuing expansionist ambitions, and also at his increasingly high-handed behaviour. In a rare display of assertiveness by the Board of Control, Ellenborough was duly recalled in April 1844, and replaced as Governor General by Sir Henry Hardinge. But the Sikhs in the neighbouring Punjab were angered by these recent imperial acquisitions, and in December 1845 they went to war against the British. They were defeated at the Battles of Lliwal and Sobraon early in the new year, and at the Treaty of Lahore in March 1846 surrendered their lands between the Beas and the Sutlej rivers to the British. But both sides recognized that this was no more than a temporary truce, and that the outbreak of a second Anglo-Sikh War was only a matter of time.

The Afghanistan fiasco and the annexation of Sind both demonstrated, albeit in very different ways, the limits to the power that the Colonial Secretary could exert, either to project British power overseas on the one hand, or to rein in over-zealous proconsuls on the other. The second crisis that Stanley inherited demonstrated other ways in which his capacity to mould events was seriously constrained and circumscribed. In January 1841 the attempt to bring to an end the first Opium War through negotiations between the British and Chinese governments collapsed, and during the winter of 1841–42 it was clear that British troops, under the leadership of General Sir Hugh Gough, were determined to launch further attacks on the Chinese mainland from their recently occupied base in Hong Kong. Stanley had no wish that the conflict should be prosecuted further; it had cost far too much, and he wanted no additional territorial possessions. All he sought was satisfaction for the injuries inflicted upon British subjects, and the establishment of peaceful and friendly relations with China, on such a footing as would afford permanent security against a recurrence of misunderstandings in the future. But from February 1842 the command of events effectively passed to Ellenborough, who was as much the 'man on the spot' in China as he was in India. Urged on by Ellenborough, Gough and his troops occupied Shanghai, and advanced further up the Yangtze River, inflicting considerable fatalities on the military and civilian population. Once again Stanley was dismayed at what he saw as the excessive and mistaken zeal of local imperial initiative, and at the 'most unsatisfactory war' that had resulted. 'There is,' he told Ellenborough, 'little advantage and no glory in such affairs as the wholesale slaughter, without loss on our part, of Chinese', and he was keen 'to close, whether by treaty or by retaining possession of such parts as we have got and choose to keep, this unfortunate war'.

In China as in India, Stanley and his colleagues were not keen on expanding British territory. Once again they feared the cost, and they also worried (rightly as it turned out) that any annexations would jeopardize their future relations with the emperor. Stanley preferred legitimate trade with China, based on treaty agreements, to 'the occupation of a Chinese Gibraltar or two', and he was quite prepared to give up Hong Kong. But there was strong pressure in London from military, civilian and mercantile circles that the island should indeed be annexed, while events in the Far East had taken on a momentum of their own. For in August 1842, and faced with Gough's troops who were threatening to take Nanjing, the Chinese Emperor sued for peace, and the Treaty of Nanjing brought hostilities to an end. When details of the negotiations finally reached London, Stanley's relief that the war was over was only equalled by his dismay when he learned that Ellenborough and Gough had committed Britain to taking possession of Hong Kong, while five ports, of which Shanghai was the most important, would be opened to British merchants under agreed import and export tariffs. The importation of opium was banned (but this was often disregarded) and the Chinese Emperor agreed to pay Britain a large indemnity for losses suffered by British merchants and the cost of the expeditionary force (which left lasting ill-feeling on the part of the Chinese). Stanley had no choice but to accept the acquisition of Hong Kong as an unavoidable commitment, but he was determined that it should be controlled from the Colonial Office in London rather than by the Governor General in Calcutta. He hoped that the colony would become 'a great mart for the commerce of all nations and for the extension of legal commerce with China'. British traders soon established themselves as a thriving commercial presence in Hong Kong and Shanghai, and Stanley appointed Sir John Davies to be Governor of Hong Kong because he believed him to be truly committed to ending what now became the illegal opium trade.

Stanley's anxieties about expensive and excessive colonial expansion in the Far East and South Asia were matched by his concerns over what he regarded as the misplaced enthusiasm of various British-based companies for the systematic colonization of the Antipodes, which seemed to him a simplistic and suspect panacea for the domestic poverty and unemployment that were such marked characteristics of the late 1830s and early 1840s. But the colonization companies were another powerful London lobby, and since it took nearly six months for mail to travel one way between the capital and the Antipodes, Stanley's ability to keep up with what was happening on the other side of the world, and his capacity to influence events and developments there, were perforce limited. There were also particular problems, both at the British and Antipodean ends, and even in between. Some of the colonization companies were dubious enterprises, and poor Britons were particularly susceptible to being defrauded of their life savings, while many of the ships carrying passengers across the seas to their new lives were unsatisfactory in their accommodation and inadequate in their provisioning. In New Zealand the Treaty of Waitangi had guaranteed the Maori possession of their lands, and conferred upon them the rights and privileges of British subjects; but as increasing numbers of settlers began to pour in, it became ever more difficult for the Colonial Office to honour and enforce these guarantees. Indeed, Stanley went so far as to denounce the New Zealand Company to Peel because of the 'perpetual small trickery which from first to last has characterized their proceedings'. It was no better in Australia, where the wild economic fluctuations meant that arriving emigrants were often confronted by land prices far higher than they had been led to expect in London, while the colony of South Australia, which had only been established in 1836, was on the brink of insolvency, and Van Diemen's Land remained a violent penal colony, where the well-meaning Lieutenant Governor, Sir John Franklin, found it hard to keep order.

The Colonial Secretary responded as vigorously as circumstances allowed to these accumulated Antipodean problems. His Colonial Passengers Act of 1842 regulated the carrying of emigrants to the Antipodes, and the Australia and New Zealand Act of the same year sought to do the same for the sale of colonial lands; and while steering these measures through the Commons, Stanley took the occasion to criticize the colonization companies for distributing misleadingly optimistic propaganda about the opportunities that awaited emigrants to the southern hemisphere. The harsh realities of colonial existence, he firmly believed, should not be concealed from those Britons who were desperate for a better life. He also sought to stabilize the finances of both South Australia and Western Australia, by giving them additional funding, and by reluctantly recognizing that in their still-early years they would be a costly burden to the mother country. By contrast, New South Wales was a more mature settlement, and Stanley carried legislation setting up a more popular constitution and giving political rights to those entitled by wealth to a say in their government. But there were serious limits as to what he could achieve. Although he recalled Sir John Franklin from Van Diemen's Land, the colony remained depressed and disorderly: half of the adult male population were former convicts, while the two largest towns, Hobart and Launceston, were notorious for their lawlessness and vice. Nor was Stanley much more successful in New Zealand, where he appointed Robert Fitzroy – who had previously captained HMS *Beagle* on Charles Darwin's voyage to the Galapagos Islands – as Governor, with instructions to ensure that the terms of the Treaty of Waitangi were strictly observed. Like Stanley, Fitzroy came to distrust the New Zealand Company, and in supporting the claims of the Maori to their lands he antagonized the settlers. By 1844 there were violent clashes between the indigenous peoples and the recent arrivals, and Fitzroy was forced to call in troops from New South Wales to restore order. In London the New Zealand Company kept up its unrelenting pressure on the Colonial Secretary to support the settlers, while the missionary societies protested that their behaviour was illegal.

The British settlers in South Africa and Canada also gave Stanley a succession of imperial headaches. In the aftermath of the Great Trek, which began in 1835, as the Boers sought to throw off British rule in Cape Colony, some 10,000 of them had settled in Natal, where they soon clashed with the British traders who were already living there and also with the Basutos and Zulus, the indigenous tribes. The British settlers, who were supported by their kith and kin in Cape Colony, demanded that Natal be formally annexed so as to maintain order, keep the peace between the Boers and the native tribes, and secure a vital base on the long sea route from Britain to India. As in the Far East and South Asia, Stanley had no wish to extend Britain's colonial possessions in South Africa; but the combination of strategic considerations and local pressure meant that by the end of 1842 he had reluctantly annexed Natal. The problems that the Colonial Secretary faced in Canada were of a different and more diverse order. In the aftermath of Lord Durham's report into the rebellions there of 1837, the Whig government had passed legislation uniting (French-dominated) Lower and (English-dominated) Upper Canada, and establishing a legislature with a nominated council and an elected assembly. But these reforms did little to assuage the bitterness and resentment that still lingered after the rebellions: indeed, the forced union of what now became Quebec and Ontario provinces only served to increase animosities and tensions. Two successive Governors, Sir Charles Bagot and Sir Charles Metcalfe, were sent out with express instructions to try to promote harmony and conciliation, but they made scarcely any headway. 'It is hard,' opined *The Times*, 'to proceed upon a system of construction by conciliation, when you have none but antagonistic elements to deal with.' Stanley was further involved in North American affairs with the passing of the Canadian Corn Laws Bill in 1843, which slightly reduced the tariff on imported Canadian wheat, while increasing it on wheat imported from the United States via Canada.

Like Lord Aberdeen, Stanley was eager to conciliate the United States, and he had vigorously defended the Ashburton-Webster Treaty from the attacks of Lord Palmerston. To the same end he quashed a proposal that Britain should annex territory in Upper California so as to constrain American expansion south into Mexico, on the by now familiar grounds that 'the formation of new and distant colonies' would 'involve heavy direct and still heavier indirect expenditure, besides multiplying the liabilities of misunderstandings and collisions with foreign powers'. (Ellenborough, recently returned from India and installed as First Lord of the Admiralty, was predictably, but ineffectually, enthusiastic.) But there remained one outstanding Anglo-American issue, and that was the settling of the western boundary between the United States and Canada from the Rocky Mountains to the Pacific coast. The British wanted it to run along the forty-ninth parallel; but in November 1844 James K. Polk had been elected American president on the intransigently annexationist slogan '54 40 or fight'. For much of 1845 there were revived fears that there might be an Anglo-American war, and this, combined with the deteriorating situation in Ireland and the continuing distress in the manufacturing districts, meant the British government was eager to settle. So, despite the president's belligerence, were the Americans, and it was eventually agreed that the forty-ninth parallel would form the boundary, with a deviation to leave the whole of Vancouver Island in British possession. News of this agreement, which had again been negotiated in Washington, reached London on the morning of 29 June 1846, just in time for Peel to announce it in his final speech as prime minister. 'By moderation [and] by mutual compromise,' he told the Commons, the British and American governments had 'averted the dreadful calamity of war between two nations of kindred origin and common language, the breaking out of which might have involved the civilized world in general conflict'.

FAMINE, ANXIETY – AND OPTIMISM

It was fitting that Peel's government fell in the aftermath of the repeal of the Corn Laws and the successful negotiation of the Oregon Treaty: for peace at home and amity abroad had been the guiding themes of his administration. Peel had come to believe that getting rid of agricultural protection would remove the last legitimate reason why the middle and working classes might deplore aristocratic government, and in July 1846 the Anti-Corn Law League was indeed wound up. And although foreign and imperial affairs had never been high priorities for him, Peel strongly preferred Britain to be at peace than at war. The price paid for these achievements was that he split the Conservative Party – which he himself had done so much to rebuild and render electable in a post-Reform Act world – for a generation. The ardent protectionists, now led by Lord Stanley and Disraeli, never forgave him for his double betrayal over Catholic Emancipation and the Corn Laws (though Stanley had been in favour of emancipation and Disraeli was not in the Commons in 1829). But Peel had also left behind a coterie of young, ambitious and devoted followers, among them Aberdeen and Gladstone, who could no longer find a home in the Tory Party, but were also uncomfortable with Whiggism, let alone with the radicalism of Cobden and Bright. The result was that for the time being, two-party politics was in abeyance: the protectionist Tories were only able to form minority governments; while alliances between the radicals, those who called themselves Liberals, the Whigs and the Peelites were intrinsically unstable. As for Peel himself, he never held office again, but during his last years he occupied a unique position in British public life. He gave essential support in the Commons to the Whig government that succeeded him, and was idolized by many members of the middle and working classes. His death in July 1850 was greeted with an extraordinary outpouring of popular grief, which was a measure of his success in relegitimating aristocratic rule and in healing the social divisions of the country.

However, when Lord John Russell succeeded Peel in the summer of 1846, that outcome was by no means a foregone conclusion. There were fewer Whigs and Liberals in the Commons than there were Tories and Conservatives, and in the short run Russell's government could only survive because of the deepening split on the opposite side of the house between the Peelites and the protectionists among the Tories and Conservatives. As befitted a Whig administration put together by the son of a duke, Russell's cabinet was very patrician indeed. Eight of its sixteen members sat in the upper house, among them such authentic grandees as Lords Lansdowne, Minto, Grey and Clarendon. As an Irish peer, Lord Palmerston continued to sit in the Commons, where one of his cabinet colleagues was Lord Morpeth, heir to another Whig magnifico, the Earl of Carlisle. The Home Secretary, Sir George Grey, was another landed, north-country baronet (and a kinsman of Lord Grey of the Reform Bill), and so was the Chancellor of the Exchequer, Sir Charles Wood, later ennobled as Viscount Halifax. But for all their shared Whiggery, this was not an easy cabinet for Russell to manage. Despite his intelligence, strong sense of history and capacity for hard work, he was a less commanding figure than Peel: distinctly short of stature, excessively thin skinned, lacking in charm or social graces, and often unpredictable in his political behaviour. At the general election held in the summer of 1847 the Whigs and Liberals together won a small majority over the combined Tory Peelites and protectionists, and since he and his colleagues were less disunited than the opposition, Russell managed to stay in power until February 1852. But he failed in his attempts to win over the Peelites on the one hand or the increasingly assertive radicals on the other.

The measures passed by the Russell government can be seen as a consolidation and enhancement of Peel's earlier legislative agenda; but they were also the result of temporarily shifting political allegiances, or of the momentum of administrative developments within individual departments. It was, for example, the disruption of party alignments that allowed the passing of another Factory Act in 1847, which reduced the time that women and adolescents could work to ten hours a day. This was carried by a temporary alliance of Whigs and Tory protectionists against the Peelites and the Liberal manufacturers, who feared that shorter working hours would mean a further slump in profits. In the same year the replacement of the Poor Law Commission established in 1834 by a Poor Law Board headed by a president was largely a piece of bureaucratic tidying-up. The Public Health Act of 1848 owed much to the inexhaustible energy of Sir Edwin Chadwick, but Russell was a strong supporter. Although the measure was watered down in the Commons, it was the first piece of national legislation ever passed concerning the well-being of the population as a whole, and it set up a General Board of Health with an inspectorate to oversee the implementation of sanitary policy by town councils or newly established local boards of health. In 1849 the government repealed the Navigation Acts, thereby ending the last vestiges of imperial protection; and this, combined with the permanent repeal of the Corn Laws in the same year, represented the final establishment of free trade, which no amount of opposition from diehard Tory protectionists would be able to overturn. Just as Peel had increasingly come to depend on Whig votes to ensure the passing of his more controversial measures, so Russell and his colleagues often needed Peel's support to get their own legislation though.

Yet such reassuring continuities cannot disguise the fact that the closing years of the 1840s were so vexed, disturbed and unstable that it looked for a time as though Peel's valiant and self-sacrificing efforts at social amelioration had failed. Once again the British economy turned down sharply: the investment bubble of the 'railway mania' had burst late in 1845, and railway shares soon collapsed; and there was a further financial crisis in 1847, which resulted in the return of severe unemployment in the textile industries of the north and across the Midlands. As had happened in previous severe downturns, the depression in investment and industry, and in trade and exports, was coincidentally intensified by another run of bad harvests, as the price of corn soared to 100s a quarter. This in turn meant that there were massive imports of grain, which resulted in a loss of gold reserves, credit stringency and business failures; and at the same time, with the tax base much reduced, government revenue plummeted. Here was a crisis at least as bad as that which the Peel administration had inherited at the end of 1841, and it was intensified by developments on the continent, where the response to the same severe and widespread economic recession was a year of revolutions in 1848 more reminiscent of 1789 than of 1830. In France the monarchy of Louis-Philippe was overthrown; in Italy there were revolutions in many of the petty kingdoms and princely states; in Vienna the Habsburg Emperor was forced to abdicate in favour of his nephew, the young Franz Joseph; and in Prussia the king for a time feared for his throne. In the midst of all this disruption and revolution, with traditional authority seemingly collapsing across much of Europe, two young subversives, Karl Marx and Friedrich Engels, published their *Communist Manifesto*. It called upon the new and disenchanted industrial proletariat to rise up everywhere and overthrow the established order, concluding with the famous rallying cry *Proletarier aller Länder, vereinigt euch!*, popularized in English as 'Workers of the world, unite!'

The shock of these disruptions was widely felt across the United Kingdom, and also across the British Empire more broadly. In Ireland the agricultural crisis had begun earlier and cut much deeper, but it was during Russell's ministry that the full impact of the Potato Famine was manifest. There had been blighted harvests in 1845 and in the following year, followed by a decline in the incidence of the disease in 1847, but it returned with renewed virulence in 1848 and 1849. This meant that, initially, it was Peel's problem, and in addition to repealing the Corn Laws, he sought to ameliorate the rapidly deteriorating conditions in Ireland by a combination of public works (financed from local rates and a central loan fund) and emergency food provision (imported Indian meal, food depots and lowered prices). But the official machinery to deal with economic catastrophe and starvation on this scale simply did not exist. In governing circles humanitarian impulses clashed with the widespread belief that subsidized improvement projects financed from London were ideologically unacceptable and/or did not work; while a serious downturn in government revenues as a result of the renewed recession made it harder than ever to justify – or to deliver – large-scale relief and welfare spending. Such were the Irish problems and challenges that Russell's government inherited, with which they had inadequate resources to deal, but which they also severely underestimated. For an amalgam of motives – inadequate funds, ideological commitment, dislike of the Irish – government intervention was significantly reduced. Public works' schemes were replaced with outdoor relief, sweeping schemes for railway construction were not adopted, and plans for land reclamation were abandoned. The Encumbered Estates Act of 1849 attempted to revive Irish agriculture by allowing heavily indebted landowners to sell on advantageous terms, but it achieved little. In the end the Malthusian crisis of the Potato Famine was solved by Malthusian means: starvation, death and emigration.

In May 1847, in the very midst of the Great Famine, Daniel O'Connell had died. While he lived, Irish discontent had focused on campaigning for the repeal of the Union, and had held aloof from the Chartist movement in Britain; but now he was gone, the increased outrage provoked by the Great Famine and what was deemed to be the British government's callous and inadequate response to it, meant that for a brief time Irish protesters did embrace Chartism. Indeed, in April 1848, *The Times* would claim that the movement had now become 'a ramification of the Irish conspiracy'. But that was hardly the whole truth of things, for the severe economic downturn at the end of the decade also stimulated the British Chartists to what would prove to be a final effort. Since the defeat of the second petition in 1842, the movement had lost much of its energy and focus, and become sidetracked into the issue of co-operative landownership schemes, where working men might live dignified lives on the land rather than be enslaved by modern machines. But the economic crisis of 1847–48 turned Chartism back to political agitation, and to the six original points of its founding document. As news was received of the revolutions in France and Italy in the early spring of 1848, the Chartists began to organize a third petition, this time with five points, and with the ballot being dropped. A new convention was called to London, and plans were laid for a mass demonstration on Kennington Common and the presentation of another petition on 10 April. But while the authorities took formidable precautions to protect property and preserve law and order in the metropolis, the London Chartists were badly organized, heavy rain on the day meant that the crowd was smaller than anticipated and soon dispersed, and for a third time parliament dismissed the petition (which allegedly contained many forged signatures) with no great difficulty. Compared to what was happening across much of Europe in 1848, the last gasp of Chartism proved a very damp squib indeed.

Here, it seemed, was a vindication of Peel's policies of free trade and social amelioration, which had consolidated and extended the reforms pursued by the Whigs during the 1830s. 'We may now thank God,' wrote one country squire in April 1848, after the Chartist demonstration had passed off peacefully, 'that the Reform Bill has been passed and the Corn Laws repealed.' Moreover, this sense of relief was accompanied by a growing feeling that, having escaped the revolutions by which so much of continental Europe had been convulsed, the United Kingdom was indeed a special and unique place. But only up to a point: for poverty, unemployment and discontent remained widespread in Britain; the Irish Potato Famine was a demographic and environmental disaster on a scale without parallel anywhere else; and as Lord Stanley had come to appreciate and fear, the colonial empire seemed increasingly difficult to manage and to control. For while the United Kingdom may have avoided revolution at home in 1848, it certainly experienced a series of imperial crises abroad. Throughout his time as Colonial Secretary, Stanley had achieved no more than limited success in reining in the annexationist demands of many pressure groups in London, and the expansionist instincts of many soldiers and proconsuls on the periphery. The result was that by 1848 the British Empire was visibly and increasingly suffering from the problems associated with imperial overload. There was an excessive reliance, especially in India and southern Africa, on multi-ethnic peasant armies, financed out of unpopular local taxation. The spending by Britain itself on the civil administration and military garrisons of empire reached unprecedented heights at this time, when the public finances were a constant source of anxiety. But this expenditure seemed unavoidable because, with the exception of Canada and New South Wales, which possessed rudimentary elected assemblies, Britain's empire was ruled more by authority and force than by consent.

At home the British may have prided themselves on their record of peaceful constitutional progress, but in the empire it was a very different story. Many of the ingredients that produced such a volatile situation across much of Europe in 1848 – among them civil disobedience, fiscal crises and an overstretched and alien military presence – were also present across large parts of the greater British imperial world. Moreover, one of the Whig government's solutions to the renewed Chartist threat and the unrest in Ireland induced by the Great Famine was to increase the numbers of prisoners who were transported to the colonies, especially to Australia and the Cape. The result was a succession of imperial crises and confrontations. There were riots and rebellions in Malta, the Ionian Islands and Ceylon. In Canada the barricades went up in

Montreal and the parliament building was burnt down. In the Cape, New South Wales and Van Diemen's Land there were campaigns of civil disobedience, as white settlers resisted imperial attempts to increase the numbers of convicts who were being transported. In New Zealand there were fears that their settlement might be used as a new convict colony, and one observer claimed that the people of Wellington had been turned into Chartists as a result. There were also serious fiscal crises, as in Jamaica and British Guiana, where the colonial governments were forced to suspend all public expenditure as their local economies collapsed in the aftermath of the global commercial crisis of 1847. Small wonder that British imperial statesmen at home, and British colonial governors abroad, found themselves drawing parallels between these volatile disturbances on the periphery of empire and the American Revolution and contemporary events in Europe. So did the colonial protesters and opponents of British imperial rule, who deployed rhetoric and arguments that drew explicitly on the revolutionary language of 1848. There might be widespread relief that Chartism had finally fizzled out, but neither at home nor abroad did the 'Hungry Forties' end altogether well.

Yet like the tumultuous 1830s, the decade defied any simple characterization across the whole ten years. Anyone reading Edwin Chadwick's *Report*, Thomas Carlyle's *Past and Present* or Disraeli's *Sybil* would indeed conclude that the early 1840s were hard and harsh times, especially in the industrializing north and Midlands, and many of the publications that poured from the presses during the second half of the decade were yet more wide-ranging in their denunciations of the ills of contemporary Britain. Charlotte Brontë's *Jane Eyre* and her sister Emily's *Wuthering Heights* (both 1847) offered trenchant critiques of gender relations, patriarchal society and class inequality. Charles Dickens's *Dombey and Son* (1848) denounced child cruelty and arranged marriages for financial gain. Charles Kingsley's *Yeast* (1848) was equally critical of the social system that condemned so many agricultural labourers to poverty. Elizabeth Gaskell's *Mary Barton* (1848) drew renewed attention to the plight of the industrial working classes in Manchester. In *Vanity Fair* (1848) and *Pendennis* (1848–50) William Makepeace Thackeray assailed the greed, idleness, snobbery, deceit and hypocrisy that he believed were still so marked among the upper classes. Nor were such criticisms confined to works of fiction. John Stuart Mill's *Principles of*

Political Economy (1848) showed scant sympathy for the iron laws of supply and demand, even as he analysed them brilliantly, and Mill would soon be recognized as the foremost intellectual champion of the working class. In the aftermath of the final Chartist debacle, Kingsley was one of the founders of Christian Socialism, campaigning against the excesses of individualism and inequality. And in *The Seven Lamps of Architecture* (1849) John Ruskin developed the earlier views advanced by Pugin, claiming that the buildings in the new industrial towns and cities were devoid of vitality and spiritual content.

At the same time there were other perspectives and developments suggesting a very different view. The Great Conservatory that Joseph Paxton completed for the Duke of Devonshire at Chatsworth in 1840, made entirely of iron and glass, might have lacked spiritual content, but it certainly did not lack for vitality or originality. Two years later Alfred Tennyson published *Poems* in two volumes, including 'Locksley Hall' and 'Ulysses', which established his public reputation and made him the obvious successor to the ageing Wordsworth as Poet Laureate. The early 1840s witnessed the advent of the postage stamp (supposedly invented by Sir Rowland Hill), the commercial Christmas card (popularized by Sir Henry Cole) and photography (largely thanks, in Britain at least, to William Henry Fox Talbot), which between them advanced both communication and the cult of domesticity. The launch of Brunel's SS *Great Britain* in 1843, and the construction of Stephenson's Britannia Bridge that was begun soon after, excited popular admiration and widespread wonder, the one acclaimed as 'the largest and most magnificent specimen of naval architecture that ever floated', the other lauded as a 'magnificent and spectacular novelty' and as a fitting 'monument to the enterprise and energy of the age in which it was constructed'. And in 1848, in the immediate aftermath of the last Chartist demonstration, William Holman Hunt, John Everett Millais and Dante Gabriel Rossetti founded the 'Pre-Raphaelite Brotherhood'. The Pre-Raphaelites sought to recover the spiritual and creative integrity that they believed art had lost since the time of the Renaissance, by embracing both romanticism and medievalism, and by emphasizing the personal responsibility of individual artists to determine their own ideas and methods of depiction. Although the Brotherhood itself did not last long, the Pre-Raphaelites and their followers would remain a major artistic force in Britain until the end of the nineteenth century.

Yet the figure who stood out during the 1840s as most opposed to the doubts, anxieties and concerns expressed by many contemporary commentators and artists was Thomas Babington Macaulay. He had forged a brilliant reputation as a scintillating essayist in the *Edinburgh Review* and as a spellbinding Commons orator at the time of the Reform Bill. From 1834 to 1838 he had been a member of the Supreme Council of India, where he became an enthusiastic supporter of the reforms of Lord William Bentinck. In 1835 Macaulay published his influential *Minute on Education*, urging the need to inculcate the most able South Asians with the superior learning of the West, so as to produce 'a class of persons, Indian in blood and colour, but English in tastes, in opinions, in morals and in intellect', who would mediate between the British proconsular elite and the majority of the population. On his return to the United Kingdom he became a full-time writer and in 1842 produced *Lays of Ancient Rome*, a series of narrative poems recounting heroic episodes from early Roman history, teaching the values of courage, patriotism and self-sacrifice, which would be learned by heart by many generations of public schoolboys. Macaulay served briefly in Russell's government as Paymaster General, but was compelled to resign after little more than a year, having lost his seat at the general election of 1847. The next year he published the first two volumes of his *History of England*, which was an instant best-seller. It justified the Glorious Revolution of 1688, both on its own merits and because of the constitutional blessings and material prosperity that it later made possible. It was a work suffused by national optimism, proclaiming that 'the history of our country during the last hundred and sixty years is essentially the history of physical, of moral, and of intellectual improvement'. His critics found the work materialist and complacent; but for Macaulay, at least, the 1840s had ended rather better than they had for many other Britons.

6

Great Exhibition, Half Time, 1848–52

Despite Macaulay's firm belief that he knew the precise direction in which history was headed, and notwithstanding his unshakeable confidence in both himself and his country, the years since the French Revolution, the Act of Union and the wars against France had been neither stable nor easy for the leaders and peoples of the United Kingdom. Unlike much of the continent, Britain may have avoided major political upheaval between the 1790s and the 1840s, but these were, nevertheless, exceptionally challenging decades for those in authority, and they were also disruptive and disrupted for virtually everyone else. But suddenly, and at exactly the turn of the half-century, it seemed as though the sky had cleared, the clouds had dispersed, and the weather had calmed, as the slump of the late 1840s proved to be the last in the sequence of vicious economic downturns that had begun in the aftermath of Waterloo. As long-term industrial growth became more pronounced and significant than short-term cyclical fluctuations, it was scarcely surprising that the Chartist disturbances of 1848 were the last of their kind, and thereafter working-class agitation and discontent gave way to a more reformist stance and assimilationist activity. At the same time the picture also seemed to brighten internationally and imperially. Despite occasional short-term stresses, the Anglo-French and Anglo-American agreements held; the avoidance of the continental traumas of 1848 was a significant boost to British national morale; most of the imperial disturbances of the same year would soon be settled; and there would be further annexations, especially in India, before the end of the decade. By the early 1850s, then, the mood of the country seemed to have lightened and lifted, which meant that the publication of Macaulay's incorrigibly optimistic *History of England*, of which the third and fourth volumes appeared in 1855, could not have been better timed.

The political figure who was the chief beneficiary and in some ways the architect of this sudden transformation was Lord Palmerston, who had again taken over the Foreign Office when the Whigs returned to power following the fall of Peel's government in the summer of 1846. Among the cabinet that Russell had put together, Palmerston was the most dominant, experienced, difficult and (eventually) disruptive figure. He had been in public life since the 1800s, which was a span that none of his colleagues could rival, and he had made a great (but also controversial) reputation as a long-lasting and assertive Foreign Secretary during the 1830s. Now he was back, and he was determined to undo what he had constantly criticized as Lord Aberdeen's supine policy of accommodation and appeasement. The other major player in dealing with the United Kingdom's relations with the outside world was the third Earl Grey, who served as Colonial Secretary throughout Russell's administration and had recently succeeded to the title on the death of his father, the second earl, who had passed the Great Reform Act. Once again the management of foreign and imperial policy was an aristocratic avocation, and that was one significant continuity between the government of Peel and that of Russell. But there the similarities ended, for Palmerston lost no time in returning to his old policy of belligerence – and bluff. In 1847 a Portuguese Jew named Don Pacifico, who had been born in Gibraltar and was thus a British subject, but who was living in Greece, had been the victim of an anti-Semitic attack, during the course of which his house in Athens had been ransacked. The Greek government refused to compensate him, whereupon Don Pacifico appealed to London for help, and early in 1850 Palmerston sent British ships to blockade Greece, so as to extract compensation and redress for him by the threat or if needs be by the use of force.

Not surprisingly, Palmerston's action was highly controversial. To his friends and admirers in parliament and beyond, it was a firm and necessary assertion of British global power, and a corresponding recognition of Britain's responsibility to support and defend its subjects, wherever in the world they might live; but to his enemies, who included radicals, Peelites, Whigs and Liberals, it was a strident and unjustified display of British belligerence, which demonstrated contempt for the new international order of peace and amity they hoped was coming into being in the aftermath of Britain's recent espousal of free trade. In mid-June 1850 parliament debated Palmerston's actions for five days and nights, and his 'gunboat diplomacy' was sternly criticized by Cobden and Gladstone, Peel and Aberdeen. But Palmerston defended himself against their motion of censure in a bravura speech that concluded with a panegyric on the United Kingdom's domestic stability and assertive global reach. The country, he insisted, had avoided the continental revolutions of 1848 because its social structure uniquely combined respect for the established order with widespread opportunities for self-help, so that all Britons might plausibly aspire to better themselves and rise up in the social scale:

> We have shown the example of a nation in which every class of society accepts with cheerfulness the lot which Providence has assigned to it; while at the same time each individual of each class is constantly trying to raise himself in the social scale, not by violence and illegality – but by persevering good conduct and by the steady and energetic exertion of the moral and intellectual faculties with which his creator has endowed him.

To this remarkably stable yet fluid social structure, Palmerston went on, was allied an equally unrivalled capacity to project national power overseas, and it was therefore, he concluded, up to parliament to decide:

> whether, as the Roman, in days of old, held himself free from indignity when he could say Civis Romanus sum; so, also, a British subject, in whatever land he may be, shall feel confident that the watchful eye and the strong arm of England will protect him against injustice and wrong.

Palmerston had spoken 'from dusk to dawn', and although there would be two days of debate still to come, the opposition had been effectively routed; with emphatic Tory support, he duly won his Commons victory by 310 votes to 264, and Don Pacifico eventually obtained compensation from both the Greek and the British governments. But although his speech brought him triumphant parliamentary vindication, the two propositions on which Palmerston's closing remarks rested were highly fanciful and tendentious. His complacent and self-satisfied depiction of the social structure of the United Kingdom hardly represented the truth of things during the first half of the nineteenth century. For it was far from clear that the majority of Britons had accepted 'with cheerfulness' the harsh lot which Providence had generally assigned them, or that the opportunities for rising in the social scale were as considerable and as abundant as he suggested. These were neither the experiences nor the views of (for example) the maimed and crippled soldiers returning to poverty and unemployment after Waterloo; or of the hand-loom weavers who had been made idle and redundant by technological advance; or of the industrial workers so often unemployed when the economy had once again turned down; or of the women and children who had been employed in the factories and down the mines; or of the Scottish peasantry evicted during the Highland clearances; or of the Irish labourers who had been lucky enough to survive but forced to emigrate in the aftermath of the Great Famine. Nor, despite his closing boast, could the 'strong arm of England' protect *all* British subjects, wherever in the world they might be, 'against injustice and wrong', as had been humiliatingly made plain by the first Afghan War, and as would again be brutally demonstrated at least twice over during the 1850s. Despite his rousing rhetoric, Palmerston's picture of a United Kingdom that was stable, happy and contented at home, and also invincibly assertive overseas, was as exaggerated and oversimplified as Macaulay's *History*.

THE NATION AT MID-CENTURY

Nevertheless, Palmerston was one of the earliest figures in British public life who understood that part of the job of politicians, operating in a new world where the public and the electorate mattered more than they had before 1832, would be to tell the people of the United Kingdom what their country looked like in terms of its domestic social structure, and how it was faring in terms of its relations with other nations and other parts of the world. As the peroration to his 'Don Pacifico' speech suggested, Palmerston knew how to convey information on these subjects to powerful and persuasive rhetorical effect, but such knowledge was also far from comprehensive or complete, and was (as Palmerston's comments showed) easily distorted by political exigencies and selected on the basis of partisan considerations. Yet it was undeniably the case that by the middle of the nineteenth century a great deal more detail was known about the circumstances and peoples of the United Kingdom than had been true fifty years before. The census of 1841 had requested much more information than the previous four that had been held at decennial intervals beginning in 1801, and it was for the first time supervised across the whole of the United Kingdom by a central authority. The 1830s and 1840s had also witnessed an unprecedented number of parliamentary inquiries and royal commissions into the changing circumstances of British social, urban and industrial life, and there had been many unofficial investigations into various aspects of 'the condition of England' question. One such study, first undertaken during the late 1830s, had been by the civil servant and statistician G. R. Porter, and his preface to the 1851 edition of his book, revealingly entitled *Progress of the Nation*, rivalled Macaulay or Palmerston in its optimistic self-congratulation:

> It must at all times be a matter of great interest and utility to ascertain the means by which any community has attained to eminence among nations. To inquire into the progress of circumstances which has given pre-eminence to one's own nation would almost seem to be a duty.

In the spring of 1851, less than a year after Palmerston had delivered his 'Don Pacifico' speech, the next national census was held, which was even more thorough than that which had been undertaken ten years before. For not only did it record the names and numbers of all household residents in the United Kingdom on the night of 30 March; it also sought more precise details concerning their place of birth, age and occupation, as well as new information regarding their marital status, their relations to the head of the household and their disability (if any). In addition, the government conducted a census of attendance at religious services, which was carried out with reasonable thoroughness in England and Wales, but less so in Scotland, and it was not undertaken at all in Ireland (perhaps because the authorities had no wish to know just how widespread and popular the Roman Catholic religion was among the majority of the inhabitants). This religious census was an official inquiry about the confessional life of the nation for which there was no precedent; it was exceptionally revealing, even as it was also flawed and incomplete; and nothing like it has been undertaken since. At the same time a third census was carried out with the aim of eliciting some basic information concerning the provision of education in the United Kingdom at mid-century, by seeking to ascertain how many boys and girls were being taught in classes on a specified day, and the type and size of the schools they were attending. As with the religious census, there was no precedent for such an inquiry, and no such investigation has been carried out again. Taken together, these three surveys represented a novel and unique attempt to ascertain what the social, religious and educational structure of the United Kingdom actually looked like (as distinct from what Palmerston claimed it was); and the conclusions drawn were less obvious and more ambiguous than at first sight the figures suggested.

The most significant generalizations that emerged from these mid-century inquiries were that the United Kingdom had both changed and not changed across the half-century since the first census had been undertaken for England and Wales in 1801; and that it was becoming an increasingly varied and diverse nation in terms of its people and places, economy and sociology, and patterns of worship and education. In 1851 the population of England and Wales stood at almost eighteen million; that of Great Britain was twenty-one million; and that of the United Kingdom of Great Britain and Ireland was more than twenty-seven million. A century earlier these would have been unimaginable figures or, if any thought had been given to the possibilities of such a prodigious explosion in numbers, it would have been with the warning, as Malthus had later delivered, that such an increase would be unsustainable given current levels of natural resources and economic activity. But the growth in the population of Great Britain since the late eighteenth century had been remarkable: in the half-century since 1801 the inhabitants of England and Wales had almost doubled in number, as had the population of Scotland, increasing from roughly one and a half million to just below three million. Ireland had shared that growth until 1841, but the ensuing Great Famine meant that by the end of that decade its demographic history had become very different from the rest of the United Kingdom; for it had experienced a catastrophic population decline, as a result of high mortality rates combined with high levels of emigration, from more than eight million in 1841 to six and a half million ten years later. This was a demographic history unlike that of mainland Great Britain, and unlike that of any western European country, and for the remainder of the nineteenth century the population trajectories of Great Britain and Ireland would be very different, the former continuing to grow, the latter in seemingly inexorable decline.

These figures also made clear the growing dominance of England vis-à-vis the rest of the United Kingdom, for it contained significantly more people than the populations of Wales, Scotland and Ireland combined. Moreover, while on 'the Celtic fringe' the majority of people still lived and worked in the countryside, it was in 1851 that for the first time the majority of the population of England was shown to be town and city dwellers, rather than country folk. Indeed, with the possible exception of the Low Countries, England was the most urbanized country in the world. But once again these aggregate figures concealed many significant local variations. England may have been a rapidly urbanizing nation, with London as the unrivalled global metropolis, and with Manchester as the 'shock city' of the 1840s, which lent some credence to Robert Vaughan's contemporary pronouncement that this was 'pre-eminently the age of great cities'. Yet at mid-century, and throughout the 1850s, more English men, women and children were living in towns such as York, Lincoln, Worcester and Hereford than they were in Leeds, Liverpool, Manchester and Birmingham. In Scotland, by contrast, only one-third of the population was classified as urban in 1851, but there was nevertheless a very marked concentration of town dwellers along the Edinburgh-Glasgow axis from the Firth of Forth to the River Clyde. In Wales, where the total population was only slightly in excess of one million, the largest town was Merthyr Tydfil, but it had fewer than 50,000 inhabitants, while Cardiff and Swansea came a long way behind. And in Ireland, which was even less urbanized than Wales, the population of Dublin bucked the national trend of steep decline by increasing, although only from 233,000 in 1841 to 254,000 ten years later.

In 1851, then, the United Kingdom as a whole was far from being a preponderantly urbanized nation and, despite the extraordinary changes that had taken place in some sectors of the economy since the late eighteenth century, it was not an overwhelmingly industrialized nation either. Even in Great Britain, which was economically far more advanced than Ireland, the largest occupational category in 1851 remained agriculture (mostly men) at slightly below two million, followed by domestic service (mostly women) at just over one million. Next came cotton textile workers at a little above half a million (almost equally divided between men and women), then building craftsmen of every kind (preponderantly male), then general labourers (ditto), then milliners, dressmakers and seamstresses of whom there were about one-third of a million (overwhelmingly female). By contrast, there were fewer than 300,000 men and women employed as woollen workers, while the sum total of coalminers was only a little in excess of 200,000. Taken together, these occupational statistics, which were the most detailed that any census had yet provided, yielded some interesting comparisons and conclusions. Agriculture employed more workers than textiles and heavy industry combined, there were more domestic servants than people employed in the cotton and woollen industries, the number of men working as blacksmiths was greater than the number employed in iron works, and more men worked with horses on the roads than with steam engines on the railways. Even in the new industrial sectors of the British economy, the units of production were relatively small (the average cotton mill employed fewer than twenty people) and steam power had only been applied to a limited number of industries. It was, then, appropriate that Britain (but not Ireland) at mid-century should be known as 'the workshop of the world' rather than 'the factory of the world'.

Even if it was workshop-based rather than factory-based, such precocious and pioneering economic pre-eminence also came at a very high domestic cost. The inhabitants of Ireland, where there had been no such progress and improvement, had suffered grievously because their economy had not industrialized, and famine and starvation had been the terrible result. Many impoverished survivors had left for the United States; but others had crossed the Irish Sea to Britain, which meant that by 1851 there were substantial Irish Catholic communities in such port cities as Glasgow (where they formed 20 per cent of the population) and Liverpool (where almost 33 per cent were Irish). Across much of Britain the fate of Ireland had been avoided precisely because of industrialization and accelerated economic growth, which had made it possible to sustain massive increases in population in ways that Malthus could never possibly have imagined. The result, during the first half of the nineteenth century, was that real wages at best held up, but they did not get much better. Such benefits as there had been from industrialization might have kept income levels stable, or prevented a disastrous decline, but they had not increased them. During its first phase the industrial revolution was far from being the harbinger of what would later be termed trickle-down economics. For most ordinary Britons the fruits of such progress and improvement as there had recently been had not yet appeared. Writing in 1848 in *Principles of*

Political Economy, John Stuart Mill made this point unambiguously – and quite correctly:

> Hitherto, it is questionable if all the mechanical inventions yet made have lightened the day's toil of any human being. They have enabled a greater population to live the same life of drudgery and imprisonment, and an increased number of manufacturers and others to make fortunes. They have increased the comforts of the middle classes, but they have not yet begun to effect those great changes in human destiny, which it is in their nature and their futurity to accomplish.

Indeed, it seems likely that during the first half of the nineteenth century, inequality in the United Kingdom was increasing, which was scarcely what Palmerston had claimed in his 'Don Pacifico' speech.

As someone who was in many ways an eighteenth-century Whig, Palmerston was not much of a religious enthusiast, and his invocation of 'Providence' in his Don Pacifico peroration was more a rhetorical device than an indication of any deeply held Christian faith. Yet in this Palmerston was something of an exception. For as the fierce debates over issues such as Catholic Emancipation and the Maynooth grant made plain, to say nothing of the spiritual disruptions provoked by the Oxford Movement, and the anxieties engendered by what was believed to be the ever-growing influence and popularity of Methodism, religion remained at the centre of the public life of the United Kingdom in the mid-nineteenth century, and of the private consciousness of many people. But no one knew just how central it actually was, and the purpose of the religious census of 1851 was to try to find out. The chosen method of doing so was to establish the extent of what was termed 'religious accommodation', and to record all those who attended church services in England and Wales, and also in Scotland, on Census Sunday, which was 30 March. Since the resulting figures recorded the size of church congregations, it was likely that some people had attended more than one service on that particular day; and no attempt was made to seek the religious affiliation of all the people of Britain, whether they had attended church on that day or not. As a result, the figures and tables proved very difficult to interpret when they were finally published in 1854. But it was generally agreed that in the case of England and Wales only half of the population who could have gone to church on that particular Sunday in fact did so; that half of those attending (which meant one-quarter of the whole population) were Anglicans who worshipped at the established Church of England; but that the other half (and thus another quarter of the whole population) attended dissenting chapels or (in much smaller numbers) Catholic churches instead.

What, exactly, did these numbers signify about the strengths and weaknesses of the Christian religion as it was observed (or not observed) by the inhabitants of the United Kingdom during the middle of the nineteenth century? It was not easy to be sure. From one perspective the results were deeply depressing. Fully one half of the population of England and Wales, which probably meant the overwhelming majority of those who were in the lower reaches of the social scale, were insufficiently engaged with religion to attend church at all; while of the other half who did the Church of England could claim no more adherents than could the combined dissenting sects. Self-evidently, the Elizabethan ideal of total religious conformity across the whole of the nation had long since ceased to be a reality, if it ever had been. The appeal of the established church, which was still believed by some of its adherents to embody the spiritual life and conscience of the whole country, did not in reality extend beyond a quarter of the population; while the massive expansion of Methodism since the late eighteenth century meant that, taken together, the many dissenting sects came very close to equalling the Church of England in their popular appeal. The most pessimistic conclusion drawn about the confessional life of the nation was that of Horace Mann, the official who compiled the statistics and wrote the religious census report, when he observed that, with half of the population staying away from any form of official Sunday worship, 'a sadly formidable portion of the English people are habitual neglecters of the public ordinances of religion'. Yet it was also possible to view these figures in a more optimistic light. Despite the disruptions and upheavals of the last half-century, a large part of the population had maintained its habits of religious practice, the Church of England remained much the largest single denomination, its attendances were more evenly spread across the country than any of its competitors, and in 1851 Anglican clergymen conducted more than four-fifths of all marriages.

The most even-handed verdict on the religious census was that the state-supported Churches of England and of Wales that had been created in Tudor times no longer provided a unifying and all-encompassing official religion for the majority of the population: by mid-century they had become essentially 'voluntary institutions' in terms of attendance at Sunday worship. But for the established church, as for the dissenting creeds (and as, indeed, for the structure of the British economy), such national aggregate figures concealed very significant local variations. The Anglican Church was much stronger in the south of England and in much of the Midlands than it was in the north and west: from Kent to Dorset, Hampshire to Worcestershire, were to be found the oldest and richest dioceses, which would soon be memorably depicted in Anthony Trollope's Barset novels, the first of which, *The Warden*, would appear in 1855, the year after the findings of the religious census were eventually published. By comparison, such dioceses as Durham, Carlisle, Lincoln and Chester were larger, poorer, and had always been less fully assimilated into the post-Reformation Anglican community; while the established church in Wales was a minority creed largely confined to English-speaking landowners, professionals and their associates. Patterns of dissenting worship were equally localized. Those Nonconformist denominations of seventeenth-century origins – namely the Congregationalists, Baptists, Presbyterians and Quakers – were strongest in a belt running from East Anglia through the south Midlands and on to the West Country; and there was also a noteworthy Presbyterian presence in Northumberland on the Scottish border. By contrast, Methodism was very strong in Wales, where it was the mass religion of native speakers, and also in England's eastern counties extending from Northumberland to Lincolnshire. Looked at another way, those counties of England and Wales where population growth and economic change had been most rapid in recent decades were places where the Church of England was weakest and nonconformity was strongest; while counties such as Dorset and Suffolk, which experienced little change and at a slow pace, were those where the established church still held up best.

Yet for all their religious diversity and religious indifference, England and Wales remained the overwhelmingly Protestant nations that they had been since the time of Queen Elizabeth. In both countries, Roman Catholics formed only a tiny minority. Some were recusant gentry who had survived the Reformation, especially in the north and west of England; but by mid-century, and uniquely among the religions of England and Wales, the majority of Catholics were recently arrived from Ireland.

Despite their imperfections the Scottish figures revealed a very different religious picture. The rate of church attendance on the morning of Census Sunday was notably higher than in England and Wales, and the largest denominational bodies north of the border were not Anglican but Presbyterian. The fact that a higher proportion of Scottish people worshipped on Sundays, and that they attended the established Presbyterian Church, reflected the different religious history that had unfolded in Scotland since the Reformation. Yet Scotland, like Wales and England, remained an overwhelmingly Protestant nation, with Catholics in a small minority. Although there had been no religious census undertaken for Ireland, patterns of religious life there remained very different. It was estimated that at least 80 per cent of the population was Catholic, priests played a much more important part in all aspects of people's lives than the Protestant clergy did in Britain, and rates of church attendance were higher as a result. The established Anglican Church had few adherents beyond Ulster except among landowners and Dublin-based professionals; and in Ulster many Protestants, being of Scottish descent, were Presbyterians rather than Anglicans. Taking the United Kingdom of Great Britain and Ireland as a whole, therefore, adherents to the established Churches of England, Wales and Ireland were a minority of all churchgoers, and they formed an even smaller minority of the whole population.

Here, then, was a second paradox revealed by the fact-finders of 1851. For just as the United Kingdom was an industrial nation where industry remained a minority activity, so it was also an outwardly Christian nation where the majority of the population declined to worship in its established churches. But these inquiries also yielded up a third paradox. The United Kingdom might be the most advanced and modern nation on the globe, but it was also a nation where the majority of the population were not only under-employed, under-nourished and largely indifferent to religion, but under-schooled as well. Of the three censuses carried out in 1851, that concerning educational provision was the most error-prone, since many schools had declined to provide information, and the attendance figures seem to have been significantly inflated. Mann himself concluded that the average child spent nearly five years at school: in reality this meant that upper- and middle-class children spent six years being educated, whereas children of the labouring classes attended for only a little over four years. This enabled him to conclude, with a kind of wilful optimism, that 'very few children are *completely* uninstructed' (his emphasis). But it seems clear that Mann grossly inflated the attendance figures, suggesting they were in the region of 80 per cent, whereas other contemporary evidence suggests that they were probably no better than 50 per cent; and whereas virtually all schools taught reading, 30 per cent did not teach writing, and more than 33 per cent did not teach arithmetic. British levels of literacy were probably higher at mid-century than they had been fifty years before, and they were also better than in contemporary Japan or Australia or France. But recent research suggests that they were well below the levels attained by the United States, the Netherlands, Sweden and Germany at mid-century.

These retrospective comparisons bear out the anxieties expressed by many contemporary commentators, namely that British workers were among the least well trained, and that British manufacturers were among the worst educated, in western Europe. This meant that while the United Kingdom might be mass-producing larger numbers of inexpensive goods than any other country in the world, they were often of inferior quality compared to the better finished and better designed artefacts being turned out on the continent, especially in France and Prussia. One early attempt to improve the aesthetics of British manufactured goods had been made by a House of Commons Select Committee, appointed in 1835, to investigate 'the best means of extending a knowledge of the fine arts and of the principles of design among the people (especially among the manufacturing population of the country)'. Convinced that 'to us, a peculiarly manufacturing nation, the connection between art and manufactures is most important', and concerned that many European governments were devoting more resources than the United Kingdom to the education of their artists and artisans, the Committee recommended the establishment of schools of design to give effect to a comprehensive programme of art education; and by mid-century there were twenty such academies in London and in the great provincial towns and cities. But it soon became abundantly clear that the schools were failing to attain the objectives for which they had been founded, namely 'the improvement of art in manufactures', and this was becoming a serious source of anxiety for the 'first industrial nation'.

Despite the fact that its literacy levels were lower than some of the nations of western Europe and North America, the United Kingdom at mid-century nevertheless boasted a substantial reading public, with a broad appetite for a print culture that at its best was probably the most sophisticated in the world. Since the Whig government had significantly reduced the Stamp Duty on newspapers in 1836, their circulation had grown dramatically: the total number sold across the whole of the British Isles in 1829 had been thirty-three million; by 1850, in England alone the figure was just double that. In that year *The Times*, which was easily the largest national daily, was selling almost 40,000 copies every morning, and it was widely regarded as being uniquely influential (Palmerston's links with its editor, J. T. Delane, were especially good, as both Russell and the queen would discover to their cost). Some semi-political weeklies, begun in the 1840s, were also doing particularly well: by mid-century *Punch* (founded in 1841) was selling 30,000 copies a week, *The Illustrated London News* (established in 1842) 100,000 and *News of the World* (set up in 1843) 56,000. The great periodicals that had been brought into being during the first third of the nineteenth century, namely the *Edinburgh Review*, the *Quarterly Review, Blackwood's* and the *Westminster Review*, and that appeared at three-monthly intervals, continued to carry articles on all the major political and public issues of the day, putting forward very different interpretations and contrasting viewpoints, and they still attracted high-quality contributors and a wide readership. But it was the novel that was regarded, in the words of another quarterly, the (theological and religious) *Prospective Review* (founded in 1845) as 'the vital offspring of modern wants and tendencies', and works of fiction remained an integral part of the United Kingdom's mid-nineteenth-century cultural landscape.

The 'condition of England' novels that had been such a marked feature of the 1840s continued to appear, albeit on a diminished scale, among them Dickens's *Hard Times* (1854), Kingsley's *Alton Locke* (1850), with its sympathetic treatment of Chartism, and Elizabeth Gaskell's *North and South* (1855), which looked at mill owners and mill workers in the fictional industrial town of Milton (based on Manchester). But on the whole the novelists of the 1850s, responding to the changed mood of the decade, would explore different issues and broaden their range. Dickens produced his most autobiographical work, *David Copperfield* (1849–50), and then turned to wider social and economic considerations: the delinquencies of the legal profession in *Bleak House*

(1852–53), debt and prison in *Little Dorrit* (1855–57), and wealth and poverty in *Great Expectations* (1860–61). Thackeray, who had never been much interested in urban squalor and industrial conditions, remained fascinated by social climbing, snobbery and hypocrisy in *The Newcomes* (1855), in which he was one of the earliest writers to use the word 'capitalism'; but he then moved to fantasy in *The Rose and the Ring* (1855), and to history in *The Virginians* (1859). With *The Warden* (1855), Trollope began his extended treatment of clerical life, which would take him six volumes to complete, and in a very different register he produced *Castle Richmond* (1859–60), which combined a deeply empathetic description of the hunger and death wrought by the Great Famine with a jarring providentialist interpretation of those tragic Irish events. Charles Kingsley turned to historical novels in *Hypatia* (1853), and *Westward Ho!* (1855), and wrote a children's book about Greek mythology, *The Heroes* (1856). By the end of the 1850s a new generation of novelists were beginning to command acclaim and attention, among them Wilkie Collins with *The Woman in White* (1859) and George Meredith with *The Ordeal of Richard Feveral* (1859) and *Evan Harrington* (1860).

These high-quality newspapers, periodicals and novels were at the apex of a teeming, interconnected and bewilderingly varied literary world of authors, editors and publishers. The novels by Dickens, Trollope, Thackeray and Collins were usually serialized before they appeared between hard covers; Dickens was the owner and proprietor of one new weekly magazine, *All the Year Round* (1859), and Thackeray was the first editor of its rival, *The Cornhill* (1860); while many of these authors and their publishers were members of the Athenaeum. But most of the mid-nineteenth-century literary output catered to a broader and less sophisticated public, in the form of 'penny dreadfuls', as well as 'improving' journals, and also 'sensation' literature, Gothic fantasies and plagiarized fictions galore. One figure who straddled many of these literary worlds was Henry Mayhew, by turns a journalist, playwright, social researcher and reformer. He was born in London in 1812, one of seventeen children, and would later become a friend of Thackeray's. Mayhew was the co-founder of *Punch* and an early contributor to *The Illustrated London News*. But he was most famous for the publication of *London Labour and London Poor* in three volumes in 1851, which derived from articles he had earlier published in the *Morning Chronicle*, and offered a vivid and often harrowing picture of the lower social echelons of the great metropolis, based on interviews he had conducted with market traders, street entertainers, labourers, sweatshop workers and prostitutes. Mayhew described the lives of such people in the minutest detail, and in so doing made plain how precarious were the existences of many dwellers in what was, at that time, the greatest and richest city in the world. The result was an unforgettable picture of what another contemporary described as 'the inferno of misery, or wretchedness, that is smouldering under our feet'.

One of the reasons for London's generally degraded environment was that its government and infrastructure were wholly inadequate for a city with a population that had grown from one million to over two million during the first half of the nineteenth century. There was no central authority, such local power as existed was dispersed between a large number of vestries, parishes and commissioners, and this chaotic structure of competing and overlapping jurisdictions helps explain why raw sewage still flowed untreated into the Thames, with serious consequences for the health of the population. In 1848 there was another outbreak of cholera, which claimed 14,000 lives in London alone, and ten years later the metropolis would be affected by what was termed the 'Great Stink'. That summer the stench from the sewage in the Thames was so bad that the curtains in the recently completed House of Commons had to be soaked in bleach so MPs could carry on with their business. The creation of the Metropolitan Board of Works in 1854 was an attempt to establish a city-wide authority, but as with the General Board of Health its powers were limited, and there would be no London equivalent to the grand rebuilding schemes begun at that time by Baron Haussmann for Napoleon III in Paris and by the Emperor Franz Joseph in Vienna. The great architectural and town-planning initiatives that had been conceived by the Prince Regent and executed by John Nash had long since been completed. By contrast, mid-Victorian London was a deliberate statement against royal absolutism and state intervention. The great aristocratic estates, extending from Bloomsbury to the edge of Hyde Park, were redeveloped for residential accommodation by speculative builders, of whom the most famous was Thomas Cubitt. And the railway termini at Euston (1837 but expanded in 1849), Waterloo (1848), King's

Cross (1852), Paddington (1854) and Victoria (1860), were monuments to the power of corporate capitalism rather than local or national government.

London, so this argument ran, might not be grand or splendid, but at least its people were not cowed or coerced by a despotic, intrusive, repressive regime. As one contemporary explained, 'The public buildings are few, and for the most part mean . . . But what of all this? How impressively do you feel you are in the metropolis of a free people!' The idea that stench, shabbiness and squalor somehow proclaimed British freedom would scarcely have convinced those unfortunates who were the subject of Mayhew's investigations. On the other hand, the initial fears that Robert Peel's metropolitan police force would resemble a continental gendarmerie had proved groundless, and by mid-century policemen were being more appreciatively regarded as 'peelers' or 'bobbies'. Moreover, London's reputation as a destination that was open to all, especially those fleeing from persecution elsewhere in Europe, remained strong. Indeed, the city's appeal to exiles and emigrés was at its peak in the immediate aftermath of 1848, when it provided a safe haven for many agitators and revolutionaries who had failed in their bids to change the world, among them Louis Blanc from France, Karl Marx from Germany, Giuseppe Mazzini from Italy, the Scalia brothers from Sicily and Lajos Kossuth from Hungary. There was also a significant community of Russian exiles in London, led by Alexander Herzen, and of Poles including Stanislaw Worcell, who preferred to live in freedom than suffer under Tsarist autocracy. Many of these exiles endured lives of hardship and monotony, often centred on Soho and Leicester Square (although Marx preferred to spend long hours in the reading room of the British Museum). At the same time London also continued to appeal, as it had done since 1776 and 1789, to those rulers and reactionaries who had failed in their attempts to prevent revolutionary change, most famously King Louis-Philippe from France and Prince Metternich from Austria.

That the United Kingdom kept its borders open, welcomed European exiles of varied political views, and let them take up long-term residence in London, was a great source of national pride, and this particular sense of self-congratulation was especially marked in the aftermath of 1848, when Britain had not only avoided revolution itself, but had also provided a second home for those who had tried and failed to bring about revolutions elsewhere. As *The Times* observed early in 1853:

> Every civilized people on the face of the earth must be fully aware that this country is the asylum of nations, and that it will defend the asylum to the last ounce of its treasure and the last drop of its blood. There is no point whatsoever on which we are prouder or more resolute.

But not everyone saw the welcome extended to thwarted revolutionaries as a disinterested display of British liberalism, tolerance and cosmopolitanism. On the contrary, by harbouring such dissidents and agitators, the United Kingdom was not so much acting high-mindedly as 'the asylum of nations', but was instead proclaiming its outright hostility to many of the reactionary regimes that had survived the revolutionary onslaughts of 1848. As King Leopold of the Belgians explained to his niece Queen Victoria, offering a very different opinion from that of *The Times*, the prevailing view across much of Europe was that 'in England a sort of menagerie of Kossuths, Mazzinis . . . etc is kept to be let occasionally loose on the continent to render its quiet and prosperity impossible'. During the era of Macaulay and Palmerston neither of these views was entirely surprising.

GLOBAL HEGEMON?

Such was the varied, paradoxical, old-and-new, rich-and-poor, rural-and-urban, agrarian-and-industrial, religious-and-secular, English-and-Celtic, Protestant-and-Catholic nation that Lord John Russell and his Whig colleagues were governing at mid-century; and they did so from a capital city that was the greatest and most prosperous in the world, but was also polluted, fog-bound and fever-infested, even as it was a beacon of hope for exiles and escapees fleeing their homeland or other countries on the continent. Since taking office, Russell had presided over what might best be described as a period of low-energy Peelite consolidation in terms of domestic legislation. But in September 1850, a few months after the 'Don Pacifico' affair had shown the continuing power of popular patriotism, there was a sudden re-emergence of popular Protestantism, occasioned by the decision of Pope Pius IX to re-establish the Catholic hierarchy in England, which had disappeared at the time of the Reformation when bishops had been replaced by vicars general. The proposal to strengthen the influence of the papacy over the affairs of the United Kingdom provoked an outburst of confessional antagonism that reaffirmed the importance of religion in the public consciousness, even as the findings of the census to be conducted in the following year would suggest that loyalty to the Church of England as an institution was in decline. The scheme originated with Nicholas Wiseman, who was the head of the Catholic Church in England and had recently been made a cardinal. Before taking it forward he had consulted not only with the pope, but also with Russell himself who, as both an Erastian (believing the church should be subordinate to the state) and a Whig (committed to religious toleration), had long supported the removal of Catholic 'disabilities' (legal restrictions and limitations).

But in proclaiming the new religious hierarchy, which would perforce be subservient to Rome, Wiseman had unwisely employed intemperate language, asserting that 'Catholic England' would again become a 'source of light and vigour', a phrase which seemed to imply that the papacy was seeking to interfere in the affairs of a sovereign and Protestant state where it was widely believed the pope should claim no such jurisdiction. In the light of recent controversies and protests concerning Catholic Emancipation in 1829, and the grant to the Maynooth seminary in 1845, to say nothing of the popular expressions of anti-popery that had been manifest in the Gordon Riots of 1780, this reassertion of papal power seemed needlessly and unwisely provocative. The Roman Catholics might be a small minority in Great Britain, but their numbers were growing rapidly in towns such as Liverpool and Glasgow in the aftermath of the Great Famine, and the fact that they were an overwhelming majority in Ireland only inflamed Protestant passions still further. Moreover, the 1840s had witnessed several conversions of prominent High Church Anglicans and members of the Oxford Movement to Catholicism, most famously John Henry Newman himself in 1845. He was not the first Anglican priest to go over to Rome during that decade, but his decision to convert encouraged others to follow, among them the hymn-writer and theologian Frederick William Faber, and Henry Wilberforce, the youngest son of the renowned Evangelical and anti-slavery campaigner. Such apostasies held out the alarming prospect of an enfeebled and retreating Church of England being undermined and overwhelmed by a reinvigorated and assertive Catholicism, determined to undo and reverse the Reformation; and these anxieties would be intensified by the subsequent conversions of the Archdeacon of Chichester, Henry Edward Manning, and the Archdeacon of the East Riding of Yorkshire, Robert Wilberforce, Henry's eldest brother.

The children of William Wilberforce were never the same happy family after they had gone their separate religious ways. (The middle brother, Bishop Samuel, remained a staunch supporter of the Church of England, and would be a fierce critic of the evolutionary theories of Charles Darwin.) But at a time when the dangers of a resurgent Catholicism seemed very real, and when there was a (no doubt exaggerated) Protestant panic about those going over to Rome, it was scarcely surprising that the pope's decision to appoint the first English bishops in the autumn of 1850 led to widespread popular protest against what was termed 'Papal Aggression'. In creating the Archdiocese of Westminster at the apex of this new Catholic hierarchy, and in establishing bishoprics with territorial designations that had not been used by Church of England, the pope had sought to sidestep Anglican sensibilities; but in this he failed, not least because Russell abandoned what seemed to have been his earlier support for the scheme and instead threw his weight behind the hostile popular protests that ensued. In November 1850 he made public a letter he had written to the Bishop of Durham, in which he denounced the pope's actions, attacked his attempts to 'fasten his fetters upon a nation which has so long and so nobly vindicated its right to freedom of opinion', and also condemned the Oxford Movement for reintroducing 'the mummeries of superstition' to Anglicanism. In the following year Russell passed an Ecclesiastical Titles Act, making it illegal for Catholic bishoprics to bear the same names as those already employed by the Church of England. But it seriously damaged his reputation as a Whig and a reformer, and it was opposed in the Commons by Catholic Irish MPs, by Peelites like Graham and Gladstone, and by radicals such as Bright and Cobden. It was also an unnecessary measure, since in naming his bishoprics the pope had been careful to avoid doing what the Act outlawed, and in any case, it would be repealed (by Gladstone) in 1871.

The anti-Catholic stance that Russell finally seems to have adopted, in contradiction to his earlier position on the issue, undoubtedly resonated across much of Protestant Britain, which may be why he took it up, in an effort to upstage Palmerston; but it won him few political friends at Westminster. As an attempt to reassert his authority over his wayward cabinet colleagues, it met with little success, while his abandonment of the historic Whig principles of Erastianism and religious toleration seemed an opportunistic repudiation of his dynastic heritage and birthright. Yet this was not the whole truth of things, for Russell, like many of his ducal forebears, had always been opposed to the spectre of Catholic despotism and ultramontanism, whereby the inhabitants of a particular country were in thrall to a distant, alien and foreign religious authority in Rome, and in the febrile atmosphere of the late 1840s and early 1850s this might not have seemed a wholly unfounded anxiety. In any case, while he was inveighing against 'Papal Aggression', Russell was also pondering a further extension of the franchise, despite having earlier averred that the settlement of 1832, of which he had been one of the chief architects and protagonists, must be considered 'final'. In the autumn of 1850, in another attempt to outflank Palmerston, who was opposed to any further enlargement of the electorate, Russell drew up a plan to do just that, and in February 1851 he suddenly informed the Commons that he would bring in new reform proposals early in the following year. But he failed to obtain a majority in support, and promptly resigned the premiership. There followed a period of high-political confusion similar to that which had occurred on Peel's earlier resignation in 1845, at the end of which Russell returned to power, but with his political authority further weakened as a result.

These political perturbations over foreign affairs and parliamentary reform, the one suddenly blowing up out of nowhere and vanishing as quickly, the other the by-product of personal rivalries, political torpor and party confusion, were the first in a series of crises and controversies that would dominate British public life for the next decade and a half, both inside parliament and across the nation. But of the two, it would be foreign affairs that would be the more important. This was partly by default, as the great domestic issues that had energized politics and galvanized, then disrupted, party loyalties from Catholic Emancipation to the repeal of the Corn Laws, had largely subsided, which meant renewed efforts to reform parliament again went nowhere. But it was also because the 1850s and early 1860s would witness the zenith of British global hegemony and its many-sided engagements with the wider world, thanks to the unrivalled position that it had by then established as the most innovative, advanced, sophisticated and prosperous economy on the planet. In domestic terms, there were undeniable limits to British industrialization at mid-century; but in international terms, the United Kingdom was far ahead of all its competitors – so much so, indeed, that they scarcely seemed competitors at all. Britain's output and consumption of coal was more than ten times that of France and six times that of those countries which would eventually become imperial Germany, while its output and consumption of iron was, respectively, three and four times greater than theirs. Steam engines may not have been all that widely used, but at mid-century they provided 1.29 million horsepower in Britain, compared to 370,000 horsepower in France and 260,000 horsepower in Germany. And more railway lines had been constructed in Britain than in France and Germany combined, even though France was, and Germany would soon become, larger territorial units.

These figures must also be set in an even wider perspective. By the middle of the nineteenth century, Britain produced half of all the world's pig-iron output, and the same proportion of the world's coal, while its looms and spindles consumed almost half of the raw cotton output of the globe, and its share of total world manufacturing production reached its highest level. Put another way, this meant that if the British index of industrialization in 1851 was set at sixty-four, its closest continental competitors were almost pitifully lagging behind: Belgium was next on only twenty-eight, Switzerland scored twenty-six, France was at a mere twenty, with Germany at but fifteen. In the same year, British per capita gross domestic product was 65 per cent higher than that of Germany and 30 per cent more than either the United States or the Netherlands. This remarkable industrial superiority meant that British engagement with the wider world was greater and more extensive than that of any other power. The primary markets for its manufactured goods remained Europe and North America, but it was also the pre-eminent exporting nation to Latin America, the Levant and China; and it was importing raw materials and foodstuffs from around the world. By mid-century Britain was also exporting capital on an unprecedented scale: in the decade after Waterloo, it had invested £6 million annually overseas, but by the 1850s that figure had increased fivefold. In part as a result of the Great Famine, it was also exporting more people as well as more money: in the 1840s, 1.6 million Britons headed to the Antipodes, British North America and the United States; during the next decade, the figure would increase to 2.3 million, and in the twenty-one years after 1850 an annual average of more than 200,000 migrated abroad. And with such extensive movements in raw materials, manufactured goods, overseas investments and emigrating Britons, it was scarcely surprising that 60 per cent of all the world's ocean tonnage was registered in the United Kingdom.

Part cause, part consequence of this mid-century consolidation of national productivity and international pre-eminence was an increased and intensified commitment to the doctrine of free trade. It might have been controversial when Peel repealed the Corn Laws and Russell repealed the Navigation Acts, and Tory protectionists were never reconciled to it. But free trade soon became an article of faith among most Victorian politicians, bankers, businessmen and manufacturers, and would remain so until the end of the nineteenth century – and in some quarters well beyond. To be sure, there was an element of humbug and hypocrisy in all this: it was easy for the British to urge equality of opportunity and 'a fair field and no favour' for everyone who was competing for business in all the markets of the world when they knew they were going to win the lion's share of the orders. But there was also a sense in which free trade *did* seem to some to be a genuinely secular-cum-religious crusade, globalizing the messianic doctrines and beliefs of the Anti-Corn Law League. In the words of one British Governor of Hong Kong, offering a novel but widely shared interpretation of the New Testament: 'Jesus Christ is Free Trade, and Free Trade is Jesus Christ.' But this slogan also papered over a multitude of different views as to how this free-trade heaven should be realized on earth. Such peace-loving internationalists as Cobden and Bright believed that free trade would of itself produce equitable relations and peace and concord throughout the world. But many businessmen did not share this piously optimistic view, and they supported Palmerston's efforts to send gunboats to open up markets in far-off places where free trade could only be imposed by force.

Yet such hegemony had its limits: for despite the United Kingdom's industrial and financial pre-eminence, and its unrivalled global economic reach, the mid-century British state lacked the capacity to project power overseas on an equivalent scale. Its undoubted economic strength did not directly translate into equivalent military might. On the contrary, the prevailing ideology of laissez-faire and the long fiscal shadow cast by the spiralling debts that had piled up during the Revolutionary and Napoleonic Wars meant the pressure to reduce public spending continued unrelentingly into the era of Peel and Russell (and, indeed, Gladstone). For decades after 1815, expenditure on the army and navy was held to an absolute minimum, to no more than 2 to 3 per cent of gross national product, which was less than it had been for much of the eighteenth century, and less than it would be again for much of the twentieth. Indeed, during the intervening period, the United Kingdom was probably less effectively 'mobilized' or resourced for war than at any time since the early Stuarts. This meant that as a European military power the British army at mid-century (65,000 troops) could not begin to compete with Prussia (127,000), France (324,000), the Habsburg Empire (400,000) or Russia (900,000), and that the transformative modernization of the British economy in the aftermath of Waterloo had not been accompanied by equivalent improvements and innovations in the army. Not surprisingly, then, there was little British appetite or capacity for military intervention on mainland Europe, although such strategically vital peripheries as the Iberian Peninsula, Belgium and the Dardanelles remained constant concerns.

In any case, most British soldiers were deployed beyond Britain's shores and beyond Europe, in the empire, where they were also supported by the Indian army, whose 300,000 troops were available for imperial service at no cost to the British taxpayer. But although the Indian army saw action in East Africa, the Middle East and as far away as China, Britain's overseas military and proconsular officials were constantly complaining that the forces they commanded were inadequate given the size of the territories they were expected to control. By contrast, the Royal Navy did assert a form of global dominance, since it remained larger than those of the next three or four navies combined. Its major fleets in the North Sea and the Mediterranean gave the United Kingdom some political leverage on the edges of the continent; and beyond Europe the navy kept the sea lanes open for trade, suppressed piracy and intercepted slaving ships, held together Britain's maritime empire, and supported the men on the spot in their dealings with indigenous rulers and peoples. But like the British army, the Royal Navy had been steadily reduced since 1815, and since its wooden sailing ships were relatively inexpensive to construct, it was for the time being possible to maintain Britain's world position on a low-budget basis. Moreover, the international situation was far less competitive than it had been before 1815: France and Spain were no longer fighting Britain in global wars for empire; Russian rivalry in Asia only gave rise to occasional alarms; and a Prussian-dominated Germany was still two decades in the future. As the foremost industrial and financial nation on earth, which also benefited from the great-power vacuum existing in many parts of the world, mid-nineteenth-century Britain could indeed operate as a global hegemon on the cheap.

Such were the strengths and also the weaknesses of the United Kingdom's international position at the middle of the nineteenth century, and they help explain why Palmerston on his return to the Foreign Office was eager to preserve the territorial balance created by the great powers at the Vienna settlement of 1815. He hoped to maintain continental alliances so as to avoid committing Britain to expensive and demanding military entanglements, and he recognized that there were considerable limitations on the nation's capacity and willingness to intervene overseas. In practice, this meant it often seemed impossible to discern any consistency in what Palmerston was doing: he might appear coolly rational and self-interested; he could embrace liberal principles and causes with widespread popular support; he would get on with despots and do business with them if it suited him; or he would bluster and bluff and bully, which sometimes worked but other times did not. As for the colonies, there had been no clear pattern or rationale to the extensive annexations that had taken place since Waterloo, and this would not change for the remainder of the century. Governments were generally reluctant to assume additional responsibilities or take on more territories, because they invariably cost money; but there was also a wary recognition that the commercial expansion of Britain across and around the globe, combined with the escalating scale of overseas emigration, might create demands for official intervention across the oceans which, however unwelcome, could not in the end be avoided. Trade and commerce, investment and settlement, did not automatically mean imperial expansion: sometimes they did, sometimes they didn't. But once colonies had been annexed, even with the greatest of reluctance, it was generally recognized they had to be retained. As Russell told Lord Grey in 1849, perhaps thinking of

1776, 'the loss of any great portion of our colonies would diminish our importance in the world, and the vultures would soon gather together to despoil us of other parts of the Empire'.

These were the constraints on Britain's global hegemony that set the limits to what Palmerston and the Foreign Office and Grey at the Colonial Office could do and could not do. On the periphery of Europe, Palmerston's early attempt to intervene in the vexed dynastic politics of the Spanish royal house, by championing a potential suitor for Queen Isabella against the candidate put forward by neighbouring France, was not a success. Nor did he wield significant influence across the North Sea in the first Schleswig-Holstein crisis of 1848–49, when German and Danish nationalists clashed, since he was in no position to deploy British troops in the area. When the European heartlands erupted in revolution in the spring of 1848, Palmerston eagerly embraced the constitutionalist rhetoric that won him the plaudits of continental liberals; but his main aim was to prevent large-scale war and any changes in the European balance of power that would be seriously inimical to Britain's interests. Hence his efforts to ensure that France should not gain influence at Austria's expense; hence his initial support of the Hungarian uprising against Austria led by Kossuth; but hence, too, his later approval of autocratic Russia's intervention to suppress the revolt on Austria's behalf; and hence his relief that moves towards the creation of a greater, unified Germany came to nothing. There were also the gesture politics at which Palmerston excelled, and which made him so popular with the press and the public: his defence of Don Pacifico, his tacit support for Kossuth when he came to Britain in 1851, and his acceptance of an address in which the Emperors of Austria were denounced as 'odious and detestable assassins'. And despite Palmerston's denunciations of his predecessor at the Foreign Office, the Clayton-Bulwer Treaty of 1850, by which Britain and the United States agreed to desist from colonizing Central America, was in the conciliatory tradition of Lord Aberdeen's earlier negotiations.

At the Colonial Office, Lord Grey had to cope with the levels of unrest that had rocked so many parts of the British Empire in 1848. Although he was a cautious policymaker and innovator, it was during this time that the new policy was initiated of granting a substantial measure of political autonomy to the colonies of settlement, which was very different from the intransigent and fatal stance that the British government had adopted towards the thirteen American colonies in the 1770s. This was accomplished by establishing 'representative' government in the form of elected bodies, and then instituting 'responsible government' by making executive councils answerable to these legislative assemblies. By such means it was hoped to retain these later settlement colonies within the empire, and the granting of responsible government also made them liable for their own military and administrative expenses. To British policymakers and proconsuls, it seemed an appealing proposition to concede a certain amount of autonomy to the colonies over their home affairs, while retaining control of foreign policy, and at the same time reducing the burden on the British exchequer. In addition to Grey himself, the most important proponent of this new approach was Lord Elgin, who was Governor General of Canada between 1847 and 1854, and who introduced responsible government in the colony in 1848. Nova Scotia and New Brunswick followed suit in the same year, Prince Edward Island in 1851, and Newfoundland three years later. During the 1850s responsible government would be established in every Australian colony except Western Australia, and also in New Zealand, although it was not always clear whether it was the governor on behalf of the British monarch, or the prime minister on behalf of the colony, who had oversight of the indigenous peoples. The same uncertainty applied in Cape Colony, where the constitution granted in 1851 was also a move in the direction of the new practice of responsible government.

But as imperial ties were being loosened in some parts of the British Empire, imperial acquisitions were also being made elsewhere, as the expansionist impulses and acquisitive zeal of the men on the spot triumphed over the general reluctance of the Colonial Office to extend its dominions still further. In the case of South Africa the grudging British annexation of Natal had merely encouraged further settler intrusion into the neighbouring region beyond the Orange River, thereby raising tensions between the British and the Boers, and also between the European whites and the indigenous peoples. In February 1848 the Governor of Cape Colony, Sir Harry Smith, peremptorily proclaimed British sovereignty over what now became the Orange River Settlement, much to Grey's dismay. Indeed, three years later Grey informed Smith that 'the ultimate abandonment of the Orange Sovereignty should be the settled point in our policy'. But Smith's insubordinate acquisitiveness was as nothing compared to that of Lord Dalhousie, who was Governor General of India from 1848 to 1856. He arrived to find the Sikhs once more in rebellious and threatening mode; but within a year they had been decisively beaten and the whole of the Punjab annexed. And in 1852 he supported the commodore of the British squadron on the Irrawaddy at Rangoon in his disagreement with the indigenous governor, thereby inaugurating the second Anglo-Burmese War, which resulted in another British victory and another British annexation, this time the province of Pegu. Dalhousie also invoked what was termed the 'doctrine of lapse' to annex seven princely states during his governorship, including Arcot, Awadh and Jhansi, where the rulers had no legitimate heirs. At the same time he began to plan for the construction of a comprehensive Indian railway system, which would be built by private enterprise, but with the Indian government guaranteeing a 5 per cent return to investors.

This was scarcely the mid-century imperialism of laissez-faire or of free trade in action, but rather the imperialism of assertive British state interference, for both Sir Harry Smith and Dalhousie had always been strong believers in the merits, necessities and possibilities of annexation, expansion and intervention. Before he took over the Orange River Settlement, Smith had previously battled (and beaten) the Sikhs in India in 1846, served in South Africa during the 1830s, and fought with Wellington at Waterloo. Dalhousie was also a protégé of Wellington's, and like his proconsular predecessor Lord George Bentinck, he believed that British rule in India should be concerned with acquisition and annexation, progress and modernity, intervention and investment, technology and reform. But this was not how Britain's imperial mission was necessarily viewed from either the Colonial Office or the Foreign Office, still less from 10 Downing Street; while those many manufacturers, traders, bankers and financiers involved with Britain's global commercial networks had very differing views as to the nature and priorities of their country's engagement with the wider world. And their interests were very different again from the British emigrants who were settling in Canada, South Africa, Australia and New Zealand in increasing numbers; *they* were more eager to acquire land, and less concerned about the rights of indigenous peoples, than were the official representatives of the British government. But the figure most preoccupied with linking the United Kingdom with the world beyond, and who captured the imagination of the British people at this time, was neither a politician nor a proconsul, soldier, trader, manufacturer, banker or emigrant: but the Scottish missionary, explorer and anti-slavery campaigner David Livingstone, who set out on his epic expedition to cross Africa in 1852, and who would be a great public figure by the end of the decade. On his death in 1873 his heart would be buried in Africa, and his body in Westminster Abbey.

EXHIBITING BRITAIN, ENCOMPASSING THE WORLD

Such was the varied condition of the United Kingdom and the British Empire at mid-century. There might be more evidence than before to support Palmerston's optimistic depiction of Britain, at the end of his 'Don Pacifico' speech, as a country that was at ease with itself, and as being pre-eminent among the nations of the world. Making essentially the same point, albeit less aggressively, a Methodist factory owner and former campaigner against the Corn Laws, Absalom Watkin, observed that 'Never have I seen clearer evidence of general well-being. Our country is, no doubt, in a most happy and prosperous state. Free trade, peace and freedom.' Macaulay could scarcely have put it better. But this was only a partial view, contradicted by many other contemporary writers, especially Marx and Mayhew. Moreover, the economic recovery from the downturn of 1848 was by no means yet certain, and while the worst of the Great Famine in Ireland might be over, its baleful effects would be long felt. And the majority of the British population was under-nourished, under-educated and possessed only a limited acquaintance with the basic tenets of the Christian religion. This might be a version of 'freedom', and the British state was undoubtedly less intrusive and less authoritarian than many of its European counterparts; but such liberty clearly had its price, and the fact that so many people were deciding to leave Britain and (especially) Ireland in the hope of making a better life overseas offers another serious qualification to Palmerston's Panglossian peroration. One such emigrant was Dickens's fictional creation Mr Micawber, in *David Copperfield*. Constantly in debt, but always hoping that something might eventually 'turn up', he emigrated to Australia where his fortunes did indeed improve, and he became a local worthy and mayor of his town.

These paradoxes and anxieties were vividly (and in some cases unconsciously) displayed in a remarkable and original scheme that resulted in the most significant and defining (and variously understood and interpreted) occasion in the history of the United Kingdom between the Battle of Waterloo in 1815 and Queen Victoria's Diamond Jubilee in 1897. The idea that eventually resulted in the Great Exhibition of the Works of Industry of All Nations originated late in 1844 with a certain Francis Wishaw, secretary of what would soon become the Royal Society of Arts. Concerned about the inadequacies of workers' education and British standards of design, he believed there would be great public benefit if British industrialists and manufacturers, following recent continental precedents in Berlin and Paris, could be persuaded to put their products on display; this in the hope of increasing public and consumer interest in their work and wares, and of encouraging friendly competition between them, thereby improving the overall standards of workmanship and design. But there was scarcely any support for Wishaw's idea from the manufacturers he approached, and it was not until the matter was taken up later in the decade by Henry Cole that it gained traction and momentum. Cole was by turns a civil servant, a journalist, an historian, and a passionate believer in progress, commerce, freedom of thought and free trade. He was also very well connected, a tireless campaigner for the causes in which he believed, a member of the Society of Arts, and seriously interested in the subject of industrial design. In 1847, and again in the following year, Cole put on exhibitions of British-made goods that were more enthusiastically supported by industrialists and also began to arouse a certain degree of public interest.

Determined to build on this initial success, Cole lobbied members of Russell's government, and his own friends in the Commons, which resulted in the setting up of a parliamentary select committee to explore standards and methods of industrial design; in 1849 he obtained the support of Queen Victoria and Prince Albert for a third exhibition, and it was their patronage that helped ensure the unprecedented and enthusiastic involvement of many manufacturers. Thus encouraged, Cole persuaded the Society of Arts to sponsor what would be the first international exhibition of manufactured goods, which would be held in London in 1851. He gained the support of Prince Albert and Sir Robert Peel, and in January 1850 a Royal Commission was established to oversee the staging of the exhibition, to superintend the planning and the buildings required, to encourage manufacturers from overseas to display their products, and to devise a scheme for the impartial awarding of prizes for the best exhibits. The Commissioners included Prince Albert, the Duke of Buccleuch and Earl Granville; Lord John Russell, Lord Stanley, Sir Robert Peel, Richard Cobden and Mr Gladstone; the presidents of the Royal Society, the Royal Academy, the Geological Society and the Institute of Civil Engineers; the chairmen of the Manchester Chamber of Commerce and of the East India Company; the architect Charles Barry, the Leeds wool manufacturer John Gott, and two of the nation's greatest bankers, Thomas Baring and Lord Overstone. They came from varied geographical, social and occupational backgrounds, but most of them were wealthy and well educated, and laid claim to a wide experience of politics and public service. Finance, industry and commerce were as well represented as the landed interest, the majority of the Commissioners had supported measures of political reform and were committed to free trade, progress and improvement, and only four of them were protectionists.

The ideological centre of gravity of the Commissioners was thus emphatically more Peelite, Whiggish, Liberal and even radical than it was Conservative, Tory or protectionist. Most of them were seriously interested in science and the arts; they were generally in favour of individualism, competition and free trade; they preferred to emphasize what different social groups within the United Kingdom had in common rather than their competing sectional interests; they understood Christianity in a broad and undogmatic manner that might further encourage a sense of national solidarity across the British Isles; and they believed that international amity and the brotherhood of mankind were economically possible and religiously essential. As befitted the transitional nature of the mid-century British economy, they understood 'industry' to mean not just machine-made artefacts, but anything produced by honest toil and manual labour; and they hoped that the exhibition, in displaying the highest-quality goods produced by all the advanced nations of the world, would serve to alert British manufacturers to the shortcomings of many of their products, and to the educational deficiencies of many of their workers. Yet for all that the Commissioners had in common in terms of unspoken assumptions and a shared outlook, it was far from clear, at the beginning of 1850, that they would be able to organize and finance the greatest international exhibition that the world had ever seen by the late spring of the following year. Where would the money come from? Would the British public be interested? And would industrialists and manufacturers from the United Kingdom and around the world accept the invitation to exhibit their goods and products in London? The road from January 1850 to May 1851, when the Great Exhibition was due to open, would be far from smooth, and the schedule was exceptionally tight.

During the first half of 1850 the Commissioners made scarcely any substantive progress. The original idea had been to obtain private sponsorship for the exhibition, in exchange for a share of the profits; but this was soon deemed inappropriate for such a major national event, and the Commissioners decided to look for more broad-based funding from the public instead. The subcommittees of the Commission, charged with particular responsibilities, did not always see eye to eye, and Cole's zealotry and determination sometimes turned out to be counter-productive. There was widespread initial opposition, both among the public and in parliament, to the decision to site the exhibition in Hyde Park, on the grounds that it would damage the trees, the flowers and the lawns, and that it would lead to an influx of domestic and foreign undesirables at the very heart of the city; and it took time to overcome this early resistance. Moreover, the original proposal for a brick building that would be four times the length of Westminster Abbey and topped by a dome that would be larger than St Paul's in London or St Peter's in Rome was met with widespread scorn and ridicule, even though much of it had been designed by Brunel. By the summer of 1850 the whole scheme seemed truly in danger of foundering, and the death of Sir Robert Peel in early July robbed the Commission of its most esteemed and committed supporter. Yet Peel's passing seems to have galvanized the Commissioners into a belated but more determined form of action. By the end of the year they had agreed on a building of revolutionary design, to be constructed of iron and glass, which had been proposed by Joseph Paxton, and was developed from the hothouses he had recently created for the Duke of Devonshire at Chatsworth. The Commissioners also succeeded in obtaining the necessary financial support, in large part from the City of London, but also from the great industrial towns of the Midlands, the north of England, south Wales and the Forth-Clyde region of Scotland.

Having finally settled on the site, the building and the funding, the Commissioners were overwhelmed with offers of objects for display; when it opened the Crystal Palace (as Paxton's building became known) would house more than 100,000 exhibits sent in by 14,000 individual and corporate lenders, selected by hundreds of committees from Britain, its colonies and other countries. As one contemporary observed, the alphabetized list of exhibits stretched all the way from 'Absynthium', provided by a Sardinian, to 'Zithers' sent in by two Viennese manufacturers, and the exhibits were divided into four categories: raw materials (minerals, metals, chemicals, food); machinery (ranging from railway engines via industrial equipment to military engineering and agricultural implements); manufactures (cotton, leather, clothing, cutlery, jewellery, glass, ceramics); and the fine arts (painting, architecture, sculpture, mosaics, enamels). To the relief of the Commissioners, the Great Exhibition was opened by Queen Victoria on time on 1 May 1851 in the presence of a glittering array of national worthies and international visitors; more than 20,000 people mobbed the Crystal Palace on the first day alone; thousands more lined the streets of London to catch a glimpse of the queen and Prince Albert. During the next five months more than six million people paid to visit what many considered the eighth wonder of the world, and by the time the exhibition closed in October 1851 it had probably been seen by one-fifth of the entire population of Britain. There had never been anything quite like it before in the whole of human history: *The Times* described the opening as 'the first morning since the creation of the world that all peoples have assembled from all parts of the world and done a common act'; and such a display could only have occurred in the United Kingdom – and it could only have happened in London.

For most among the millions of visitors it was the sheer, miraculous abundance of the *things* that were on display, drawn from all the four corners of the globe, that was so extraordinary and unforgettable: 'All of beauty, all of use,' Tennyson wrote in his 'Exhibition Ode', 'That one fair planet can produce.' Indeed, some visitors were bewildered and astonished, not only by Paxton's innovative and futuristic building, but also by the overwhelming quantity and variety of the goods inside it; and one admitted to being in a 'state of mental helplessness' before such an embarrassment of riches. But it was the people who crowded into the Crystal Palace who were themselves at least as extraordinary as the artefacts on display, as the millions of visitors came from all parts of the United Kingdom, and from a remarkably wide range of social backgrounds. Most of those who showed up at the beginning were from among the traditional titled and territorial classes, and the professional, entrepreneurial and financial bourgeoisie; but towards the end of May 1851 'shilling days' were introduced, and thereafter, the majority came from those much lower down the social scale. Many were from London, but great numbers also journeyed in from the provinces on excursion trains that the recently constructed railway network now made it possible to provide. Indeed, it was claimed that as a result more Britons were moved about in 1851 than ever before in the nation's history, and they had travelled on the trains and visited the Crystal Palace in a friendly and harmonious way, which belied the social tensions and conflicts of earlier decades and recent years; 'never before,' John Tallis claimed in his three-volume history of the Great Exhibition, 'had there been so free and general a mixture of classes as under that roof'. According to one contemporary, 'the exceeding popularity of the exhibition eventually became its greatest wonder'. As *The Times* put it a few days before the Crystal Palace closed on 11 October 1951, 'the people have now become the Exhibition'.

Yet although the great display of 1851 undoubtedly engendered a widespread sense of wonder among those millions who visited, it also sent out many other messages and varied signals. As befitted the hybrid and transitional nature of the mid-century British economy, where handicraft, workshop and factory production co-existed, domestic 'industry' was exhibited in many different guises, many of the artefacts on display were the work of individual craftsmen, and despite the modernity of Paxton's palace its 30,000 glass panes had been hand-blown rather than mass-produced. In its conception and execution, the Great Exhibition was never just, or even primarily, a display of a triumphant, modernizing, factory-based manufacturing. Those who were the least enthusiastic about the exhibition were the farmers and the agriculturalists, which helps explain why many more exhibits came from England than from Wales (with the exception of the southern coalfields), or from Scotland (apart from the industrial districts of the central Lowlands), or from Ireland (where only Belfast, Dublin and Cork contributed more than a few items). The exhibition was primarily intended as a place of instruction and education, and to be 'more of a school than a show', by introducing a wide public audience to the marvels of manufacture and the miracles of industry; but given the prevailing levels of working-class schooling and literacy, it is unlikely to have achieved that aim. It was also meant to display examples of best practice, from which British manufacturers might learn; but for many of them, the Great Exhibition was more of a marketing opportunity than a learning exercise, and by the time the Crystal Palace closed most of the goods on display had, indeed, been sold.

The Great Exhibition was also, appropriately, about the Greater Britain beyond the United Kingdom, for among the displays located at its very heart were those drawn from what the *Official Catalogue* described as 'British Possessions in Asia' (India and Ceylon), Europe (Malta, Gibraltar and the Channel Islands), Africa (the Cape of Good Hope, Mauritius and the Seychelles), the Americas (Canada and the Caribbean Islands) and the Antipodes (Australia and New Zealand). Here again the messages sent out were ambiguous. The very phrase 'British possessions' suggested a greater degree of metropolitan control and imperial coherence than were in fact the case; although the exhibiting of the Koh-i-noor Diamond, which had been taken by the British after their recent conquest of the Punjab, was a spectacular trophy from Dalhousie's recent aggressive annexations in India. Most of the colonies were depicted as sources of raw materials that were exported to Britain (among them timber and furs from Canada, wool from Australia, and cotton from India), and also as markets to receive British manufactured goods in return. Some parts of the empire also used the opportunities presented by their participation to present a particular version of their own histories and hopes: the Australian colonies, for example, had transformed themselves from convict settlements to a vital component in the empire of the future; while the Governor of Cape Colony hoped that the South African exhibits might encourage more Britons to emigrate there. More generally, the Great Exhibition brought the British Empire 'home' to the British capital and to the British people, in novel ways, by putting it on display within the Crystal Palace, thereby enabling visitors to encounter far-off places they had never seen and in all likelihood would never see. For Britons who might have had little or no connection with the empire, the Great Exhibition made clear that it was a vital component of the nation's wealth, power and global prestige.

Yet the Great Exhibition also aroused considerable controversy, concerning both its ethos and displays, and its critics assailed it from a variety of contradictory perspectives. The young William Morris found the Crystal Palace and everything inside it to be 'woefully ugly', and John Ruskin, who was no admirer of such newfangled materials as iron and glass, was of the same view. So was Augustus Pugin, whom the Commissioners generously allowed to install a Medieval Court in the Gothic Revival style, which celebrated traditional craftsmanship and pre-factory methods of production, and was thus a reproach to almost every other exhibit on display. In a similar vein, Ralph Nicolson Wornum, the art historian and later Keeper of the National Gallery, criticized both manufactures and manufacturers for their superficiality, incongruity of style and excess of ornamentation: 'The paramount impression conveyed to the critical mind,' he wrote, 'must be a general want of education in taste.' But the anti-industrial critics did not have it all their own way. One anonymous author, writing in the *Eclectic Review*, thought Pugin's sham Medieval Court to be a 'poor, meretricious make-believe', with its 'false outlines and incongruous ornament', which was decidedly inferior to modern machinery where form followed function. Meanwhile, the Scottish engineer and inventor James Nasmyth thought the very idea of trying to improve the quality of British goods so that they could rival the best produced on the continent was mistaken and misguided – and unnecessary. British manufacturers, he insisted, would always be 'beat by foreigners in respect to abstract perfection', and so instead of trying to compete with them on their own terms they should concentrate on 'quantity and fair average excellence'. 'Quantity,' he insisted, perhaps a touch optimistically, 'is what England will ever be supreme in. Economic production is our forte, and ever will be.'

Thus regarded, the Great Exhibition might seem to have been an essentially secular enterprise (none of the original Commissioners was an ordained clergyman), primarily concerned with celebrating progress, proclaiming improvement and validating private enterprise. But while the United Kingdom might be in the vanguard of modernity, the exhibition also demonstrated that there was already serious international competition. As the Commissioners had recognized from the outset, some of the best-designed goods and the most advanced industrial products had been manufactured in Europe or North America, and in areas such as hydraulic or electrical machinery, which required the application of advanced scientific knowledge, Britain was noticeably lagging behind the United States, to the dismay of observers like Lyon Playfair and Charles Babbage. Beneath the widespread sense of self-congratulation, there was thus some justified anxiety about the United Kingdom's future economic prospects. Nor was this the only way in which the Great Exhibition expressed what might be termed the contradictions of mid-nineteenth-century progress. For, as was made plain by the concern over the religious census and the popular outbursts against the restoration of the Catholic hierarchy, Britain was still in many ways a faith-bound society, and from its planning stages there was much religious comment and controversy about the Great Exhibition. To its critics, the Crystal Palace was a latter-day Tower of Babel, where hedonism, godlessness and the worship of material things were rife, and which deserved, and would receive, godly punishment and retribution. But to its admirers, it was a divinely inspired undertaking: at the opening ceremony the Archbishop of Canterbury had prayed for the Almighty's blessing on the exhibition, massed choirs had sung Handel's Hallelujah Chorus, and it was confidently proclaimed that 'we are carrying out the will of the great and blessed God'.

This conviction that God was somehow on the side of the Great Exhibition also fed easily into its overarching theme of pacifist internationalism. Prince Albert believed it would be a festival of 'peace, love, and ready assistance, not only between individuals, but between the nations of the earth', and Richard Cobden thought it would 'break down the barriers that have separated the people of different nations, and witness the universal republic'. But in the era of 'Don Pacifico' this Christian internationalism also morphed into a more specific insistence that the Great Exhibition proclaimed the United Kingdom was providentially destined to global pre-eminence as the instrument of divine purpose. Such was the view of Charles Kingsley:

> The spinning jenny and the railroad, Cunard's liners and the electric telegraph, are to me . . . signs that we are, on some points at least, in harmony with the universe; that there is a mighty spirit working among us . . . the Ordering and Creating God.

But this, too, was a contested verdict, and there were others, among them Charles Babbage, who put the Great Exhibition in a broader, more sceptical, better-informed and less complacent perspective:

> It is a grave matter of reflection, whether the Exhibition did not show very clearly and distinctly that the rate of industrial advance of many European nations, even of those who were obviously in our rear, was greater than our own; and if that were so, as I believe it to have been, it does not require much acumen to perceive that in the long race, the fastest sailing-ships will win, even though they are for a while behind.

'The Roman Empire,' Babbage concluded, in appropriately and ominously Gibbonian terms, 'fell rapidly because, nourishing its national vanity, it refused the lessons of defeat and construed them into victory.'

AFTERMATH AND AFTERSHOCKS

The final day of the Great Exhibition was 11 October 1851, when more than 50,000 people visited the Crystal Palace; those who remained until closing time at five o'clock in the afternoon sang the national anthem; then there was widespread cheering and the visitors made their way to the exits to the pealing of bells. But what was the legacy of this unprecedented display? As befitted an exhibition that from the planning stage had been invested by different people with multiple meanings, its long-term impact was correspondingly varied. For many Britons, and for the remainder of the nineteenth century, the Great Exhibition would become increasingly bathed in a nostalgic glow: its successor, the International Exhibition held at South Kensington in 1862, would seem lacklustre by comparison; and ten years later, Jane Budge observed that 1851 had been 'the brightest year of all Victoria's reign, when . . . the world was at peace, and England, prosperous at home and honoured abroad, saw men of almost every nation gather in friendly rivalry to her shores'. More prosaically, the Crystal Palace itself was dismantled, and re-erected, with embellishments and additions, in the south London suburb of Sydenham, where it would play host to more exhibitions, as well as concerts and sporting events, until it burned down in 1936. More lastingly, the surplus that the Great Exhibition accumulated, close to £200,000, would be used to acquire land in South Kensington on which would be constructed the Victoria and Albert Museum, the Natural History Museum and the Science Museum, as well as the Imperial College of Science and Technology and the Royal

Albert Hall. This institutionalized commitment to science, the applied arts, and education and design was the Prince Consort's abiding legacy to his adopted land, and no member of the British royal family since has made so many-sided a contribution to the cultural and intellectual life of the United Kingdom.

Soon after the Great Exhibition had come to its triumphant close, international affairs again intruded in British public life, as events in France would have major repercussions on domestic politics. In the aftermath of the fall of the July Monarchy in the revolutions of 1848, Louis-Napoleon, nephew of the former French emperor, had been elected President of the French Republic; but in December 1851 he carried out a coup d'état against the constitution, abolished the republic, and declared himself to be the Emperor Napoleon III in direct succession to his uncle. Without asking the permission of the cabinet or the queen, Palmerston immediately expressed the British government's 'entire approbation' of Louis-Napoleon's unconstitutional actions. This seemed an uncharacteristically supine action by a Foreign Secretary who prided himself on being generally opposed to continental despots, while the unhappy memories of the long, hard wars against the first Emperor Napoleon still lingered in Britain in the popular mind and in political circles. But Palmerston believed Napoleon III to be an anglophile and thought he would be more friendly to Britain than any alternative French regime. Nevertheless, the queen and Prince Albert were outraged: they disliked what they regarded as their Foreign Secretary's high-handed behaviour combined with his loose eighteenth-century morals, and they urged Russell to dismiss him, which he duly did before the year was out. Russell and Palmerston, whom Queen Victoria would later describe as those 'two terrible old men', had never got on, and this was a chance for the prime minister to reassert his waning authority over his cabinet. But not for long, for Palmerston was determined to have his 'tit for tat' with Russell, and in February 1852 the government was defeated in the Commons, on a motion proposed by Palmerston, whereupon Russell resigned.

The end of Russell's Whig administration would usher in a new and very different era of parliamentary politics, in which the relatively clear party divisions between Whig-Liberals and Tory-Conservatives, which had held up for much of the 1830s and 1840s, would no longer be so strongly marked. Nor was this the only sense in which, precisely at mid-century, the old world seemed to be giving way to the new. For just as the Great Exhibition and the Crystal Palace seemed to point towards the future, in which the second half of the nineteenth century might be very different from the first half, so the great men of the past, who had linked the United Kingdom to those stern and demanding late eighteenth- and early nineteenth-century years of turmoil, battle and war, were leaving the stage. Both Whig leaders of the 1830s had already quit the mortal coil: Earl Grey, who had passed the Reform Act, had died in 1845, and Lord Melbourne, his successor, had followed three years later. William Wordsworth, who had welcomed the new dawn of the French Revolution in the 1790s with naive and youthful ardour, but who had long since embraced more conservative views, died in April 1850, and his extensively revised autobiographical poem, *The Prelude*, on which he had been working since 1798, would finally be published a few months later. Soon after, Sir Robert Peel died, widely revered as the greatest statesman of his time, to whom statues were put up across the country, paid for by the pennies donated by grateful working men. And in the closing weeks of 1851 the country's greatest painter, Joseph Mallord William Turner, whose extraordinary oeuvre captures so much of the mentality and materiality of the United Kingdom as it changed and evolved from the 1790s to the 1840s, breathed his last. In his final years he had still been pushing the boundaries of his art in such new, audacious and semi-abstract directions that Queen Victoria would later tell her grandson and the future King George V that he was mad.

All these men had been alive at the time of the French Revolution, but on 14 September 1852 two men died who had entered the world more than forty years apart: Augustus Welby Pugin, who had only been born in 1812, and the venerable Duke of Wellington, who had long preceded him in 1769. Like Turner, Pugin was an undoubted genius, but pace Queen Victoria, he was much the closer of the two men to madness, and he had descended into complete insanity just before his death. His Medieval Court at the Crystal Palace had been much criticized by Evangelicals for importing popery and superstition into the very heart of the Great Exhibition, and one Anglican clergyman visiting from Liverpool claimed that it was 'filled with Babylonish garments and Tractarian toys'. Two months after Pugin's death the new Palace of Westminster, which in its ornamentation and iconography was so much his creation, and which was the very antithesis of Paxton's modernistic Crystal

Palace, was opened. By then the Duke of Wellington was still unburied, and he would not be interred in St Paul's Cathedral, next to Nelson, until eight weeks after his death. Fresh from the success of the Great Exhibition the year before, Prince Albert was determined that the great duke should be awarded a funeral so spectacular that it took two months to organize, and to construct the overweight and over-embellished 'funeral car', which would be drawn by twelve horses, but which was so heavy that it frequently got bogged down in the London streets on its way to St Paul's on the day of the funeral. As a ceremonial the duke's obsequies were rather a shambles; but it was the greatest display of public mourning between Nelson's funeral in January 1806 and Gladstone's in May 1898, and one and a half million people turned out to watch. And in his first major task as Poet Laureate, Alfred Tennyson composed a majestic funeral ode, which caught the sense of epochal transition as it lamented 'the last great Englishman is low'.

With the exception of the crisis years during the passing of the Great Reform Bill, Wellington had been the nation's most venerated and iconic figure since Waterloo: venerated because the splendour of his military deeds were widely remembered and gratefully recognized; and iconic because in paintings, prints and engravings, and on many consumer objects, he had become the most visually depicted Briton ever to have lived. His last service to the state of which he was both pillar and ornament had been in the year before his death. Worried by the sparrows that were nesting in the Crystal Palace, Queen Victoria had asked the duke's advice: 'Sparrowhawks, Ma'am', was his reply, and the unwanted birds were duly dispatched. But the excessively ornamented funeral car, reviled by Thomas Carlyle as an 'incoherent huddle' of drapery, flags and emblems, suggested that, in inadvertent corroboration of one of the didactic aims of the Great Exhibition, British design did indeed need improving. A second funereal solecism was committed by Tennyson when he described Wellington as 'the last great Englishman': for he was Anglo-Irish in his ancestry, and a fervent believer in the greatness of Britain and its empire. But the Poet Laureate was correct in wondering how the nation would manage without its one undoubted hero, who had for so long been the strongest living link with the Napoleonic Wars, and with the triumphs that in retrospect increasingly outweighed the tribulations of those times. That was certainly Queen Victoria's anxiety, which she shared with many of her subjects, namely that the age of giants was over and the era of pygmies was at hand. This was an unnecessarily gloomy apprehension, for there would be no lack of heroic figures during the years ahead (the queen and the Poet Laureate among them, to say nothing of Mr Gladstone).

But in the short run the anxious mood was well justified, for while the remainder of the 1850s would witness continued social stability and growing material prosperity, British pre-eminence as the global hegemon, and the very core of Victorian religious beliefs, would both be shaken to their foundations. Already some contemporaries were wondering whether the United Kingdom could hope to hang on to its far-flung empire, with its varied territories and governmental structures, that was run on a shoestring, and held together by sentiment and bluff at least as much as by power and force. And what of the American republic, rapidly developing as a land-based empire extending from the Atlantic to the Pacific? 'The superiority of the United States to England,' warned *The Economist* at just this time, 'is ultimately as certain as the next eclipse.' As the largest English-speaking nation overseas, the great republic was already vastly superior in people and resources to those of the scattered British communities in Canada, in the Antipodes and in South Africa. This was one of many reasons why politicians, policymakers and pundits in London would become increasingly concerned with the future of these colonies of settlement, and from the 1860s onwards they would move ever higher up the agenda of imperial preoccupations and priorities, eventually overtaking India itself. The consolidation and defence of these dispersed communities, initially in the case of British North America, then in Australia, and finally (and least successfully) in South Africa, would become a major task of imperial statecraft, which in turn would lead to further calls to draw the whole of the settlement empire together in some greater form of imperial federation.

By mid-century the evolving and expanding British Empire was raising a further difficulty for those in the imperial metropolis, which was not geopolitical but religious. Ever since the establishment of the American colonies, many Britons had regarded their worldwide imperium as the providential instrument and Protestant expression of God's divine purpose. Yet in practice, the majority of its subjects increasingly espoused very different faiths: there were Catholics in Ireland and Canada; there were adherents of the Dutch Reformed Church in

South Africa; and the teeming millions who lived in British India were overwhelmingly Muslims and Hindus. Moreover, these inhabitants of South Asia constituted by far the largest community in the empire, and missionary efforts to convert them to Christianity had generally proved unavailing. In terms of its aggregate population, therefore, the British Empire was no longer a Protestant imperium at all, and there was no realistic likelihood that it ever would be again. This not only eroded the belief that the British were pre-ordained to rule the globe and in the process convert the 'heathen' to the superior religion of the Christian faith: it also fed the growing mid-Victorian doubts about the very validity of the Bible itself, on which Tennyson meditated in *In Memoriam*, published in 1850. For while he emphatically reaffirmed the importance of the Christian religion, and reasserted his belief in the existence of an afterlife, he also recognized that it was no longer possible to maintain the literal truth of the Bible's account of the Creation and the Flood. There might, Tennyson concluded, be grounds for faith in humankind, but it was no longer certain just where the balance lay between doubt and belief, anxiety and hope:

> I falter where I firmly trod,
> And falling with my weight of cares
> Upon the great world's altar-stairs
> That slope thro' darkness up to God,
>
> I stretch lame hands of faith, and grope,
> And gather dust and chaff, and call
> To what I feel is Lord of all,
> And faintly trust the larger hope.

7
Equipoise and Angst, 1852–65

Lord John Russell's administration had been an unconscionable time a-dying, and it had taken twelve months from February 1851 for it finally to expire. It was effectively brought down by Palmerston, whom Russell had recently dismissed, who took his revenge by joining with the Tory opposition and carrying an amendment to the Militia Bill in the Commons against the government. Russell resigned for a second time, and the fourteenth Earl of Derby (as Lord Stanley had become the previous year on inheriting the title from his father) formed the first of three short-lived minority Conservative administrations over which he would preside across the next sixteen years. Although almost entirely forgotten today, Derby was one of the great figures of nineteenth-century British public, social and cultural life: he was a fine debater, a classical scholar of note and a significant patron of the turf; he was also an authentic grandee, with very rich, coal-bearing estates in Lancashire, and leader of the Conservative Party for an unrivalled span of twenty-two years. But he had not started out on the right in politics. As Lord Stanley, he had entered parliament in 1820 as a Whig, and he had held junior office under Lord Grey at the time of the passing of the Great Reform Act; but he had broken with the government over its reforms to the Church of Ireland and he had resigned in 1834. He subsequently embraced Peel's new brand of reforming Conservatism, and became Colonial Secretary in the government Peel formed in 1841. But once again Stanley fell out with his erstwhile leader, resigning in protest against the repeal of the Corn Laws, and from 1846 he was the head of what was essentially the Tory protectionist rump, since such devout and devoted Peelites as Graham, Herbert, Aberdeen and Gladstone were determined to keep their distance.

Derby was thus a very seasoned – albeit somewhat maverick – politician when he formed his first administration. But this could not be said of his colleagues, for most of the talent and experience in the Conservative Party of the 1840s had resided with the Peelites, and once they had departed there was little left except the landowners, agriculturalists and protectionists who were an increasingly marginalized and fanatical sect, and although they would formally abandon protection in 1852, they would go down in posterity as 'the stupid party'. Not surprisingly, Derby found it difficult to form a cabinet from such unpromising and untried material. Very few of his colleagues would have held office had it not been for the Tory split in 1846, and the effective departure of the Peelites. Derby's administration would be remembered as the infamous 'Who?, Who?' government, so called because that was the question repeatedly asked by the old and deaf Duke of Wellington upon being told the names of the many obscure cabinet appointees. The Home Secretary, Spencer Walpole, bore a distinguished eighteenth-century name, but he had only sat in the Commons since 1847. The Foreign Secretary, Lord Malmesbury, admitted his only qualification for the post was that he had edited the correspondence of his grandfather, who had been ambassador to Spain, Russia, Prussia and France. And the Colonial Secretary was Sir John Packington, an unknown Worcestershire country squire and baronet whose chief claim to fame was that he had married three times. But the most extraordinary – yet also the most portentous – appointment was that of Benjamin Disraeli as Chancellor of the Exchequer. A Jew by birth who had converted to Anglicanism, he had attended neither public school nor an ancient university, and his reputation was that of a shameless adventurer, with no obvious political principles, and as a writer of somewhat pretentious but derivative fiction, in particular with *Sybil* (1845), his attempt at a 'condition of England' novel.

The Peelites (and especially Gladstone) disliked Disraeli because he had made his parliamentary reputation in the late 1840s by his savage, brilliant and wounding attacks on their leader, putting into unforgettable (and unanswerable) words the loathing and rage of the Tory agriculturalists and High Anglicans who believed that Peel had betrayed them, not once (over repeal) but twice (over Catholic Emancipation as well). Yet the churchmen and the protectionists did not like Disraeli either: he was exotic, he owned no country house or land until he was in his forties, he had no family links to the aristocracy, and he was an almost complete outsider. So when the Tories split in 1846, the protectionists had initially preferred Lord George Bentinck to be their Commons' leader; and he was followed by another patrician, Lord Granby, who would later become the sixth Duke of Rutland. Only in February 1852, with the formation of Derby's first ministry, did Disraeli become the undisputed head of the Conservatives in the Commons, and he would remain their leader until he departed for the Lords in 1876. But the post of Chancellor of the Exchequer was one for which he was completely unfitted: his personal finances had always been chaotic, and mastery of the intricate details of Treasury business and public accounts was not his forte. Moreover, the Tory rank and file wanted protection restored, but this was impossible for a minority government facing a free trade majority in the Commons. The resulting budget was simultaneously a bundle of expedients but also too clever by half: in an effort to placate the agricultural interest, Disraeli reduced the malt tax; to try to appease the radicals he distinguished between 'earned' and 'unearned' categories of income tax; and there were many sleights of hand in the figures he presented. Gladstone, who had never forgiven Disraeli for his attacks on Peel, responded by delivering a powerful speech that cut the budget to pieces. It was defeated by 305 votes to 285, and Derby's government resigned after less than a year in office.

POLITICS IN FLUX

The clash between Gladstone and Disraeli over the budget of 1852 was the beginning of a celebrated personal rivalry and intense political animosity that would last until Disraeli's death in 1881, and it would become one of the defining features of the public life of the time. Both had started out as Conservatives, but although Disraeli had broken with Peel, he would remain a loyal Tory, convinced of the importance of maintaining party distinctions; whereas Gladstone, who was abidingly devoted to Peel, was beginning a political journey that would take him in a very different direction, towards what would soon become the Liberal Party and eventually (at least according to Queen Victoria) to a kind of 'half-mad' radicalism. Yet this is to anticipate, since for the best part of the next decade and a half neither Gladstone nor Disraeli would be the foremost political leaders of the day. Wellington's generation might have moved on, but for the remainder of the 1850s, and during the first half of the 1860s, all the prime ministers had been born in the closing decades of the eighteenth century, and by the standards and statistics of the time they were exceptionally long lived: Aberdeen lasted from 1784 until 1860; Palmerston had been born in the same year, and would go on until 1865; Russell, the greatest survivor of them all, had been born in 1792 and would live on until 1878; while Derby had entered the world in 1799 and would die the youngest of the four in 1869. Palmerston would be the first ever prime minister to be in office at the age of eighty, and Russell would outlive him by six years. By contrast, Disraeli had been born in 1804, Gladstone five years later (the same year, 1809, as Charles Darwin), and both Queen Victoria and Prince Albert in 1819. So whereas Aberdeen, Palmerston, Russell and Derby were from the last generation of the eighteenth century, Disraeli, Gladstone, Queen Victoria and Prince Albert (and Darwin) were from the first of the nineteenth.

At the beginning of the 1850s, Victoria and Albert were still only in their early thirties. She had become queen at the age of eighteen, and they had married when they were barely twenty. Yet for different but equally compelling reasons neither of them was overawed by the fact that the majority of the British politicians with whom they were dealing were not only much older and more experienced in the affairs of the world, but (at least compared to the queen) also much better educated. In Victoria's case, this was because she was headstrong and wilful, and although Albert would regard it as his main task in life to cure her of these temperamental shortcomings, he never in fact succeeded. Hence in part the queen's adolescent infatuation with Lord Melbourne, and her determination to keep him in office, and to keep Peel out, at the time of the so-called 'Bedchamber Crisis'. But although the prince would also come to possess strong feelings for or against the public men with whom he dealt, the reasons were significantly different: for Albert, as befitted the son of a ruling German Protestant prince (albeit a minor one), possessed a strong sense of the royal prerogative; he envisaged the role of the monarchy in British public life as being fundamental rather than ornamental; he wanted a crown above party, not so it should be marginal and neutral but so it could be disinterestedly involved in affairs of state; he was eager to be actively engaged with ministers in their governing and their decision making; and he thought it entirely appropriate that he and the queen should conduct their own foreign policy by direct correspondence with their royal relatives who occupied many of the thrones of Europe. Moreover, Albert's success in working with many of the leading public and political figures of the day over the Great Exhibition seemed a convincing vindication of this view.

In terms of the flux and flow of politics, the times were also propitious for an exertion of royal power, even though it was never as decisively or as successfully wielded as Victoria and Albert might have wished. To be sure, the passing of the Great Reform Act had given a notable impetus to party organization, with the establishment of the Reform Club and the Carlton Club; but these were neither catering to nor mobilizing a mass electorate. In the aftermath of the Great Reform Act only one in seven of the adult population of the United Kingdom had the vote (in England and Wales the figure was one in five): no women possessed the franchise, having been explicitly excluded, and the English electorate was probably a smaller proportion of the population in the 1830s and 1840s than it had been during the late seventeenth and early eighteenth centuries. Although most of the worst corruptions and abuses had been eliminated, landed influence remained strong in many boroughs and counties: Buckinghamshire was Tory on account of the influence wielded by the Duke of Buckingham, just as Bedfordshire was Whig because of the sway exerted by the Duke of Bedford. Moreover, successive *general* elections, and notwithstanding that adjective applied to them, remained agglomerations of particular contests often fought on local issues. A high proportion of seats, sometimes more than half, were not contested; and despite Peel's Tamworth Manifesto, national party programmes were of little significance to ministers or MPs or voters. As Peel discovered when repealing the Corn Laws, what was popular in the country was not necessarily popular with many of the party rank and file in the Commons.

So while the identity and organization of the two major parties had become more marked after 1832, every ministry thereafter, and especially in the years from 1846, faced the problem of the (eventually uncontrollable) indiscipline of some of its supporters. This had been true of Melbourne, Peel and Russell, all of whom had lost their Commons majorities by the end; and from 1846 to 1865 only Palmerston's administration, formed in 1859, would retain its rank-and-file support intact until the dissolution in July 1865. But it was not just that governments found it hard to retain the loyalty of their ostensible but often erstwhile followers; it was also that after 1846 neither of the two largest parties ever won enough seats to command unaided the votes they needed to survive in the House of Commons. The Tories (or Conservatives), led by Derby and Disraeli, were the larger of the two, but their numbers fluctuated between 230 and 280, and before Derby's death they never achieved a clear majority. And although the Whigs were merging almost imperceptibly with those who were calling themselves Liberals, they were never as large a Commons' contingent as the Tories, and in any case their two leaders, Russell and Palmerston, were often rivals and opponents more than allies. That left three other groupings: the radicals, along with the Peelites and the Irish, who exerted a disproportionate influence just because neither the Tories nor the Whigs could deliver a Commons majority on their own. All three groupings disliked the Whigs, for being too grand and cliquish, for living too well, and for being lukewarm about further reform; but their shared disdain for the Conservatives, who remained strongly committed to the Church of England, was even greater. This meant these additional groups were more likely to support the Whigs than the Tories, and explains why Lord Derby would lead three minority governments; but sooner or later the backing of these groups for the Whigs also invariably fell away.

These, then, were the mid-century years in which the House of Commons, rather than the still very restricted electorate, made and unmade governments, and it was this unusual parliamentary fluidity and uncertainty that gave Victoria and Albert considerable scope to interfere. They did so immediately following the fall of Derby's first minority Tory government in December 1852, by asking Lord Aberdeen to form a coalition consisting of Peelites, Whigs, Liberals and radicals, the very groupings who had just brought the Derby administration down. Aberdeen, had been in public life for the best part of forty years, he had been the acknowledged leader of the Peelites since the split of 1846, and he would be the first Scottish prime minister since Lord Bute in 1763. Earlier in the year he had struck a deal with Russell, which had paved the way for the coalition, and which indicated that the Peelites were moving in a leftward direction. Even though Aberdeen's own followers amounted to scarcely forty MPs, the Peelites did very well out of the coalition arrangements: six of them were in the new cabinet, which was exactly the same number as the Whigs, who formed a much larger parliamentary phalanx, and there was only one radical. Russell was given the Foreign Office; and since Aberdeen and Palmerston had never agreed on international affairs, the new prime minister was eager to confine Palmerston to domestic matters by sending him to the Home Office. Aside from the premiership, the most senior office given to a Peelite was the Exchequer, to which Aberdeen appointed Gladstone, fresh from his parliamentary triumph in destroying Disraeli's budget. Disraeli had already had his say about this contrived cabinet even before it formally came into being, when he observed in one of his budget speeches that 'England does not love coalitions', by which he meant coalitions that kept the Tories out, which Aberdeen's emphatically did.

As if in corroboration of Disraeli's sceptical dictum, the Aberdeen coalition was intrinsically unstable from the very beginning. There was general Whig resentment that the Peelites, though so few in number, had obtained so many cabinet posts, but there were also long-standing personal animosities. Because of their previous clashes and disagreements, relations between Palmerston and Russell, and between Palmerston and Aberdeen, were never good, and the negotiations between Aberdeen and Russell, which had led to the creation of the coalition, had left Russell unhappy and resentful: in part because he felt the Whigs had been short-changed, and in part because he had annoyed Aberdeen by prevaricating over which offices he himself should accept. The two men also viewed the coalition very differently. Aberdeen had set out to construct an integrated government from the materials available, and he regarded Russell as one among many plausible claimants who had to be accommodated. Russell, on the other hand, saw the union between himself and the prime minister as an alliance of equals, and gave more weight than he should have done to Aberdeen's vague reassurance that at some future date he would resign the premiership in Russell's favour. Matters were made worse by Russell's subsequent erratic behaviour. In addition to the Foreign Office, he had taken on the Leadership of the House of Commons, which Aberdeen had been determined to keep out of Palmerston's hands. In February 1853, having held the post for less than three months, Russell resigned from the Foreign Office, and Lord Clarendon was appointed in his stead. Thereafter Russell held a succession of undemanding offices, and he became increasingly frustrated and resentful that Aberdeen showed no sign of relinquishing the premiership to him. As a result of this unstable triangle of incompatible personalities represented by Russell, Aberdeen and Palmerston, the past, present and future prime ministers, there was never any realistic likelihood the coalition would settle down.

Nevertheless, in terms of domestic issues, it could claim some achievements to its credit, especially during the early years, of which Gladstone's first budget was the most notable and portentous. Since 1846 the Peelites had repeatedly denounced what they regarded as the financial irresponsibility of the intervening administrations: in Russell's case because Peel's hard-won budgetary surplus had been squandered on what seemed to them to be misguided and costly relief programmes in the aftermath of the Great Famine; and in Derby's case because of Disraeli's unsound and incompetent budget. As the heir of Peel, and a zealous believer in the cause of financial probity and fiscal responsibility, Gladstone was determined that *his* first budget, presented in April 1853 in a speech lasting four and a half hours, would mark a return to prudent and expert management of the public purse. Unlike any of his predecessors, he sought to look at and to plan ahead for the next seven years. He reduced the tax on tea and abolished the excise duty on soap, thereby promoting the righteous causes of temperance and cleanliness. He extended the income tax to Ireland, which had previously been exempted, in return for the cancellation of the remaining Great Famine debts, and the payment threshold was lowered from £150 to £100. But these immediate extensions and increases were combined with a commitment to successive annual reductions in the income tax from the existing rate of 7d in the pound to its complete abolition by 1860, thereby giving posthumous vindication to Peel's earlier claim, on reintroducing the tax in 1842, that it was merely a temporary measure. This was a brilliant stroke, made possible by the buoyant economy and burgeoning government revenues of the early 1850s; and with this budget Gladstone established his reputation as the most significant politician among the younger generation, with a high-minded sense of purpose, a grasp of complex detail, and a command of the Commons that none of his contemporaries could rival.

Perhaps unexpectedly, since it was not his obvious metier, Palmerston proved himself a vigorous force at the Home Office, carrying a series of reforming and codifying measures that attempted to mitigate the worst consequences of industrialization and population growth, which would have been unthinkable in the days of Lord Liverpool's administration (to which Palmerston himself had belonged). He passed a new Factory Act that removed many of the loopholes in earlier legislation, and made it illegal for young people to work between the hours of six p.m. and six a.m.; a Truck Act that prohibited employers from paying their workers in goods instead of wages or compelling them to buy goods only in company shops; a Smoke Abatement Act that sought to stem the growing air pollution in towns and cities; a Vaccination Act that made the practice compulsory for all children; and a Merchant Shipping Act that consolidated previous legislation and improved standards of safety at sea. Palmerston was also a notable penal reformer, instituting the first juvenile-reform schools and separate prisons for young offenders, reducing the length of solitary confinement in prison from eighteen to nine months, and abolishing the last vestiges of the system of transportation, where the only remaining destination was Van Diemen's Land (recently renamed Tasmania). Of even greater long-term significance was the publication, in 1853, of what was known, after its co-authors, as the Northcote-Trevelyan Report, which would eventually result in the creation of a professional domestic civil service, recruited and promoted on the basis of merit rather than, as had hitherto been the case, on the basis of patronage, connection, inertia and seniority. The Report recommended that future appointments be made via competitive examination, and that promotions should also be made on the grounds of ability. These proposals would not be fully accepted until 1870; but they eventually resulted in the creation of a new mandarin caste: university educated, literate, incorruptible and apolitical, and pervaded by a tone of impartial public-spiritedness.

These measures, although of undoubted significance, and representing a substantial advance on what had gone before, were, nevertheless, more matters of legislative consolidation, administrative progress and technocratic competence than they were the outcome of great causes or crusades (though Gladstone did succeed in investing his budgets with a considerable degree of heroic moral grandeur). As such, they were appropriate for a period in which public engagement with domestic politics was significantly less extensive or impassioned than it had been for much of the 1830s and 1840s, and when party divisions were also less clear-cut, not least because the great domestic causes of earlier decades were conspicuously absent. These were, then, the 'anti-reforming times' that Gladstone would later describe (and regret). But had Russell had his way, that epithet would not have applied, at least to the years of the Aberdeen coalition. As his disenchantment and frustration grew, Russell again took up the cause of parliamentary reform, proposing a very modest additional enfranchisement. Although he had earlier insisted that the Great Reform Act, which he had himself introduced into the Commons, had been the final settlement of the issue, he now became the first front-bench politician to take up the cause again, and he duly proposed two bills, in 1852 and in 1854. But neither among the public nor in parliament was there any appetite for such a measure, and both bills were dropped in the face of opposition and apathy. Nor were these the only parliamentary casualties: in 1854 proposals concerning the Poor Law, education, the politics and the admission of Jews to parliament also failed to make it onto the statute book. As was so often true for governments during these years, the Aberdeen coalition's support in the House of Commons was fading fast.

Part cause, part consequence of the coalition's visible loss of political momentum was the intensification of the personal antagonisms involving Palmerston, Russell and Aberdeen. In December 1853 Palmerston resigned in protest against Russell's recent espousal of parliamentary reform, to which he himself was strongly (but also opportunistically?) opposed. He got a generally good press, and his resignation was also taken, rightly or wrongly, as signifying his strong support for a firmer anti-Russian policy in the crisis that was developing in eastern Europe. Eventually, but with little enthusiasm, Aberdeen was obliged to take Palmerston back, but he now allied himself with Russell against the prime minister, urging a more robust and belligerent policy towards Russia than that which Aberdeen was pursuing. Between the summer of 1853 and that of 1854 the coalition gradually drifted into war with Russia, and there was strong public support for more determined and resolute action, and thus also for Palmerston himself. But a succession of military disasters late in 1854 gravely weakened the standing of Aberdeen's government in parliament and also with the public. In January 1855, and by now wholly alienated and disaffected, Russell resigned from the cabinet, whereupon the rest of his former colleagues followed suit. Queen Victoria refused to accept their resignations, and sent

Russell himself a stinging rebuke. But this merely delayed the inevitable fall of the Aberdeen coalition. Later in the month the radical MP John Arthur Roebuck proposed the appointment of a committee (over which he would himself eventually preside) to inquire into the conduct of the Crimean War. It was carried by an overwhelming vote of 305 against 148, and it spelt the end of the coalition government. Aberdeen's reputation never recovered, and he would not hold office again.

It was now up to the queen and Prince Albert to appoint Aberdeen's successor, and they were determined it should not be Palmerston, whose private morals and public unreliability they alike detested. She turned a second time to Lord Derby, but he gave up when Palmerston and Gladstone both refused to join a Tory-dominated ministry. Russell, to whom the queen turned next, fared even less well; in the light of his part in the fall of the Aberdeen coalition, no politician of substance was willing to serve under him. And so with regret, the queen turned to Palmerston, who formed an administration closely resembling its predecessor, minus Aberdeen and Russell. Within less than a month the leading Peelites – Gladstone, Herbert and Graham – resigned, ostensibly because Palmerston had refused to veto Roebuck's inquiry into the war, more substantively because they were uncomfortable serving under a prime minister whose policies they in many ways disliked. Palmerston now headed a purely Whig-Liberal government, but it was not expected to last: the prime minister was already in his early seventies, John Bright thought him an 'aged charlatan', while Disraeli dismissed him as an 'impostor', a 'pantaloon', who was 'very deaf, very blind, and with false teeth'. Yet Palmerston would live for nearly eleven more years, and he would be prime minister until his death, only interrupted by the brief interlude of 1858–59 when there would be another minority Tory government led by Lord Derby. Palmerston's progressive rhetoric abroad (albeit often bluffing) and support for limited reform at home meant he could usually count on more support than the Conservatives, while his position in the country was one of unrivalled popularity thanks to his assiduous cultivation of the press and his generally strong rapport with public opinion, which saw him as vigorous and manly rather than old and failing. More than any other politician, Palmerston was both the embodiment and the beneficiary of what was undoubtedly, in domestic terms, the settled and relatively quiet times that have been called 'the age of equipoise'.

As prime minister he presided over the same sort of practical, but limited, reform that had earlier characterized his years at the Home Office. The stamp duty on newspapers was abolished, which led to a further sharp increase in their circulation; British companies were given limited liability status, which meant they could expand, by issuing shares for sale, without liability for any company debts or losses being incurred by shareholders; the Metropolitan Board of Works was established to oversee the sewers and drainage of London; and counties that had hitherto failed to establish a police force were now compelled to do so. The Matrimonial Causes Act of 1857 made it possible to obtain a divorce without resorting, as had previously been the case, to special parliamentary legislation, and was deplored by Gladstone for bringing divorce within reach of people of all classes; while the Obscene Publications Act of the same year, banning literature deemed liable 'to deprave and corrupt', tended in the opposite direction by proclaiming a new and more restricted kind of public morality. But as with the Aberdeen coalition, foreign affairs soon intruded. In October 1856 the Chinese authorities at Canton had arrested a small sailing vessel called the *Arrow*, which was flying the British flag. Dissatisfied with the apology and restitution that were offered, the Governor of Hong Kong ordered the bombardment of Canton, and when news reached London in February 1857, Palmerston, perhaps hoping to repeat his earlier 'Don Pacifico' triumph, naturally supported the Governor's decision; but there was no equivalent victory for him in parliament, where Cobden's motion of censure in the Commons was carried by 263 votes to 247, as the radicals and Peelites joined forces with the Tories, and as Disraeli, Gladstone and Russell, as well as Cobden and Bright, all spoke against him. Palmerston's response was to appeal to the country over the head of parliament, and in the subsequent election he won a majority of eighty-five seats over all his opponents, which was the largest since the Whig triumph in 1833.

The 1857 election result was a ringing endorsement for Palmerston's activist foreign policy, one indication of which was that both Cobden and Bright, who were opposed to any such overseas, state-backed British interventionism, both lost their seats. But Palmerston's triumph was short lived. In January 1858 an Italian refugee who had previously lived in England, named Orsini, tried to assassinate the Emperor Napoleon III in Paris. The French police discovered that the explosives had been made in Britain, and Napoleon demanded the British government take action to prevent further conspiracies, a request with which Palmerston thought it reasonable to comply. He duly introduced appropriate legislation, but with the febrile state of public opinion, and the fluid state of party politics, his government went down to defeat, as the Peelites and the radicals voted with the Tories. The government resigned, once again the queen sent for Derby, who as in 1852 tried to persuade Whigs and Peelites to join his cabinet, and as before they refused. This second minority Conservative government, in which Disraeli, Walpole and Malmesbury held the same cabinet posts that they had in 1852, lasted slightly longer than its predecessor, but still less than a year and a half. This time, some domestic legislation was passed: the property qualification for MPs was abolished; after a decade-long campaign, practising Jews were finally admitted to parliament; and there were some changes in the structure of local government and the administration of public health. At the same time, and perhaps in the hope of outflanking the Whigs and their supporters, Disraeli decided to take up the cause of parliamentary reform, introducing a modest bill that would lower the county franchise, extend the vote to those with certain specific qualifications, and offer a very small measure of redistribution. But in March 1859 the measure was defeated in the Commons by 330 votes to 291, and at the ensuing general election, which was notable for the high level of uncontested constituencies, the Conservatives made some gains but still remained a minority.

Once more, Palmerston would be the beneficiary, though he was not Queen Victoria's preference: she sent for Lord Granville, a well-connected Whig, but he was unable to form a government, and so the sovereign reluctantly turned to Palmerston for the second time. By then, his parliamentary position had been much strengthened, as the result of a famous meeting that had taken place in London in June 1859 at Willis's Tea Rooms, at which Whigs, Liberals, Peelites and radicals all pledged to work together. Palmerston and Russell agreed to bury the hatchet and to serve under each other as the queen chose, and Gladstone finally decided to ignore the continuing siren songs of the Tories (Derby had offered him posts in 1852 and again in 1858) and instead, and irrevocably, threw in his lot with what now became formally known as the Liberal Party. To Victoria's disappointment, Palmerston formed his second administration, which in some ways resembled the Aberdeen coalition, but with a more radical tinge: Lord John Russell (who would be created an earl two years later) went back to the Foreign Office, and

Gladstone returned to the Exchequer. Moreover, it was no longer a coalition, but something more closely approximating to an authentically Liberal government with the former Peelites more or less fully assimilated. Once again, there were some cautious reforms, concerning food and drink adulteration, bankruptcy and insolvency, police and public health in Scotland, and contagious diseases, and there was a further consolidation of the criminal law. This was not a government, as the future Lord Salisbury would later claim, which accomplished 'that which is most difficult and most salutary for a parliament to do – nothing'. And its record also contradicted Palmerston's own assertion that 'we cannot go on adding to the statute book ad infinitum'. On the other hand, Russell's reform bill of 1860, which closely resembled those he had vainly put forward in 1852 and 1854, made no progress, and no one in the cabinet or the Commons seemed much concerned, let alone excited by it.

There was one other area where these years witnessed some moderate reform, or at least interest in reform, and that was education. In part as a result of the establishment of University College and King's College in London, and in part in response to the disruptive effect of the Oxford Movement, a royal commission had been set up in 1850 to investigate the ancient universities of Oxford and Cambridge, which were widely regarded as being too narrow and restricted – socially, intellectually and religiously. The eventual result was legislation passed in 1854 and 1856 bringing limited change to both places, by opening up teaching posts to some degree of competition, and by allowing dissenters to study for first degrees; this in turn helped improve educational standards and led to a broadening of the curriculum, in the direction of such subjects as science and history, though more significant reforms would have to wait until 1871. Soon after, four more Royal Commissions were established: Newcastle, on the elementary education provided for the poorer classes of England and Wales (1858–61); Clarendon, on the major public schools, namely Eton and eight others (1861–64); Taunton, on the schools not investigated by Newcastle or Clarendon (1864–67); and Argyll, on Scottish schools (also 1864–67). (No such inquiries were undertaken for Ireland.) As with the reform of Oxford and Cambridge universities, these were essentially bipartisan investigations, and they made extensive recommendations concerning the involvement of local and central government in elementary education, and the reform of the management, curriculum and endowments of public schools and grammar schools; but it would not be until the end of the decade that this resulted in legislation.

In retrospect, but also at the time, the most significant personal question hanging over British politics between 1846 and 1859 had been: what would happen to Gladstone? Would he, as Derby certainly wanted, and as even Disraeli was prepared to contemplate, return to the Tories? Or would he continue his leftwards journey and throw in his lot with the Whigs and the Liberals? He did not like Disraeli, but he did not like Palmerston either. Yet looking to the future, he had better prospects of being Liberal leader once Russell and Palmerston had passed on than he did of leading the Conservatives once Derby had died, because that task would surely fall to Disraeli. But even though he had made his choice in 1859, throwing in his lot with the Liberals, Gladstone was not a happy member of Palmerston's government. The Crimean War had been expensive, and a succession of foreign crises in the early 1860s (ditto) meant that Palmerston made repeated demands to increase the military budget. Indeed, between 1861 and 1865 a higher proportion of central government expenditure was devoted to defence than at any other time of peace during the whole of the nineteenth century. This in turn made it impossible for Gladstone to phase out the income tax by 1860, as he had promised to do in 1853, and on several occasions he contemplated resignation. But thanks to a buoyant economy, Gladstone did succeed in lowering the income tax on incomes over £150, from 10d in the pound in 1861 to 5d in the pound five years later. He also abolished the duty on paper in 1861, and his lengthy, commanding budget speeches restored his parliamentary reputation after his wilderness years of 1855–59. Meanwhile, Palmerston's parliamentary position seemed increasingly unassailable, as he managed to keep almost everyone happy and took care never to antagonize more than one group at a time. As a result, and uniquely among all governments that had been in office since 1832, his Commons majority remained intact for the entire duration of parliament.

WORLD POWER AND ITS LIMITS

The high politics of these years thus present an intriguing paradox: on the one hand, an extraordinary cavalcade of larger-than-life personalities, with constantly shifting tensions, animosities, betrayals, friendships, loyalties and alliances; but on the other, the substance of public business was generally parochial, humdrum, low-key and unexciting (with the exception of Gladstone's heroic budgets). At one time or another, cabinets were mooted that contained virtually all the main political players, regardless of their vexed personal relations or differing party affiliations. The domestic measures passed by the Derby, Aberdeen and Palmerston administrations were virtually indistinguishable from each other: careful, cautious, codifying and consolidating, but little more. In the same way, both Russell and Disraeli took up the cause of moderate parliamentary reform, but although their proposed changes were far from fundamental, neither of them got anywhere with their schemes, yet it did not seem to matter. Pace Palmerston, the statute book was actually being added to, not least by himself, but certainly not ad infinitum; and pace Gladstone, the times were not entirely anti-reforming, but nor were they ardently in favour of large-scale change or transformative interventions. Indeed, the majority of Britons had no wish that the state should be doing more than it was. In some ways, indeed, the most significant political event of these years may have been the death of Prince Albert in 1861, allegedly from typhoid fever, or perhaps from a chronic bowel condition. His disconsolate widow abandoned herself to secluded grief, and this, combined with the increasingly dissolute behaviour of the Prince of Wales, would by the end of the decade precipitate what Gladstone would call the 'great crisis of royalty'. Moreover, had Albert lived, it is difficult to believe that the queen would have taken against Gladstone so strongly and so intransigently as she soon would do, or that she would have been so beguiled by, or become so infatuated with, Disraeli. Most nineteenth-century Britons died at about Albert's age (he was forty-two), but in his case the political consequences were considerable.

Since domestic politics was conspicuously lacking in great causes and great issues, foreign affairs assumed a more than usually prominent part in the public business of the time almost by default. Aberdeen's government fell in 1855 because of its military incompetence, and Palmerston's was defeated two years later because he misjudged the political and public mood over the *Arrow* affair. But it was also true that during these years international developments and global events were of markedly greater significance than they had been in earlier decades. From the late 1840s to the mid-1860s there unfolded a succession of interlinked crises: the European revolutions (1848), the Taiping Rebellion in China (1850–64), the Crimean War in Russia (1853–56), the Indian Rebellion (1857–58) in South Asia and the American Civil War (1861–65). They were the direct successors to the earlier global crisis of the late eighteenth and early nineteenth century, and all of them carried profound implications for Britain as the pre-eminently engaged world power at mid-century. The revolutions of 1848 had been the result (among other things) of further crises for many of Europe's fiscal military states, and they had convulsed not only much of the continent but also many realms of the British Empire. The Taiping Rebellion was a further indication of the weakness of the Qing Dynasty, and this offered both challenges and opportunities for British merchants who were eager to increase their Far Eastern trade. The Crimean War had come about because the British government had felt reluctantly compelled to support the Ottoman Empire and take up arms against the Russians, who seemed a constant threat to the British position in South Asia. The Indian Rebellion was an unprecedented challenge to British power on the subcontinent, and for a time it seemed as though the rebels might win. And the American Civil War meant there were some difficult choices for the British to make between supporting the northern or southern states.

Across the 1850s and early 1860s these successive crises in different parts of the world, all of which impinged significantly on British power and material interests, developed in interconnected and overlapping ways, and this in turn meant that it rarely fell to one single administration to deal with them in their entirety. To be sure, between 1852 and 1865 only three people held the post of Foreign Secretary, but they never did so for a long and sustained spell. Lord Malmesbury occupied the office in the two minority Conservative governments, while Lord John Russell and Lord Clarendon took it in turns in both the Aberdeen and the Palmerston administrations. Malmesbury was far from distinguished, and Russell was erratic, but Clarendon was a much safer pair of hands. As before, the conduct of foreign policy was widely regarded as an aristocratic preserve, and Malmesbury, Russell and Clarendon were all authentic grandees. But for much of the time, Palmerston was a – perhaps *the* – major figure in the handling of international relations, both as Aberdeen's Home Secretary and when he was prime minister. At the same time, the often unsettled affairs of the British Empire were overseen by no fewer than ten Colonial Secretaries, between 1852 and 1865, who rarely served for more than a year. With the exception of Russell, who held the office briefly between February and July 1855, and who had previously occupied the same post during the last two years of Melbourne's government, none of them were political figures of the first rank. Under these circumstances, the making of foreign and imperial policy was less disciplined and deliberate than it had often been during the 1820s, 1830s and 1840s. It was never clear just what position Palmerston would take, especially regarding imperial annexations or withdrawals, and as so often, they were determined by an unpredictable amalgam of administrative momentum, the views of the permanent officials in London, and the actions of the men on the spot.

The first pressing issue with which the Derby and the Aberdeen governments had to deal was the problem of re-establishing stability in the settler colonies in the aftermath of the disruptions associated with 1848. Following the recent Canadian precedent, the preferred solution was to extend responsible government to other parts of the settler empire. A New Zealand Constitution Act had already been passed in 1846, but Governor George Grey had opposed the division of the country into separate European and Maori districts. He feared it would place the 13,000 settlers in an unacceptable position of dominance over the 100,000 Maori, and on his advice the measure was indefinitely suspended. Grey drafted alternative legislation, and most of his proposals were embodied in the New Zealand Constitution Act of 1852, passed by the first Derby minority government. It established a General Assembly (essentially a parliament), consisting of an appointed Legislative Council and a nominated House of Representatives, and the country's domestic affairs would be conducted by an Executive Council, appointed by the Governor. Although the franchise was property based, and confined to male settlers, the qualifications were comparatively low, a much higher proportion of the (admittedly small number of) New Zealand white men had the vote than did their counterparts in the United Kingdom, and the franchise would be further extended in 1860. But parliament and the Executive Council were still subordinate to the Governor, and when the first General Assembly convened at Auckland in 1854 it called for the establishment of responsible government. The Colonial Office offered no objection, and in April 1856 Governor Thomas Gore Browne laid out the procedure for its operation. In matters assigned by the legislation of 1852 to the General Assembly he would accept the advice of responsible ministers, whether he agreed with them or not; but imperial interests, foreign policy and Maori affairs remained the preserve of the British government.

By this time the same constitutional provisions were also being extended to the Australian colonies. The discovery of gold in Victoria and New South Wales in 1851 had led to an unprecedented influx of immigrants, and this, combined with the effective ending of penal transportation soon after, provided a further incentive to constitutional progress and consolidation. Between 1855 and 1859 responsible government became the established practice in New South Wales, Victoria, Tasmania, South Australia and finally in Queensland. By the end of the 1850s, which had seen the continuous and cumulative implementation of responsible government, regardless of which party was in power in London, the British settler colonies in the Antipodes and North America had become essentially autonomous in the conduct of their domestic affairs (although London retained control of foreign policy). Here, as elsewhere in the empire, timely reform played a major role in pacifying the political radicalism and the popular unrest of the late 1840s; and as in New Zealand, the new elective franchise embraced a higher proportion and more diverse social range of the adult male population than was contemplated in the United Kingdom, not only compared to the provisions of the Great Reform Act, but also compared to the schemes more recently proposed by Russell and Disraeli. (Although the new Cape Colony constitution of 1851 only prescribed representative rather than responsible government, the franchise was again much wider than in the United Kingdom, and had been devised so as to ensure that Africans as well as settlers could vote.) During the 1850s there may not have been much appetite among MPs for the further reform of the Westminster parliament, but there was considerable enthusiasm for constitutional change and the extension of the franchise in the settler colonies.

This was a noticeably coherent and successful policy, regardless of who was in charge of the Colonial Office, namely the abandonment of expensive direct administrative responsibilities, which were transferred to the colonies themselves, and their replacement by looser imperial controls, along with significant constitutional innovation and liberalization. This combined economy at home with arm's-length consolidation overseas, and it was not meant to prefigure the break-up of the settler empire. Occasional outbursts, such as Disraeli's in 1852 when he declared such colonies were 'millstones round our necks', were convincingly contradicted by the reality of what was actually happening. Elsewhere, there was no such clear-cut pattern, as annexation in some areas was accompanied by the giving back of territory elsewhere. In 1852 the tiny Bay Islands off the Caribbean coast of Central America were acquired by the Colonial Office – but without the knowledge of the Foreign Office. In India, Dalhousie continued his vigorously expansionist policy of invoking the doctrine of 'lapse' to acquire additional princely states. Further east, where the Chinese Emperor was vainly trying to put down the Taiping Rebellion, the British bombarded Canton in the aftermath of the *Arrow* incident, thereby initiating the second Opium War (1856–60), which would eventually result in the destruction of the imperial summer palace in Beijing, the granting of further trade and jurisdictional concessions by the Chinese government, and the annexation by the British of additional Hong Kong territories. But in the Mediterranean the Ionian Islands were made over to Greece in 1862; in Africa the Orange River Settlement, only annexed in 1848, was given up six years later; and although the British acquired Lagos in 1861, their subsequent failed expedition against the Ashanti led a parliamentary select committee to deplore all 'further extension of territory or assumption of government'.

As the mix-up between the Colonial Office and the Foreign Office over the Bay Islands suggests, there was no consistent policy, and London was in no position to make or impose one. In the case of the Indian princely states, the annexation of Lagos and the bombardment of Canton, it was the local men who drove imperial acquisitiveness and military aggression forward, with the British government following very sceptically and reluctantly behind. The United Kingdom may have been the global hegemon, but from the official vantage points of Westminster and Whitehall the projection of British power in many parts of the world was reluctantly undertaken. There was no general wish to interfere in the internal affairs of foreign states, and less was achieved by way of annexation and intervention than the men on the spot would have wished. The Opium Wars and the burning of the summer palace in Beijing were a bitter (and still unforgotten) humiliation for the Chinese; but beyond a certain amount of increased trading, the British made little impact on a vast country that was being convulsed by the Taiping Rebellion. Naval squadrons blockaded the Brazilian coast in the 1850s, and patrolled the West African coast, with the aim of stopping the still vigorous slave trade; but they produced little by way of results. In the case of Japan, the British joined in the international bid to open up the country in 1858, which was led by Commodore Perry of the United States, but the first British minister, Sir Rutherford Alcock, thought such intervention would have little significant effect. The Anglo-Persian War of 1857–58 was initiated from Calcutta rather than from London, and was more motivated by fear of Russian expansionism than any wish to dominate Persia. And although the British might have forcibly opened the decaying Ottoman Empire up to international trade, they also regarded it as a key counterpoise to Russian expansionism on land and by sea, which meant that the Foreign Office had no wish to see it weaken still further.

Such was the context – of an imperial policy that was far from coherent, and of a limited inclination and capacity for wider global intervention – in which the United Kingdom nevertheless went into battle with Russia in the Crimea during the mid-1850s, its only major continental land war between 1815 and 1914, and the sort of conflict it was widely believed the Vienna settlement of 1815 had rendered impossible. But Russia was not only, looking west, a member of the Concert of Europe; it was also, looking east, one of the four great Asiatic empires with which, by virtue of its global reach, Britain interacted during the nineteenth century in a complex, interconnected, geopolitical entanglement. The decay of Mughal authority in India, which the East India Company had helped bring about, had provided an opening for further British intervention and expansion in South Asia, and this in turn had led to the promotion of the narcotics trade with China, and the ensuing Opium Wars and Hong Kong annexation. But as the British became more entrenched in India, they also came increasingly to fear (and to exaggerate?) the threat represented by Russia's aggressive policy towards Persia and Afghanistan, and also by its predatory maritime ambitions vis-à-vis the Black Sea and the Mediterranean. In addition, Russia was the least popular of the European powers as far as British public opinion was concerned, because of its autocratic government, and its savage treatment of those liberals who vainly campaigned for domestic reform (often from their exiled homes in London); while its behaviour in Asia seemed scarcely consistent with its responsibilities as a member of the Concert of Europe. As a result, the British were generally inclined to support the decaying Ottoman Empire as an essential counterweight to Russia. The earlier interventions in the 1830s and 1840s had been undertaken with the aim of promoting the modernization of the imperial government in Constantinople, and also of improving Anglo-Ottoman trade, thereby, it was hoped, halting decline and rejuvenating the ailing but indispensable empire.

But in the summer of 1853 this intrinsically unstable stand-off between the two empires was thrown into jeopardy, when Russia sent troops into the principalities of Moldavia and Wallachia in the Balkans (roughly the equivalent of modern Romania), which were formally under Ottoman suzerainty. The ostensible grounds for this invasion were that the Orthodox Christians living there needed protection from Ottoman intolerance and persecution; but it was clearly an act of territorial aggression intended to precipitate the complete collapse of Ottoman authority throughout the Middle East. In defence of his realms the Ottoman Sultan declared war on Russia in October, but the following month the Russians destroyed the Ottoman fleet near the Black Sea port of Sinope. The British press and public clamoured for military action against Russia; but Lord Aberdeen had no wish to go to war, and was much criticized for what seemed to be his policy of drift and indecision, while Palmerston conveniently escaped censure because he had recently resigned from the government, in opposition to Russell's espousal of parliamentary reform. In December, Britain and France reluctantly agreed that they must co-operate to save the Ottoman Empire from destruction. In January 1854 the British and French Mediterranean fleets entered the Black Sea and in March the two countries finally declared war on Russia. The Anglo-French strategy was to fight a limited engagement, with the intention of capturing Sebastopol, Russia's prime Black Sea port in the Crimea, so as to make it impossible for the Russian fleet to enter the Dardanelles and capture Constantinople. But because Sebastopol was heavily fortified, a naval bombardment would be insufficient to ensure its capitulation, so British and French troops were dispatched to the Crimean Peninsula, with the aim of increasing pressure on the Russian port from the landward side. Initially the strategy seemed to work, as the allies defeated the Tsar's troops at the Battle of Alma in September 1854, and the fall of Sebastopol seemed imminent.

But thereafter things began to go badly wrong for the British and French forces. Two subsequent engagements, at Balaclava and at Inkerman (in October and November 1854 respectively), were inconclusive: the allies failed to defeat the Russian troops, but the Russians failed to expel the British and the French from the peninsula. The Battle of Balaclava would become infamous for the futile and costly charge of the Light Brigade of British Cavalry, while at Inkerman the British again sustained heavy losses. But there was worse to come in the form of the winter that, as Napoleon had discovered forty years before, was invariably Russia's most reliable and deadly ally. To the deaths and casualties arising from the incompetence of the senior British commanders, Lords Raglan and Cardigan, were added the scourges of cholera, dysentery and other virulent diseases, which reached epidemic proportions and threatened to wipe out the allied forces completely. Thanks to the support of her friend Sidney Herbert, who was Lord Aberdeen's Secretary at War, the zealous, energetic and well-connected Florence Nightingale went out to Scutari, tended to the sick and wounded, and would acquire a legendary (and perhaps inflated?) reputation as 'the lady with the lamp'. Despite the shortcomings of the British high command, and the lack of food, fuel, medicines and other supplies, the allied forces gradually gained the upper hand during the early months of 1855, and they finally conquered Sebastopol in September. Protracted diplomacy then ensued, eventually resulting in the Peace of Paris of 1856. Russia lost any rights it had claimed in the two Balkan principalities; it was compelled to accept the complete neutralization of the Black Sea and the destruction of all its naval bases there; and the continued existence and integrity of the Ottoman Empire was guaranteed.

In military and diplomatic terms this was a successful outcome for Britain: the Russians had been confined and humiliated (for now), and the Ottoman Empire had been preserved (for now). Even as the Aberdeen coalition lost public credibility, some of the most glaring defects in the British war effort were being remedied. The Secretaryship of State that had combined the Colonies and War was divided, and a new consolidated War Office was created. By early 1855 the worst supply problems had been overcome, and the lines of command between London and the Crimea improved. The costs were considerable: spending on the army and the navy tripled from £15.3 million in 1853 to £46.7 million in 1856, which made it impossible for Gladstone to fulfil his pledge to abolish the income tax. Instead, he met the cost of the war by increasing taxes rather than by extensive borrowing, and with a buoyant economy there was more than enough money available. Palmerston also came out well from the war, largely because he was lucky enough to take over as prime minister at just the point, in February 1855, when the tide began to turn in favour of Britain and France. But there were serious criticisms, especially from the radicals like Roebuck, who believed the war had been incompetently prosecuted by an aristocratic leadership that was not up to the job, and from John Bright, who insisted that the war, however much supported by the press and public opinion, was simply wrong. The United Kingdom may have been on the winning side, but there was no Wellington and no Waterloo, and the war in which, as Alfred Tennyson put it, 'someone had blundered' (and, indeed, many more than merely one), scarcely enhanced Britain's lustre or prestige as the greatest power in the world. And despite the undoubted – if belated – improvements, there were more deep-rooted problems at the heart of the United Kingdom's civilian establishment and in its armed forces, among them aristocratic amateurishness and privilege, military incompetence, administrative muddle, and perhaps, also, a too restricted franchise. Sooner or later these problems would have to be dealt with – though not, it seemed, while Palmerston was still alive.

In any case, there was almost immediately a more urgent challenge that the British military and the British Empire had to face, much further away in South Asia. The United Kingdom had fought the Crimean War to support and strengthen one alien empire and to cow and constrain another so as to sustain the Asiatic balance of power in what were widely (but not universally) agreed to be Britain's best geopolitical interests; but the so-called Indian 'Mutiny' (or 'Great Rebellion') was a savage battle for the very survival of the United Kingdom's imperial and mercantile presence on that vast continent, and thus for its continued existence and credibility as the pre-eminent global power. The stakes were at least as high as they had had been at the Battle of Waterloo, and the outcome would once again be a very close-run thing. The origins of what is also referred to, from different perspectives, as the Indian Rebellion or the first War of Independence, were many, varied, and in some ways deep-rooted and of lengthy gestation, but there were more immediate causes as well. At the most general level, there was pent-up but growing opposition to the increasing intrusiveness of the British presence; this had been an economic disaster for the Indian cotton industry, which virtually collapsed in the face of the massive importation of British manufactured textiles. Moreover, the determination of the government in Calcutta, urged on by Macaulay, to impose the English language and British values in ignorance or defiance of the many, venerable and sophisticated indigenous cultures also caused great offence. And the aggressive annexation of princely states, culminating in Dalhousie's acquisition of Awadh, seemed a further indication of the United Kingdom's hostile and predatory intentions; and so did such modernizing reforms as the abolition of suttee and the construction of railways, both of which seemed further indications of British contempt for indigenous ways and customs. The result was that by mid-century there was growing hostility to the British presence, and especially to its prime agent and representative, namely the rich and rapacious East India Company.

Yet at mid-century this British imperial venture remained largely a matter of private enterprise, and neither in London nor Calcutta was there adequate official oversight of the Company's affairs. As its fiat extended across the subcontinent, the Company had been able, with the grudging consent of the Mughal Emperor in Delhi, to collect land revenues, and much of this money was siphoned off for the benefit of its employees or its British shareholders. This large and growing income had also enabled the Company to build up its own private army of nearly a quarter of a million men, and to hire from the crown between 20,000 and 30,000 British troops to stiffen its Indian sepoys and act as a check on unrest in the army. Although it had lost some of its trading monopolies in 1813 and 1833, the Company was still exporting opium to China in large quantities. The culture of this bizarre company state, masquerading as an arm of the British government, was more characterized by a concern for jobs, for perquisites and for plunder than by any higher sense of imperial mission or purpose; and its employees were a selfish and self-perpetuating oligarchy, ever eager to demand more territorial expansion, and more trade in opium. Even as the British government, acting via the Board of Control, regretted or deplored such pressures, the Company wanted more opportunities, more revenue and more corruption. But while this was a rapacious regime, it was also a brittle and vulnerable one. Much of the subcontinent still lay outside the Company's control, in the form of autonomous princely states, with their own revenues and private armies. Its borders with Persia, Afghanistan and Tibet were far from secure, and behind them lay the perennial Russian threat. And the whole Company presence was perforce an imperialism of occupation but not of settlement. As one Governor General put it, 'The Government of India, unlike the colonies of the Crown, has no element of national strength on which it can fall back in a country where the entire English community is but a handful of strangers.'

Lord Dalhousie, whose high-handed actions had caused so much offence, left India in 1856, and the accumulated dissatisfaction with what the British had been doing in the country finally boiled over in the following year. One reason was that it marked the hundredth anniversary of the British victory at the Battle of Plassey in 1757, and to disenchanted South Asians a century of British misrule was more than long enough. Moreover, in January 1857 it was rumoured that the grease being used for a new type of rifle cartridge introduced by the Company army was made of fat from cows (sacred to Hindus) and from pigs (unclean to Muslims). Since cartridges had to be bitten into by troops prior to inserting them into their rifles, both Hindus and Muslims were equally affronted at this apparent lack of respect shown to their religions. During the early months of 1857 sepoys rebelled in many parts of northern India, most significantly at Meerut. They captured Delhi, where the last (and largely nominal) Mughal Emperor proclaimed his support for the insurgents, and much of the Upper Ganges valley and upland Central India fell to the mutineers. Some ruling princes such as Nana Sahib also joined in the rebellion; and for a time it appeared as though the Company's power might be broken. A brutal, bitter war ensued, with terrible atrocities perpetrated on both sides (though the London newspapers unsurprisingly gave more attention to Indian than British savagery): at Cawnpore, Nana Sahib ordered the slaying of hundreds of British men, women and children; in retaliation, the British killed everyone suspected of being a rebel when they won the town back. But the British were lucky in that the Great Rebellion was far from being a focused, co-ordinated or pan-Indian phenomenon: whatever the general level of discontent, the subcontinent was just too big for that. In retaliation, they assembled an army in the recently conquered Punjab; they recaptured Delhi by the end of 1857; and additional troops, belatedly sent from Britain, fought their way up the Ganges, exacting savage reprisals. Resistance smouldered on into 1859, but the Great Rebellion had been broken.

This eventually successful outcome helped restore national and military morale that had been severely dented by the disasters of the Crimean War, but the Great Rebellion was the greatest nineteenth-century crisis the British Empire faced. The government in London played little direct role, the initial British military response was slow, and the medical back-up was no better than in the Crimea. Moreover, the failings of the East India Company could no longer be ignored, and in 1858, once the Rebellion was largely over, Palmerston announced that the Company would be abolished. His administration fell before the necessary legislation could be passed, but it was carried by Lord Derby's second minority government in a significant display of bipartisanship. All the Company's territories and property were transferred to the British government, the Board of Control was abolished, and a new cabinet post was created, that of Secretary of State for India. As the representative of the monarch, the Governor General, who would be responsible to the Secretary of State, was upgraded to the exalted rank of Viceroy, and in future all British territories would be administered by an Indian Civil Service, whose high-minded and disinterested ethos was very different from that which had prevailed before. The Company's sepoy army was put under crown control and drastically reduced in numbers, while the proportion of British troops was substantially increased (again at the expense of the Indian taxpayer). Henceforward, there would be more investment in railways, but the wider aim of 'modernizing' Indian society, which had been so appealing to the likes of Bentinck, Macaulay and Dalhousie, was largely relinquished. The Mughal Emperor was deposed and the last of his line was tried for treason and exiled to Rangoon. In an accompanying proclamation, which further reversed previous policy, Queen Victoria declared that Britain no longer harboured any more territorial ambitions in India, and would in future respect and safeguard the rights of the native princes, most of whom had been loyal in 1857. This meant that henceforward British government in South Asia would be more on the side of tradition than of modernity.

One extraordinary scheme, initially mooted earlier in the 1850s, to which the Great Rebellion gave brief validation and reinforcement, was the construction of what was then by far the world's largest ship, initially named *Leviathan*, subsequently the *Great Eastern*, which would turn out to be the last and most unsuccessful brainchild of Isambard Kingdom Brunel. His proposal was to construct a vessel so large that it could sail non-stop around the Cape of Good Hope to India, carrying sufficient coal to power it for the whole journey, and enough troops, horses and artillery to reinforce the Indian army so substantially that any revolt on the subcontinent would be crushingly suppressed. Construction had begun in 1854, and it took more than two years to build a ship that was nearly 700 feet long, which was six times larger than any vessel then afloat and would not be surpassed in size for almost sixty years. But the launch on the Thames at Millwall, in November 1857, was a failure, it took another year to get the ship into the water, and the company that had been incorporated to finance the venture ran out of money. The first time the *Great Eastern* was in open sea it was plagued by a series of freak accidents and fatalities, the few transatlantic crossings that it made were blighted by further mishaps, and it never did undertake (nor could it have undertaken) any of the journeys to India that had been its initial raison d'être. Underpowered and ungainly, it would be Brunel's most spectacular financial and engineering failure, only briefly redeemed when the vessel was used to lay the first transatlantic cable in 1866. But Brunel would know nothing of this, for he had died in 1859, in part from the strain, overwork and worry that the *Great Eastern* had caused him. He was, in a way, the most conspicuous British victim of the Great Rebellion.

By contrast, and more on account of luck and good timing than anything he did, Palmerston came out well from the Crimean War and Great Rebellion, and the same was true of the Italian Risorgimento, even though the United Kingdom was more an onlooker than an active participant, and despite the fact that Napoleon III was in many ways the prime mover. Although Britain and France had been allies during the Crimean War, relations between the two countries had turned sour by the end of the 1850s. This was partly in the aftermath of the Orsini affair; partly because Napoleon increased spending on the French navy and coastal defences; partly because this gave rise to the revival of hysteria about another possible cross-Channel invasion; and partly because of a residual British dislike of the French, especially when they were again being led by someone ominously named Napoleon. The result was a widespread (but ephemeral) outbreak of febrile francophobia, yet a further increase in government spending on defence, including the construction of forts still visible on the south coast, and the semi-spontaneous establishment of the 'volunteer movement' that by 1862 numbered more than 150,000 men. Meanwhile, Napoleon had appointed himself the patron and supporter of the developing nationalist movement for Italian unification, and to this end he went to war with Austria in 1859 in the hope of expelling the Habsburg Emperor from the north of the peninsula. His victory helped pave the way for Italian unification, largely brought about by Giuseppe Garibaldi and the Count of Cavour in 1860. Palmerston mistrusted Napoleon's intentions, yet his government gave general support to the liberal movement for Risorgimento, though it did nothing more, and as a result probably received greater credit among the Italians, and also in the United Kingdom, than it deserved. However implausibly, Italian unification was viewed as a particularly Palmerstonian triumph: a victory for liberal principles and constitutional monarchy over French, Habsburg and papal absolutism.

In the case of the American Civil War, which broke out in 1861, the official British attitude was one of strict neutrality between the North and the South; there was nothing that the United Kingdom could do or should do militarily to intervene in a domestic quarrel half a world away, but for good reason, political and popular opinion was deeply divided. From one perspective, the South was a quasi-aristocratic society, and as such it was greatly admired by many British patricians and landowners, and it was also the United Kingdom's most significant supplier of raw cotton. Moreover, for much of the duration of the conflict, Palmerston, Russell and (especially) Gladstone were generally sympathetic to its cause. On the other hand, the South was also a slave-owning society, and Britain had abolished slavery throughout its empire thirty years before. This meant that radicals such as John Bright naturally espoused the cause of the North, and so did many Lancashire workers, despite (or because) the North's blockade of the South meant they were unemployed as there was no American cotton available for them to spin or to weave. There were also two embarrassing diplomatic incidents: late in 1861 the Union navy had stopped a British steamer, the *Trent*, and removed two Confederate agents, who had been heading across the Atlantic to Britain with the aim of putting the Southern case; and in the summer of 1862 the Southern sloop-of-war, the *Alabama*, which had been constructed on the Mersey, was allowed by the authorities to sail across the Atlantic, despite strong Northern objections, where it destroyed considerable quantities of Union shipping. These two incidents might have led to a British recognition of the South's independence and autonomy, and also to the outbreak of war between the United Kingdom and the Northern states. But in the end, flexible diplomacy, combined with a growing recognition of the stature of Abraham Lincoln and the justice of his Emancipation Proclamation, meant conflict was averted, which was just as well given the complete triumph of the North by the spring of 1865.

In reality, there was little Britain could do regarding the American Civil War, and as the Italian Risorgimento had already shown, the same was essentially true in Europe. In 1863 another rebellion had broken out in Russian Poland, which had been despotically ruled since the partitions of the late eighteenth century. The revolt was again suppressed with great harshness and Britain joined France and Austria in delivering vigorous protests at Russia's brutal actions. But for all its liberal posturing, there was nothing that the Palmerston government could do, since it lacked the political will and the military capacity to fight for the liberation of Poland. Indeed, the only such continental conflict that Britain had successfully engaged in had been that for Spain during the Napoleonic Wars, and it was only the continued presence of the Royal Navy in the North Sea, the Atlantic and the Mediterranean which perpetuated the illusion that the United Kingdom was a military power of European significance. This detachment (or, alternatively, weakness) was made abundantly plain in 1864 when the vexed issue of the Schleswig-Holstein duchies surfaced once again. During the previous crisis of 1848–49 Palmerston had wielded little influence; this time around, when Prussia was under the leadership of the ruthless and calculating Bismarck, his capacity to intervene was even further reduced. The Iron Chancellor was determined to achieve German unification under Prussian hegemony, and along the way he was set on provoking a war with Denmark in order to gain the port of Kiel. British public opinion strongly supported Denmark against such Prussian bullying, but Palmerston refused to join Napoleon III in either convening a European peace conference or going to war with Prussia. As with Poland, strong British words meant nothing; the policy of bluff no longer worked. Bismarck got his war, his military victory and the Schleswig-Holstein duchies, and there was nothing that Palmerston could do about it.

This was the first portent of a new European world of *Realpolitik*, of national aggression, and of blood and iron, which would be very different from that in which the ageing prime minister had lived and worked. Palmerston tried to laugh the whole matter off by asserting that only three people had ever understood the Schleswig-Holstein question properly: one of them was dead, the second was mad, and he himself was the third, but he had forgotten what it was all about. This was an unconvincing position to adopt, not least because the war between Denmark and Prussia had also put the widowed, grieving queen in a particularly difficult position. In 1859, and in a match that had been vigorously promoted by Victoria and Albert, their eldest daughter, Princess Victoria, had married Crown Prince Frederick of Prussia. The object of this matrimonial diplomacy was to cement a perpetual alliance between Britain and Prussia, the two great Protestant nations bordering the North Sea, and with the further hope that the princess's liberal instincts and high-minded cleverness might in the fullness of time moderate the authoritarian and militaristic culture of the Prussian court and people. But four years later, in 1863, the Prince of Wales had married Princess Alexandra, who was the daughter of the King of Denmark, and someone of such ravishing beauty and sweet temper that it was hoped, vainly as it turned out, she might restrain and moderate her husband's increasingly wayward morals. With a Prussian son-in-law and a Danish daughter-in-law, the queen was thus placed in an impossible situation once their two countries went to war against each other: here was an early indication that relations between the different nations of Europe could no longer be carried on by the personal connections among the cosmopolitan cousinhood of continental royalty in the way that Victoria and Albert had hoped and believed.

MATERIAL COMFORTS, MENTAL PERTURBATIONS AND CULTURAL ENRICHMENT

With so much excitement during these years being generated by what were often the vexed and violent politics of international relations and imperial affairs, it was scarcely surprising that there was little need or appetite for the same sort of excitement in the content and subject matter of politics at home. Underpinning this widespread domestic sense of relative calm and tranquillity was an era in which, unlike the 1830s and 1840s, long-term growth and improvement more obviously outweighed the countervailing vicissitudes and cyclical fluctuations, and these unprecedentedly prosperous and stable years may indeed have witnessed the summit of the nation's global economic pre-eminence. To be sure, there were exceptionally high food prices in 1853–54 (when the Crimean War cut off supplies of Russian corn); there was the cotton 'famine' of 1862–65, which led to great hardship in Lancashire (as a result of the American Civil War); and there was a commercial crisis and financial panic in 1857 (coinciding with the Great Rebellion). But these were nothing like the disruptions that had been caused by the severe downturns of the late 1810s, the 1830s and the 1840s, and throughout the 1850s and the early 1860s all the indicators of the nation's economic performance pointed strongly and seemingly inexorably upward. Between 1851 and 1861 the population of Great Britain grew from twenty-one million to more than twenty-three million (although that of Ireland continued its post-Famine decline), and by 1871 it would reach twenty-six million. Despite the gloomy prognostications of the protectionists, the repeal of the Corn Laws had not spelt the ruin of British agriculture, and the subsequent era of 'High Farming' saw the sector more prosperous than at any time since the early years of the nineteenth century. The mileage of railway track doubled, and remained the longest in extent of any country in Europe, while receipts from goods and passenger traffic trebled. Exports were growing at the rate of 11 per cent a year, which was almost double that of previous decades; the United Kingdom possessed two-thirds of the world's capacity for cotton factory production; and it still accounted for half the global output of coal and iron. Exports of woollen manufactures almost doubled, imports of raw materials from Australia and New Zealand grew almost fourfold, and despite the crises in the Crimea and South Asia, trade with the rest of the world continued to grow.

During these mid-Victorian years, it seemed as though progress and prosperity had become systemically embedded, along with domestic stability and growing international trade and overseas investment. In essence, the economic success of mid-Victorian Britain lay in its ability to make cheap cloth, iron, steel (in 1856 Henry Bessemer patented a new process for the mass production of steel from molten pig iron) and machinery for much of the rest of the world, and to transport them to overseas customers in British-built and British-insured ships. The fact that the incomes of landlords, businessmen, farmers and artisans all rose together, and showed substantial gains, helped underpin the mood of domestic optimism (even as that mood was being simultaneously undermined by events in the Crimea and India). According to the tax returns, the gross value of income from lands, houses, mines and other forms of real estate assessed under Schedule A increased from £85.8 million in England and Wales in 1843 to £131.3 million in 1865, and in Scotland during the same period from £9.4 million to £16.2 million. The income reported under Schedule D (profits from trades and professions) rose from £52.7 million in England and Wales to £95.96 million during the same period, and in Scotland from £4.8 million to £9.8 million. With a buoyant agricultural sector, and with their revenues from real estate, coalmines and docks greater than ever, the aristocracy and gentry were unprecedentedly prosperous, and grandees such as the Duke of Devonshire, the Marquis of Bute and the Earl of Derby were prodigiously rich. Bankers were likewise making high profits, especially from arranging international loans: hence the purchase by the Rothschilds of their broad acres in Buckinghamshire, and their construction of such grand palazzos as Mentmore Towers and Waddesdon Manor. Middle-class businessmen also enjoyed stable prosperity for the first time during the nineteenth century. As a result some sought to establish a new type of industrial relationship with their workers, among them Titus Salt, who in Bradford, West Yorkshire, built an entire estate for his employees, called Saltaire, complete with public baths, a library and a chapel; others, like Joseph Gillott of Birmingham, became enthusiastic collectors of contemporary British art.

With very rare exceptions, those who made their money in provincial industry – such as Josiah Wedgwood, Richard Arkwright, Jedediah Strutt and Thomas Cubitt – were generally less well off, even as late as mid-century, than the greatest British landowners or the London-based investment bankers. Since the scale of British industry remained on the whole small, the dimensions of British industrial fortunes generally remained correspondingly limited. Not surprisingly, then, and hence the continued laments of men like Cobden and Bright, politics during the 1850s and 1860s remained largely the preserve of the aristocracy and gentry, not only in the cabinet and the House of Lords but also in the Commons. From this perspective, the most significant middle-class development during the age of equipoise was the gradual solidifying of the closest Britain has ever come in recent times to a national intelligentsia: in some ways Coleridge's 'clerisy' a generation after he had first coined that concept of learned and literary people wielding cultural authority. It was partly that there was a fully fledged reading public, eager and able to purchase the newspapers and periodicals, the works of history, theology, contemporary comment, fiction and poetry, which such men (and they were nearly all men) produced. It was partly that intermarrying families like the Darwins, Stephens, Arnolds, Huxleys, Stracheys and Trevelyans, who overlapped across the worlds of industry, land, politics, the church, academe and the British and Indian civil service, were comfortably off so that many of them had the time to think and the leisure to write. It was partly that being an intellectual had also become a career open to talents, as exemplified by such self-made figures as Carlyle and Dickens. And it was partly that this largely London-based world was institutionalized and interconnected through the mid-Victorian proliferation of dining clubs and literary societies, of which the Athenaeum, whose heyday was the third quarter of the nineteenth century, was the most famous and significant.

It was not only the aristocracy, bankers, businessmen and intelligentsia who had never had it so good. As the mid-Victorian economy expanded at a rate to keep pace with continued population growth (except, still, in post-Famine Ireland), more jobs were created in the industrial sector even as the (undeniably prosperous) agricultural sector continued to decline in relative importance. Between the early 1850s and the late 1860s the number of engine makers, shipwrights, railway workers and printers doubled, and the trend in real wages was emphatically upward from the early 1850s to the late 1860s. The quality of life in many of the industrial towns and cities remained dreadful, with continued overcrowding and environmental degradation, while malnutrition remained rife and life expectancy low; but the average full-time hours worked began to fall, and housing began slowly to improve (though again Ireland was the exception). It was these gradual changes for the better, combined with a recognition that the confrontations of the 1830s and 1840s had achieved little, that explains the transformation in working-class politics from agitation to accommodation. In some industries skilled workers created trades unions of a new type, with a national membership and a national organization based in London, boasting considerable funds. The Amalgamated Society of Engineers, founded in 1851, was just such a union, and it provided the model for such later leaders as George Odger and Robert Applegarth, who were hostile to sectional agitation or strike action. A further indication of this new cult of respectability was the establishment, in 1853, of the United Kingdom Alliance, a national temperance organization, with strong working-class support. Such developments drove Friedrich Engels to despair: 'The English proletariat,' he wrote to Karl Marx five years later, 'is becoming more and more bourgeois.' The Reverend S. G. Osborne, a kinsman of the Duke of Leeds, saw things rather differently: the British, he observed in 1864, were 'a people at peace among ourselves'.

It was scarcely coincidence that at almost exactly the same time, between 1857 and 1861, Henry Thomas Buckle, the son of a wealthy London shipowner, published what were meant to be the first two instalments of his *History of Civilization in England*, which was much closer to Osborne than to Engels in its optimistic world view and in its self-satisfied sentiments. These books were meant to be but the prologue to a much larger work, comparing civilization in England, and greatly to its advantage, with those of France, Spain, Scotland, Germany and the United States; but Buckle died in 1862, having only completed a fragment of his intended magnum opus, with but one chapter on the history of England itself. Like many mid-Victorians, he did not see barbarism and civilization as the Gibbonian polar opposites, pitted against each other in primordial and perpetual conflict. On the contrary, he believed that they were the extreme ends of a linear continuum, that human history demonstrated 'the progress from barbarism to civilization' along that scale, and that it did so especially and exceptionally in the case of England itself which, by comparison with any other nation, represented the apogee of human accomplishment and achievement thus far. Buckle's volumes sold well, and made a huge impact on the general public. They were much admired by mid-Victorian, middle-class secularists, radicals and liberals, among them Charles Darwin, T. H. Huxley and John Stuart Mill, who liked Buckle's bold interpretations, vigorous prose and confident belief in the progress and superiority of England; but he was fiercely condemned by conservative religious and academic writers, among them Lord Acton and Bishop Stubbs, who thought Buckle was insufficiently respectful to the Christian religion and to the detail of historical evidence, and also by Scottish Presbyterians, who were as hostile to him as they felt he was hostile to them and their nation. Either way, Buckle's *History of Civilization* is one of the key works for understanding the mood and mind of the mid-Victorian generation.

Buckle offered a long-term perspective on how the English had come to reach what seemed to be such a uniquely advanced state of civilization. During the same decade another writer was engaging with the same subject, although from a more prescriptive and didactic perspective, as he sought to discover what sort of people had, by their own effort and character, helped bring about such a state of affairs, and to exhort others to emulate these examples, which would be to their own and to the nation's benefit. His name was Samuel Smiles, and in 1859 he published his most famous work, *Self-Help*. Appropriately, and unlike Buckle, Smiles was a self-made man from Scotland, who during the 1840s had become closely associated with the industrial economy of provincial England. He lived much of his life in Leeds, where he witnessed at first hand the failure of popular radical agitation. He also acquired there a detailed knowledge of the self-improving world of friendly societies and mechanics institutes, and he made his living by working as a manager for a succession of railway companies. One result of these experiences and engagements was *Lives of the Engineers*, published in several volumes between 1861 and 1862, celebrating such iconic figures as John Smeaton, John Rennie, Thomas Telford, Matthew Boulton and James Watt, and George and Robert Stephenson, and showing by their examples what character and application might achieve. But *Self-Help* was his most famous book, emphasizing the importance of the application of good character to the problems and possibilities of daily life, which he regarded as the key to individual self-fulfilment and general social improvement. Collective agitation, he had come to realize, accomplished little; individual striving and endeavour, on the other hand, based on strength of character and laudable ambition, might accomplish much. Within the first year the book had sold 20,000 copies, it was translated into many languages, and Smiles would follow it up with other studies, appropriately entitled *Character* (1871), *Thrift* (1875) and *Duty* (1880).

From their different perspectives, Buckle and Smiles sought to show how the United Kingdom had achieved the supreme position it had attained among the nations and the civilizations of the world, and how future generations of ordinary people might strive not only to improve themselves, but also thereby to enhance the condition of the country still further. Yet there was a tension during these years between the belief in progress and freedom, individualism and liberty, as propounded by Buckle and Smiles, and the reality of growing government intervention, albeit on a scale less marked and less controversial than it had been during the 1830s and 1840s. In this debate, which broadened into wider issues concerning the liberty of the individual, the role and reach of the state, and the means and processes whereby societies developed and evolved, the mid-Victorian intelligentsia was much involved, among them Herbert Spencer, who was an engineer by training, an agnostic in religion, and by turns a philosopher, biologist, anthropologist, sociologist and liberal political theorist. In 1851 he had published a book entitled *Social Statics*, in which he exalted the rights of the individual against government interference of all kinds, and claimed that the whole tendency of the biological development of the human species was towards greater individuation. Indeed, he went so far as to predict that humanity would eventually become completely adapted to the requirements of living in society, and that the state would wither away altogether. Spencer would later expound a more fully developed theory of evolution in *First Principles of a New System of Philosophy* (1862), in which he urged that all structures in the universe developed and evolved from a simple, undifferentiated homogeneity to a complex, differentiated heterogeneity. Here, he believed, was a universal law, applying as much to the stars and the galaxies as to biological organisms and human social organization.

In different but related ways, Buckle, Smiles and Spencer were all interested in the subject of freedom and, either implicitly or explicitly, in its relation to government; but it was another member of the mid-Victorian intelligentsia, John Stuart Mill, the intimidatingly well-educated son of James Mill, who engaged with these issues most directly. By this time Mill junior had rejected much of his father's doctrine of utilitarianism, he had married Harriet Taylor in 1851, and as the Chief Examiner to the East India Company had spent the years 1856–58 vainly advocating its continued existence despite successive governments' successful determination to abolish it. But these were also exceptionally fertile years for Mill intellectually. In *On Liberty* (1859) he asserted the individual's right to freedom, privacy and self-fulfilment against the countervailing forces of tyranny, overbearing government and (increasingly) public opinion, which he lamented (and in opposition to Spencer) were driving out variety and plurality in human existence and making all people ever more alike. In *Representative Government* (1861) he urged that the rights and duties of citizenship should be exercised without reference to property, gender, status or class, but he conceded that they could not automatically be extended to all, because they depended on a minimum degree of education and independence. And in *Utilitarianism* (1863) he expressed the hope that existing inequalities would eventually pass away, globally as well as locally, and that the 'aristocracies of colour, race and sex' would eventually disappear. Appropriately for his times, Mill never quite reconciled his commitment to privacy, spontaneity and personal freedom with his belief in the collective good, the need for political institutions appropriate for different kinds of society, and a culture of active citizenship. Reluctantly recognizing the importance of the latter, Mill would be elected as a Liberal for the Westminster constituency in the general election of 1865.

Insofar as there was a prevailing concern during these years, which preoccupied the mid-Victorian intelligentsia, and with which Buckle, Smiles, Spencer and Mill were all in their own ways engaged, it was that of explaining and understanding the causes of progress (and of retrogression) in human (and non-human) history. How did species, and societies, change and develop over time? More particularly, discussion had centred, since the 1830s, on the issue of whether such evolution was a process that was externally determined by an omnipotent 'creator', or internally, organically and biologically driven according to some (not yet discovered) natural laws. The outcomes of these debates would clearly have profound implications for science, history, philosophy, theology and politics. Charles Darwin, who was a very well-connected and well-travelled member of the mid-Victorian intelligentsia, had been brooding on these problems since his voyage on the *Beagle* during the 1830s, and somewhat belatedly in 1859 he published the most momentous and disturbing book that appeared in the United Kingdom during the nineteenth century, *On the Origin of Species by*

Means of Natural Selection. Darwin did not actually use the word 'evolution' until the book's fifth edition published in 1869; but in the context of the times he was clearly offering a new explanation as to how evolution happened, namely 'natural selection' (which Herbert Spencer would later gloss as meaning 'the survival of the fittest'). To be sure, Darwin had no satisfactory explanation for the process of 'natural selection', he was relatively tentative in his findings and his claims, and he did not present the fullest development of his ideas until he published *The Descent of Man, and Selection in Relation to Sex* in 1871. But his ideas were widely taken up and extended, not only by Herbert Spencer but by such contemporary savants as Sir John Lubbock, Walter Bagehot and T. H. Huxley.

Like many of the intellectuals of his generation, Darwin was a sceptic in matters of religion, but he was a loyal supporter of his rural parish, and he never intended his book as being primarily an anti-Christian polemic (furthermore, he would be buried in Westminster Abbey). But his belief in progress, in the struggle for existence, and in mankind's gradual development from a lowly condition to its current elevated state, clearly carried with it profoundly disturbing implications for those believers in a supreme being who had allegedly created the world in six days and made man in his own image. Within a year of the publication of *Origin of Species* these alternative interpretations were famously disputed at the University Museum in Oxford between two other members of the mid-Victorian intelligentsia: Thomas Henry Huxley, a biologist and comparative anatomist, and the most belligerent proponent of Darwin's views, and Samuel Wilberforce, the local bishop, an eloquent speaker, amateur scientist, and a son of the great anti-slavery campaigner. There is no full contemporary account of the confrontation between the two men, and many other people spoke in addition to Huxley and Wilberforce (though Darwin himself was absent on account of illness). Wilberforce criticized Darwin's theory on ostensibly scientific grounds, and he is alleged to have asked Huxley whether it was through his grandmother or his grandfather that he considered himself descended from a monkey. In replying, Huxley defended Darwin's theory and concluded by insisting that he was not ashamed to have a monkey for an ancestor, but that he would be ashamed to be connected to a man who used his great gifts to obscure the truth. This exchange may in fact never have taken place; and the

Oxford debate may have acquired a legendary status only in retrospect. But the idea of evolution, given new force and cogency by Darwin, was deeply unsettling to many educated British minds. It cast the most serious doubt yet on the central tenets of the Christian faith, and it put science and religion in an opposition that remains to this day. It was the most subversive development of these years, and its implications for society at large, for faith and for science were as far-reaching as they have proved long-lasting.

The year 1859 had been something of an *annus mirabilis* in British publishing and intellectual life, for not only had it seen the appearance of significant works by Smiles, Mill and Darwin, but it was also the year in which a learned, independent, freethinking woman named Mary Ann (or Marian) Evans published her first full-length novel. She had been born in 1819 in the English Midlands, and brought up in an intensely pious Anglican religious atmosphere; but she had also been unusually well educated for a girl of her time, and she would become fluent in French and German, and well read in science, philosophy and religion. By the 1840s she had turned against religion, and this rejection was intensified when she translated into English two German works which argued that much of Scripture was historically incredible. In 1851 she moved to London, and established herself as a single working woman in the world of radical, freethinking, metropolitan journalism; and she formed close friendships with the publisher John Chapman, with Herbert Spencer, and with the writer and critic George Henry Lewes, with whom she lived as his wife from 1853, even though he was already married. By the (double) standards of the time, this was much more outrageous behaviour than that of Lord Palmerston, and it helps explain why she published her first novel under a male pseudonym. That name was George Eliot, and the novel was *Adam Bede*. By then, she was also much caught up in the evolutionary debates that Darwin's recent publication had further intensified, and there would be some allusions to them in her next novel, *The Mill on the Floss* (1860). These two works established Eliot as one of the greatest of nineteenth-century novelists, later consolidated by the publication of *Middlemarch* (1871–72); but whereas Darwin would be buried in Westminster Abbey in 1882, despite his freethinking, Eliot had been denied such a final resting place two years earlier, for precisely that reason.

In March 1858, when Derby's second minority government was in office, when the Indian Rebellion was not yet fully overcome, and one year before Darwin's *Origin of Species* appeared, the assembled members of a metropolitan dining society known as the Club, which had been in existence since 1764, concluded during the course of their post-prandial discussion on 'the highest period of civilization' that it was occurring 'in London at the present moment'. Small wonder that H. T. Buckle's two volumes would soon find such a ready market. The Club's members, all of them men, included such figures from the political elite as Lord Aberdeen, Lord John Russell, Lord Clarendon and Mr Gladstone; worldly historians such as Lord Macaulay, George Grote, James Froude and William Lecky; poets and men of letters such as Tennyson and Matthew Arnold; and clerics such as Bishop Samuel Wilberforce (which meant T. H. Huxley was not elected until after Wilberforce's death). Many of the Club's members also belonged to the Athenaeum, and it was these sorts of associations and connections that provided the informal institutional support for the consolidating intellectual life of the metropolis during the 1850s, where the vigour of public debate contrasted strongly with the torpor of politics. Just a year before this particular dinner, in 1857, two members of the Club, Macaulay and Earl Stanhope, had been involved, with Thomas Carlyle, in founding a new London venture: the National Portrait Gallery. The aim was to display visual images of those figures who had been of the greatest significance in the nation's history. This might seem another instance of mid-Victorian self-regard and hero-worship. But from the very beginning it was also recognized that 'defects of character' should not prevent the acceptance or acquisition of a portrait if the sitter was of sufficient historical significance. In the light of Lord Palmerston's well-known moral lapses, to say nothing of George Eliot's unconventional lifestyle, it was a wise and necessary provision.

At the same time cultural life was also being increasingly enriched in the provincial towns and cities of the United Kingdom. Some of this progress was the result of local authorities taking advantage of Acts of Parliament, such as that passed in 1845 which had allowed them to levy rates to fund local museums, and that carried five years later enabling them by similar means to construct public libraries in England and Wales (a provision extended to Scotland three years later). The result was the establishment of museums and libraries in many of the major towns and cities of mid-Victorian Britain, with the aim of edifying and educating not just the middle classes but also the labouring classes. The creation of the Birmingham Central Library may serve as an example. Belatedly established under the legislation of 1850, construction was begun in 1862, the lending library was opened three years later, and the reference library in 1866. 'We have,' observed the Reverend George Dawson, 'made provision for our people – for *all* our people.' Private Acts of Parliament were also obtained for the establishment of cultural institutions. In 1853 the local authority in Halifax acquired the power to borrow money to construct a town hall, which would house the municipal offices, courthouse and police station, along with a fine central space named for the monarch. The result was a magnificent neoclassical structure, built to designs by Charles Barry and his son Edward Middleton Barry, which was opened in 1863 by the Prince of Wales. And even before the Central Library was constructed in Birmingham, an Act had already been obtained in 1854 'for the diffusion and advancement of science, literature and art amongst all classes of persons resident in Birmingham and the Midland counties', which led to the establishment of the Birmingham and Midland Institute. It superseded the Birmingham Philosophical Institution, which had recently closed after half a century's existence, and also contained the first museum in the town.

Birmingham and Halifax were but two of the provincial towns energetically investing in culture at this time. In Bradford another great building named St George's Hall was financed by a joint-stock company, and constructed in the neoclassical style between 1851 and 1853. It contained a restaurant 'for the accommodation of mercantile men', and a concert hall with a magnificent organ, and was intended to elevate the tone of local life by providing concerts and hosting meetings for the well-to-do and for the working classes. In Manchester the Free Trade Hall, celebrating the earlier success of the Anti-Corn Law League, was constructed by public subscription between 1853 and 1856, and became the home of the Hallé Orchestra, one of the first to be founded in the provinces, and for the rest of the century and beyond it would serve as the city's premier concert hall. A year after the Free Trade Hall was opened, Manchester hosted an exhibition entitled 'Art Treasures of Great Britain', which ran from May to October 1857, displaying more than 16,000 works and visited by more than a million people. But the most significant attempt to combine civic pride with cultural enrichment was the construction of Leeds Town Hall between 1853 and 1858, initially financed by a joint-stock company but subsequently by the town council. It was the grandest and most lavish of all the civic buildings constructed during that decade, opened by Queen Victoria and the Prince Consort, and it became the home of the Leeds Choral Society, one of many such organizations that specialized in performing oratorios (Mendelssohn's *Elijah* had first been performed at Birmingham Town Hall in 1846). Many provincial towns also played host to the annual meetings of the British Association for the Advancement of Science, founded in 1831. Huxley and Wilberforce had debated evolution at the gathering held in Oxford in 1860, and subsequent meetings took place in Newcastle, Birmingham, Nottingham, Dundee, Norwich and Exeter.

These organizations and events were intended to benefit all the classes and levels of British society that were able to avail themselves of the facilities and opportunities they were providing in unprecedented abundance. Other initiatives were particularly focused on the respectable and aspiring working classes, who by the 1850s were beginning to register real and lasting improvements in their standards of living for the first time. Pre-eminent among them were the mechanics' institutes, the first of which had been established during the 1820s, in Edinburgh, Glasgow, Liverpool, Manchester and London (the latter eventually evolving into Birkbeck College of London University). They were often supported by local industrialists with paternal and philanthropic inclinations, among them the engineers Joseph Whitworth and James Nasmyth, who were eager that their workers should understand the machinery they were operating. By the 1850s there were more than five hundred such mechanics' institutes spread across the towns and cities of the United Kingdom, and by then they were providing libraries, lectures and, occasionally, laboratories and museums to educate the minds and expand the horizons of adult workers. Just as the town halls, concert halls, libraries, cathedrals and art galleries provided the mid-century models for similar buildings in the developing imperial cities of Wellington, Auckland, Cape Town and Toronto (as well as Bombay, Calcutta and Madras), so the mechanics' institutes would proliferate across the British Empire, especially in Australia and Canada, and also the United States. And wherever they were established, throughout the English-speaking world, they provided a self-helping and high-minded alternative to what were increasingly deemed to be such wayward working-class activities as drinking, gambling and the pursuit and enjoyment of cruel sports.

During the 1850s and early 1860s much of Palmerston's popular appeal derived from the fact that he seemed, and indeed was, a Regency figure out of his time, delighting in boxing, and allegedly seducing one of Queen Victoria's chambermaids when staying at Windsor Castle. Yet in cultural terms the dominant trends of the time, influenced by a very different ethos emanating from the sovereign and her consort, were in a contrary direction, epitomized in such phrases as 'the call to seriousness' and 'the march of intellect', and above all in the concept and practice of what was termed 'respectability'. Even as church-going seemed to have declined by mid-century, such Christian virtues as kindness and decency, prudence and thrift were being more widely embraced by many who were lower down the social scale. One indication was a significant decline in the practice of cruel sports that had been so marked a feature of working-class culture during the early decades of the nineteenth century. The legislation of 1822, which had made it an offence to inflict cruelty on cattle, had been greatly extended in 1835, outlawing cockfighting and bear-baiting as well as bull- and badger-baiting. Five years later the Society for the Prevention of Cruelty to Animals obtained the royal patronage of Queen Victoria herself, and so successful were its efforts to enforce these new laws that by the 1850s most cruel sports, in both the towns and the countryside, had virtually disappeared or were in the final stages of terminal decline (with the significant exception of fox hunting, which was confined to those much higher up the social scale). At the same time the massive growth in the number of friendly societies and insurance companies (the Royal Liver was founded in 1850, the Prudential four years later) helped to mitigate such risks and hazards of working-class life as unemployment, disability and a pauper's funeral, as the habits of saving and thrift became more widespread. This was a very different world from that in which Lord Palmerston had been a young man.

THE END OF AN ERA

Ever since the fall of Sir Robert Peel over the Corn Laws in 1846, Palmerston had been the dominant figure in British politics, and as much by accident as by design he was the perfect frontman for what seemed to be the 'anti-reforming' times that followed at home, and the high point of British global hegemony that was achieved and asserted abroad. When party politics were in flux, when there was a limited appetite for further domestic reform, when foreign affairs were unusually important, when the British polity seemed uniquely liberal and stable, when the United Kingdom was engaged with more parts of the world, and when the press and public opinion were significant forces, Palmerston was the personification and the embodiment of his times in a way that was true of no other political contemporary. Perhaps appropriately, his most lasting monument would be the Foreign Office in Whitehall, which resulted from a competition for new government buildings launched during his first administration in 1856. From the outset there was controversy as to the appropriate style. Gothic had been used for the Houses of Parliament, as being the quintessential national idiom, but thanks to Pugin it had also become associated with Catholicism and 'papal aggression'. Classicism had been disdained as redolent of the French Revolution and the Prince Regent's extravagance, but perhaps it was now more appropriate for a nation with global range and reach? Sir George Gilbert Scott was entered into the competition during Lord Derby's second brief administration; but on returning to power Palmerston refused to accept his Gothic Revival design, and ordered Scott to come up with an alternative scheme 'in the ordinary Italian' style, or face the sack. Swallowing his pride, Scott grudgingly complied, and went on to produce the magnificent neo-Renaissance building that seemed the fitting headquarters of the greatest power in the world, and still adorns Whitehall today.

As *The Times* wrote approvingly at the end of 1861, 'Lord Palmerston in truth represents the precise state of the national mind in opposing unnecessary changes without setting up resistance as a principle, and in countenancing all foreign approximations to the political theories and system of England.' But these words, like the classical certainties of the new Foreign Office, belied and denied the many uncertainties of his age. The Crimean War had exposed the serious limitations of the British military, and especially its capacity to fight a land-based war either in Europe or in Asia. The Great Rebellion made plain the weaknesses of British power in India, and in its aftermath British rule became more conservative, but as a result would soon provoke an equal but opposite reaction among the increasingly well-educated urban middle classes of Calcutta and Bombay. Across the Atlantic, Abraham Lincoln had successfully preserved the United States as a single nation, which thereby made possible its late nineteenth-century development and emergence as the greatest economic powerhouse the world had yet seen. And Otto von Bismarck had embraced a blood-and-iron doctrine of *Realpolitik* that would soon make Germany the pre-eminent European power. It had already made Palmerston's policies of bluff and bluster ineffective and irrelevant, so that he had indeed become the 'impostor', the 'pantaloon' and the 'aged charlatan' whom Disraeli had earlier denounced. These international concerns, and their long-term consequences, were paralleled by domestic developments of which the same was, and would be, true. For at the end of the 1850s Darwin's *Origin of Species* offered scientific validation for the doubts about Christian conviction and biblical authority that Tennyson had poetically expressed at the beginning of the decade. Both at home and overseas, the Palmerstonian 'age of equipoise' was neither as confident in its power nor as certain in its beliefs as that phrase retrospectively suggests; and in any case, it did not last all that long.

By the summer of 1865 Palmerston had been occupying 10 Downing Street continually since June 1859 (and, indeed, he had been in office for most of the time since 1846), and he was the only prime minister since 1832 who had maintained his Commons majority intact throughout the whole life of parliament, something that neither Melbourne nor Peel nor Russell, let alone Derby, had managed before him (and nor would Derby or Russell manage to do so after him). Queen Victoria and Prince Albert never liked Palmerston, and his parliamentary support sometimes faltered, but he never appealed to the British electorate in vain, and he did so once again in July 1865, when at the general election the government gained about a dozen seats. But before parliament met, Palmerston expired in October, his last words being, defiantly, resignedly or jokingly: 'Die, my dear doctor, that's the last thing I shall do!' In truth, this event had been long awaited, with anxiety and apprehension by some, and with eager anticipation by others, and the reaction to Palmerston's passing among members of the political class mingled regret at the close of what had been a remarkable career and a remarkably stable period with enthusiasm for the opportunity this created to get the country moving forward again. There was a need,

Richard Cobden wrote in February 1864 (and he too would die the following year), for 'great reforms which are necessary to avert great disasters in this country', and Palmerston was not the man for that task. 'Our quiet days are over,' agreed Sir Charles Wood, who had been Russell's Chancellor of the Exchequer, speaking at Palmerston's funeral, but with rather different emphasis; there would, he regretted, be 'no more peace for us'.

Palmerston had died at just about the right moment, outliving his own time, both domestically and internationally, but not by all that much. At home there was a growing sense that there should be, and must be, a new round of the sort of fundamental reform – of education, of the armed services and of the franchise – that had not been seen since the 1830s and 1840s, as evidenced by the establishment of the (middle-class) Reform Union in March 1864 and the (working-class) Reform League a year later. Gladstone had found himself increasingly out of sympathy with Palmerston during the final years of his government: the continued high costs of the premier's foreign policy had again obliged him as Chancellor to abandon his earlier hopes of abolishing the income tax; and he also became increasingly unsympathetic to Palmerston's opposition to any further instalment of parliamentary reform. From the early 1860s Gladstone began to make public speeches in the provinces, and his encounters with members of the respectable working class led him to proclaim in the Commons in May 1864 that 'every man who is not personally incapacitated by some consideration of personal unfitness or of political danger is morally entitled to come within the pale of the constitution'. At the general election held in the following year he ceased to solicit the votes of (as he saw it) the reactionary electors of Oxford University, whom he had represented since 1847, and went instead to the more congenial constituency of South Lancashire, where he came among working people 'un-muzzled'.

As Lord Derby conceded, this meant Gladstone was now increasingly 'the central figure in our politics', his importance was 'far more likely to increase than to diminish', and this certainly betokened the advent of a new era in public life. With Palmerston dead, Russell must surely soon follow; and in which direction, thereafter, would Gladstone take the Liberals and the Whigs? Nor was that the only question waiting to be answered. What sort of foreign policy would or could replace the Palmerstonian bluff and bluster that had recently left the

United Kingdom as little more than an ineffectual onlooker in Europe? Nor had the recent efforts of successive governments to consolidate and tranquillize the United Kingdom's far-flung realms been wholly successful or sufficient. The Colonial Laws Validity Act, passed in 1865, recognized the recent establishment of self-government in the settler colonies, and sought to spell out and define the relationship between the imperial parliament in London, which retained ultimate authority, and the colonial legislatures, which within these limits had full domestic law-making powers. But the Canadian and Australian colonies remained separate and dispersed; in New Zealand relations had deteriorated between the settlers and the Maori; while in South Africa contacts between the British, the Boers and the indigenous peoples continued to be troubled and vexed. And in October 1865, the very month of Palmerston's death, the brutal suppression by Governor Edward Eyre of a relatively minor native revolt in Jamaica, which he implausibly likened to the Great Rebellion, outraged and polarized domestic opinion. Carlyle, Ruskin, Tennyson and Dickens supported the Governor's action; Darwin, Huxley, Bright and John Stuart Mill were fiercely critical. In more ways than one, and both at home and abroad, there were challenging and troubling times ahead.

8

Leaping in the Dark, 1865–80

At the general election held in the summer of 1865 the Liberals, as the party had been known since 1859, were returned with a commanding position in the Commons, having won majorities in Ireland, Scotland and Wales and, more unusually, in England, too. This was Palmerston's last electoral triumph, but since he had died before the new parliament could meet, the queen sent for Earl Russell (as Lord John Russell had become in 1861), who formed an administration at the end of October very like the one it had succeeded. The only major new appointment was that Lord Clarendon replaced Russell as Foreign Secretary, but this indicated continuity rather than change, for Clarendon was returning to the very office he had occupied under Aberdeen and again under Palmerston during the mid-1850s. Yet it was also widely believed – or feared – that continuity was no longer enough in what had suddenly become the brave new post-Palmerston world. Russell himself, by this time in his mid-seventies, must have known that this second administration would be his last, and he resolved to carry a significant measure of parliamentary reform, partly because he thought this was what the mood and needs of the country now required, and partly because it would help him regain his earlier reputation as the champion of progressive causes. In this late-life endeavour, Russell's chief ally was Gladstone, who continued as Chancellor of the Exchequer, but who was in a more powerful position than ever as he also became the Liberal leader in the House of Commons. It could only be a matter of time before he succeeded Russell at the head of the whole party; he represented a much more populous constituency than before, with a large working-class element; and he was moving rapidly from being a late-Peelite technocrat to being the champion of a new brand of more democratic, inclusive and energized Liberalism.

Between 1848 and 1865, and as Russell had discovered to his cost, there had been little public or political interest in reopening the issue of parliamentary reform, which was widely believed to have been 'finally' settled back in 1832. Since then, the number of voters in the United Kingdom had risen in absolute terms as the population had grown, but in relative terms there had been scarcely any significant increase: in the early 1860s, 18.4 per cent of the adult males in England and Wales had the vote, whereas in Ireland and Scotland the number was 13.4 per cent in both countries. But with Palmerston gone, the reform of parliament and the further extension of the franchise assumed a new urgency: Russell was eager to force the pace, Gladstone had come to believe that respectable working-class men ought to be given the vote, and the anomalies of the system seemed increasingly indefensible. For example, Honiton, Totnes, Wells, Marlborough and Knaresborough, containing fewer than 23,000 people between them, returned as many MPs as the 1.5 million inhabitants of Liverpool, Manchester, Birmingham, Leeds and Sheffield. Accordingly, in March 1866 Gladstone introduced a relatively mild reform bill in the Commons (Russell being in the Lords), which would have extended the franchise to those who could meet particular rental qualifications, and those with savings-bank deposits of a certain amount. But the precision of Gladstone's statistics provoked rather than allayed anxiety, and his fellow Liberal MP, Robert Lowe, delivered a series of bitter speeches denouncing the extension of the franchise to those whom he deemed and damned as unfit for the vote. Many Palmerstonians were dismayed by any reform proposals emanating from their leaders, and they were willing to support the Liberal rebels, led by Lowe and Lord Elcho. The result was that as the reform measure made its way through the Commons, Russell's seemingly impregnable majority vanished, and in June 1866 his government was defeated by 315 votes to 304.

Russell resigned, and Derby formed his third Tory administration. Like its two predecessors, it was a minority government (as most disaffected Palmerstonians soon returned to the Liberal fold). As in 1852 and 1858, Disraeli was Chancellor of the Exchequer, and by now also the unchallenged leader of the Conservatives in the Commons. His main aim was to pass *any* major piece of legislation in the hope of showing that the Tories were again fit to govern, which they had not been since 1846. Despite the earlier defeat of Gladstone's scheme, reform remained high on the political agenda, in part kept there by the riots in Hyde Park in favour of such a measure that took place later in the summer of 1866, which caused damage to railings and flowerbeds, but were nothing like as threatening as the organized and nationwide popular protests of the 1830s and 1840s. Disraeli accordingly resolved to introduce his own reform bill, hoping he could sell it to Conservative backbenchers, disaffected Palmerstonians and the Liberal rebels alike, by presenting it as an essentially conservative measure, with 'safeguards' designed to prevent the excesses of democracy rather than provide, as Gladstone's scheme had seemed to do, a road map leading directly towards them. Disraeli introduced his bill in March 1867, and during the next two months breathtakingly abandoned all the 'safeguards', and cynically accepted a series of amendments, the effect of which was to make the bill much more radical than Gladstone's. But the Tories, who had had enough of their near-permanent opposition status since 1846, were willing to stay with Disraeli, and this, combined with the temporary support of sufficient disgruntled Liberals, meant he was able to get the measure through. This was a virtuoso parliamentary triumph, mingling desperation, audacity and opportunism, and Disraeli duly succeeded Derby on his retirement as Conservative leader and prime minister in February 1868. But for Gladstone, who had been comprehensively outmanoeuvred in the Commons, it was 'a smash without example'.

LIBERAL ZENITH, CONSERVATIVE COMEBACK

During the final parliamentary stages of what became the Second Reform Act, Derby famously observed that the Conservatives were 'making a great experiment and "taking a leap in the dark"'. Indeed, their bill as eventually passed was a much bigger leap, and to a much darker destination, than Derby or Disraeli had initially envisaged or intended. Indeed, Lord Cranborne, the future third Marquis of Salisbury and a rising figure in the Tory Party, was so outraged that in March 1867 he resigned his post as Secretary of State for India in protest; Robert Lowe, who had been thwarted in his attempts to prevent any extension of the franchise, feared that the dreaded democracy had arrived and with it the end of the world; while Thomas Carlyle likened the passing of the bill to 'shooting Niagara', and deplored the idea that 'any man' was now deemed to be 'equal to any other'. But as had been the case in 1832, these anxieties were exaggerated, for while Disraeli abandoned many of the 'safeguards' in his bill, the measure as eventually carried (and there was corresponding legislation for Scotland and Ireland passed in 1868) brought about less change than some had hoped for and others had feared. In terms of the franchise, the most significant change was the giving of the vote to all borough householders, whether owners or not, in England, Wales and Scotland (but not Ireland), who paid their own rates; there was also some lowering of the voting qualifications in the counties. The effect of this enfranchisement was to add approximately one million voters to the electorate, mainly concentrated in the large towns. Of all males aged twenty-one and above, slightly more than one-third now enjoyed the vote in England and Wales, just under one-third in Scotland, and less than one-sixth in Ireland. This meant that in no part of the United Kingdom was the electorate doubled, and in Ireland it was hardly increased at all.

The Second Reform Act may have widened the franchise more considerably than Derby and Disraeli had originally intended, but it was scarcely a radical measure, as evidenced by the fact that John Stuart Mill, who was briefly an MP between 1865 and 1868, could only muster seventy-three votes in favour of female suffrage. The same was true regarding the redistribution of seats: of the fifty-two that were taken from the small boroughs in England and Wales, seven were given to Scotland, twenty-five were reassigned to the English counties, and a mere nineteen were transferred to such large towns as Leeds, Manchester and Birmingham. At the same time the number of MPs returned by Greater London was only increased from eighteen to twenty-two. As such, the Act favoured the country over the town, and the agricultural sector over the industrial, and at the same time the number of voters in the boroughs was increased by a much greater percentage than the number of voters in the counties. Since the big boroughs tended to be Liberal strongholds, Disraeli was happy to write them off, and to expand their electorates; whereas the county constituencies were more likely to vote Conservative, which explains why their electorates were not increased so significantly. Most of the new voters were confined within the largest towns; many small boroughs with populations below

20,000 continued to return MPs; and at least eighty constituencies were still controlled by aristocratic patrons. This was neither a national nor a uniform structure of representation, and nor – pace Cranborne, Lowe or Carlyle – was it remotely democratic. According to Reform League agents, the election of 1868 was as corrupt as its predecessors, not surprisingly, given that there was still no secret ballot. From a partisan perspective Disraeli had sought to retain and to strengthen those aspects of the electoral system that favoured the Conservatives rather than the Liberals; and either way it continued to privilege the landed interest. Country gentlemen and aristocratic relatives were still the largest group in the Commons, great territorial magnates formed the overwhelming majority in the House of Lords, and they would continue to dominate cabinets until the end of the century.

In terms of the functioning of the electoral and parliamentary system in the aftermath of the legislation of 1867–68, the so-called 'leap in the dark' did not, in many ways, turn out to be so disruptive or destabilizing after all. A generation later, this was 1832 over again. As so often, the ruling elite of magnates and gentry had had to make concessions and compromises, but as a result they had retained much of their power and influence. Yet as had been the case with the Great Reform Act, so in 1867 the further reform of parliament did carry with it significant consequences and developments that were unintended and unexpected. Although the public agitation of 1866–67 was less energized and ominous than that which had taken place between 1830 and 1832, the advent of the Reform Union and the Reform League had betokened a new level of popular engagement with parliamentary politics, and also portended some significant developments in terms of party-political organization. Even before the Second Reform Act was passed, the Birmingham Liberal Association had been founded by a group of radical, largely Nonconformist figures, led by the rich and retired screw manufacturer, Joseph Chamberlain, who would later go into local politics as a radical Liberal and be a renowned and reforming mayor of Birmingham. The aim of the association was to create a significant local organization, known as the 'caucus', which would ensure that voters be encouraged to vote Liberal, and it was soon replicated and followed by similar organizations in other parts if the country. The Tories responded in turn by creating, in 1867, and at Disraeli's behest, what soon became known as the National Union of Conservative

Associations, which held annual conferences and also sought to mobilize voters in the constituencies.

All this lay in the future, although in 1867 it was not a far-off future. Of greater immediate significance was the change in the personnel and generations of the leaders of both political parties. Palmerston was gone, Russell quit public life having yet again failed to pass an additional measure of parliamentary reform, and Derby retired early in 1868, one of the great public figures of the nineteenth century, but who never formed a Conservative government with a Commons majority. With his departure the stage was set for the major parliamentary battle between their successors, Gladstone and Disraeli, and in the short run the laurels had gone to Disraeli. As Russell's lieutenant, Gladstone had been unable to carry a Liberal Reform Act; and having succeeded Russell as Liberal leader, he had then failed to prevent Disraeli from passing what became a yet more radical measure. Disraeli, on the other hand, had successfully thwarted Gladstone, had driven his own measure through, and had been the first to obtain the keys to 10 Downing Street. But as soon as the Second Reform Act was passed, the Liberals were again in a position to unite around Gladstone, who now took command of his party, both in parliament and in the country at large, and began to fashion it into a new progressive instrument. He cultivated organized dissent, which was becoming known as the 'Nonconformist conscience', and in the Commons he carried the abolition of compulsory church rates against the Disraeli government; and although Gladstone had earlier defended the importance and necessity of a state-supported church, he took up the cause of disestablishing the Church of Ireland. He also made overtures to the trades unions and the Reform League, who would supply him with important local information for the impending election.

So while Disraeli, as leader of the Conservative Party, was enjoying his first but very temporary occupation of 10 Downing Street, Gladstone as leader of the opposition had begun to reposition and project himself as the most significant and charismatic Liberal statesman of the age, following and emulating the earlier examples of Lincoln in the United States and Cavour in Italy. Although he had twice faltered over parliamentary reform, no one doubted Gladstone's energy as a man of business or his competence as a man of government. He was also acquiring a substantial popular following as 'the people's William', and he was increasingly insisting that God was on his side as well. Moreover, by taking up the policy of Irish Church disestablishment, he found an issue that enabled him to mobilize and hold together (at least for a time) the many different and disparate elements of the political left, both in the country and in the Commons. Most Irish MPs were delighted to support the withdrawal of state support from what was widely regarded in their country as an alien and much-disliked Anglican church. Since many radicals were freethinkers, they were also generally in favour, albeit for very different reasons; so were the Whigs, who had always believed in religious toleration; and so was the 'Nonconformist conscience', which was generally hostile to any form of state-sponsored religion. During the closing stages of Disraeli's government, Gladstone had successfully carried three Commons resolutions, urging that the Church of Ireland should be disestablished, and no longer enjoy its privileged position of being supported by the British state. In the ensuing general election he won a clear mandate to implement such a measure, as the Liberals obtained 384 seats and the Tories 274.

In explaining the Liberals' electoral victory, Gladstone recognized the growing importance of organized dissent, and of Scotland, Wales and Ireland, when he observed 'our three *corps d'armée*, I may almost say, have been Scotch Presbyterians, English and Welsh nonconformity, and Irish Roman Catholics'. In terms of the 1868 election this analysis was undoubtedly correct. Religion, and its relation to the state, had been an important issue since 1829, both for those who wanted that relation changed, and for those who did not. This was something that Gladstone understood (by contrast, Disraeli did not), both substantively and in terms of its political implications for those wanting change. He was obliged to justify his own volte-face, having previously advocated the continued existence of a state church, and this he now did at great length. Gladstone's analysis also rightly appreciated the dependence of the Liberals on the votes of 'the Celtic fringe', where the party regularly obtained a majority, while only occasionally accomplishing the same feat in the English constituencies. In future the Liberals would need the votes on the margins of the United Kingdom to enable them to dominate England, whereas the Conservatives would use their power base in England to try to dominate the United Kingdom. This in turn may help explain why Gladstone had suddenly become so concerned, not just with Irish Church disestablishment in particular, but with Hibernian matters in general. For as he put it at just this time, his new and self-appointed mission, which would preoccupy the rest of his very long political life, was 'to pacify Ireland'. In this endeavour he was sure that God was on his side: 'The Almighty,' he noted, on becoming prime minister, 'seems to sustain and spare me for some purpose of His own, deeply unworthy as I know myself to be.' Disraeli, on entering 10 Downing Street for the first time, had seen things rather differently: 'Hurrah!' he had exclaimed. 'I have climbed to the top of the greasy pole.'

In some ways the cabinet that Gladstone formed in December 1868 did not differ markedly from the Whig-Liberal administrations of the Palmerston era, and the continued significance of patrician personnel was a powerful riposte to those who believed the Second Reform Act had ushered in the end of the world. Of his cabinet's fifteen members, six were peers (including a duke and four earls), while another of them, Lord Hartington, was heir to the Duke of Devonshire. Lord Clarendon returned to the Foreign Office for the fourth time; when he died in 1870, Gladstone replaced him with Earl Granville, who had previously been his Colonial Secretary. The Lord President of the Council and the Lord Privy Seal were, respectively, Earl de Grey and the Earl of Kimberley. The Duke of Argyll was Secretary for India, and Lord Hartington was Postmaster General. But other ministers came from very different social backgrounds, and in this regard Gladstone's first cabinet was at least as much about change as it was about continuity. Robert Lowe, who had been so vehemently opposed to both Gladstone and Disraeli's reform proposals, was Chancellor of the Exchequer: he was a barrister. George Goschen was President of the Poor Law Board: he was a banker. Edward Cardwell was Secretary for War: like Gladstone, he was the son of a Liverpool merchant and a former Peelite. John Bright was President of the Board of Trade: he was a radical, a Quaker and a Rochdale mill owner, and one of the first Nonconformists to sit in a British cabinet. Another was W. E. Forster, who joined in 1870 as Vice President of the Board of Education: he, too, was a Nonconformist manufacturer. Two of Gladstone's colleagues had also lived and worked for a time in the settler empire: Robert Lowe in Sydney during the 1840s, and H. C. E. Childers, First Lord of the Admiralty, in Melbourne during the 1850s.

Having mobilized public opinion behind a clearly articulated set of proposals, having won a Commons majority of more than one hundred, and having formed a cabinet that would be more radical and more collegial than might have been expected, Gladstone duly embarked on a programme of legislation the like of which had not been seen since the Whig reforms of the 1830s. But why, beyond the attraction of Irish Church disestablishment both to the Liberal MPs and their supporters across the United Kingdom, was Ireland at the top of his list? Part of the answer was the general state of depression and discontent that continued to exist long after the Great Famine was over. But the more specific form this discontent and dissatisfaction with British rule took was the establishment in Dublin in 1858 of the Irish Republican Brotherhood or, as they were more popularly known, the Fenians. They enjoyed substantial emotional and financial support from the large Irish Catholic population by this time established on the east coast of the United States, and by the mid-1860s the Fenians could boast a membership of more than 50,000, largely recruited from the lower levels of Irish agrarian society. In January 1867 a group of Fenians arrived in Britain, determined to wage a guerrilla campaign against the British state; in September they attacked a prison van in Manchester, rescuing two of their comrades and killing an unarmed police sergeant; and in December they blew out a wall in Clerkenwell prison, where they sought to free their chief of arms procurement, killing twelve Londoners and injuring many more. In the same year the Fenians attempted to rise in rebellion in Ireland and their American collaborators also vainly tried to invade Canada. The perpetrators of the Manchester and Clerkenwell outrages were apprehended, and several of them were sentenced to what would be the last public hangings in Britain.

Although there was widespread popular outrage at these acts of violence, the Fenians undoubtedly reinvigorated Irish nationalism, and their subversive activities convinced Gladstone that he must, indeed, seek to 'pacify' the country, in the hope of retaining the Union. The act to disestablish the Church of Ireland, passed in the summer of 1869, was his project from the very beginning; it was his administration's first major legislative enterprise, and he was personally responsible both for drafting it and for carrying it through parliament. In substantive as well as political terms, the case for disestablishment was overwhelming: in 1861 only 12 per cent of Ireland's population of almost six million were members of the Church of Ireland, whereas nearly 80 per cent were Roman Catholics. As such the Church of Ireland was one of the most potent symbols of Protestant England's unwanted supremacy over Catholic Ireland. Disestablishment was thus a relatively easy way of removing a central grievance of the majority of the Irish people, and to assure them of the Westminster parliament's goodwill, while making no substantive political concession to them. The leaders of the Church of Ireland recognized that the bill was bound to become law, and disagreement and negotiations centred on getting the best financial terms. The Irish Church was duly disestablished, and in future bishops would no longer be appointed by the crown on the advice of the prime minister, but by a church synod. The Church of Ireland was also disendowed, in that it would receive no further income from the British government. But it retained property and other assets worth in excess of £10 million, and there were additional sums made available for such 'non-religious' causes as poor relief and agricultural improvement.

Gladstone's second piece of Irish legislation, with which he was as closely identified as church disestablishment, concerned the land – a subject by which he would be much preoccupied for the next twenty years. But although the alien, Protestant, Anglo-Irish landowners were as much anathema to many Irish people as the established church had been, the land question was a more complex and difficult matter to resolve. Most Irish tenants wanted to own their land, but the interests of those with large holdings, and who were thus better off, were very different from those with smallholdings, who were often very poor; and while Gladstone, as a firm believer in the preordained rural hierarchy, was determined to sustain and support the landlords, many of them were absentee, irresponsible and with precarious finances. These were not easily reconciled circumstances or aspirations; moreover, any government interference in the relations between landlord and tenant would be seen as a deeply subversive assault on the sanctity of private property and freedom of contract. The Irish Land Act that was passed in the summer of 1870 limited the powers of landlords to evict tenants, guaranteed compensation to departing tenants for any improvements they had made on their property, and provided government loans with which tenants could purchase their holdings. Gladstone was pleased with the measure, but it was much less successful than Irish Church disestablishment in that it failed to achieve its purpose: its main provisions were easily evaded by landowners, who could raise rents high enough to allow evictions, or impose restrictive covenants that effectively nullified the Act's compensation clauses, while the money and terms offered to those tenants who wished to purchase their holdings were insufficient. As a result the measure failed to deal with the shortcomings of Irish landlords, the problems and grievances of Irish tenants, or the structural problems of the agricultural sector.

These two Irish reforms were very much Gladstone's own. But there were other issues where evidence had been accumulating and momentum building up during the later Palmerston years, that the Liberal government also sought to address. One of them concerned the inadequate provision of mass elementary education. In 1861 the Newcastle Commission had complacently reported that 95 per cent of children aged eleven or less were attending some sort of school. But other, private surveys undertaken later in the decade suggested that almost one-half of those aged between five and thirteen were still not receiving serious elementary education. All these statistics were intrinsically unreliable; but those that posited lower rather than higher levels of education seemed the more convincing; and it was also widely felt that the North's victory over the South in the American Civil War, and Prussia's recent military triumphs over Austria and France, were in part explained by the fact that their troops were better educated than those they vanquished. Moreover, the enfranchisement of many members of the urban-based, male working class in 1867 meant they needed to be better schooled if they were to wield their votes in a responsible and well-informed way. This meant, as Robert Lowe sardonically put it, that there was now a pressing need to 'educate our masters'. But in addition the structure of education in the United Kingdom was at best confused, at worst chaotic, and it was riven by religious and sectarian strife: there were state-aided, non-denominational, 'British' schools, and there were also voluntary, so-called 'National' schools, which were denominational, and almost entirely free of state support. But however they were funded, and whatever their religious affiliation (or none), the elementary education system of the United Kingdom was clearly inadequate to the needs of the world's most advanced industrial society.

However, it was easier to diagnose the limitations of the system than to come up with a prescription that could be afforded, made to work, and would simultaneously satisfy both those who wanted elementary education to remain largely controlled by the churches and those who wished it to become primarily secular. In 1869 two organizations were founded to promote these very different objectives: the (Anglican) National Education Union, supporting the existing denominational system, and the (Nonconformist) National Education League, which advocated universal and non-sectarian schooling, and which had developed from the Birmingham Education League, another radical initiative set up two years earlier by George Dixon and Joseph Chamberlain. Not surprisingly, the Education Act that eventually emerged in 1870, steered through by W. E. Forster, satisfied neither pressure group. To be sure, it did seek to create enough school places for all children in England and Wales between the ages of five and thirteen (a separate measure was passed for Scotland in 1872; there was no corresponding legislation for Ireland). The 'voluntary' schools were to be given maintenance grants from the central government, equivalent to the amounts privately subscribed; and in areas where voluntary effort made inadequate provision, or where a majority of ratepayers requested it, school boards were created, which would be elected, and given the powers to levy rates, build new schools and appoint teachers. But problems arose when it came to trying to resolve the sectarian and religious issues. In board-run schools religious education was ostensibly limited to the Bible and hymn singing; but in practice denominational religion could still be taught. This left Nonconformists with an abiding sense of grievance, and many who had voted for Gladstone in 1868 would not do so in 1874 for this very reason.

In truth, Forster's Act satisfied neither Nonconformists nor Anglicans, and it established a hybrid system of elementary education that extended state control but also encouraged the continued existence and, indeed, the expansion of voluntary, denominational schools. Yet in terms of increasing educational provision, the Act was of lasting importance. All the great cities, and many lesser towns, elected school boards, and by 1880 they would be overseeing more than 3,000 schools, which were generally better equipped than the denominational schools. In 1870 London was spending £1.6 million on education; by 1885 the sum would have risen to £5.1 million; and elsewhere local-authority expenditure also increased considerably. Gladstone's government also turned its attention to the universities of Oxford and Cambridge, building on the legislation of the 1850s, which had ended the Anglican exclusivity of the undergraduate body. The University Test Act, passed in 1871, opened up academic posts, previously confined to Anglicans, to Nonconformists and Jews (and, effectively, to unbelievers) – though only to men. By then, and thanks to an Order in Council of June 1870, appointment to the higher levels of the civil service had also been reorganized, as the old system of nomination was done away with, to be replaced by competitive examinations and what was termed 'open competition'. This was a significant advance on the Northcote-Trevelyan reforms of 1855, changing the basis of recruitment from patronage to something approaching merit (albeit as defined by the ability to pass a certain form of examination). Henceforward the higher echelons of the civil service would increasingly be dominated by those deemed to be 'the best men', namely graduates of Oxford and Cambridge, many of whom would come from Balliol College, of which Benjamin Jowett became Master in 1870, and which would become the unrivalled breeding ground for Whitehall mandarins and imperial proconsuls.

These changes in elementary education, higher education and civil-service recruitment were attempts to remedy some of the shortcomings of the mid-Victorian state apparatus that had become increasingly apparent during Palmerston's final years; the same was true of reforms to the military undertaken by Edward Cardwell at the War Office, which were a belated effort to deal with the inadequacies of the army that had been so embarrassingly and disastrously displayed during the early phases of the Crimean War. In 1869 Cardwell abolished flogging in peacetime, and in the following year he passed legislation making it possible for men to enroll for shorter periods, and placing the commander-in-chief (the recalcitrant Duke of Cambridge) under direct War Office control. Influenced, as were the advocates of educational reform, by the extraordinary success displayed by Bismarck's troops in the Franco-Prussian War, Cardwell determined to improve the British army's competence, skill, efficiency and adaptability, and to this end he sought to abolish the purchase system and also to introduce what was termed 'territorialization'. The case against purchase was that it effectively restricted the personnel in the higher ranks in the army to aristocrats and landowners who could afford to buy their commissions, and as the mistakes in the Crimea had made plain, many of them were very poor officer material. But the opposition to this measure in the Lords was so bitter that abolition was eventually secured in 1871 by royal warrant rather than by legislation. As for 'territorialization', Cardwell's aim was to associate particular regiments and Volunteers with specific localities, by dividing the United Kingdom into sixty-six districts, thereby exploiting county identities, all with the aim of improving recruitment.

While these measures were being passed, Gladstone was also much preoccupied by what he called 'the great crisis of royalty', brought on by Victoria's continued grief-stricken seclusion since Prince Albert's death in 1861, which meant she rarely appeared in public and spent most of her time isolated at Windsor, Osborne and Balmoral; and also by the stench of scandal that was beginning to adhere to her increasingly wayward son and heir, the Prince of Wales. In *The English Constitution* (1867) Walter Bagehot memorably described the hapless royal pair as the 'retired widow' and the 'unemployed youth', and public dissatisfaction reached its peak during Gladstone's first administration, further encouraged by the overthrow of the Emperor Napoleon in France and the subsequent establishment of the Third French Republic. By the early 1870s republican clubs had proliferated across the United Kingdom, and some advanced Liberals, such as Joseph Chamberlain and Sir Charles Dilke, made no secret of their anti-royalist views. All this caused Gladstone much concern. He was a firm believer in the British monarchy as an essential national institution, and he grieved that it seemed more unpopular than at any time since the late Hanoverians. Accordingly, he repeatedly sought to persuade the queen to come out of her widowed seclusion, and to show herself more conspicuously and more frequently to her subjects; and he also devised a scheme to give employment to the Prince of Wales. Gladstone's idea was to establish a royal residence in Ireland, which the prince would occupy for part of the year as an essentially ornamental Viceroy, leaving the detailed administration of the country to be headed by a deputy who would replace the Lord Lieutenant. Gladstone hoped that such a scheme would not only provide the Prince of Wales with something worthwhile to do, but would also help bind the Irish people more loyally to the crown.

Gladstone and his colleagues had thus energetically embraced a remarkable agenda of activity, to which should be added the Ballot Act of 1872 that introduced secret voting at the polls, a Licensing Act that was a nod to the temperance lobby, and legislation aimed to strengthen the rights of trades unions. But there were significant limitations to what the prime minister and his cabinet achieved. The disestablishment of the Irish Church was a major symbolic gesture, but little more, while the Irish Land Act made scarcely any impact on landlords or tenants or on agriculture. Despite its undoubted significance, Forster's Education Act resulted in widespread dissatisfaction, and by alienating so many Nonconformists it undermined the Liberal coalition that Gladstone had created in the country and in the Commons; and the government made no attempt to deal with secondary education. The reform of civil-service recruitment was only implemented slowly, and the Foreign Office did not embrace 'open competition' until after the First World War. Cardwell's abolition of purchase in the army did little to change the social composition of the officer corps, for he failed to increase pay or reduce expenses, which meant that in practice rich aristocrats and gentry were still the only people who could afford these commanding positions. The prime minister's efforts to solve 'the great crisis of royalty' got nowhere, because the queen had become enchanted with Disraeli and had come to loathe Gladstone. Accordingly, she refused to accede to his well-meaning but tactless exhortations that she appear in public, and she vetoed the scheme for a royal residence in Ireland. And while the Ballot Act was a major piece of legislation, which transformed voting procedures and the electoral system, there was no general enthusiasm for it, and little political credit to be gained from it.

As is almost invariably the case with any long-lived reforming administration, the Liberal unity, energy, momentum and enthusiasm that had been so marked in 1868 and immediately after had largely dissipated within four years. At the same time a succession of scandals concerning appointments and financial matters cast doubt on the administration's claims to either probity or competence, as Gladstone showed himself obstinate and high-handed in making ill-judged recommendations for judicial and ecclesiastical appointments, and was subjected to harsh criticism in the Commons. Between 1868 and 1874 the Liberals lost thirty-two seats to the Conservatives at by-elections (while only ten constituencies changed hands in the opposite direction), a sure sign that the government was losing its hold on the country. In a further effort to 'pacify Ireland', and also in an attempt to recover the political initiative, Gladstone introduced an Irish University bill early in 1873. Irish Catholics had long argued that they lacked acceptable higher-education provision, since Trinity College, Dublin, was staunchly Protestant, and a bastion of the Anglo-Irish Ascendancy. The aim of

Gladstone's bill was to establish a new and alternative Irish university, which would be primarily for the country's majority Catholic population, and the prime minister believed he had secured the necessary support of the Roman Catholic hierarchy in Ireland and thus of Irish MPs in parliament. But Gladstone was mistaken, for he was denounced by the Roman Catholic Church on the grounds that the proposed new university would have no religious tests whatsoever, and in March 1873 an unholy alliance of Conservatives and Irish Catholic MPs defeated the second reading by three votes. Gladstone's cabinet thereupon resigned, and the queen invited Disraeli to form what would have been yet another minority Conservative government.

During the first years of Gladstone's ministry the Tory opposition had made little headway: the political initiative lay emphatically with the government, Disraeli was criticized by his own followers for having taken what seemed to be the failed gamble of the Second Reform Act, and for much of the time he was unwell. But by 1872 the government's growing difficulties, allied with an improvement in Disraeli's health, meant he had recovered something of his fighting spirit, and during the first half of that year he had delivered two unusual speeches, at Manchester and the Crystal Palace, which were landmarks in the Conservative revival. Unlike Gladstone, Disraeli was not in the habit of addressing large public audiences, which may be another reason why these two addresses made such an impact. He denounced the radical wing of the Liberal Party as a threat to the established institutions of the country; he criticized the Liberals as a whole for being in misguided thrall to 'cosmopolitan' and 'continental' ideas; and he dismissed Gladstone and his tired cabinet colleagues for resembling a 'range of exhausted volcanoes'. It was, he insisted, the Conservatives who were the 'national' party, pledged to uphold the constitution of the country, determined to improve the condition of the people, and committed to supporting and strengthening the empire. In truth, these two speeches were high on phrases but low on policies; yet they resonated and caught a mood, and reinforced the impression that the Tories were coming in with the tide. But Disraeli wanted power on his own terms, and preferred to bide his time, forcing the government to endure a further period of unpopularity. So when Gladstone and his colleagues resigned in March 1873, he declined to accept the queen's commission to form yet another minority Tory government, on the grounds that the Liberal defeat had been inflicted by a temporary and unusual alliance of Irish and Conservative MPs remote from the normal workings of the parliamentary system.

Disraeli's refusal obliged Gladstone to resume office, in seriously weakened circumstances, with the Liberals' political fortunes clearly in decline. Late in 1873 Gladstone tried to counter the government's unpopularity, and to give a new impression of energy and purpose, by shuffling his cabinet: John Bright, who had resigned on health grounds early in 1871. was brought back as Chancellor of the Duchy of Lancaster, and Robert Lowe was moved to the Home Office, with Gladstone once again taking the Exchequer. A final piece of legislation, the Judicature Act, was also passed, which consolidated the many contradictory and overlapping jurisdictions of the time, and created the High Court of Justice and the Court of Appeal. But it was to little avail, and early in 1874, and to the surprise of both parties, Gladstone suddenly decided to hold a general election. The Liberals were indeed exhausted after six years in office, their party was not well organized, and Gladstone failed to come up with the right policy at the right time, as he had so successfully done in 1868. All he could offer, in what Disraeli derided as a 'prolix' election address, was a renewed undertaking to abolish the income tax. Having previously denounced Gladstone and his colleagues for resembling 'extinct volcanoes', Disraeli now changed tack, and upbraided them for having been over-active purveyors of 'incessant and harassing legislation'. By contrast, the Conservatives offered to put an end to this law-making frenzy, to defend the constitution and the Church of England, and to display 'a little more energy in foreign affairs'. On both sides, in fact, the general election of 1874 was a subdued and negative affair.

But it yielded a decisive outcome. For the first time since 1841 the electorate had returned a substantial Conservative majority: there were 350 Tories, 245 Liberals and 57 Irish Home Rulers (a new group, led by the Protestant Isaac Butt, who were no supporters of the Fenians, but were disappointed at the limited nature of Gladstone's attempts at Irish 'pacification'). This gave Disraeli a Commons majority just short of fifty, but in England the Conservatives were ahead by one hundred seats. They were still a minority in Wales, Scotland and Ireland; but the recent emergence of the Home Rulers was more of a threat to the Liberals than to them. The Tories did spectacularly well in the English county seats, and also made significant gains in the larger boroughs. Gladstone immediately resigned not only the premiership but also the leadership of the Liberal Party (where he was succeeded by Hartington in the Commons and Granville in the Lords), and during the next few years his parliamentary appearances would be intermittent and unpredictable. Meanwhile, Disraeli's persistence had been abundantly rewarded, and he set about creating the first Tory cabinet since Peel's that had a reasonable prospect of serving out the full parliamentary term. Half its members were peers, and a majority came from the traditional titled and territorial classes. The Foreign Secretary was the fifteenth Earl of Derby, son of the former Conservative leader, who had held that post in the minority government of 1866–68. The Secretary for India was the former Lord Cranborne, who had recently succeeded as third Marquis of Salisbury, now reconciled to Disraeli's leadership, and the Colonial Secretary was Lord Carnarvon. The Chancellor of the Exchequer was Sir Stafford Northcote, a West Country baronet, and the Postmaster General was Lord John Manners, son of the Duke of Rutland and a friend of Disraeli's since his Young England days. The one notable non-landed appointment was that of R. A. Cross, a Lancashire banker and solicitor, to the Home Office.

As prime minister in 1868, Disraeli had only enjoyed the illusion of power; but in 1874 he obtained the substance of it as well. With a united cabinet, a secure parliamentary majority in both houses, an opposition in defeated and chaotic disarray, and a monarch who greatly preferred him to Gladstone, Disraeli could have carried whatever programme he wished. Yet he had little idea what to do with power now he had finally obtained it. This was partly because it had come to him too late: he was sixty-nine when he became prime minister for the second time, and he was tired and frequently unwell. Moreover, and unlike Gladstone, Disraeli was never a man of government, for he was neither a commanding administrator nor a creative legislator: most of his political life had been spent in opposition, he had never headed a major department for a long period, and the only significant measure he had carried was the Second Reform Act, which was more the expression of desperate opportunism than the outcome of any long-pondered strategy or carefully considered scheme. Nevertheless, at their first cabinet meeting his colleagues expected Disraeli to reveal detailed plans, giving legislative substance to the general proposals he had outlined in his speeches at the Crystal Palace and at Manchester. But he had no such agenda, and after Gladstone's six years of frenzied activity, no such inclination, so for the remainder of 1874 there would be no burst of legislative energy comparable to that which his predecessor had unleashed in 1868. A Factory Bill was already in preparation, which further reduced working hours, and the new government took it over; a Licensing Act, which modified the Liberals' unpopular statute, did not involve much planning or preparation; and the Public Worship Regulation Act, designed to curb the growing practice of what were termed Anglo-Catholic liturgical 'excesses' in the Church of England, was passed at the initiative of the Archbishop of Canterbury, and also to please the queen, but turned out to be a damp squib.

Only in 1875, when the new ministers had found their feet, did Disraeli's government carry any serious programme of social legislation, and most of it was the work of Cross, the Home Secretary. There were two measures concerning trades unions, which capitalized on the widespread dissatisfaction with the recent Liberal legislation in this field: the Conspiracy and Protection of Property Act, and the Employers and Workmen Act. The first provided a less restrictive limitation on peaceful picketing, the second removed the privileges previously enjoyed by employers vis-à-vis their workers respecting breaches of contract; and the effect of these two measures was to do away with the old Combination Laws, strengthen the legal position of the trades unions, and recognize the rights of workers to strike and to collective bargaining. The Public Health Act consolidated and clarified the complex mass of sanitary legislation that had been passed since the 1840s, including that of Gladstone's government, and it would remain the basis of legislation in this area until the 1930s. The Sale of Food and Drugs Act forbade the use in food or drugs of anything 'injurious to health', again codifying earlier piecemeal legislation, and it remained the principal measure on that subject until 1928. The Artisans Dwellings Act made powers available to all boroughs regarding slum clearance and the rehousing of their inhabitants, which had previously been restricted to a few large cities; as a result, local authorities across the United Kingdom were enabled to provide improved housing at the expense of their ratepayers. The same year also saw the passing of another Factory Act, which further sought to protect women and children against exploitation, and at the Exchequer, Northcote carried a Friendly Societies Act, which offered additional legal safeguards for working-class efforts at savings and insurance.

No previous Tory government had done so much for social reform within a single year; but as with Gladstone's earlier endeavours, the Tories achieved significantly less than was sometimes claimed for them. With the exception of the trades union legislation, most of the Acts laid down guidelines that were enabling rather than mandatory, which meant the appropriately designated local authorities could either adopt or ignore them – and most chose to ignore them. The Artisans Dwellings Act, for example, gave new powers to municipal authorities to draw up improvement schemes, but they were costly and thus unattractive to ratepayers; by 1881 only ten of the eighty-seven towns in England and Wales to which it applied had availed themselves of its provisions (of which Joseph Chamberlain's Birmingham was the most famous and atypical example). Likewise, the Sale of Food and Drugs Act did not compel local authorities to appoint the chemical analysts who alone could have provided the necessary evidence concerning adulteration. The Public Health Act was full of detail, but it was essentially a consolidating measure, lacking any new ideas or principles, while the legislation concerning friendly societies did little substantive to render them more financially secure. This was, then, not so much a coherent programme of social reform as a series of rather haphazard bills, the timing of which were dictated by circumstance rather than by any coherent philosophy; and it certainly did not amount to a British equivalent of the state social-welfare scheme that Bismarck was simultaneously enacting in Germany. Moreover, Disraeli showed little personal interest in the measures, and it would be a mistake to see this legislation as representing the long-awaited fulfilment of his Young England dream of a paternal aristocratic government assuming appropriate responsibility for the welfare of the workers.

Like the Liberal measures that had gone before, these Tory social reforms did not represent a shift away from the prevailing ideology of laissez-faire to a new era of state intervention. As Disraeli himself observed in 1875, 'Permissive legislation is the characteristic of a free people.' A Conservative backbencher put it more graphically in the same year, when he extolled the 'suet-pudding legislation' as 'flat, insipid, dull', but also as being 'very wise and very wholesome'. There were more measures to come, but they added little to the limited achievements of 1875. In the following year a Rivers Pollution Act was passed (one of the earliest statutes concerning the environment), along with an Education Act (which compelled parents of children aged between five and ten to send them to school), and also a Merchant Shipping Act (which aimed to ensure all such vessels should be seaworthy); but the first of these laws was ineffectual, the aim of the second was to prevent not promote the establishment of school boards in the countryside, and the third was poorly drafted and again lacked powers of compulsion. The Prisons Act of 1877 was a rare piece of legislation in that it centralized the whole penal system under a Prison Commissioner, and similar legislation was also passed for Ireland and Scotland; while a Factory Act carried the following year was another essentially consolidating measure. In any case, by this time, Disraeli's involvement in the domestic affairs of his administration had lessened even further. For much of 1876 he was afflicted by bronchitis, asthma and gout, and there were complaints from MPs that he was no longer vigorous enough to conduct the government's business in the Commons. The criticisms were unanswerable, and Disraeli's solution was to stay on as prime minister but to quit the Commons and enter the Lords, which he duly did in August 1876 as Earl of Beaconsfield.

GENDER AND RACE, CULTURE AND ANARCHY

It gave Queen Victoria enormous pleasure to confer an earldom on Disraeli, for he was by a substantial margin the favourite prime minister of her widowed years. Gladstone had intimidated her, with his lengthy prime-ministerial memoranda, and he had treated her as though she were a venerable institution rather than a flesh-and-blood being. Disraeli, by contrast, flattered her, flirted with her, and although he invariably showed her all the deference that he believed to be her due, he engaged with her as a lonely, bereaved woman (he was by now a widower himself). Indeed, it was in his relations with his sovereign that Disraeli's delight in the politics of gesture and phrase-making reached its zenith. He wrote Queen Victoria the most extraordinary letters that any prime minister can ever have written to a sovereign; he called her the 'faery queen' and attributed to her far greater constitutional powers than she possessed. Having won the queen's confidence and something close to her love, he was far better placed to solve the 'great crisis of royalty' than Gladstone could ever have been. In truth, the tide of public opinion had already begun to turn away from republican agitation and back towards the monarchy, with the service of thanksgiving that had been held to celebrate the recovery of the Prince of Wales from typhoid in 1872. But Disraeli, seizing the moment, encouraged (rather than cajoled) the queen to begin appearing in public again, one indication of which was that she attended the state opening of parliament for him in 1876, again the following year, and also in 1880, something she had never done for Gladstone. She was also, like many of her subjects, turning away from the liberalism and internationalism of the 1850s and 1860s towards the conservatism and imperialism that would be the hallmark of the closing decades of her reign; and Disraeli was both the agent and the beneficiary of this late-life, right-royal transformation (of which it is difficult to think Prince Albert would have approved).

It was one of the many ironies of the nineteenth-century United Kingdom that while its head of state for almost two-thirds of the time was a woman, it was extremely difficult for anyone else of the queen's sex to play any significant part in the public or political work of the British nation and empire. One other female who did make a significant impact was Florence Nightingale. Her medical activities in the Crimea may have been exaggerated, both by herself and the media, but she was beyond question a major force in the subsequent establishment of nursing as a profession, due to her book, *Notes on Nursing* (1859), and the founding of the Nursing School at St Thomas's Hospital in London the following year, the first of its type and purpose in the world. Thereafter, Nightingale remained an important public presence, through her writings on a broad range of national and imperial health matters, and via her lobbying and close connections with such well-connected and influential figures as Sidney Herbert and William Jowett. Another such person was Angela Burdett-Coutts, who inherited a huge sum from her grandfather, built up a significant collection of nineteenth-century art, and numbered among her friends both Charles Dickens (who dedicated *Martin Chuzzlewit* to her) and the Duke of Wellington. Her life's work was philanthropy, as she gave away more than £3 million of the fortune she had inherited: supporting homes for young women who had been forced into crime and prostitution; endowing the bishoprics of Cape Town and Adelaide; donating funds to prevent cruelty to animals and children; and pioneering social housing. She also worked with Florence Nightingale, helping her in her efforts to promote nursing, and was actively interested in improving the condition of indigenous Africans.

These three women may have had more in common with each other than with most others of their sex, as in some ways their class undoubtedly trumped their gender. They were all unusually well born and well off: Victoria was a regnant queen, Nightingale's parents were prosperous gentry, and Burdett-Coutts inherited a substantial fortune from her grandfather. They lived well beyond the average age of nineteenth-century women: Victoria died at eighty-one, Nightingale made ninety and Burdett-Coutts reached ninety-two. They were also much honoured: Queen Victoria became an empress, Florence Nightingale was the first woman to be appointed to the Order of Merit (in 1907, the next was Dorothy Hodgkin in 1965), and Angela Burdett-Coutts was made a peeress in her own right (in 1871, again a very unusual form of recognition for a woman). Yet even they, or perhaps especially they, found it difficult being a woman in a man's world, and to reconcile their public and private roles. All the politicians with whom Victoria dealt were men, and much better educated than she was; yet she also took pride in being a soldier's daughter, and she did not like pregnancy or motherhood, but nor did she think that women should have the vote. Nightingale never married, and like the queen she often used the threat, or the reality, of physical or mental illness as a way of disconcerting men with whom it might otherwise have been difficult to deal – or to control. While playing lady bountiful was an entirely acceptable female role, Burdett-Coutts shocked polite society by marrying, at the age of sixty-seven, her American-born secretary, just twenty-nine years old; as a result she was compelled to forfeit three-fifth's of her income to her sister according to a rogue clause in her step-grandmother's will forbidding her from marrying a foreign national.

The Prince of Wales and future King Edward VII preferred his women to be biddable, beddable and beautiful, but as the eldest son of Queen Victoria he also had no difficulty in recognizing an alpha female when he met one. The Order of Merit he bestowed on Florence Nightingale was in his personal gift as the sovereign, while he is reported to have described Angela Burdett-Coutts as being 'after my mother the most remarkable woman in the kingdom'. Yet the fact remains that during the reign of Queen Victoria virtually all women were second-class citizens, in both law and custom. The Great Reform Act was the first piece of legislation concerning the franchise that had explicitly declared that no women were entitled to vote in parliamentary elections. But they were expected to marry, and failing to do so carried a social stigma at least until the outbreak of the First World War. At her wedding, a woman became legally subordinate to her husband, and her property became his to dispose of as he wished, unless the bride's rights had been protected by a trust, an arrangement that only the wealthy could afford. Until the Matrimonial Causes Act was passed in 1857, divorce was impossible without a private Act of Parliament, and even after that legislation it remained much easier for men to divorce their wives than for women to divorce their husbands. Husbands could divorce their spouses on the grounds of a single act of adultery on their part, but family law did not even admit that male adultery was an offence. Here was an example of what was still the prevailing double standard: it was acceptable for men to have sexual relations with many women, both before and during marriage; but it was not acceptable for women to behave in the same way; or, if they did so openly, they would be shunned by polite society.

As during the early decades of the century, the conventional ideology of gender relations was built around the notion of 'separate spheres': the man went out into the worldly public domain of politics, government, business, the professions or labouring work, while the woman remained at home, bearing and bringing up the children, overseeing the domestic arrangements, and creating a warm and nurturing environment. In sum, and in the words of the poet Coventry Patmore, she was supposed to be 'the angel in the house'. But the reality was more complex, and generalizations are exceptionally difficult. Some mid-Victorian wives felt confined and subordinated by the institution of marriage and the demands of domesticity; others relished the opportunities provided by hearth and home and family, and enjoyed fulfilling sexual and emotional relations with accommodating and supportive husbands. Although the education of middle- and upper-class women was almost invariably more limited than that of their spouses (compare Victoria and Albert), many of them developed a range of complementary domestic 'accomplishments', such as reading, embroidery, painting and piano playing, often to high and satisfying levels; while Jane Austen, the Brontë sisters and George Eliot were pre-eminent practitioners of that quintessential nineteenth-century British art form, the novel. In any case, the boundaries between the private and public spheres were never as clear or as complete in practice as the prescriptive literature on the subject suggested. During the mid-Victorian period roughly one-third of the labour force was female, and while the largest occupational group were domestic servants, who were by definition employed in the private sphere, the textile industry was also a significant employer of (usually unmarried) women.

This period also witnessed significant changes in the legal and political status of women, although as so often it was those who were higher up the social scale who benefited most. The Women's Protective and Provident League, founded in 1874, encouraged females to organize in trades unions, and such associations were soon recognized by the Trades Union Congress; but such women were only a tiny minority in the whole unionized labour force. The Married Women's Property Act of 1870 permitted female spouses to retain their earnings if they worked, and after 1882 they could keep property they owned before marriage separate from that of their husbands. The Municipal Franchise Act of 1869 gave female ratepayers the vote in local elections; women could, from the outset, be elected to the school boards established by Forster's Education Act; and from 1875 they could become Poor Law Guardians. Many middle- and upper-class women also involved themselves in charitable and philanthropic work, though none on the scale of Angela Burdett-Coutts. Among them was Octavia Hill, who moved to London from Wisbech in Cambridgeshire in 1852 and became a friend and admirer of John Ruskin. One of her interests was to provide better working-class housing in the great metropolis, and Ruskin later made over to her part of a legacy that enabled Hill to buy properties in Marylebone. She improved them and installed tenants who were casual or seasonal labourers rather than respectable artisans. Hill believed that good-quality accommodation and moral improvement went together. Tenants were expected to be regular and sober in their habits, and she did not tolerate disruptive behaviour or being in arrears with the rent. She also sought to make more general improvements to the urban environment, and would later be one of the founders of the National Trust.

At the same time educational opportunities were increasing for women. Forster's Act brought more and better elementary schooling to girls as well as to boys, but the main improvements were enjoyed by the middle classes. From 1850 to 1890 Frances Buss was headmistress of the recently established North London Collegiate School, which pioneered secondary education for girls, especially from the lower middle classes. Buss became an acknowledged authority on female improvement: she believed it was as important to educate girls as boys; she advocated competitive examinations for girls on the same basis as for boys; and she urged women to take up the new opportunities for involvement with school boards and local government from the 1870s. Higher up the social scale, Cheltenham Ladies' College was founded in 1853, the Girls' Public Day School company was inaugurated in 1872, and an increasing number of endowed grammar and proprietary schools for girls followed. In many such schools, the curriculum focused on 'domestic' and 'feminine' subjects, but there was clearly some increased scope for book-learning. In turn this made possible a significant breakthrough in higher education, albeit only in the first instance for a tiny minority, with the foundation of Girton College (1869) and Newnham College (1871) in Cambridge, and Lady Margaret Hall (1878) and Somerville College (1879) in Oxford. This development was deemed sufficiently novel and noteworthy (and in some quarters subversive) that Gilbert and Sullivan would satirize it in *Princess Ida* (1884), whose eponymous heroine has founded a female-only university, which taught that women are superior to men.

Such improvements in the prospects of women fell far short of granting them the vote in parliamentary elections. In January 1867 the first organization to campaign for votes for women was established as the Manchester Women's Suffrage Committee, which was quickly joined by similar associations in London, Birmingham and Bristol. The London branch provided a degree of national co-ordination, and one of its most prominent members was Millicent Fawcett, who had helped found Newnham College. Her sister was Elizabeth Garrett Anderson, who had qualified as a doctor at the University of Paris in 1870, and became the first female member of the British Medical Association three years later. The two sisters had been much influenced by the writings of John Stuart Mill, who as an MP made a sustained argument in favour of women's suffrage during the debates over the Second Reform

Act. There was, he insisted, no 'adequate justification for continuing to exclude an entire half of the community, not only from admission, but from the capability of ever being admitted within the pale of the constitution'. But his efforts were to no avail and in 1869 he published *The Subjection of Women*, setting out his arguments more fully and systematically. 'The legal subordination of one sex to another,' he insisted, 'is wrong in itself.' It was 'one of the chief hindrances to human improvement', and 'ought to be replaced by a system of perfect equality'. 'Under whatever conditions,' he concluded, 'and within whatever limits, men are admitted to the suffrage, there is not a shadow of justification for not admitting women under the same.' But neither parliament nor public opinion in general shared that view, and the cause of female suffrage did not prosper in the 1870s.

Instead, women with ambitions to be involved politically took up campaigns on single issues, as their predecessors had done during the 1820s and 1830s. The most famous (and, ultimately, successful) of these campaigns was that against the Contagious Diseases Acts, which had been passed in 1864, 1866 and 1869. The aim of these measures was to reduce the amount of vice and immorality in garrison towns throughout the United Kingdom by requiring prostitutes to be registered, and by allowing special police to arrest such women, who would then be compelled to undergo a medical examination for venereal disease. If found infected, women could then be consigned to hospital for treatment under conditions resembling incarceration. Such measures were deemed to be an infringement of individual freedom, since prostitutes ought to be free to choose that occupation if they wished; and since it was only women not men who were subjected to compulsory and humiliating medical examinations, this was a further indication that the double standard remained very much the norm. The campaign against this legislation was led by the National Anti-Contagious Diseases Association and the Ladies' National Association; the latter was led by Josephine Butler, who wrote pamphlets on social questions and sought to increase opportunities for women in higher education and the professions. In 1871 she told the LNA that 'all men, however pure their conduct' were 'depravers of society, who hold the loathsome and deadly doctrine that God has made man for un-chastity and woman for his degraded slave'. The Contagious Diseases Acts would eventually be repealed in 1885.

Four years earlier, another remarkable woman of British nationality had died in London. She was named Mary Seacole and had been born in Kingston, Jamaica, the daughter of a Scottish army officer and a free black woman. Seacole spent the 1840s managing the family hotel in Kingston and the early 1850s travelling in Central America. During that time she acquired some basic nursing skills, treating patients in the cholera epidemic which ravaged Jamaica in 1850, and again in Panama soon after. When the Crimean War broke out in 1853, Seacole determined to travel to Britain to volunteer as a nurse. Her application to the War Office was refused, but she travelled to the Crimea anyway, and early in 1855 she set up the 'British Hotel' behind the lines, which offered 'comfortable quarters for sick and convalescent officers', and she also eased the suffering of some of those who had been wounded on the battlefield. Seacole returned to London towards the end of 1856, and the following year published her autobiography, *Wonderful Adventures of Mrs Seacole in Many Lands*. She spent much of the 1860s back in Jamaica, but often in financially straitened circumstances, and the Seacole fund was set up in London, whose patrons included the Prince of Wales, the Duke of Edinburgh and the Duke of Cambridge. By 1870 she was back in London, and was soon established on the periphery of the royal court: a nephew of Queen Victoria, Count Gleichen, carved a marble bust of her in 1871, which was exhibited at the Royal Academy, and she became the personal masseuse to the Princess of Wales, who suffered from a stiff rheumatic knee.

There is still no consensus as to whether Seacole was, or was not, the 'black Florence Nightingale', and she would soon be forgotten after her death in 1881, but she was clearly a recognized and accepted figure in London during the last decade of her life. Yet by then racial attitudes in Britain were undoubtedly beginning to coalesce and harden; for while there was a gradual move towards increasing opportunities for women during these years, attitudes towards race generally shifted in the opposite direction. The mid-Victorians certainly thought they were the most civilized society the world had ever seen; but they had also believed in the essential unity, equality and perfectibility of mankind, which meant that if other peoples and other races were given a helping hand, by missionaries, imperial interventions and provisions, or the reform of 'native' customs, then they, too, might improve and advance to higher stages of civilization (this was generally the view of David Livingstone).

But by the 1860s and 1870s there was a growing belief in a racial hierarchy that was fixed and immutable, which meant there was little realistic prospect that those who ranked low on it could be raised up or advanced. This change of view was partly because of a growing misapplication of Darwin's ideas concerning evolution: he made no claims about the superiority of one race over another, but others argued that this was precisely what the notion of the 'survival of the fittest' implied. 'Physical science,' Charles Kingsley noted in 1871, 'is proving more and more the immense importance of Race; the importance of hereditary powers, hereditary organs, hereditary habits, in all organized beings, from the lowest plant to the highest animal.'

There were other influences at work by this time, not least the widespread proliferation of views about the nature of race and the importance of racial hierarchies, which owed little to Darwin or to those who referenced but misrepresented him. Even as Marx and Engels had declared in their *Communist Manifesto* of 1848 that the history of human society was the history of class struggle, other contemporaries, ranging from Benjamin Disraeli to the Scottish doctor and publicist Robert Knox, were insisting that, on the contrary, it was 'race' that was 'everything'. The rise of interest in anthropology only served to reinforce this view, as exemplified in the words of Dr James Hunt, President of the Anthropological Society of London, who declared that there were 'about six races below the negro, and six above him'. African explorers, who were especially active by this time, took a similar view. Richard Burton frequently compared Africans with animals, and felt that what he believed to be the arrested mental development of Negroes showed they belonged to the 'lower breeds of mankind'. Sir Samuel Baker, reporting on the Lake Albert and Nile basin, noted in his journal in 1863 that 'Human nature viewed in its crude state as pictured amongst African savages is quite on a level with that of the brute, and not to be compared with the noble character of the dog.' This was the time at which Charles Kingsley concluded that he no longer believed in the intrinsic unity and equality of mankind: 'I have,' he noted, 'seen also, that the differences of race are so great, that certain races, e.g. the Irish Celts, seem quite unfit for self-government.' From this perspective, the Catholic Irish were as much alien and racially inferior as the Negroes of Africa or the West Indies. Many British politicians had scarcely any first-hand knowledge of Ireland or its people: Gladstone only visited it once, Disraeli never.

These were significant shifts in attitudes, which at least in part were also responses to events in the empire. The Great Rebellion had been a massive blow to British confidence in its imperial mission in South Asia. Fifty years later the Duke of Argyll could still recall 'how sick we felt with anxiety, how alarming the prospect appeared, and how all our flowers had lost their glory', on first hearing the news of the uprising. The most lasting change in attitudes that resulted was a growing hostility to South Asians, where the earlier stereotype of the 'mild Hindu' was replaced by a belief in his deceptiveness and cruelty. As the first post-Mutiny Viceroy, Lord Canning, put it, 'the sympathy which Englishmen . . . felt for the natives has changed to a general feeling of repugnance', There were similar changes in attitudes to parts of the settlement empire. During the 1860s the earlier localized conflicts between the settlers and the Maori in New Zealand escalated into full-scale land wars on North Island, as 18,000 British troops, supported by artillery, cavalry and local militia, battled 4,000 Maori warriors. The Maori successfully resorted to guerrilla tactics, but the British eventually won thanks to their superior numbers and weaponry, and large areas of land were confiscated from the Maori as punishment for their rebellion. But once again, the natives were 'misbehaving', and as a result of these long and costly conflicts, relations between New Zealand settlers and the United Kingdom became for a time so embittered that by 1869 the *New Zealand Herald* was openly expressing the view that it would have been better if France rather than Britain had been the first European power to plant its flag on the islands.

These growing imperial anxieties and changing racial attitudes were further intensified because of the debate on the controversial actions taken by Governor Edward Eyre in suppressing the rising in Jamaica in October 1865 – a small-scale eruption that he had mistakenly likened to the Great Rebellion, with the result that 439 individuals were killed, 600 were flogged and 354 were court-martialled. Opinion in Britain was deeply divided about Eyre's actions, especially among members of the new intelligentsia: some condemned him as a vengeful and brutal racist, whereas others praised him as a heroic defender of Christian civilization. The Eyre controversy also spilled over into the parliamentary debates on what would eventually become the Second Reform Act. The Governor's supporters believed that giving the vote to uneducated British men would result in the same sort of 'horrors' and atrocities that they insisted had been committed by the unlettered Jamaican natives, whereas those who opposed Eyre's actions wanted to extend the British franchise because it was the right thing to do, and because a refusal to do so might encourage just the sort of violence that had recently taken place in Jamaica. A royal commission eventually produced a cautious report, criticizing Eyre for the 'positively barbarous' floggings, but praising him for containing the disturbances. But *The Times* caught the public mood when it declared that the Jamaican rebellion was 'more in the nature of a disappointment' than the Great Rebellion. For whatever hopes there might have been that 'the Negro could become fit for self-government' had now been gainsaid and belied by events in Jamaica. 'Alas,' it concluded, 'for the grand triumph of humanity and the improvement of races.'

The combined result of changes in thinking about racial categories and hierarchies, and the anxieties and disillusion engendered by the Great Rebellion, the Maori Wars and the Governor Eyre controversy was undoubtedly a hardening of racial attitudes from the 1860s onwards. The earlier, optimistic view that indigenous peoples could be 'improved', and brought up to Western standards of living and civilization, including assuming responsibility for their own self-government, was largely abandoned. People of colour were disdained as savages or children or, if they were educated, were regarded as being completely untrustworthy. As Lord Kimberley, the Colonial Secretary, put it in 1873, 'except in quite subordinate posts, we cannot safely employ natives'; and, he added, 'I would have nothing to do with the "educated" natives as a body.' Hence the growing social distance between the British officials and military that opened up in India and the colonies of rule from this time onwards. But this was not the whole of the story. In some ways the modernizing impulses of empire continued: the universities that had been founded in the 1850s in Calcutta, Madras and Bombay (and also Sydney, Melbourne and Toronto) had not only been established decades before they would be inaugurated in Leeds, Liverpool, Manchester and Birmingham; they also continued to turn out those very educated Indians whom the imperial authorities increasingly came to dislike and distrust as 'infernal baboos'. The British would also continue their construction of railways, docks and harbours, in South Asia and latterly in Africa; there would be continuing medical research into tropical diseases (Sir Robert Ross proved that malaria was carried by mosquitoes in 1897); and many of the largest and most advanced British companies, in shipping and in railways, would be imperial ventures. Not for nothing would Lenin later describe imperialism as 'the highest stage of capitalism'.

Nevertheless, from the 1860s onwards, there was also a pronounced turning away, on the part of those governing the empire, from a belief in progress, improvement and modernization, and a rediscovery and renewed appreciation of traditional social structures. 'The policy of suppressing or suffering to go to ruin, all the aristocracy and gentry of India,' wrote Sir Charles Wood in the aftermath of the Great Rebellion, 'is a mistake. We must be stronger with the natural chiefs and leaders of the people.' Lord Kimberley agreed. Having listed those 'natives' with whom the British should have nothing to do, he added, 'I would treat with the hereditary chiefs only, and endeavour as far as possible to govern through them.' Lord Lytton, Disraeli's Viceroy, made the same point: 'If we have with us the princes,' he insisted, 'we have with us the people.' Henceforward, in India and in those large parts of Africa that would soon be acquired, this would be the policy and technique of imperial government: through traditional structures of royal and aristocratic authority that the British came to believe resembled those which existed in the United Kingdom itself. For a nation that sought to govern its empire on a limited budget and with a minimum of personnel, this was the only way to proceed, and in any case, patrician Britons felt at ease collaborating with high-born princes and emirs and sheikhs. Just as gender was undercut by class, so was race; and just as prescriptions about 'separate spheres' were significantly modified in practice, so the shoestring management of empire increasingly required accommodation with indigenous rulers and native princes. The problem was that having 'the princes' with them would be no guarantee that the British would have 'the people' with them, too.

These contemporary issues of gender, race and empire were being explored by the new generation of novelists who had begun to make their mark during the 1860s. In both *The Woman in White* (1859) and *Man and Wife* (1870) Wilkie Collins criticized the partiality of the law, which continued to make the position of women inferior to that of men, especially in the case of money. In *Middlemarch* (1872) George Eliot explored, among other contemporary issues, the problems and constraints facing intelligent and ambitious women through the unhappy and unfulfilled life of her heroine, Dorothea Brooke. In marrying the elderly Reverend Edward Casaubon, Dorothea condemned herself to a loveless marriage with a man who resented her youth, enthusiasm and energy, and she soon discovered that his own researches, with which she had hoped to help him, were years out of date and would never be published. In *Far from the Madding Crowd* (1874) Thomas Hardy offered a much darker vision of rural life in his fictional county of Wessex than Trollope had conveyed in his Barset novels, as he explored the themes of love, honour and betrayal through the life of Bathsheba Everdene. Although rich, proud, beautiful and independent-minded, and faced with a succession of suitors, Bathsheba's relationships almost always went wrong, either because she was too proud to persevere in them or because the men were disloyal and untrustworthy. By then, Collins had also published his most famous novel, *The Moonstone* (1868), the title referring to a precious gem, modelled on the real Koh-i-Noor diamond, which had been illegally acquired, and brought to Britain, by thc unplcasant Colonel Herncastle through deception and murder, but which by the end of the story would be returned to India. Although widely regarded as the first major detective novel, *The Moonstone* also depicted British rule in South Asia in a far from positive or benevolent light.

Yet the most significant cultural engagement with current events during these years came not so much from these novelists, resonant though their writings were, but from the poet and critic Matthew Arnold, who did not share the interests of Collins, Hardy and Eliot in gender and race. His father, Thomas Arnold, had been headmaster of Rugby School from 1827 to 1842, and had been a reforming and transformative figure: he improved discipline and teaching, made the chapel the centre of school life, insisted that the formation of a manly and Christian character was more important than academic success, and inspired the foundation of other public schools committed to the same ideals (among them Marlborough, Lancing, Wellington and Haileybury). Matthew Arnold was educated at Rugby School and Balliol College, Oxford, but on the early death of his father he was compelled to earn his living, which he did for much of his life as a state-employed inspector of schools. He began to publish poetry in the early 1850s, and he was twice elected Professor of Poetry at Oxford, in 1857 and 1862. Much of his poetry engaged with a world very different from that of his father, by exploring the contemporary corrosion of faith by doubt. This was especially so in the case of 'Dover Beach' (1857), where he depicted a nightmarish world of uncertainty from which the old religious verities had receded as a result of the undermining of the biblical basis of Christian religion by Darwinian theories of evolution. Arnold's anxieties were further fuelled by the (relatively minor) riots during the crisis over the Second Reform Bill, and by the Fenian agitations in Ireland and Britain, and it was in this pessimistically preoccupied mood that he set about writing his major work of social criticism, *Culture and Anarchy*, which was first serialized in the *Cornhill Magazine* in 1867–68, and subsequently published in book form in 1869.

Arnold was unimpressed by the reformist energy of Gladstonian Liberalism and by the Christian engagement of the 'nonconformist conscience'. As he saw it, contemporary politics were heading seriously in the wrong direction by embracing further parliamentary reform, the Church of England no longer offered conviction or solace, while utilitarianism was mistakenly preoccupied with materialism and the pursuit of individual wealth. Where, then, in such a shifting and deteriorating world, in which neither politics nor religion could provide definitive answers, and made worse by the prevailing philistinism of the Victorian middle classes, might certainty still be found? Arnold's answer was to propose 'culture as the great help in our present difficulties', and as the only plausible antidote to the anarchy that seemed increasingly to threaten society. By engaging with 'the best which has been thought and said in the world', he urged, and 'turning a stream of fresh thought upon our stock notions and habits', it should be possible to 'do away with classes' and 'make all men [sic] live in an atmosphere of sweetness and light'. From one perspective *Culture and Anarchy* was an impassioned assertion of the primacy of culture in life; but from another it was so vague in its pronouncements as to be of little practical value. Yet either way, it came close to arguing that high literary culture was virtually a substitute for organized religion, which Arnold elsewhere described as no more than 'morality touched by emotion'. Although he retained some sort of religious sensibility, he doubted the existence of God, and could not accept the Bible as fact. 'The story is not true,' he later wrote of its account of Jesus; 'it never really happened.' Darwin had undermined religion by science; Arnold did the same with culture. For all the Liberal optimism of Gladstone's government, doubt was increasingly the mood of the times.

INTERNATIONALISM VERSUS IMPERIALISM

Just before Disraeli assumed office as Chancellor of the Exchequer in Derby's government in July 1866, Bismarck's Prussian army defeated the forces of the Habsburg Emperor at the Battle of Königgrätz. Soon after, Disraeli offered the House of Commons a florid, complacent and rather fanciful account of Britain's pre-eminent place in the contemporary world, not only in regard to relations with its near neighbours, but also in terms of its global reach and engagement. It was not, he observed, that the country declined 'to interfere in the affairs of Europe' because it had 'taken refuge in a state of apathy': on the contrary, it was 'as ready and as willing to interfere as in the old days, when the necessity of her position requires it'. Indeed, he went on, there was 'no power that interferes more'. England, he insisted, 'interferes in Asia, because she is really more an Asiatic power than a European [power]'; and, he continued, she interfered 'in Australia, in Africa and in New Zealand' as well. This was an early example of Disraeli's appropriation of the so-called 'Palmerstonian mantle' in foreign and imperial policy. Yet the realities of Britain's overseas engagements were rather different. By the mid-1860s, as the continental balance of power began to shift, Britain's capacity to 'interfere' in European affairs, which had never been all that great since 1815, was significantly less than it had been even in 'the old days' of Palmerston's prime, as evidenced by his inability to wield any effective influence over the Schleswig-Holstein issue; while in imperial affairs, virtually all administrations since the end of the Napoleonic Wars had only very reluctantly acquired more territory, with its attendant risks, responsibilities and expenses. But despite these undeniable realities, Disraeli sought to perpetuate the illusion that the United Kingdom could command the continent when the need and the necessity arose, and he would meet with some success in this endeavour between 1876 and 1878, while at the same time holding out the prospect of a growing imperial engagement with and military assertiveness in other parts of the world.

There was one major piece of colonial legislation and imperial consolidation that was carried out by the minority Conservative government led by Derby and subsequently Disraeli, namely the British North America Act of 1867. That measure created the Canadian Confederation, and thus what would become the senior settler dominion of the British Empire, by fusing together the maritime provinces of New Brunswick and Nova Scotia (yet leaving out Prince Edward Island, which would not join until 1873, and Newfoundland, which would retain its separate existence until 1949) with what had been since 1840 the single colony of Canada, but which was now again divided into the separate provinces of Quebec and Ontario. There were many reasons for what would in retrospect be this pioneering and prototypical piece of imperial consolidation that would later be followed in Australia and South Africa. One was the destabilizing impact of the American Civil War, combined with fears that the triumphantly preserved Union might cast predatory eyes on the scattered and vulnerable British realms that lay to the north, or that it would continue as a base from which the Fenians might launch further attacks. A second was that there was a greater likelihood London investors might put up money for a trans-Canadian railway, to rival the one begun by the United States in 1863, if there was a united political authority supporting the enterprise. A third was that a Canadian confederation held out the prospect of establishing a trans-continental dominion, which western provinces might join (Manitoba did so in 1870 and British Columbia, on the Pacific coast, followed suit in 1871). And finally, but (as it turned out) vainly, the Derby-Disraeli government hoped a Canadian confederation might reduce demands on the imperial exchequer, by moving some of the costs of imperial defence from the United Kingdom to the newly consolidated dominion.

The survival of the American Union, the Confederation of Canada and the soon-to-be-completed trans-continental railways that would extend from coast to coast, constituted a major geopolitical reorientation in the western hemisphere. They consolidated two vast, land-based nations that were prodigiously endowed with agricultural potential and mineral resources, opening up the Midwest and the prairies to European markets, and increasing the American and British presence on the Pacific rim and the seas beyond. The American transcontinental railroad was completed in 1869 (and British Columbia only joined the Confederation on condition that an equivalent Canadian line would also be constructed); and in the same year, on the other side of the world, a similar geographical reorientation took place with the opening of the Suez Canal in Egypt, which dramatically shortened the sea route to India. For Britain, more than any other European country, the

Canal would be the vital link with the subcontinent, in terms of people and trade. From the very beginning British ships had made up more than three-quarters of the tonnage that used it, providing corroboration for Disraeli's claim that Britain was also an 'Asiatic power'. But in Europe, meanwhile, and in defiance rather than corroboration of Disraeli's words, Germany was unified under Prussian leadership in the aftermath of the Franco-Prussian War, with Britain as little more than an impotent bystander. The result was the most significant shift in the European balance of power since 1815, as Bismarck's *Reich* superseded France as the pre-eminent continental force. Like the post-Civil War United States, this new, consolidated Germany was a massive, land-based empire, with a large population and many natural resources. And, like the United States again, during the closing decades of the nineteenth century it would mount an ever-increasing challenge to the industrial supremacy of the United Kingdom.

By the time the Canadian Confederation was a reality, and German unification had been completed, Gladstone's first government was in charge of Britain's foreign and imperial policy. As before, the Foreign Office and the Colonial Office were both held by aristocrats: Lords Clarendon and Granville in the former case, Lords Granville (who transferred from the Colonial to the Foreign Office on Clarendon's death) and Kimberley in the latter. Apart from his brief and somewhat eccentric period as Special Commissioner of the Ionian Islands in 1858–59, Gladstone had little direct experience of foreign affairs, since like Peel before him, his portfolios and preoccupations had been essentially domestic. But he did have views, as when in 1862 he had briefly expressed support for the slave-owning South in the American Civil War; and they later became more inflected by moral and religious convictions about liberty and freedom. Well versed and deeply immersed in the continent's history, languages and culture, Gladstone was a strong believer in the Concert of Europe and in liberal internationalism; and since he was equally convinced of the need to keep government spending down, he was also a determined opponent of imperial expansion and Palmerstonian belligerence and bluff. Hence his support, during the 1860s, for Italian unification, the liberal internationalists' cause par excellence. Hence his dislike of German unification under Prussian leadership, which was by no means a liberal or internationalist enterprise, and his doubts about Bismarck's 'scrupulousness and integrity'. And hence his support of Lord Granville, in his efforts to impose further measures of imperial retrenchment in the aftermath of Canadian Confederation, by withdrawing the garrisons of British troops from Canada, Australia and New Zealand; moves that Disraeli, beginning to sense the potential in playing the imperial card, denounced as a misguided attempt to break up the British Empire.

At the Foreign Office, Lord Clarendon lacked the prime minister's element of idealism in his thinking on international affairs, but he shared Gladstone's view on the need for peace and co-operation in Europe, along with an aversion to assuming any additional treaty obligations. He sought, unavailingly, to damp down the flames that would eventually burst forth in the Franco-Prussian War, although Bismarck would later claim, perhaps disingenuously, that had Clarendon lived, the conflict might have been averted. Clarendon's successor, Lord Granville, also adhered to the policy of patient diplomacy, the pursuit of peace, and the avoidance of any more European alliances. This meant that, after registering some concern about the continued neutrality of Belgium, which it was obliged by a treaty of 1839 to uphold, Britain had stood on the sidelines as the Franco-Prussian War was fought out. Meanwhile, Granville was much more active, and much more successful, in improving the United Kingdom's relations with the United States in the aftermath of the Civil War. The unresolved dispute over the *Alabama*, the British-built Confederate cruiser, which had done so much damage to Union shipping, was referred to international arbitration, and in 1872 the government agreed to pay a substantial sum in compensation to the United States. In the same year an issue that the earlier treaty of 1846 had not clearly resolved, concerning British and American claims to the San Juan Islands in Puget Sound, was also finally settled. As a result, Anglo-American relations were put on a new and better footing, which had the further advantage of allaying British anxieties concerning Canada's security, while the settlement of these differences by arbitration was held up as an exemplary way to conduct international affairs.

By staying out of European conflicts and cultivating good relations with the United States, Gladstone's foreign policy owed more to Lord Aberdeen's view of international affairs than it did to Palmerston's, or Disraeli's, and this was scarcely surprising given how close Aberdeen and Gladstone had been in the government Peel had formed in 1841. Yet when it came to imperial affairs, there was no such clear-cut contrast: for as so often in the past, the desire for retrenchment and the wish to avoid assuming any new responsibilities, ran up against countervailing forces requiring expenditure or annexation that could not always be resisted. In his best-selling book, *Greater Britain*, published in 1868, the young radical, republican politician, Sir Charles Dilke, urged that the colonies of settlement should free themselves from British imperial rule. For the time being they had no wish to behave in such a manner, even in New Zealand, where self-government had been granted in 1856, and where there had been serious criticism of Britain at the time of the Maori Wars; but there was no real desire for independence. And although, in the aftermath of the post-Rebellion proclamation, further annexations were in abeyance in India for the time being, the impossibility of letting go, combined with reluctant acquisitiveness, remained in practice the policy elsewhere, as evidenced by the example of Africa. In 1870, while still at the Colonial Office, Lord Granville declared his willingness to give the Gambia to France; but nothing ever came of the proposal. Two years later the British government acquired, for cash down, a string of Dutch forts along the Gold Coast, and in 1873–74 it felt compelled to mount a full-scale war against the King of the Ashanti, who was threatening these recently acquired holdings from the north; and it was as the leader of the British troops in that campaign that General Garnet Wolseley first came to public notice in Britain.

Despite such unavoidable exceptions, Gladstone's administration of 1868–74 was generally characterized by a commitment to reform at home along with the pursuit of tranquillity abroad; but in the case of Disraeli's government that followed, the priorities were exactly the opposite. He largely left domestic matters to his colleagues, while he concentrated his declining energies and intermittent attention on foreign and imperial affairs, appropriately and aristocratically assisted until 1878 by the fifteenth Earl of Derby at the Foreign Office and Lord Carnarvon at the Colonial Office. Although he was in fact a far less cosmopolitan and well-travelled figure than Gladstone, Disraeli had been (at least retrospectively) prescient when he had sensed in the early 1870s that nationalism and imperialism would supersede liberal internationalism, and as a past master of theatrical gestures and vague yet memorable phrases, he thought himself ideally equipped to play the role of world statesman. Such gestures and phrases were much in evidence when, in late 1875, Disraeli purchased for the British government from the bankrupt Khedive of Egypt the shares he owned in the Suez Canal for £4 million. Disraeli masterminded the acquisition in secret, with the direct help of the Rothschild Bank, and only then did he seek the approval of the cabinet and the Commons. Since the Canal was the imperial lifeline between Britain and India, Disraeli was determined that the French, who had built the Canal and were major shareholders, should not consolidate their interest by getting their hands on the Khedive's holdings as well. Despite perceptions to the contrary, the purchase did not give Britain complete control of the Canal; but it was a remarkable deal, and a dazzling coup, which would also turn out to be a good investment.

The following year Disraeli took another step towards enhancing the centrality of the empire in British life, by associating it more closely with the British monarchy, when he carried the Royal Titles Act. Queen Victoria was additionally proclaimed Empress of India, which meant that henceforward she would no longer be merely Victoria *Regina* but Victoria *Regina et Imperatrix*. Although this was rightly regarded as another quintessentially theatrical Disraelian gesture, the queen herself put pressure on her prime minister to pass the measure, not least because she had no wish to be outranked by her eldest daughter Victoria, who as Crown Princess would one day become Empress of Germany. But there were broader explanations, too. Since the deposed Mughal Emperors were deemed to have vacated their imperial Indian throne in the aftermath of the Great Rebellion, it seemed both possible and appropriate to declare that the British crown was their legitimate successor, and to adjust and augment its title accordingly. Moreover, in setting the British sovereign at the apex of the South Asian social hierarchy, the act further reinforced the earlier commitment to the rulers of the Indian princely states that had been made in the proclamation of 1858. The measure also provided further proof of Disraeli's claim that Britain was an 'Asiatic power', since China, Japan, Russia, Persia and the Ottoman realms were all ruled by emperors, and so it seemed right and, indeed, essential that India should be ruled by a super-sovereign of similar standing and stature. It further signalled that Britain was serious and determined to maintain its position on the continent, and that it would stand firm against any Russian threats to the buffer states that lay between. And finally, Disraeli justified the legislation on the grounds that it was 'only by the enhancement of titles' that it was possible to 'satisfy the imagination of nations' (whatever that meant).

Although Gladstone had resigned the leadership of the Liberal Party after his election defeat, and had intended to devote the rest of his life to theology and scholarship, he remained a member of the House of Commons. He criticized both of the Conservative government's measures, the purchase of the Suez Canal shares to no great effect, but the augmentation of the queen's title with substantial Liberal support, since Disraeli had unwisely neglected to consult the opposition over a matter which, because it concerned the monarchy, he certainly should have done. These actions did to some extent signal a more assertive imperial policy, but in other ways it was continuity that was much more apparent. Just as Gladstone had acquiesced in the punitive expedition against the Ashanti, which was in practice largely decided on by Kimberley at the Colonial Office and Cardwell at the War Office, so Lord Carnarvon annexed Fiji in 1874, without Disraeli and his colleagues knowing much about it, on the grounds of the island's chronic instability, where both the local rulers, and the increasingly assertive British representatives, needed to be brought under some sort of control. But Carnarvon was another reluctant annexationist and he urged the first colonial government to avoid 'all possible expenditure that is not necessary'. Equally Gladstonian in its drive for imperial retrenchment was the attempt made in 1875 by Sir Stafford Northcote, the Chancellor of the Exchequer, to abolish the British garrison in Hong Kong, since it would save the taxpayer £100,000 a year. But, like many attempts to reduce the cost of imperial defence and administration, nothing came of this proposal.

The greatest overseas challenge that Disraeli faced concerned the complex and interconnected geopolitics of the British, Habsburg, Ottoman and Russian Empires. On the whole, and as had been the case during the Crimean War, it remained British policy to support the continued existence of the declining Ottoman Empire as an essential counterweight to Russia, which was (perhaps exaggeratedly) regarded as an ever-present threat to Britain's Indian Empire. During the mid-1870s there were a series of uprisings against Ottoman rule in the Balkans, which were looked upon with favour by the neighbouring empires of Austria-Hungary (which was anxious to extend its influence in that region) and Russia (which saw itself as the leader of 'pan-Slavism'). In May 1876 the Ottomans suppressed a nationalist uprising in Bulgaria with extraordinary ferocity, and an estimated 12,000 rebels were killed. Disraeli dismissed the massacres as 'largely invention', and completely misjudged the reaction of the British public. A campaign was soon mounted, led by journalists and churchmen, condemning the killings, and especially the vengeful treatment meted out to those of the Eastern Orthodox faith, and in late summer Gladstone himself joined in. In September he published a pamphlet, entitled 'The Bulgarian Horrors and the Question of the East', denouncing the behaviour of the Ottoman authorities and urging them to leave 'bag and baggage' from 'the province they have desolated and profaned'. Within three weeks Gladstone's pamphlet sold 200,000 copies; he returned to the political stage mounting an energetic public campaign on behalf of the persecuted Christian minorities; provincial, radical and Nonconformist opinion rallied to him; and there were five hundred demonstrations against the government's pro-Ottoman policies.

By the end of 1876, the Bulgarian agitation was beginning to lose momentum; but it had significant political consequences. Disraeli's government was seriously wrong-footed by the strength of the public reaction against the Ottomans, and it never regained its earlier confidence, while Gladstone increasingly came to feel that he must return to front-line politics to mount a moral campaign against the iniquities of what he would call 'Beaconsfieldism' (derived from the title Disraeli had recently taken on being made a peer). Meanwhile, Russia declared war on the Ottomans early in 1877 and by January 1878 the Tsar's army was at the gates of Constantinople. This was not acceptable to the British government, the Royal Navy was ordered to the Bosphorus, and troops were moved from India to the Mediterranean. At this point Carnarvon (who was anti-Ottoman) and Derby (who was anti-intervention) resigned from the Colonial and the Foreign Office respectively. Disraeli did not mind much and replaced them with Sir Michael Hicks Beach and Lord Salisbury. But for a time another war between the United Kingdom and Russia seemed highly likely, the government budgeted an extra £6 million for military expenditure, and popular opinion now reversed itself, and turned from being strongly anti-Ottoman to being vehemently against Russia. 'We don't want to fight,' ran the music-hall song of the time, 'but by jingo if we do, we've got the ships, we've got the men, and we've got the money too.' The Russians, the song went on, 'shall not have Constantinople'. In the event there was no such war, since the Ottomans and Russia came to terms in March 1878, and at the insistence of Austria-Hungary a

European conference was convened that would be held in Berlin later that year, at which all remaining issues would be settled.

The Congress of Berlin was the most extended gathering of Europe's senior statesmen since the Congress of Vienna had met over sixty years before, and it was a measure of its success that there would be no major war between the European powers for another generation and more. Most of the contentious subjects had already been settled beforehand, and they were essentially ratified at Berlin; they included the boundaries of what were recognized as the independent Slavic states in the Balkans, the continued independence of the Ottoman Empire with its capital at Constantinople, and the effective checking, once again, of the latest attempt by Russia to penetrate the Balkans and the Middle East. More particularly, Britain gained control from the Ottomans of the island of Cyprus, in the belief that it would furnish a base from which power could be exercised in the eastern Mediterranean. During the course of these negotiations the United Kingdom was represented by Disraeli, Salisbury and Lord Odo Russell, the first British ambassador to Berlin. According to Salisbury, Disraeli often had only 'the dimmest idea of what was going on', because of his 'deafness, ignorance of French and Bismarck's extraordinary mode of speech'; and there can be no doubt that the reason he confined himself to speaking in English was not to assert his nation's interest, but because he was incapable of conversing in any other language. Yet Bismarck himself took a very different view of Disraeli's performance: 'The old Jew,' he said admiringly, 'there is the man!' Not surprisingly, Disraeli agreed, and on his return to cheering crowds in London he claimed he had secured 'peace with honour', and he told the queen that she was now 'the arbiter of Europe'. Both were characteristically memorable and magniloquent phrases; the first was plausible, the second was not.

THE FALL OF 'BEACONSFIELDISM'

In domestic terms the climax of Disraeli's government had been 1876, with his reincarnation as Earl of Beaconsfield, and the queen's as Empress of India ('one good turn deserves another', the cartoonists noted); while in foreign affairs it came two years later, with his triumph at the Congress of Berlin. Thereafter, the political climate turned against him, in part because the late 1870s witnessed an acute commercial and industrial depression, marking the abrupt end of the mid-Victorian boom years, and also giving an early warning that the 'workshop of the world' would be facing much stiffer economic competition. 'There are symptoms which suggest,' warned J. A. Froude in 1870, 'misgivings as to the permanency of England's industrial supremacy.' Between 1877 and 1879 unemployment increased from less than 5 per cent to more than 11 per cent; there was a spate of bankruptcies of which that of the City of Glasgow Bank in 1878 was the most sensational; and 1879 was one of the worst years of the second half of the nineteenth century for business and manufacturing. These setbacks also coincided with a depression in many parts of the agricultural sector, which was bad news for a Tory party that was so heavily reliant on rural support. Four wet summers during the late 1870s were the harbinger of much greater troubles, as the opening up of the American prairies, in the aftermath of the Civil War, combined with new agricultural machinery and the fall in freight rates caused by the change from sail to steam, created a global market in agricultural produce and a corresponding collapse in agricultural prices. In 1877 English wheat had averaged 56s 9d a quarter; the following year it had plummeted to 46s 5d a quarter, and it dropped steadily thereafter. This cheap imported grain intensified the impact of the rainy summers: across the broad arable acres of the United Kingdom, farmers' incomes dwindled, and so did the rents they could pay their landlords.

Such a severe downturn seemed to provide belated vindication for those who had argued, Disraeli among them, that the abandonment of agricultural protection by Peel in 1846 had been a great mistake. But while virtually all the European nations now imposed tariffs on imported foodstuffs, Disraeli refused to follow suit. Agriculture was no longer as important a sector of the United Kingdom economy as it had been in the 1840s, and a tax on food would not be well received by the urban working classes, especially at a time of industrial depression and rising unemployment. But this did not play well with the squires or in the shires, which still remained the Tory heartlands; while the depression in industry and trade also turned many of the recently enfranchised workers against the government that had given them the vote. The effect of the agricultural depression in Ireland was even more disastrous, negating such limited economic recovery as there had been since the

Great Famine. Tenant farmers could not pay their rents, landlords with precarious finances found their incomes drastically reduced, the number of evictions soared, notwithstanding Gladstone's legislation of 1870, and violence escalated. The Irish Home Rule Party, which had first appeared in parliament in 1874, had previously followed such a coy and conciliatory policy that it seemed to represent no real threat to the Union. But in 1875 a young Protestant landlord named Charles Stewart Parnell was returned to Westminster at a by-election as MP for County Meath, and he was determined to take a more aggressive stance against landlords and against the Union, which would intensify protests in Ireland and lead to major disruptions in parliament. The 'Irish question' was back, and it would not go away for the rest of the nineteenth century or, indeed, for the whole of the twentieth century – and beyond.

At the same time the imperial situation turned against Disraeli, too, as his fine phrases and theatrical gestures were drowned out by serious military and diplomatic reverses, as Britain 'interfered' in Asia and Africa, but to no good purpose and at a high cost in terms of domestic repercussions and international prestige. In South Africa the already tense and intractable relations between the British (in Cape Colony and Natal), the Boers (in the Transvaal and the Orange Free State) and the native peoples (especially the powerful kingdom of the Zulus), had been rendered more fraught and difficult by the discovery of gold and diamonds in the late 1860s. In 1877 the Colonial Secretary Lord Carnarvon appointed Sir Bartle Frere as Governor and High Commissioner to British South Africa, and simultaneously annexed the Transvaal, much to the Boers' rage and resentment. A determined expansionist, Frere strongly supported this London initiative, but in defiance of the Colonial Office he also sought the further extension and consolidation of British power at the southern end of Africa. To that end he tried to suborn the Zulus, but in January 1879, at the Battle of Isandlwana, the army of Chief Cetshwayo inflicted a devastating and humiliating defeat on British forces – an almost unprecedented military triumph of black men over white men. Disraeli's government reacted with feeble indecision, neither supporting Frere unconditionally nor recalling him. Eventually, in July, Cetshwayo was defeated and captured at the Battle of Ulundi, and his kingdom was broken up. But with the Zulu threat removed, the Boers in turn felt confident enough to reject the recent annexation of the Transvaal by London, and in February 1881 they would inflict their own defeat on the British at the Battle of Majuba Hill, thereby regaining independence for the Transvaal, subject to what was vaguely described as British 'suzerainty'.

This South African debacle was seriously damaging to the standing of a government that had sought to promote the cause of British imperialism so vigorously and to identify the Conservative Party with it; but it was made worse by the simultaneous unravelling of the administration's policy in South Asia. In 1876 Disraeli had appointed Lord Lytton as Viceroy, and he duly presided over a traditional-cum-invented durbar at which Queen Victoria was proclaimed Empress of India. But like many Britons before him, Lytton was also worried by the dangers of Russian incursion into Afghanistan, especially when the Emir received a delegation from the Tsar in the summer of 1878, but stopped a British mission at the border. Lytton duly issued an ultimatum, which was ignored, and three British armies then invaded. A year later the Afghans ceded control of the passes linking their country and India to the British, and accepted British direction of foreign policy and a resident British minister at Kabul. Disraeli himself seems to have had no fully worked-out policy towards Afghanistan, although it is far from clear that he wanted his Indian Viceroy to be this assertive. And once again the vigorous imperial policy espoused by the man on the spot backfired: in September 1879 the newly appointed British minister in Kabul was killed by mutinous Afghan soldiers, along with all his staff. To be sure, these further reverses on another distant frontier of the empire were soon avenged, as General Roberts would lead British troops from India into Kabul the following month. But once again the government's policy of imperial adventurism had seemed incoherently formulated and incompetently executed.

These imperial misadventures and misjudgements intensified the disenchantment with Disraeli's government resulting from the severe economic downturn, and they played into Gladstone's hands. Although disappointed by his failure to rouse the Liberal Party more effectively over Bulgaria, he was convinced that one more campaign would rid the British nation and its empire of the evils of 'Beaconsfieldism'. In January 1879 he announced that he would give up his Greenwich seat and contest the Midlothian constituency at the next general election; and later that year he began a speaking campaign in Scotland the like of which had never been seen before in British politics. In a succession of addresses, which attracted unprecedented press and public interest, Gladstone attacked 'Beaconsfieldism' as a 'whole system of government' that was rotten in every aspect. In particular, he accused Disraeli of presiding over an administration that had been fiscally reckless with its massive increase in military spending, and which had failed to adhere to the cardinal principles of foreign policy, namely the promotion of peace, the support of the Concert of Europe, and the avoidance of 'needless and entangling engagements', both on the continent and far beyond. To be sure, Gladstone had nothing to say about the recession, and he had no remedies to offer at a time of widespread distress. But nor did Disraeli, who kept aloof throughout most of the campaign, and had little to offer by way of a policy or a programme. The final result stood that of the 1874 election on its head: the Conservatives dropped from 352 seats to 239, while the Liberals rose from 243 to 351 (and the number of Irish Home Rulers increased from fifty-seven to sixty-three). As Gladstone vividly and vengefully observed, the fall of 'Beaconsfieldism' was like 'the vanishing of some vast magnificent castle of Italian romance'.

It was an accurate remark, and also an apt analogy, for while writers such as Wilkie Collins, George Eliot and Thomas Hardy were engaging with contemporary issues like gender, sexuality and race, the years from 1865 to 1880 were the heyday of the 'political' novel, too, in which several writers explored not only the reality, but also the romance, of power. Why was this? It was partly because popular engagement in politics had evolved from the agitation of the 1830s and 1840s, over parliamentary reform and slavery, the Corn Laws and Chartism, via widespread veneration for Peel and then, for every different reasons, support for Palmerston, to admiration for Disraeli and Gladstone; this meant they were political celebrities and public figures in a way that had never been quite true of any of their predecessors, and which would be recognized in the great portraits that Millais would paint of Gladstone in 1879 and Disraeli (albeit unfinished) in 1881. But there was more to it than that, as the political parties were increasingly organizing, and in so doing involving more people. In 1870 Disraeli had established the Conservative Central Office under John Gorst, who revived local Tory associations, and this played a major part in their election victory of 1874. Two years later Joseph Chamberlain gave up the mayor's parlour in Birmingham and at a by-election was elected one of the town's MPs. The following year, in an effort to move the party in a more radical direction, he established the National Liberal Federation, which owed much to the Birmingham Liberal Association and the National Education League, and sought to bring together numerous Liberal associations all over the country and standardize their organization. As a result of these developments, the public were now involved in party politics as never before.

So it was scarcely surprising that these years witnessed the heyday of political novels, since readers were also often voters. In his later works Anthony Trollope abandoned the ecclesiastical intrigues of Barchester and produced his six political 'Palliser' novels (originally known as his 'parliamentary' fictions). Set during the earlier period of fluid party politics in the aftermath of the repeal of the Corn Laws, Trollope centred his stories on the eponymous Plantagenet Palliser, later Duke of Omnium, whose family wielded significant political influence in Barsetshire, and who would eventually become prime minister of a coalition including both Whigs and Tories. Equally engaged, but less even-handed, was George Meredith in *Beauchamp's Career* (1875), which sympathetically described the life of the young idealist Nevil Beauchamp, who went into politics as a self-proclaimed radical, thereby alienating many of his staunchly Tory upper-class relatives, satirized by Meredith as selfish and degenerate. Yet the most remarkable political novel of the period was that which Disraeli began to write in the aftermath of his electoral defeat. It would be his last work, entitled *Falconet,* and the central character was modelled on Gladstone, here depicted as the prince of prigs and held up to continuous odium and ridicule. It was a fitting finale for a politician who had always blended fact and fiction in his politics no less than in his novels, so much so that an imaginary 'Lord Beaconsfield' had originally appeared in *Vivian Grey* (1826), Disraeli's first novel. But regrettably he would not live to complete his last one – although Gladstone was probably relieved.

9
'Disintegration' Averted?, 1880–95

Queen Victoria was devastated by the news of Disraeli's electoral defeat. She could not bear it that their close relationship was at an end, and she regarded with horror the prospect that Gladstone might yet again return to power. His attacks on 'Beaconsfieldism' had, she believed, been irresponsible and unpatriotic in the extreme, and she declared she would 'sooner abdicate than send for or have anything to do with that *half-mad fire-brand* who would soon ruin everything, and be a *Dictator*' (her emphasis). Having received Disraeli's resignation with the greatest of reluctance, she looked to Lord Hartington, as the Liberal leader in the Commons, and to Lord Granville, who held the same position in the upper house, to save her from this dreaded fate. But they were in no position to do so, for Gladstone had already made it clear to them that he must return to power, and that he would only do so as prime minister. Accordingly, Hartington refused the queen's invitation to form a government, and along with Granville, he told her that she had no choice but to send for Gladstone. The electorate had given its verdict, and in the light of it the queen could no more keep Disraeli in than she could keep Gladstone out. How different from those far-off days when she had been able to exert herself during her first years as sovereign to sustain Lord Melbourne and exclude Sir Robert Peel, when Prince Albert had made such strong claims for the active role the British monarchy should play at the very heart of the British government, or when, more recently, Disraeli had flatteringly attributed to her far greater powers than in practice she by then possessed.

It was often said that Gladstone was 'terrible on the rebound'; and just as he had recovered from the political setbacks of 1866–67 to win victory in 1868, he had now emerged from semi-retirement to unite the all too fissiparous forces of Whigs, radicals, Irish and Nonconformists in another Liberal electoral triumph. Despite the queen's visceral opposition, he formed his second administration in April 1880, with as secure a Commons majority as in his first, and he did so at the advanced age of seventy. To begin with he assumed the extra burden of being Chancellor of the Exchequer, and he was also Leader of the House of Commons, responsible for arranging its parliamentary timetable. In addition, since both the Foreign Secretary (Lord Granville, once more) and the Colonial Secretary (Lord Kimberley again) were in the upper house, much of their business in the Commons also fell to him. As these appointments suggest, Gladstone's second cabinet was in many ways a decidedly traditional one. There were four other peers, including two very grand Whigs: the Duke of Argyll as Lord Privy Seal, and Earl Spencer as Lord President. The Home Secretary, Sir William Harcourt, was of highborn lineage and would eventually inherit estates at Nuneham in Oxfordshire. Having declined the queen's invitation to form his own government, Hartington again served under Gladstone, as Secretary of State for India. So the patricians were still in the majority, and the Whigs were strongly represented, while the radicals continued to be few in number. The aged John Bright was back as Chancellor of the Duchy of Lancaster, and W. E. Forster returned to the cabinet as Chief Secretary for Ireland; but the most significant new face around the table was that of Joseph Chamberlain, the former mayor of Birmingham and creator of the National Liberal Federation, who combined radical politics with fierce ambition, and now became President of the Board of Trade after only four years in the Commons.

Although in many ways Gladstone's second cabinet resembled his first, it would turn out to be much less harmonious than its predecessor, and within five years it would all but have broken up amidst deep personal and political recriminations. Yet none of this could have been foreseen in the spring of 1880, when the Liberals' future again seemed as bright as the Conservatives' prospects seemed gloomy. Many Tories feared that the victory of 1874 had been merely a flash in the pan, and that the normal Whig-Liberal dominance had been restored. For another twelve months Disraeli continued to lead the party from the Lords, but on his death in April 1881 the Conservatives followed the precedent that the Liberals had earlier set on Gladstone's (first) 'retirement', by dividing the party leadership between Sir Stafford Northcote in the Commons and the Marquis of Salisbury in the Lords. Neither of them promised well: Northcote would be no match for Gladstone, whose private secretary he had once been; while Salisbury was completely out of sympathy with the enlarged electorate that the Second Reform Act had brought into being. In 1880, then, it would have seemed inconceivable to virtually everyone that it would turn out to be more than twenty-five years before the Liberals would win another landslide victory at the polls, and during that quarter century the Conservatives would successfully establish themselves as the natural party of government. As Disraeli had once observed, the 'vicissitudes of British politics' were indeed inexhaustible.

GOOD TIMES, BAD TIMES

In 1883, when the Conservatives' electoral prospects still seemed gloomy, Lord Salisbury gave vent to these anxieties in an essay published in the *Quarterly Review*, apocalyptically headed 'Disintegration'. As his title suggested, he feared the British nation and empire were facing a potentially catastrophic crisis, because many of the people who had been given the vote in 1867 were, he believed, ignorant, envious, antagonistic and unpatriotic (as Robert Lowe had argued at the time). And in response to this recently (and wrongly) widened franchise, the Liberal Party of Gladstone (and now of Chamberlain) had embraced ever more radical politics, and thus represented an unprecedented challenge to the established order, not only at home but also overseas. Urged on by their leaders who ought to have known better, Salisbury feared that antagonistic social groups were mobilizing against each other for class war, that the traditional, propertied hierarchy was in jeopardy, and that the possessions of 'churchmen, landowners, publicans, manufacturers, house-owners, railway shareholders [and] fund-holders' were all at risk. Moreover, he went on, such partisan animosities not only portended domestic collapse: they would also erode the will to rule, both in the United Kingdom and overseas. Ireland, he feared, was 'the worst symptom of our malady', for to dissolve and abandon the Union would not only be intrinsically disastrous, but would also mean that 'all claims to protect or govern anyone beyond our own narrow island were at an end'. The repudiation of great-power obligations overseas, and the weakening of the will to rule, Salisbury pessimistically concluded, would invariably follow, and with it the 'loss of large branches and limbs of our Empire'.

During the early 1880s Salisbury's sombre analysis concerning the bleak electoral prospects for the Tories, and his fear that the United Kingdom and the British Empire might break up, were widely shared by many Conservative peers, MPs and voters; and also by Queen Victoria, who dreaded that if the Liberals had their way, with Gladstone back in power and 'un-muzzled', she might be reduced to the impotent and degraded status of a mere 'democratical monarch'. These were the gloomiest prognostications by any Tory leader since the Duke of Wellington had predicted in the early 1830s that parliamentary reform would spell the end of civilization as he knew it. Yet Salisbury, like Wellington before him, grossly exaggerated the threats the nation faced. At home, Gladstone may have expressed support for 'the masses' against 'the classes', but he was as eager to preserve the established social order as Salisbury; both of them were in fact 'out and out inequalitarians', and Salisbury's fear of the ill-educated and irrational mob would soon be matched by Gladstone's anxieties about the corrupting power of plutocracy (which had been brilliantly explored and exposed by Trollope in 1875 in his greatest and angriest novel, *The Way We Live Now*). Neither wanted Ireland to leave the Union: their disagreements would be over the best means to achieve what was in fact an agreed end, the Conservatives (as in the past) preferring coercion, the Liberals (ditto) conciliation. Although both men were reluctant to take on any more territory, or to spend any more money on maintaining Britain's position as the pre-eminent global power, neither wanted the British Empire to break up, and both were reluctantly forced to assume greater imperial responsibilities. And, pace Salisbury, there were many signs that the condition of the United Kingdom was improving and consolidating, rather than disintegrating and declining.

One indication was that between 1871 and 1891 the nation's population had grown from less than thirty-two million to nearly thirty-eight million, even as Ireland's numbers continued to decline. With the possible exceptions of Belgium and Holland, the United Kingdom was the most densely populated and heavily urbanized nation on the planet, with more than two-thirds of Britain's inhabitants (though not those of Ireland) living in urban areas. Underlying this increase was continued economic growth, not so much in the case of textiles where, judged by the size of its labour force, expansion had climaxed by the 1850s, but rather in the case of heavy industry. In 1851 coalmining had employed 216,000 people, but thirty years later the figure was 495,000 and still rising; and in iron and steel and engineering the increase was even more rapid, owing to the stimulus of such new inventions as the Bessemer process. As a result, railway mileage in the United Kingdom continued to grow, linking the most distant parts of Ireland to Dublin and of Scotland and Wales to London; and the majority of the world's merchant shipping was both built and registered in Britain. England's industrial heartlands remained in the Midlands, the northeast and northwest; but there were rapidly developing and increasingly integrated industrial economies in the Scottish Lowlands (coalmining and shipbuilding), in south Wales (coalmining and steel) and in northern Ireland (textiles and shipbuilding), centred respectively on Glasgow, Cardiff and Belfast; and even as nationalist demands intensified in Ireland, and began to stir in Scotland and Wales, these three areas would remain closely tied to the Union. In terms of raw materials and manufactured goods, the United Kingdom continued to prosper as an export-dominated economy: in the early 1880s almost half the value of all British exports was in textiles, and nearly another quarter in coal, iron and steel or engineering.

Britain's share of global manufacturing output reached its high point at just short of 23 per cent in 1881, when it also produced 44 per cent of the world's exported manufactured goods – astonishing figures for a country that, in terms of its landmass, was smaller than Spain or France or Germany, let alone Austria-Hungary, Russia or the United States. During the 1880s three and a half million Britons left their homelands for the United States (70 per cent), for British North America (11 per cent), for the Antipodes (11 per cent) and for South Africa (2.5 per cent). From 1880 to 1893 the numbers leaving the United Kingdom never fell below 200,000 a year, and they would peak at 320,000 in 1893. At the same time Britain was exporting capital abroad in unprecedented quantities: investing in Indian, colonial and foreign railway companies; in government bonds, both inside and outside the empire, which often funded railways or other kinds of infrastructure; and in overseas companies controlling utilities such as gasworks or waterworks, or banks, real estate, mines and plantations. The result was a doubling in the value of Britain's overseas assets from less than £1,000 million in the early 1870s to around £2,000 million by 1900 – a mountain of overseas wealth that no other Western nation came close to rivalling. Just as the mining and manufacturing regions of the United Kingdom made it the world's pre-eminent industrial power, so the City of London consolidated its position as the world's indispensable financial power, in terms of banking, insurance and shipping, along with commodity trading and the Stock Exchange, and the number of people employed increased by more than a quarter between 1881 and 1901.

This was scarcely a nation on the brink of 'disintegration', as the industrial economies of Ireland, Scotland and Wales were increasingly anglicized and assimilated, and as the finance and service sectors in the City of London complemented the manufacturing and exporting sectors in Britain's industrial regions. Moreover, growing national prosperity meant the standard of living of many of its inhabitants was visibly improving. Aristocrats such as the Bedfords, Cadogans, Derbys and Northumberlands were richer than ever, with substantial incomes drawn from mining royalties and urban real estate. Plutocrats like the Guinness brothers (in brewing), Weetman Pearson (international contracting) and Alfred Harmsworth (newspapers) were beginning to make their own fortunes that rivalled even the most Himalayan aristocratic accumulations. The middle-class professions were expanding with unprecedented rapidity, the number employed in public service and the professions increased from fewer than 600,000 in 1871 to more than 800,000 twenty years later; and many of the foremost politicians of the next generation would be lawyers, among them H. H. Asquith, David Lloyd George, Edward Carson, Rufus Isaacs, John Simon and F. E. Smith. There was also a notable growth in office-based, white-collar, lower-middle-class employment; the brothers George and Weedon Grossmith memorably satirized the foibles and petty snobberies of such people when they created Mr Pooter in *The Diary of a Nobody* (1892). But the greatest improvements were probably registered by the better-off members of the working class, for whom, across the last quarter of the nineteenth century, real wages increased by one-third, and many families were able to include meat, bacon and eggs, and tea with sugar, in their diet for the first time.

The quality of life was also improving. Epidemics of cholera and typhus largely ceased, the death rate began to fall, and the numbers in receipt of poor relief declined. Thanks to improved policing methods, violent crime, so vividly evoked in the novels of Dickens, diminished dramatically, and Sherlock Holmes, who made his first appearance in the late 1880s, had good reason to lament that there was often little serious work for him to do. Part cause, part consequence of these developments was that the uniformed constabulary, who had initially been distrusted as state-sponsored snoopers, had by now acquired a more positive reputation as the avuncular and incorruptible embodiment of benevolent authority (and had been gently mocked as such by Gilbert and Sullivan in 1879 in *The Pirates of Penzance*). Another form of paternalism was embraced by such high-minded employers as the soap magnate William Lever, who founded the model village of Port Sunlight in 1889 on Merseyside, where he provided high-quality accommodation for his workforce, and the Quaker chocolate manufacturer George Cadbury, who began his own settlement at Bournville on the southern edge of Birmingham four years later. Lever's soap and Cadbury's chocolate were two of the many products that were now reaching a mass market, as a result of the late nineteenth-century 'retailing revolution', which saw the rise of shops with multiple branches, such as Home and Colonial Tea, Maypole Dairy, Liptons and Boots. In 1880 there were forty-eight such firms with fifteen hundred branches between them; by 1895 there were two hundred of them with more than six thousand branches. There were also the Co-operative Wholesale Societies, with their working-class membership, which combined low prices for foodstuffs with a strong moral fervour. By 1881 there were half a million members, and annual turnover was £15 million; by 1891 those figures had jumped to more than one and a half million members, and a turnover of £50 million.

There were other indications that the quality of life was improving for many lower down the social scale. Following Forster's Education Act, standards of literacy generally improved, and by the close of the nineteenth century many members of the working class were reading a broader range of books and newspapers than their predecessors had ever done. The 1880s and 1890s also witnessed the widespread proliferation of the music hall as a new form of mass, popular entertainment, with acts and artistes purveying a unique amalgam of suggestiveness and knowingness, comedy and jingoism. By the 1890s there was a vogue for 'Imperial tableaux', in which the colonies were depicted as being loyally subservient to the mother country, and they were interspersed with 'tableaux vivants', in which female performers often simulated nudity by wearing flesh-coloured body-stockings. Many of the halls had by then been grouped into highly profitable theatre chains that were nationwide in their scope, and in London 45,000 people a night were crowding into its thirty-five largest halls. For those who preferred more bracing forms of recreation, bicycling became the new craze, thanks to the development of the safety machine and the invention of the pneumatic tyre in the late 1880s. By 1891 there were more than 5,000 manufacturers providing this cheap, self-propelled transport, offering unprecedented opportunities for city dwellers to ride out to the country. Since easy terms and cheap second-hand machines were readily available, cycling was embraced by better-off working men (and women) as well as the middle classes. Indeed, the most pronounced social change of these years was the creation of a new sort of working-class culture, which was in many ways profoundly conservative, as would be evidenced by the number of pubs in working-class districts named 'The Earl of Beaconsfield' or 'The Lord Salisbury'.

This sense that the 1880s were a prosperous and stable decade, rather than one in which the forces of disintegration had triumphed, was well articulated in the celebrations marking the Golden Jubilee of Queen Victoria, who had reigned for fifty years by 1887. And it was coincidentally appropriate that this unprecedented royal and popular spectacle was staged during Salisbury's premiership: for it was essentially a festival of those Disraelian values of monarchy, hierarchy and empire, all sanctified by a service of thanksgiving held at Westminster Abbey. In British terms there had been nothing like it since the Golden Jubilee that George III had attained in 1809, but there was scarcely anyone alive who could remember that far-off occasion. It was also a celebration of Queen Victoria's unrivalled position as the doyenne of European royalty, and many of her continental relatives visited Britain to take part in the observances (among them Crown Prince Frederick of Prussia, resplendent in a white uniform, but already stricken with cancer of the throat that would kill him the following year after a brief reign as German Emperor and King of Prussia lasting barely a hundred days). In addition, the Jubilee provided an opportunity for the crafting of a structured and self-satisfied national narrative, extolling half a century of remarkable progress, politically, socially, economically and culturally, and not only domestically, but imperially and internationally as well. And it would serve retrospectively as the prototype of what would become a grand sequence of late nineteenth- and early twentieth-century royal extravaganzas, in which the governing elite rediscovered the capacity to organize great state pageants, and the British people rediscovered their liking for them.

But the mood of national self-congratulation associated with the Golden Jubilee was not universally shared. The year before, Alfred (by now Lord) Tennyson had offered a very different perspective, by producing a sequel to his earlier poem, *Locksley Hall*, which he had written in 1835 and published seven years later. Although the late 1830s and early 1840s had been depressed and difficult times, Tennyson's original poem had furnished a confident, early-Victorian vision of a future that could not fail to be an age of wonder; but in 1886, in *Locksley Hall Sixty Years After* (his arithmetic was somewhat awry), he offered a very different and much more somber vision. In retrospect, Tennyson insisted, that ardent, optimistic belief in a world that would be forever moving forward and getting better had been deeply misplaced, for Britain was now menaced by an irresponsible democracy, by the Irish troubles at home, and by Russian threats abroad, and it was governed by politicians who no longer told the truth, but pandered and lied to an ignorant mass electorate, thereby further demoralizing the nation and undermining its achievements. This was Arnold's *Culture and Anarchy*, or Salisbury's 'Disintegration', in verse; and as the most famous and influential poet of his day, Tennyson's words undoubtedly carried weight. But Gladstone, who had given Tennyson his peerage in 1884, was stung by them, and published a rejoinder entitled ' "Locksley Hall" and the Jubilee: A Critique of Lord Tennyson's Poem', in the *Nineteenth Century*, a monthly literary magazine founded in 1877, where he rehearsed the optimistic narrative of the last half-century, concluding that 'men [sic] who lived fifty, sixty and seventy years back, and are living now, have lived into a gentler time'.

Yet even Gladstone was compelled to concede that there were many faults in the state of contemporary British society. In the year of the Golden Jubilee the United Kingdom was in the midst of a lengthy economic downturn, lasting from 1873 to 1896, as the mid-Victorian boom was followed by what late Victorians would call the 'Great Depression'. Economic growth was slower than at any time since the industrial revolution; the belief that progress in Britain would be permanent and self-sustaining received a serious jolt; and there were especially bad years of high unemployment, not only in the late 1870s, as Disraeli had discovered to his political cost, but also in 1886–87 and again in 1893–94. To be sure, the depression was primarily one of 'prices, profits and interest', and that was in some ways beneficial, for it was the downturn in prices that led to the increases in real incomes that helped fuel the rise in working-class spending on consumer goods. But these same developments, in significant part arising from the worldwide reduction in agricultural prices resulting from the globalization of the market in arable crops and meat, hit the rural economy, and rural society, very hard. Farmers' profits fell, landlords' rents fell, and so did the value of their broad acres. Arable agriculture suffered more than livestock farming, but many landowners saw their incomes cut by as much as a third, while the exodus from the country to the town intensified. After the prosperous mid-Victorian years of 'High Farming', the rural sector entered a deep depression from which it would not recover until after the Second World War. In the long run the repeal of the Corn Laws had turned out as badly as Peel's critics had feared and forecast, but renewed demands for protection made no headway, and the Royal Commission on Agriculture, appointed in 1881, merely confirmed the depression but offered no palliatives or cures.

There were also growing doubts and anxieties concerning the nation's once unchallenged industrial sector. For even as the United Kingdom's manufacturing reached its peak output as a proportion of world production and exports, there were ominous signs that its global pre-eminence could not be maintained in the face of the competition being mounted by the recently unified Germany and the reconstructed United States, both of which seemed set to surpass Britain, not only in manufacturing output but also in manufacturing innovation. During the early 1880s the United Kingdom had produced one-third of the world's steel, which was substantially ahead of the United States, and almost double that of Germany. But by the early 1890s, America had significantly overhauled Britain, Germany was catching up fast, and the *Reich* would be ahead by the end of the decade. There was a further problem, in that the 'Great Depression' was a global phenomenon, and from the late 1870s to the 1890s most European nations, as well as the United States, responded by imposing tariffs to protect their own industries and to keep out foreign imports. But the United Kingdom remained wedded to free trade as an article of quasi-religious faith, and refused to retaliate in kind, and many manufacturers feared their largest export markets, which remained in Europe and the United States, would be cut off, even as the unprotected British market seemed increasingly flooded with cheap manufactured goods from abroad. These anxieties were articulated but were not allayed by the Royal Commission on the Depression in Trade and Industry (1885–86), which praised the Germans for their enterprise and warned of the 'increasing severity' of their competition, both at home and abroad, and these concerns would be reiterated by E. E. Williams in his alarmist book *Made in Germany* (1896).

These industrial and international worries fed into domestic concerns about the general health and well-being of the United Kingdom, for while the standard of living and the quality of life were improving for many people, this was not true for everyone. On the contrary, there was abundant evidence that, a century after the industrial revolution had begun, unacceptable levels of poverty continued to co-exist with unprecedented amounts of plenty. In 1885 the Royal Commission on the Housing of the Working Classes revealed alarming evidence of gross overcrowding among the lower paid. The 1891 census showed that one-tenth of the population was living more than two per room, including one-fifth of London's inhabitants. The Royal Commission on Labour of 1891–94 calculated that the average annual earnings of adult male manual workers in 1885 had been only £60, and that over 80 per cent of them earned 30s a week or less. The Royal Commission on the Aged Poor of 1895 showed that more and more elderly people were obliged to go into workhouses. These official inquiries were complemented by investigations made by private individuals. In 1883 the Congregational minister Andrew Mearns published a pamphlet entitled *The Bitter Cry of Outcast London*, drawing attention to the 'pestilential human rookeries' of deplorable squalor in which poor people lived in Bermondsey. Two years later the crusading journalist W. T. Stead authored *The Maiden Tribute of Modern Babylon*, describing what he termed the white slave trade whereby young women were trapped in a (short) life of prostitution. And in 1891 Charles Booth produced the first instalment of his survey of *Life and Labour in London*, which concluded that 30 per cent of the inhabitants of the world's greatest city were living in poverty.

It may have been true that the national standard of living and quality of life had improved markedly across the nineteenth century, but these were disturbing revelations for those who believed that Britain had been providentially marked out as the most civilized, progressive and successful nation on the globe. So it was scarcely surprising that the 1880s were also years, as Salisbury and Tennyson feared, characterized as much by social restlessness as by social repose. In the middle of the decade, when unemployment was high, workers all over Britain took to the streets in angry demonstrations. In February 1886 there were riots in Trafalgar Square, shops in Pall Mall and Piccadilly were looted, and the Lord Mayor launched a 'ransom' fund for the unemployed. Further trouble erupted in November 1887, with another huge demonstration in Trafalgar Square on what became known as 'Bloody Sunday', and which took place despite a banning order from the Chief Commissioner of the Metropolitan Police. Between August and November 1888 five prostitutes were brutally murdered in London's Whitechapel by a sadistic killer who was given the name 'Jack the Ripper', and the photographs taken of his gaunt, emaciated and disembowelled victims remain a shocking reminder of the darker side of late-Victorian life. The following year Ben Tillett led the London dockers out on strike, and they enjoyed widespread public support in Britain and throughout the empire. The result was a famous victory of the workers over their employers, which emboldened others to strike, among them the match girls at Bryant and Mays, and the gas workers in support of an eight-hour day. In terms of organized popular protest, and in terms of investigations into the condition of the country, there had been nothing like this in Britain since the 1830s and 1840s.

These striking and protesting workers, who were especially in evidence on the streets of London in the months before and after the Golden Jubilee celebrations, had no wish to overthrow the established order: they just wanted to get their fair share. But these disturbances did portend a new phase of working-class assertiveness, known as the 'New Unionism', as semi-skilled and unskilled labouring men began to organize for the first time, in emulation of those skilled artisans who had been unionized since the 1840s. Membership of the Trades Union Congress increased from two-thirds of a million in 1886 to one and a half million four years later, while the number of unions affiliated to the TUC grew from 122 to 311. There were also the first attempts to create a new political organization to represent and further the interests of working men at Westminster, as Keir Hardie established the Independent Labour Party in 1893 for precisely that purpose. By then left-leaning middle-class intellectuals, who also sought (among other things) to improve the lot of the workers, had already set up their own associations, and 1884 had been a bumper year for such activity. H. M. Hyndman refounded his earlier Democratic Federation as the Social Democratic Federation; William Morris established the Socialist League as a dissident offshoot; and Beatrice and Sidney Webb, in alliance with H. G. Wells and George Bernard Shaw would soon join the Fabian Society, which was also begun in that year. Those who worried about 'disintegration' and class warfare regarded such developments with fear or disapproval, and they denounced all such organizations as 'Socialist'. Yet the workers and the writers had little in common, since the trades unionists wanted to obtain better pay and conditions, whereas the intellectuals wanted to change society more fundamentally, but could not agree as to how to do so.

Although the class war that Salisbury had feared did not come to pass, the downturn in the British industrial economy undoubtedly intensified animosity between workers and employers. At the same time the more severe recession in the agricultural economy increased hostilities between tenants and landlords, beginning in England and Wales, with what would become known as 'the revolt of the field'. Even before the depression began to hit the rural regions, there had been some localized attempts at trades union organization, which were consolidated and extended by a Warwickshire farm worker and Methodist preacher named Joseph Arch. In the spring of 1872 he had founded the National Agricultural Labourers' Union, with the aim of improving wages and conditions of rural workers, and within two years it claimed more than 80,000 members. In the short run the result was an undoubted improvement in the conditions of agricultural labourers, but the onset of the depression weakened the union's bargaining position and numbers declined. In response, Arch turned by the end of the 1870s to political agitation, urging that the vote should be extended from property owners to rural labourers who mostly lived in rented accommodation. It had previously been supposed that, if enfranchised, they would merely vote as their landlords told them, but the recent protests and agitations suggested that many of them were more radical in their political views. Their cause was also taken up by Jesse Collings, the radical parliamentary colleague of Joseph Chamberlain, who urged that all agricultural labourers should be given smallholdings and allotments so as to improve their conditions more significantly. During the early 1880s Collings and Chamberlain campaigned vigorously for this policy under the slogan 'Three acres and a cow'.

Underlying these campaigns was a growing hostility to landlords, who were memorably denounced by Chamberlain as a class of idle parasites 'who toil not neither do they spin'. But it was on the 'Celtic fringe', where farming had never been as prosperous, and where rural society had always been further divided on ethnic, religious and linguistic lines, that opposition to the landowners erupted violently and vociferously. This was especially so in Ireland, where impoverished Catholic tenants turned on their indebted Protestant landlords, refused to pay their rents, or simply could not afford to do so, and began waging what was termed a Land War. The landlords responded with a spate of evictions, as a result of which the violence and the intimidation escalated still further. In the Highlands and islands of Scotland there were similar disturbances, known as the Crofters' War, which recalled and harked back to the earlier violence at the time of the clearances, and this anti-landlord agitation also spread to Wales. Such antagonisms not only represented an unprecedented social threat to the established rural hierarchy: they also represented an unprecedented political threat to the integrity of the United Kingdom, as demands for the regulation and reform of landlord-tenant relations were accompanied by renewed agitation for Home Rule and the dissolution of the Union in Ireland, and there were similar demands for the first time in Scotland and Wales as well. Here, indeed, was the spectre of national 'disintegration' that Salisbury feared, and it gave added urgency to the question he had posed: did the British governing classes still possess the political will and determination to do something about it, and would they be able to hold the United Kingdom together?

That was far from clear, for during the 1880s British culture and society seemed increasingly afflicted by what was called 'decadence'. Its most famous proponent was Oscar Wilde, who had begun that decade as the leader of the 'aesthetic movement', lampooned by Gilbert and Sullivan in *Patience* (1881), but who by the end of it had become the leader of the so-called 'decadents', professing to prefer pessimism to optimism, the decayed to the living, the abnormal to the normal. They were also suspected of drug-taking and homosexuality, and they were widely regarded in strait-laced circles as degenerate and corrupt. In fact, there were never that many of them, but the anxiety and alarm the 'decadents' deliberately and undoubtedly provoked, along with simultaneous fears about the 'white slave trade', help explain the passing of the Criminal Law Amendment Act of 1885, which raised the age of consent for girls from thirteen to sixteen; in addition, as the result of an amendment carried by Henry Labouchère, it criminalized for the first time as 'gross indecency' all forms of homosexual activity, in public or in private. Hence the police raid, four years later, on a homosexual brothel in Cleveland Street in London's plush Fitzrovia district, and although the scandal was largely hushed up, it was rumoured that some of the greatest and grandest names in the land were implicated. The four plays that Wilde wrote at this time – *Lady Windermere's Fan* (1892), *A Woman of No Importance* (1893), *An Ideal Husband* (1895) and *The Importance of being Ernest* (1895) – all explored upper-class decadence: their idle, leisured characters, interested in little but social gossip; and the darker explorations of hypocrisy, blackmail, corruption and double lives. Here, indeed, was striking corroboration of those who feared, like Salisbury, that public service and the will to rule were under threat from within as well as without.

These years were also characterized by growing doubt and disbelief. During the first three-quarters of the nineteenth century most of the political elite had been steeped in religious teaching, and saw public life as a Christian avocation; and while Darwin, Tennyson and Arnold struggled with their faith, they still regarded it as a battle worth waging. But by the last quarter of the nineteenth century, doubt was triumphing over faith. Gladstone and Salisbury were both, albeit in different ways, committed churchmen, but the same was not true of their prime-ministerial successors. Arthur Balfour's first publication was entitled *A Defence of Philosophic Doubt* (1879), and two of his abiding interests were spiritualism and humanism; Lord Rosebery married Hannah Rothschild, who kept her Jewish faith and was buried in a Jewish cemetery; Asquith, though of Nonconformist parentage, preferred the Greek and Roman Classics; and after devouring *The Martyrdom of Man* (1872) by William Winwood Reade, the young Winston Churchill would abandon his Christian beliefs. In 1869 T. H. Huxley had coined the word 'agnostic' to describe someone who was an unbeliever, and in

1892 he declared that agnostic principles were 'irreconcilable with biblical cosmogony, anthropology and theodicy'. In 1876 the former clergyman Leslie Stephen (brother of Sir James Fitzjames Stephen and father of Virginia Woolf) published 'An Agnostic's Apology', reprinted in 1893, in which he argued that the only position to take regarding the possible existence of God was doubt. And Thomas Hardy's later novels, including *The Mayor of Casterbridge* (1886), *Tess of the d'Urbervilles* (1891) and *Jude the Obscure* (1895), criticized conventional attitudes to sex and marriage, and suggested it was Fate, rather than God, which determined human actions. So doubting of Christianity did these works appear that the Bishop of Wakefield allegedly burned his copy of *Jude*.

Yet although established religion was in decline, and doubt and scepticism were on the rise, there was still considerable social risk in publicly admitting unbelief; so adherence to established religion remained expected, as in the case of royal weddings, coronations, jubilees and funerals, even on the part of unbelieving politicians (and Arthur Sullivan was as celebrated a composer of hymn tunes and oratorios as he was of comic operas). Most public men were willing to conform, at least outwardly, but Charles Bradlaugh was not. Originally elected as Liberal MP for Northampton in 1880, he refused to swear (as distinct from affirm) the required parliamentary oath of allegiance on the grounds that he was both an atheist and a republican (and he was also a believer in birth control). Bradlaugh was duly expelled from the Commons and his seat was declared vacant, and like John Wilkes before him, he would be expelled several more times after wining a succession of by-elections. Eventually, in 1886, he was allowed to affirm instead of swear the oath, and take his seat, and two years later legislation was passed confirming the validity of this new arrangement. Henceforward, it was no longer compulsory for MPs to be Christians of any denomination, let alone for them to be Anglicans, as had been obligatory before 1829. Like disintegration and decadence, so doubt, agnosticism and even atheism were increasingly the mood and mores of the time. But the reaction of many voters to the economic downturn, political separatism, social conflict and growing scepticism of the 1880s would be to rally to the established order, with the aim of averting the very 'Disintegration' that Salisbury had feared. In Britain, as elsewhere in Europe, the 'Great Depression' led to a major realignment of politics and parties, and in a rightwards direction.

THE TURN TO THE RIGHT

One indication that the public mood was shifting during the 1880s was the extraordinary success of Gilbert and Sullivan's comic operas. Their second joint work, *Trial by Jury*, had premiered in 1875, but it was only three years later, with *HMS Pinafore*, that they hit their stride. From 1879 through to 1889 they produced a new operetta virtually every year, beginning with *The Pirates of Penzance* and culminating with *The Gondoliers*. Gilbert's words could be barbed and sharp, but they were often undercut by Sullivan's sparkling and sentimental melodies; and even as the operas lampooned such national institutions as the monarchy, the House of Lords, the government, the judiciary and the armed services, they also celebrated them. Like the Sherlock Holmes stories, they presented a settled world, seemingly at risk of subversion, but where the established order was miraculously preserved at the eleventh hour. In 1882, a year before Salisbury wrote 'Disintegration', the collaborators had produced *Iolanthe*, which was part Wagnerian spoof, complete with a fairy queen (a nod to Disraeli and Victoria?), part a send-up of the peerage. At one point Earl Tolloler is urged by Iolanthe to give away his Irish estates to his tenants, and at the beginning of the second act Private Willis, on sentry duty in New Palace Yard, pondered why it was that:

> Every boy and every gal
> That's born into this world alive
> Is either a little Liberal,
> Or else a little Conservative.

This was a plausible view of party politics since 1865, but by the late 1880s it would no longer be so valid. Yet it was a shrewd move for Gilbert and Sullivan to put the House of Lords on the stage, for by the middle of the decade the peers would again be at the centre of political controversy, both as legislators and landlords.

On taking office in 1868, Gladstone had declared that his mission had been 'to pacify Ireland'. Fourteen years on, that task was yet more pressing and also more daunting, not least because of the violent Land War that the tenants were waging against their landlords. But it was also because of a major change in the politics of Anglo-Irish relations. After the general election of 1880 the new leader of the Irish Home Rule Party, Charles Stewart Parnell, wanted Home Rule for Ireland *now*, not at some indefinitely defined future date, and he was prepared to resort to extraordinary means to get it. His first momentous decision as leader was to support the Land War that tenants were waging against the evictions, by refusing to pay their rents – protests which became known as 'boycotting', after the hapless Captain Boycott, the embattled land agent for Lord Erne, whose tenants in County Mayo were the first to stop their payments. But at the same time, and in what was termed a 'New Departure', Parnell and his Nationalist followers carried the fight to London, with a parallel campaign of obstruction and disruption in the Palace of Westminster itself that made it increasingly difficult for the government to conduct its parliamentary business or carry legislation. This created serious problems for Gladstone in his dual capacity as both Leader of the House and prime minister. This was made plain in early 1881, when he sought to pass another Irish Coercion Bill, which aimed to restore order by suspending habeas corpus and giving increased powers to the Chief Secretary. But the delaying tactics of the Nationalists were so successful that they kept the Commons sitting continuously for forty-one hours between 31 January and 2 February, until the Speaker ended the debate.

As had been true of many of its predecessors when dealing with Ireland, the Liberal government tried to pursue the irreconcilable policies of coercion and conciliation at the same time, and did not succeed with either, let alone both. So, having finally passed the Coercion Act, under which Parnell was subsequently arrested, Gladstone introduced another Land Bill that was much more far-reaching than the limited and ineffectual measure he had carried in 1870, which gave Irish tenant farmers fixity of tenure, and established a new type of land court to set fair levels of rent. This measure was a significant interference by government in what had previously been regarded as the free and sacrosanct contractual relations between landlord and tenant. It was clearly more concerned with the plight of the tenants than with that of the landlords, and it enraged and distressed many landowners, including some Whig grandees, who thought that Gladstone was turning against them. Meanwhile, the government had struck a secret deal with Parnell, known as the Kilmainham Treaty after the Irish prison where he was being held, whereby he would be released if he called off the boycotts and protests in exchange for further legislation cancelling the arrears of rents owed by more than 100,000 Irish tenants to their landlords. But to some of Gladstone's supporters this seemed like capitulation to violence and illegality: Lord Cowper, the Lord Lieutenant of Ireland, resigned; and so did W. E. Forster as Chief Secretary. This abandonment of Gladstone by a Whig grandee on one wing of the Liberal Party, and by a radical former businessman on the other, was a portent of more desertions to come. Cowper and Forster were replaced, respectively, by the fifth Earl Spencer, another Whig magnifico, and Lord Frederick Cavendish, who was not only Hartington's younger brother but also a relative of Gladstone's by marriage.

Many landowners, whether Tory or Whig, were deeply disturbed by what Gladstone had done, and their anxieties would soon be abundantly vindicated. In May 1882, just after the installation of Earl Spencer as the new Lord Lieutenant, a small group of Irish terrorists murdered Lord Frederick Cavendish and the Irish permanent under-secretary, T. H. Burke, in Phoenix Park. Not since the assassination of Spencer Perceval had there been such a tragic and traumatic episode in British public life; and this was much worse because it was clearly the work of conspirators and in defiance of government policy. This made any further concessions by Gladstone impossible, and his irregular dealings with Parnell were widely condemned – the more so when it emerged that he had agreed 'to co-operate cordially for the future of the Liberal Party in forwarding Liberal measures and measures of general reform', which seemed to imply that the Kilmainham Treaty was more of a shabby deal to improve the Liberal Party's prospects than it was to solve Ireland's problems. In fact, Parnell was no terrorist, and he deplored the Phoenix Park murders. But although he was released from gaol, and became a hero to many Irish Nationalists, the violence, the boycotts and the murders went on through much of 1882, and Parnell was powerless to prevent them. Using all its coercive powers, the government gradually managed to reduce the disorder and restore a semblance of peace. But once again the combination of coercion and conciliation had failed to solve the Irish question; there was growing disquiet among the Whigs about Gladstone's long-term aims; and Queen Victoria was as outraged by the Phoenix Park murders as by what she regarded as Gladstone's completely misguided handling of Hibernian affairs.

During the initial two years of his second government, Irish affairs took up so much time that there was little opportunity for the sort of wide-ranging legislation that had characterized Gladstone's first prime ministership. But some other laws were passed, and they possessed an undeniable radical tinge. The Ground Game Act gave occupiers concurrent rights with landowners to kill rabbits and hares, ending the centuries-old monopoly on game that had previously been reserved to proprietors; and the Employers' Liability Act made it easier for workers to sue employers, which signalled a further interference with freedom of contract. Flogging was finally abolished in the army and navy, completing one aspect of Cardwell's earlier reforms. The Married Women's Property Act gave them the same rights as unmarried females, which meant their possessions were no longer automatically determined to be legally owned by their husbands. The Settled Lands Act increased the ability of landowners or their trustees to sell land and other assets, and paved the way for sales of pictures, libraries and estates at the end of the decade by some of the nation's greatest and grandest families, such as the Dukes of Marlborough and Hamilton. And Jesse Collings steered through an Allotments Extension Act, which meant that by 1886 there were 400,000 such holdings under four acres. There was also some tidying-up legislation concerning patents and bankruptcy law. But the most important measure, which made a far greater impact on the workings of the electoral system than the Ballot Act of 1872, was the Corrupt Practices Act of 1883, which limited the amounts that candidates could spend, made it easier to punish those who indulged in bribery, and effectively shifted the task of canvassing from paid agents to volunteers.

Gladstone followed this with an even more audacious piece of legislation, namely the Third Reform Act. Early in 1884 the government introduced a single Representation of the People Bill, the first to deal with the United Kingdom as a whole (previously there had been separate measures for England and Wales, for Scotland and for Ireland), which aimed to create a uniform household and lodger franchise based on the one that had been introduced for the English boroughs in 1867, and which would now be extended to the counties as well. The bill passed the Commons but was rejected by the Lords, where both Tory and Whig peers declared they would only let it proceed if the government also put forward concurrent proposals for the redistribution of seats.

The Lords' rejection provoked a campaign of popular protest, encouraged by Joseph Chamberlain, behind the slogan 'Peers versus People', and this may have helped bring about the inter-party discussions that took place towards the end of 1884 to resolve the impasse. As a result, two measures were eventually passed in the following year that both extended the franchise and redistributed seats. The former, in response to Joseph Arch's campaigning, gave the vote to agricultural labourers who were householders. The latter swept away virtually all the two- and three-member constituencies, as well as more than one hundred boroughs with populations of less than 15,000: in future all counties and large towns were divided up into single-member constituencies. The under-representation of London was remedied, in part by reducing the over-representation of the southeast; but what had become the over-representation of Ireland (where the population continued to decline) remained. Equally significant was the increase in the numbers enfranchised. Thanks to the Second Reform Act, the proportion of adult males entitled to vote had grown from 16.7 per cent (1.31 million) to 30.3 per cent (2.53 million); by 1891, thanks to the Third Reform Act, it had doubled to 61 per cent (approximately six million).

Since all women were still denied the franchise (an amendment to extend it to them had been defeated by 271 votes to 135), this meant that approximately 30 per cent of the adult population of the United Kingdom wielded the vote during the next thirty years. This was still not democracy, and nor was it intended to be. But the Third Reform Act did bring into being something approaching a coherent, rational, nationwide system of voting and representation. By abolishing most of the remaining small boroughs, the Act further weakened the Whigs (who often sat for such places), while the county constituencies and the new suburban seats in the big cities would definitely favour the Tories; and since the largest relative increase in the size of the electorate was in Ireland, that was bound to benefit the Nationalists. This may explain why Chamberlain suddenly began to take an interest in Ireland, with the aim of thwarting the growing demands for Home Rule. As President of the Board of Trade he had accomplished little, and his Merchant Shipping Bill of 1884 had been withdrawn in the face of hostility from shipowners and a lack of enthusiasm for it in the cabinet. Later that year, and far exceeding his departmental brief, he took up the issue of Irish local government reform, and proposed the establishment of representative county bodies along with a 'Central Board', with the aim of fending off any demands for even greater Irish autonomy. But the cabinet was divided over the scheme, which seemed to some Whigs to make Home Rule *more* likely not less. With the government in disarray, the Conservatives won a vote amending the budget in June 1885, supported by most Irish members, while many Liberals declined to back their own front bench. Gladstone took this as a vote of no confidence, then promptly resigned, and again looked forward to retiring 'at the end of the current parliament', which had only two years to run.

To the queen's delight, Salisbury formed his first Conservative government, albeit another minority administration, although with the (perhaps unexpected) support of the Irish Nationalists, in exchange for undertaking not to renew Irish coercion. Half its members were peers, and another quarter had close landed connections, including two sons of dukes, while Salisbury was both prime minister and Foreign Secretary. One result of the temporary pact between the government and the Nationalists was the rapid passing of a more far-reaching piece of land reform, known as Ashbourne's Act, than Gladstone had achieved in 1870 or in 1881. It made available £5 million to tenants (to which the same sum would be added by further legislation passed in 1888), and as a result 25,000 farmers bought their holdings. The dismantling of the Anglo-Irish landed ascendancy and the creation of an indigenous Irish peasant proprietorship had begun: 'Disintegration' furthered by Lord Salisbury himself. Secret talks were also held between Parnell and the Irish Viceroy, Lord Carnarvon: they proved inconclusive, although Parnell's preference was that Salisbury should take up Home Rule, which he had a much greater chance of getting through the House of Lords than did Gladstone. But Salisbury had no intention of doing so, and nor, yet, did Gladstone, who was much more concerned by what he saw as the excessive radicalism of Chamberlain's 'Unauthorized Programme', which he had first announced at the beginning of 1885. Chamberlain's main targets of attack were the usual radical suspects, namely the monarchy, the Lords, the established church and the landed classes in general, and in advocating manhood suffrage, payment of MPs and graduated taxation, he was hoping to appeal to the newly enfranchised agricultural workers, and to drive the Liberal Party in a much more radical direction. But the Whigs deplored these wild proposals, Gladstone agreed with them, and Chamberlain's programme remained 'unauthorized' by the party high command.

It was in this generally confused state of affairs that parliament was dissolved in November 1885, and the first general election was held with the new franchises and constituencies established by the Third Reform Act. Gladstone expected the Liberals to gain another large majority, which might provide him with the opportunity to deal comprehensively with the Irish question, as he had recently come to believe it was now necessary to do, by combining measures of Home Rule and land reform. But he made no such public commitment, while most other Liberal candidates did not mention Home Rule at all. The results of the election were politically indecisive and arithmetically extraordinary. The Liberals lost seats compared to their showing in 1880, which suggests that Chamberlain's 'Unauthorized Programme' made little impact on the newly enfranchised voters in the way he had hoped; but with 334 MPs returned, they were much the largest party in the new House of Commons. Having obtained 250 MPs, the Conservatives had done marginally better than in 1880, but they were again a minority. Yet this seemingly clear-cut Liberal triumph was completely negated by the fact that the Irish Nationalists had increased their numbers from fifty to eighty-six seats, which meant that if Parnell's followers combined with the Tories, they effectively cancelled out the Liberal majority. So Gladstone's hopes had not been realized, and he found himself at the head of a party without the secure parliamentary majority that might enable him to carry his wide-ranging Irish measures. Moreover, many Whigs were increasingly anxious that what they still regarded as 'their' Liberal Party had become too radical during the previous five years; by contrast Chamberlain and his followers were disaffected because they had not succeeded in making the party radical enough; while the Irish Nationalists preferred the Tories.

Disappointed in his anticipated commanding Commons majority, Gladstone's initial response was to try to persuade Salisbury to stay in office and embrace Home Rule, on the grounds that the Conservatives were better placed to carry such a measure, both in the Commons (with Irish Nationalist and some Liberal support) and in the Lords (where Salisbury, unlike Gladstone, might persuade the Tory peers to let the legislation through). But Salisbury was determined not to add another Conservative volte-face to those of 1829, 1846 and 1867, and so he rejected Gladstone's overtures, dismissing his call to treat the Irish question as being above party politics, as mere (and characteristic) Liberal 'hypocrisy'. Disappointed again, Gladstone's recent conversion to Home Rule was declared publicly by his son Herbert in mid-December 1885, just as the election was drawing to a close. It was easily represented as an opportunistic bid for the support of the Irish Nationalists, who did indeed transfer their allegiance from Salisbury back to Gladstone; but, in fact, Gladstone had for some time been moving in the direction of embracing and espousing a more comprehensive settlement of Irish matters (although it never seems to have occurred to him to consult or notify any of his Liberal colleagues). Salisbury did not resign in the aftermath of the general election; but when parliament met in January 1886, his government was defeated on an amendment to the Address moved by Jesse Collings. The opposition majority of 329 to 250 almost exactly resembled the numbers of Liberals and Conservatives who had recently been returned: but the reality was very different. The Irish Home Rule Party supported the amendment; but seventy Liberals, dismayed by Gladstone's recent espousal of Home Rule, had abstained, and eighteen more (including Hartington), who were even more dismayed, had actually voted with the Conservatives.

But however ominous (from Gladstone's point of view) the voting patterns, they were enough to make it plain that Salisbury could not go on. He duly resigned, and in February 1886 Gladstone formed his third administration. There were still Whig grandees a-plenty: Earl Spencer as Lord President, Lord Granville (again) as Colonial Secretary, Lord Ripon at the Admiralty, Lord Kimberley at the India Office, and the young Lord Rosebery as Foreign Secretary. Once again the radical element in the party was fobbed off with lesser offices: A. J. Mundella at the Board of Trade, and Chamberlain at the Local Government Board (Sir Charles Dilke, who might also have hoped for office, had just been cited in a divorce case, and his political career was ruined). In terms of its personnel and political balance Gladstone's third cabinet closely resembled his second. But there were two significant and worrying differences: Hartington had refused to serve, and Chamberlain was more disaffected than ever, having been offered only another of the most junior posts. Nor were these Gladstone's only difficulties, for he also faced the entrenched opposition to Home Rule from the House of Lords and from the queen. Yet he had become convinced by recent events that it was no longer possible to hold

Ireland against its will, especially with the enlarged electorate that had been brought into being by the Third Reform Act. Coercion had self-evidently not worked, which meant all that was left was conciliation. Accordingly, Gladstone reasoned, it was time to make far-reaching political concessions, which might secure continued Irish loyalty for the Union, and also to embrace more wide-ranging land reform, which would remove the grievances that had provoked the recent violence and rent strikes. After all, self-government had been progressively given to the overseas colonies of settlement, and they were becoming *more* loyal, not less. Why should the same policy not succeed in Ireland, too?

Back in office, Gladstone immediately began work on the two measures that together he hoped would result in the 'definitive' Irish settlement, that would fully and finally 'pacify' the country, and that he now believed to be essential. The first was a Home Rule Bill, providing for a separate Irish parliament, which would meet in Dublin, and henceforward there would be no more Irish MPs elected to the British House of Commons. But the Anglo-Irish Union of 1800 would not be completely repealed, since the imperial legislature at Westminster would retain full control over all matters concerning the crown, foreign policy, defence, customs and excise, and religious establishments. The restored Irish parliament would consist of an elected lower house and a nominated upper house, sitting and voting together. Its powers would be clearly circumscribed, but it would be enabled to levy taxes, which would largely be spent as the Dublin legislature saw fit, but with a specified proportion to be disbursed for 'imperial' purposes. This was a relatively moderate scheme, of limited devolution, with the aim of keeping Ireland as an integral part of the empire; but at the same time Gladstone also proposed a far-reaching scheme of land reform, which would make available a huge sum, ranging from £50 to £120 million, from the British exchequer, enabling tenants to buy out their landlords and establishing a whole new class of peasant proprietors who ought to be grateful to the imperial authority for making possible their very existence. By March 1886 the details of these measures were made available to the cabinet, whereupon Chamberlain (and Sir George Otto Trevelyan, Secretary of State for Scotland) resigned, on the grounds that, whatever Gladstone maintained to the contrary, Home Rule would in fact spell the end of imperial unity and security, causing Britain to 'sink to the rank of a third-rate power' in the world.

At the age of seventy-six Gladstone introduced the Government of Ireland Bill into the Commons early in April, in a speech lasting over three hours, and the Land Purchase Bill followed one week later. He sought to depict Home Rule for Ireland as the next stage of imperial evolution, following the encouraging precedents set by the still-loyal settlement colonies, especially Confederated Canada. But he encountered widespread opposition. Some of it focused on the details, but there were also more general objections. For the Conservatives, Lord Salisbury declared that the Irish were incapable of self-government, and that renewed coercion and continued emigration were the only solutions to the current problems. The maverick Tory, Lord Randolph Churchill, had recently played what was termed 'the Orange card' when he urged that, in opposing Home Rule, Protestant 'Ulster will fight, and Ulster will be right'; and he would later coin the phrase 'Unionist Party' to describe the coalition of anti-Home Rule forces that was now coming into being. But it was from Gladstone's erstwhile supporters, who felt themselves deeply betrayed, that the strongest opposition came. Lord Hartington, as the leader of the Whigs, appeared at a great London gathering with Lord Salisbury to oppose the bill. John Bright, tapping into England's residual anti-Catholicism, declared that 'Home Rule' would be 'Rome rule'. And Joseph Chamberlain charged Gladstone not only with political inconsistency but also with proposing a measure that would fatally weaken both the Union and the empire. On 8 June Gladstone made the final speech in the Commons at the end of the debate on the second reading: 'Think, I beseech you, think well, think wisely, think not for the moment, but for the years that are to come, before you reject this bill.' His plea went unheeded, and the measure was defeated by 341 votes to 311, with ninety-four Liberals going into the lobby against Gladstone, while six more abstained.

The rejection of Gladstone's first Home Rule Bill was a major political event, but it was also the expression of more deep-rooted changes. The majority of great Whigs, such as the Dukes of Devonshire, Bedford, Marlborough and Westminster, who had become increasingly concerned since 1880 about where the Liberal Party was going, now deserted Gladstone, and although a minority stayed loyal, the landed establishment and the House of Lords would henceforward be overwhelmingly Conservative and Unionist. Many rich bankers and financiers based in the City of London, among them the Rothschilds (to whom Gladstone had vainly given a peerage in 1885), also abandoned their earlier liberal internationalism, and aligned themselves with conservatism and imperialism. In Scotland there was strong Presbyterian sympathy for Protestant Ulster, and a significant minority of Liberal MPs went over to the Unionist cause. And Joseph Chamberlain, the thwarted radical in the Liberal Party, led his West Midlands followers to defend the Union and embrace the empire. These different groups formed the Unionist alliance, and joined with the Conservatives in parliament and in the constituencies, where Tories and Liberal Unionists agreed not to stand against each other. These developments were well reflected in the general election that Gladstone insisted on calling immediately after his Commons defeat, and only seven months after the last contest had taken place. It was, essentially, a referendum on his Irish policy, and he lost it. The Conservatives registered major gains, winning 316 seats, while the Liberals were reduced to 191; the Irish Nationalists held steady at 85 seats, and the Liberal Unionists returned 78. This meant the anti-Home Rule MPs outnumbered those in favour by 394 to 276, and although there were Home Rule majorities in Ireland, Scotland and Wales, the vote in England was overwhelmingly in support of the Conservatives and Unionists, who between them won 332 seats to 123 for the Liberals.

Gladstone promptly resigned in July 1886, his third government having lasted barely six months; and to the queen's relief Lord Salisbury became prime minister for the second time. Despite the successes of the Conservative and Unionist alliance at the recent election, he would appoint no Whigs and no Liberal Unionists to his cabinet, which contained seven peers and seven commoners (five of whom were sons of peers or baronets). The Foreign Office was given to Sir Stafford Northcote, recently ennobled as the Earl of Iddesleigh. The most sensational appointment was that of Lord Randolph Churchill, aged only thirty-six, the darling of the National Union of Conservative Associations, and self-styled champion of something (or nothing?) called 'Tory democracy', as Chancellor of the Exchequer and Leader of the Commons. But within six months the government faced a major crisis, as in December the increasingly erratic and unstable Lord Randolph threatened to resign unless his demands for a significant reduction in military expenditure were met. Salisbury duly called his bluff, and accepted his resignation

18. Indian mutineers about to be blown apart by British cannon during the Great Rebellion of 1857. There were terrible atrocities on both sides.

PUNCH, OR THE LONDON CHARIVARI.—August 3, 1867.
REFORM
A LEAP IN THE DARK.

19. A horse with the head of Disraeli takes a leap in the dark with his reform bill of 1867, while the Liberal leaders look on askance.

20. A Victorian domestic interior by Charles West Cope, 1860. The piano was a major signifier of status and respectability.

21. Queen Victoria commissioned this painting, *No Tidings from the Sea*, by Frank Holl, in 1870, depicting the grieving family of a fisherman.

22. Melbourne from the Botanic Gardens, by Henry Gritten, 1867. It was already on the way to becoming one of the great cities of the southern hemisphere.

23. The Grand Assemblage, or Delhi Durbar of 1877, at which the Viceroy, Lord Lytton, proclaimed Queen Victoria to be Empress of India.

24. The Ottawa Parliament Buildings, constructed for confederated Canada. Troops deliver a *feu de joie* for Queen Victoria's birthday, 1868.

25. William Ewart Gladstone by Sir John Everett Millais, 1879, the year he launched his campaign against the iniquities of 'Beaconsfieldism'.

26. Benjamin Disraeli, Earl of Beaconsfield, also by Millais, completed posthumously in 1881.

27. C. J. Staniland, *The Emigrant Ship*, completed during the late 1880s, a decade of high emigration from the United Kingdom.

28. Lord Salisbury by Sarah Acland, 1894. He was as formidable as he looks in this portrait.

29. Piano transcription (1887) of Gilbert and Sullivan's *Iolanthe* (1882)

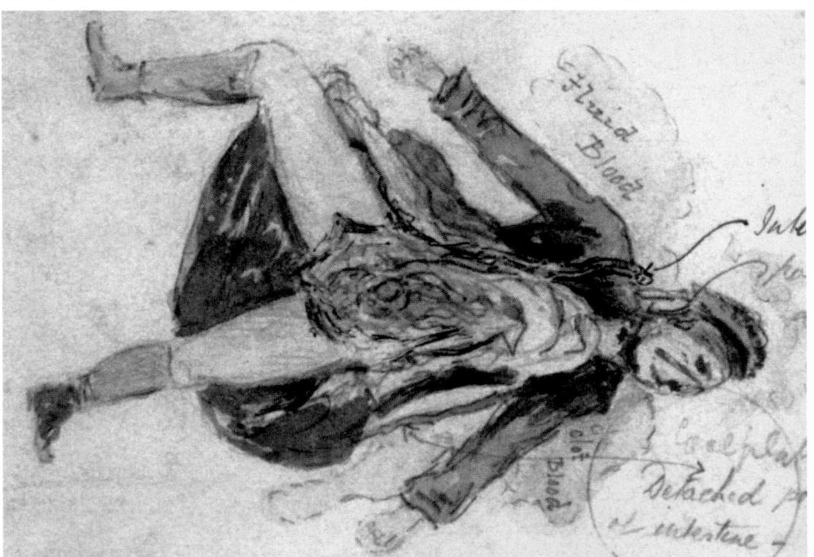

30. (*top left*) A cycling couple, *c.* 1900: the advent of the bicycle gave many women greater freedom of movement (and expression?) than ever before.

31. (*above*) Sketch of the body of Catherine Eddowes, one of the victims of Jack the Ripper, 1888. A harrowing depiction of the darker side of late nineteenth-century London life.

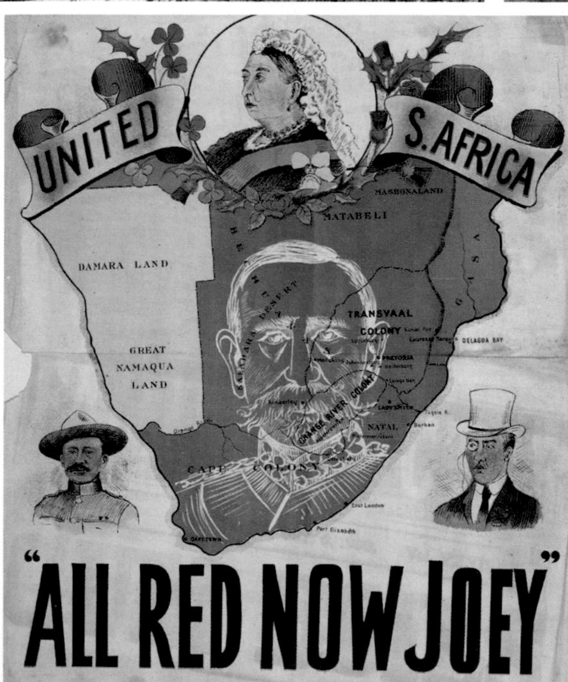

32. (*above left*) The centre of Birmingham, 1893. This grand ensemble of civic buildings was still intact during the 1950s, but not all of them survive today.

33. (*above*) A bonfire to celebrate the relief of Mafeking in 1901. Note the effigy of the Boer, *top left*, shortly to be incinerated.

34. (*left*) A poster celebrating the victorious conclusion of the South African conflict, which was often known as 'Joe's War', in which the Colonial Secretary, Joseph Chamberlain, played a major and controversial part.

35. Crowds mourning their monarch at Parliament House, Melbourne, January 1901, under a banner proclaiming 'Our Beloved Queen Gone to Her Rest'.

36. Female graduates of Glasgow University, *c.* 1900. They had studied scientific subjects, which was very unusual for women at that time.

37. (*right*) Jawaharlal Nehru as a Harrow schoolboy, aged fifteen in 1904. He would later attend Trinity College, Cambridge.

(Lord Randolph would never hold office again and died in 1895), and he appointed in his stead as Chancellor the one and only Liberal Unionist who would hold office in his administration: the banker MP George Goschen, who had been First Lord of the Admiralty under Gladstone between 1871 and 1874, but had subsequently fallen out with him. At Goschen's insistence, Iddesleigh was sacked from the Foreign Office, which Salisbury again took over himself. More significantly, Salisbury brought in his nephew, the young Arthur Balfour, to be Secretary of State for Scotland in November 1886, and he moved him to be Chief Secretary for Ireland the following year. He also promoted the non-patrician Charles T. Ritchie, a jute merchant from Dundee, to be President of the Local Government Board.

Naturally, Ireland was a major item on the new government's agenda; in the aftermath of the failure of Gladstone's Home Rule Bill, Parnell's authority significantly waned among his supporters, while the Irish Nationalists, urged on by the more radical William O'Brien and John Dillon, organized a 'Plan of Campaign', in which tenants once more agitated and protested against high rents and evictions. Salisbury's government responded with renewed coercion in the form of a draconian Crimes Act, which was the brainchild of his nephew, who thereafter would be excoriated in Ireland as 'Bloody Balfour'. In the short run it provoked more riots and violence, but by 1891 Balfour's policies seemed to be working, as unrest noticeably diminished. In addition to the Crimes Act, Balfour also passed another Land Act in 1891, which made an extra £33 million available for land purchase; but it contained so many complicated legal clauses that it did not go fully into effect until it was amended five years later. Meanwhile, Parnell's reputation also declined on the other side of the Irish Sea when, in 1887, *The Times* published letters, allegedly written by him, apologizing to his supporters for having denounced, contrary to his own beliefs, the Phoenix Park murders. But after a court case and a judicial inquiry it emerged that a disreputable Irish journalist had forged the letters, and Parnell's reputation in England was completely restored. Not for long: for in November that year he was named as the co-respondent in a divorce case brought by Captain W. H. O'Shea against his wife, Katherine. Irish Catholic clergy and English Nonconformists alike turned against Parnell as a proven adulterer, a majority of Nationalist MPs repudiated his leadership, and in October 1891 he died suddenly at the age of only forty-five. The cause of Home Rule was much weakened as a result, especially among English Liberal voters.

Like Disraeli in 1874, Salisbury sought to offer a welcome Tory respite from the heroics and histrionics of Gladstone's great causes and moral crusades. But it was not just in Ireland that Salisbury's government was in fact more active than this might suggest. At the Exchequer, Goschen lowered income tax and declined to cut military spending to the extent that Lord Randolph Churchill had intended. At the Local Government Board, Ritchie carried the Local Government Act of 1888, which was a far-reaching measure of reform for England and Wales, and similar legislation was enacted for Scotland (but not for Ireland) in the following year. These Acts established elected county councils in every shire; they created separate county boroughs for all towns with more than 50,000 inhabitants; and they rationalized the chaotic and competing jurisdictions of the great metropolis into a single London County Council (although leaving the historic City of London, with its Corporation and Lord Mayor, intact). As a result, local government, by elected representatives, was significantly strengthened: these new authorities could raise their own revenues by levying rates, and they were made responsible for policing, education and welfare; and the creation of county councils began the gradual shift of power away from the aristocracy and gentry in the countryside. At the same time, and in recognition of the depression that had hit so many farmers and landowners, the government established a Board of Agriculture, presided over by Henry Chaplin, the very embodiment of the traditional, Tory, fox-hunting squire. And in 1891 all fees in state elementary schools were abolished, and the minimum age at which children could be employed in factories was raised from ten to eleven. These largely incremental reforms helped consolidate the Conservatives' reputation as the natural, safe and reliable party of government.

Nevertheless, by the early 1890s the Salisbury government was visibly losing popularity. As with the final stages of Disraeli's second administration, the renewed downturn in the economy put the men in power on the defensive. Meanwhile, the Trafalgar Square protests, the 'Jack the Ripper' murders, the London dock strike and the 'Baring crisis' of 1890 (when the family bank nearly collapsed, having invested unwisely in unsound Latin-American bonds), all added to the general mood of anxiety and alarm; and the government suffered a succession of by-election defeats, some in normally safe Tory seats. In June 1892 Salisbury obtained a dissolution of parliament, and the general election was held in the following month. Most Liberals campaigned on the so-called 'Newcastle Programme', which had been adopted the year before and essentially meant Home Rule for Ireland, church disestablishment for Wales and Scotland, and a further extension of the franchise. Gladstone, who by this time only really cared about Home Rule, was more enthused about some parts of the programme than others. The Conservatives and Unionists, by contrast, promised to keep the United Kingdom intact, and to keep things quiet. The mood of the country was not easy to gauge; but in general, English opinion was against Home Rule, whereas the voters in Wales, Scotland and Ireland were more likely to be in favour of it. The election saw substantial gains for the Liberals, who increased their seats from 191 to 272, and significant losses for the Conservatives and Unionists, whose MPs were reduced from 394 to 314, while the Irish Nationalists held steady at 81. Disappointingly for Gladstone, the Unionists remained the largest contingent in the new Commons, and the Liberals were well short of a majority. But together with the Irish Nationalists they carried a vote of no confidence against the government, and Salisbury and his colleagues resigned in August 1892.

NEW IMPERIALISMS FOR OLD

In 1883, just one year after Gilbert and Sullivan had produced *Iolanthe*, and Salisbury had written 'Disintegration', the Regius Professor of Modern History at the University of Cambridge, Sir John Seeley, published a book entitled *The Expansion of England*, in which he famously and exaggeratingly observed that the British appeared to have conquered and settled half the world in what he called 'a fit of absence of mind'. But Seeley's work was more of a warning of challenges to come than it was a celebration of great things already achieved. For he feared that the geopolitical future lay with large, contiguous, land-based empires, held together by railways, rather than with dispersed seaborne empires, much more tenuously held together by naval and maritime connections. As for the nation states of Europe, Seeley believed that on their own they had no great future at all. Fifty years hence, Seeley predicted, the United States and Russia would have depressed even France and Germany, the pre-eminent powers of continental Europe, into second-class nations. The lesson and the warning for Britain were clear, for it, too, was in many ways much more a nation than it was an empire. Unless it gave up its insular and absent-minded ways, and consolidated its overseas realms into some sort of formal, federated structure, which might enable it to hold its own against the sprawling, transcontinental Russian and American behemoths, then Britain, too, would decline into the rank of another second-rate power. This, Seeley was convinced, was the great challenge the United Kingdom faced in the closing decades of the nineteenth century, and looking forward into the opening decades of the twentieth.

However much Seeley's warnings resonated during the 1880s and beyond, their demands for raising the imperial consciousness, and for giving unprecedented effort and money to imperial consolidation and defence, did not sit easily with the views of either Gladstone or Salisbury. Although Gladstone shared Seeley's view that the United States was predestined to overtake the United Kingdom, and had written an article in 1878 making just that prediction, he remained a committed liberal internationalist. He maintained his belief that the Concert of Europe was the best way of regulating relations between the nations of the continent, and was wary of assuming any more imperial responsibilities unless he absolutely had to. Hence his denunciation of the iniquities and excesses of 'Beaconsfieldism' during the Midlothian campaign and at the general election of 1880. And while Disraeli had indeed favoured a more forward imperial policy in Africa and Asia, accompanied by a wealth of theatrical gestures, he had scarcely met with conspicuous success; whereas Lord Salisbury would be much more circumspect, for he was almost as reluctant as Gladstone to add to the nation's imperial responsibilities. So the consolidation of the settler colonies that Seeley called for did not occur during the 1880s. Instead, and much more unexpectedly, there was a complete reversal of policy towards Africa where, except for the troublesome southern tip, some residual anti-slaving bases in the west, and occasional forays against the Ashanti, previous British governments had generally been reluctant to become involved. But from the 1880s both Liberal and Conservative administrations suddenly became convinced, albeit often reluctantly, that the continued protection of Britain's Indian Empire demanded the urgent acquisition of vast new tracts of territory across the length and breadth of Africa.

But the decade did not begin like that. Instead, it started with the sort of confusion and inconsistency that had so often characterized British colonial policy. During the Midlothian campaign Gladstone had denounced Disraeli's annexation of the Transvaal, but on assuming office in 1880 he gave the Boers the impression that this was no longer precisely what he thought, although in fact it probably was. In December the Boers rebelled against imperial rule, and in February 1881 they defeated the British at the Battle of Majuba Hill. The scale of the battle was relatively insignificant, but this military reversal, coming so soon after the earlier defeat inflicted on the British by the Zulus at Isandlwana, caused an outcry inside parliament and beyond; and it made Gladstone's desire to be generous to the Boers, which he had failed to communicate clearly to them, now appear as a capitulation in the face of military defeat. The subsequent negotiations between the British and the Boers were protracted, bad-tempered and unsatisfactory to both sides in their eventual outcome. The Pretoria Convention, signed in August 1881, recognized the 'independence' of the Transvaal Republic, but subject to continued British 'suzerainty', which was taken by the British (but not by the Boers) to mean control over foreign policy and some influence over the Boers' treatment of black peoples. But no one quite knew what 'suzerainty' actually meant or was and three years later, at the renegotiations that led to the London Convention, the word was silently removed, as part of the concessions made by the British. Yet the Boers remained alienated and dissatisfied, resenting Britain's economic power, residual local influence, and contempt for their own culture and beliefs, and they ignored the frontiers that had been agreed, expanding into 'native' lands. Nothing had been satisfactorily settled, and it could only be a matter of time before there would be further trouble.

During the Midlothian campaign another of the iniquities of 'Beaconsfieldism' that Gladstone had vehemently denounced had been Disraeli's purchase of the Suez Canal shares on behalf of the British government, and in so doing he was repeating the criticisms he had originally made when the deal had been announced in 1875. Yet having (incompletely) withdrawn from the Transvaal, he soon (incompletely) interfered in Egyptian affairs. The Khedive had been forced to sell his canal shares because his personal finances were in a chaotic state, and so were those of the whole country, and in 1879 the management of Egypt's international debts and tax revenues was put under the 'Dual Control' of British and French officials. Such humiliating foreign intervention provoked a nationalist-cum-religious backlash, led by Colonel Arabi, and the British officials in Cairo warned that Dual Control was breaking down and that Egypt's public finances were on the brink of total collapse. Since the British took four-fifths of Egypt's exports, provided nearly half of its imports, and held one-third of the country's public debt, this was cause for serious concern. Moreover, for Gladstone, public finance was both a fiscal and a moral issue (and some of his relatives were Egyptian bondholders). In May 1882 the British and French navies arrived off the coast of Alexandria to protect their financial interests, which led to an anti-Western riot in which six British-born people were killed. Soon after, the admiral in command of the British squadron was given vague orders and decided, once the French had recalled their own ships, to bombard Alexandria. The cabinet subsequently authorized the sending of troops, led by General Wolseley, and the Egyptians were defeated at the Battle of Tel-el-Kebir in September. Arabi was exiled to Ceylon, and Britain assumed indirect but virtually complete control of Egypt exercised by Sir Evelyn Baring, the British Consul General from 1883 to 1907; and despite repeated protestations that the British were on the brink of leaving, it would eventually be declared a protectorate on the outbreak of war in 1914.

In the short run, the British occupation of Egypt was extremely popular; but such aggressive and peremptory behaviour was almost completely at variance with Gladstone's much-vaunted principles of liberal internationalism, non-intervention and support for the Concert of Europe. It was also noticeably less successful in the medium term. The coercion of Egypt set a new precedent for belligerent intervention in African affairs that would soon be emulated by other powers: especially France, which felt duped and deceived by Britain's peremptory bombardment and occupation, and Germany, which under Bismarck would also be eager to join in what would soon become the clamour for colonies. For the next decade and a half Anglo-French tensions would be considerable, and Bismarck would delight in stoking them, which in practice meant that little remained of the Concert of Europe. But this was not the only adverse consequence of the Egyptian occupation, for it provoked a crisis in neighbouring Sudan, which was also nominally part of the Khedive's dominions. Even more than in Egypt, there was popular Muslim resentment at the Westernizing tendencies that had intensified since the British occupation of Egypt, and these found a focus in the leadership of the Mahdi, a quasi-religious leader who raised the standard of revolt. The Khedive organized an army to suppress the rebellion, but the Mahdi's forces destroyed it late in 1883. Gladstone's government decided to evacuate those Britons remaining in the Sudan, and Major General Charles Gordon, a strange, complex, religiously obsessed character, was put in charge. He arrived in Khartoum in February 1884, but soon found himself besieged by the Mahdi's forces. The cabinet, preoccupied with parliamentary reform, belatedly organized a relief expedition, again led by General Wolseley, but it reached Khartoum in January 1885 two days after Gordon had been killed.

Following Isandlwana and Majuba Hill, this was a third colonial reversal for the British in the space of six years, and on learning of it in February 1885, Queen Victoria's outrage echoed the popular mood. 'The news from Khartoum,' she telegraphed Gladstone, Granville and Hartington, 'are frightful, and to think all this might have been prevented and many precious lives saved by earlier action is too frightful.' Gladstone the 'Grand Old Man' (GOM) was derided and lampooned as the 'Murderer of Gordon' (MOG), while Gordon became an instant imperial martyr. But Gladstone rode out the public storm and the parliamentary censure (by fourteen votes), not least because another imperial crisis helped divert attention away from the Upper Nile. In 1880 Gladstone had sent Lord Ripon to India to succeed Lord Lytton, who had been Disraeli's Viceroy, and one of Ripon's tasks was to try to calm relations between India and Afghanistan. This he seemed to have managed, by recognizing Abdur Rahman, a nephew of Sher Ali, as Emir; and in return the Emir acknowledged British control over his foreign policy, although he was not required to accept a British resident in Kabul. But Russia's expansion in the region continued, and in March 1885 the Tsar's armies defeated an Afghan force at the Battle of Panjdeh. Here was another crisis on the frontiers of the British Empire, following hard on the news of Gordon's murder, and Granville's reaction was little short of panic-stricken: 'It is too dreadful,' he wailed, 'jumping from one nightmare to another.' But this time a better solution would be reached, although a final settlement was only brokered during Salisbury's brief minority government later in the year. The Russians gained Panjdeh, but they were thwarted in their wish to gain the important town of Herat, and Britain had not been drawn further in on the northwest frontier.

But this successful settlement, and the semi-retreat from the Transvaal, were rare examples of Gladstone's anti-expansionist policy actually working. During the five years of his second administration the pace of colonial annexation markedly increased, especially in Africa, and not just on the part of Britain. King Leopold of the Belgians had been showing an interest in the Congo region since 1876; France had begun to intervene in Senegal in 1879 and occupied Tunisia two years later; and Bismarck, eager to find places in the sun for Germany, was eyeing parts of southwest Africa, East Africa and the Cameroons. These incursions, combined with the sudden British occupation of Egypt, set off a chain reaction that (to mix the metaphor) would soon become known as the 'Scramble for Africa', as the European powers sought to partition the continent between them. The motives for this sudden burst of annexation, acquiring vast swathes of a continent about whose peoples and resources very little was known, were complex and varied. But in every case it seems as though they were driven more by anxiety and alarm than by hubris and optimism. The onset of economic recession, combined with the widespread imposition of tariffs and a growing sense of commercial competition, helped produce a climate in which it seemed ever more urgent for the European powers to make pre-emptive colonial strikes; while the pressure from the men on the spot for increased government support mounted inexorably. This was especially so in the British case, where vital strategic interests in Egypt and South Africa, both essential staging posts on the route to India, meant it seemed equally essential to forestall and keep out potential rivals, especially France, Germany and Portugal. The result was that British policymakers, regardless of party-political affiliation, became preoccupied during the 1880s with securing the headwaters of the Nile, and keeping a route open northwards from Cape Colony and Natal.

Perhaps surprisingly, many of these colonial tensions were settled relatively amicably at another Congress of Berlin, which met between November 1884 and February 1885. France, Germany and Belgium got much of what they wanted, while British claims to Somaliland (at the southern end of the Red Sea and opposite Aden), to Bechuanaland (keeping the northern route open from the Cape) and to Nigeria (protecting West African trading interests) were ratified. In the febrile atmosphere of the time these might have seemed essential acts of imperial pre-emption, but for a British government elected to rein in the excesses of 'Beaconsfieldism' this was an astonishing record. Indeed, the Governor of Cape Colony, Sir Hercules Robinson, thought it strange that the Liberals, who 'came in on the platform of curtailing imperial responsibilities should be likely to add more to them than any previous ministry'. Moreover, such unprecedented imperial acquisitiveness hardly squared with Gladstone's oft-repeated claim that, as an ardent liberal internationalist, he was emphatically on the side of 'people rightly struggling to be free'. This might have been his view of the American South (briefly), of Italy (for much longer) and (perhaps) of Ireland, but he did not think this of Germany, Egypt or other parts of Africa. Yet who was he to decide whether such struggles for freedom were right or wrong, and on what basis or criteria? The reality was that in an era of growing national rivalries, in Europe and overseas, the process of extending the British Empire had acquired an almost irresistible momentum, and there was nothing Gladstone could do to stop it, least of all in the case of Africa. By the mid-1880s the continent had become a British (and European) obsession. As Lord Salisbury later observed, when he had left the Foreign Office in April 1880 no one gave serious thought to that part of the world, whereas when he returned there in the summer of 1885 people were talking of little else.

But it was not just Africa that was partitioned during these years. In the Far East there were similar powerful pressures for Gladstone's government to annex and acquire new colonies to fend off rivals, especially the French in Indochina and the Dutch in Sumatra. In the case of Borneo the British were eager to prevent the northern part of the island falling into foreign hands, because of its strategic importance on the flank of the vital sea routes from Singapore to China. But they were also keen to limit and to outsource their responsibilities, so in 1881 the government chartered the British North Borneo Company, which would administer the area under the nominal authority of the Sultans of Sulu and Brunei (and would also provide the precedent for subcontracting British imperialism out to private enterprise in large parts of

Africa). Elsewhere in the region there were Australian anxieties about the active presence of the Dutch in the East Indies, especially in western New Guinea, which at its closest was less than a hundred miles from the Australian mainland. Accordingly, in 1883 antipodean representatives visited Lord Derby, the Colonial Secretary, in London, urging that Britain annex Samoa, New Guinea and the New Hebrides. Derby was appalled: 'I asked them,' he wrote to a friend, 'whether they did not want a planet all to themselves, and they seemed to think it would be a desirable arrangement if only feasible.' 'The magnitude of their ideas,' he concluded, 'is appalling to the English mind.' But not, it seemed, to the Australians, for despite Derby's strong opposition the Australians seized part of eastern New Guinea to forestall Dutch annexation, and declared a protectorate over it. Once again, the men on the spot had pushed the boundaries of empire forward, in successful defiance of London's non-expansionist preferences.

Even during Salisbury's brief minority government, which held office during the second half of 1885, further annexations continued, this time in South Asia. The prime minister had appointed Lord Randolph Churchill, then only thirty-six, to his first cabinet post as Secretary of State for India. Lord Randolph had denounced Gladstone's interventionist Egyptian policy, but he showed no such restraint once he had obtained high office. Eager to make a splash, he annexed Upper Burma in what was essentially another display of acquisitive pre-emption, and which displayed no respect, as the British had undertaken to do since 1858, for the rights of a native ruler. The French had been showing interest in the region from their adjacent possessions in Indochina, and King Thibaw had made friendly overtures to them, even confiscating British property and transferring it to the French. He also issued a proclamation calling for the liberation of British-controlled Lower Burma. The British response to these threats was to declare the king to be a tyrant who reneged on his treaty obligations, and in October 1885 he was sent an ultimatum. There was no reply to it, and the following month the British sent 10,000 troops to fight what would be the very brief third (and final) Anglo-Burmese War. They captured Mandalay, King Thibaw was deposed and sent into exile, and Lord Randolph presented Upper Burma to Queen Victoria as a New Year's Day present on 1 January 1886. It would be the United Kingdom's last substantial imperial acquisition on the Indian subcontinent. By extraordinary coincidence, on 28 December 1885, just four days before Lord Randolph offered Upper Burma to the Queen-Empress, the Indian National Congress was founded. Its initial aim was to lobby for increased indigenous representation in the British-controlled government of the subcontinent and in the Indian Civil Service; and it would subsequently and successfully campaign for Indian independence. One of the founders of the Congress was Dadabhai Naoroji, who would be the first Asian elected to the British parliament in 1892 where, as a Liberal, he would speak in favour of Irish Home Rule.

At least in retrospect, the early to mid-1880s were a pivotal time in Britain's imperial history, and in more ways than one. With the generally reluctant acquiescence of the men in London, the empire was expanding more rapidly than ever; but it was also facing unprecedented nationalist challenges, in Ireland, Egypt and India, which might portend an imperial 'disintegration' to parallel and reinforce the one that Lord Salisbury feared at home. So it was scarcely surprising that in 1884 the Imperial Federation League was founded to campaign for just the sort of colonial consolidation that Seeley had called for in *The Expansion of England*, published in the preceding year, and for which the Confederation of Canada might provide an appropriate model and precedent. The League spawned branches in Canada, Australia, New Zealand and the West Indies, and it was supported in Britain by Conservatives, Unionists and Liberal imperialists; but when it tried to move from vague objectives to particular policies, it fell apart and was dissolved in Britain in 1894. Yet it could claim one limited achievement, for it was the League that had called for the first conference of colonial prime ministers at the time of Queen Victoria's Golden Jubilee in 1887. The conference duly took place, attended by more than one hundred delegates, from both the self-governing and dependent colonies (although India was not represented). It was only a deliberative body, and its resolutions were not binding. But the colonies in Australia and New Zealand did agree to pay £126,000 per annum towards the cost of the Royal Navy, to help fund its deployment in the Pacific, and in return the British government agreed not to reduce the navy's Pacific Station without the consent of the colonies. The conference also approved a proposal to lay a telegraph cable between Canada and Australia, the final link in an imperial network of communication encircling the globe.

By then, Salisbury had been back in power since the summer of the previous year, and he was faced with the further intensification of imperial acquisitiveness, about which he was no more enthusiastic than he had been in earlier times. He had no great hopes for the colonial conference of 1887, nor for the further demands in respect of additional imperial expansion by which he feared he would be confronted. 'I will do my best,' he told a colleague, 'to keep my temper, but the *outrecuidance* [presumption] of your Greater Britain is sometimes trying.' And when urged by the Australians to take over the New Hebrides, his anger boiled over. 'They are,' he complained, 'the most unreasonable people I have ever dreamed of.' They wanted, he went on, for Britain to incur 'all the bloodshed, the dangers and the stupendous cost of a war with France, of which almost the exclusive burden will fall upon us'; and all this 'for a group of islands which are to us as valueless as the South Pole, and to which they are only attached by a debating club sentiment'. Salisbury was quite right to doubt the worth of such acquisitions, since the small Pacific Islands were useless as markets, produced little except coconuts, and lacked mineral resources that might be profitably extracted. Yet although he fended off the annexation of the New Hebrides for the time being, Salisbury could not prevent further imperial expansion in the Pacific. In 1888 and 1889 several uninhabited islands were annexed by the British as possible relay stations for the trans-Pacific cable that would soon be laid; and although the Admiralty thought them to be of no use, the Gilbert and Ellice Islands were declared a British protectorate in 1892, on the by now familiar pre-emptive grounds that this was to avoid 'the alternative of letting the Germans into our sphere of influence'.

But it was in Africa, in the aftermath of the Congress of Berlin, that the seemingly inexorable forces leading to further imperial expansion were at their most unrestrained. In West Africa, East Africa and South Africa there were strong pressures on the government in London from the men on the spot to safeguard what were deemed to be Britain's vital strategic and commercial interests. Salisbury's response, following the recent precedent of the North Borneo Company, which itself harked back to the example of the East India Company, was to try to extend the British Empire on the cheap; he would do this by subcontracting these areas of it out to private enterprise, in the hope that chartered companies would be able to make sufficient profit from the exploitation of natural resources to defray the expenses of administration, at no cost to the British exchequer. In 1886 the government bestowed a charter on the Royal Niger Company, led by Sir George Goldie, which established British claims to the area of the Lower Niger River, against competition from the French and the Germans. Two years later, in the aftermath of an agreement reached by the British and Germans as to spheres of influence, the Imperial British East Africa Company was given a charter, under the leadership of Sir William MacKinnon, to oversee the development and government of what would eventually become Kenya and Uganda; this included the aptly named Lake Victoria, widely believed to be the source of the River Nile. And in 1889 the fabulously rich Cecil Rhodes, who had made a prodigious fortune in diamonds and gold, obtained a charter for his British South Africa Company, which he would use to open up a British imperial route to the north, to exploit what were believed to be the rich gold deposits in Mashonaland, and to fend off Portuguese expansion from neighbouring Angola and Mozambique.

In fact, and as the far from encouraging precedent of the East India Company should have suggested, none of these three ventures worked in the way that had been intended, because in no case did the anticipated mineral resources eventuate. Even the British South Africa Company, which had Rhodes's millions behind it, failed to become a profitable enterprise; and because government oversight was minimal, the conduct of the company employees vis-à-vis the native peoples was often brutal. In the end, subcontracting the empire out to private enterprise did not pay and did not work, and sooner or later, albeit very reluctantly, the British government would have to step in, formally annexing territory, taking over the administration, and converting protectorates into crown colonies. Meanwhile, between 1886 and 1892, Lord Salisbury negotiated a series of treaties with other European powers, building on and consolidating the earlier agreements reached at Berlin, which more precisely defined the spheres of influence where the new British chartered companies would operate in Africa, and also made some particular territorial adjustments in several parts of the world. As a result, the British and French, despite Salisbury's earlier forebodings, came to an arrangement whereby they effectively shared control of the New Hebrides via a joint naval commission; and Britain gave up the island of Heligoland in the North Sea to Germany in exchange for obtaining the island of Zanzibar. Britain and Germany also agreed to divide their spheres of influence on mainland East Africa by drawing a line from the coast, south of Mombasa to Lake Victoria; while Britain and Portugal came to terms in Central Africa, as a result of which the British South Africa Company was given free rein in what became Northern and Southern Rhodesia, Nyasaland was declared a British protectorate, and Portuguese attempts to block Britain's northern expansion were thwarted.

Although by this time the Conservatives and their Unionist allies were well established as the defenders of the integrity of the United Kingdom and as the people most wedded to the cause of empire, Salisbury was in fact no more eager to increase Britain's imperial burdens and responsibilities than Gladstone had been; but he was also no more successful than Gladstone in implementing such a cautious policy. Even if no one knew what the economic potential might be of far-off Pacific Islands or large swaths of sub-Saharan Africa, or even if they doubted there might be any economic potential at all, no British government was prepared to run the risk of letting France, Germany or Portugal acquire lands which might, eventually, turn out to be of great economic value, as had been the case when diamonds and gold were discovered in South Africa. Hence the need to forestall other European powers, by pre-emptive annexation. Such views were also shared by those who formed the imperialist wing of the Liberal Party for, as Lord Rosebery insisted, acquiring more territory was a way of 'pegging out claims for the future . . . to take our share in the partition of the world which we have not forced on, but which has been forced on us'. But simultaneous developments in other parts of the empire were tending in rather different directions. In South Asia, and in response to the pressure exerted by the recently established Congress, the Indian Councils Act was passed in 1892, empowering universities, district boards, municipalities and chambers of commerce to recommend members to provincial councils (though since these bodies had very limited powers, this was scarcely a major concession). In the same year Western Australia was granted responsible government, which gave impetus to the movement urging the federation of all six Australian colonies.

For the Liberal and Conservative governments alike, there were at least two paradoxes in their imperial policies during these years. The first was that neither party, when in government, wanted to annex more territory or acquire more colonies, but that was in fact what happened: such was the disjunction and the disconnect between the 'official mind' at home and the 'men on the spot' overseas, and it was often the latter who prevailed over the former. The second paradox was that while this sudden expansion of the formal British Empire seemed to betoken an imperial nation extending its global reach as never before, in reality the opposite was the case. For this late nineteenth-century phase of British imperialism was more defensive and pessimistic than it was aggressive and hubristic, trying to preserve some of its overseas positions in the face of mounting competition from other European powers. Yet, and perhaps this is even a third paradox, this was not how it seemed to many Britons at the time. The jingoism of the music halls was also reflected in a whole new genre of imperial adventure novels. They may have owed something to Wilkie Collins's *The Moonstone*, but they offered much less nuanced accounts of derring-do and redemption on the frontiers of empire, beginning with Robert Louis Stevenson's *Treasure Island* (1883), set in the Caribbean, and H. Rider Haggard's *King Solomon's Mines* (1885), located in Africa. The popular appetite for such imperial fictions would intensify during the remainder of the century, although it remains unclear just how deeply the majority of the United Kingdom's population engaged with their empire. *They* may have been enthusiastic for it, but most politicians were not. Here was another indication of how opinions and beliefs had fragmented during these years. In that sense at least, the epithet 'disintegration' was indeed valid.

EXIT LIBERALS

On 15 August 1892, now aged eighty-two, Gladstone became prime minister for the fourth time, an event that Queen Victoria predictably greeted 'with much regret'. He was in good health for someone of his very advanced years, but his eyesight was failing, and he was the oldest man to head a government in modern British history. For the last twenty years he had repeatedly hovered on the brink of retirement, and it was only his determination to carry Irish Home Rule that had kept him in public life since 1885. But since his last definitive electoral victory in 1880, the big Commons majority he needed to have any chance of passing such a measure had repeatedly eluded him. To his great regret he now found himself at the head of a minority government, and he was only able to form what everyone recognized must be his last administration with the support of the Irish Nationalists, which gave them a comfortable, but not wholly reliable, majority over the Conservatives and Unionists. As a result of the Whig desertions over Home Rule, Gladstone's last Liberal cabinet was the least aristocratic of the nineteenth century. Lords Kimberley, Ripon and Spencer were back, as was Lord Rosebery at the Foreign Office, although he was maddeningly difficult and indecisive before finally accepting. Two other patricians who reappeared were Sir William Harcourt, who returned to the Exchequer, and Sir George Otto Trevelyan, who went back to the Scottish Office. But almost half of the cabinet consisted of middle-class professionals or businessmen, among them the former journalist John Morley, who was Chief Secretary for Ireland (as he had briefly been in 1886); James Bryce, a lawyer and academic, who was Chancellor of the Duchy of Lancaster; Henry Fowler, a Wolverhampton solicitor, who was President of the Local Government Board; and the young H. H. Asquith, who had enjoyed a brilliant career as a barrister, and was given his first cabinet post as Home Secretary.

In his final speech on the second reading of the first Home Rule Bill, Gladstone had described the opportunity of solving the Irish question as 'one of those golden moments in our history' that might never return. Now he had his second chance, and the issue of Home Rule dominated – and consumed – the eighteen months of his final period in office. To his admirers, this was an heroic cause and a noble campaign, and as such it was the appropriate culmination of the Grand Old Man's life and work; but to his detractors, Gladstone merely presented the sad and pitiful spectacle of someone desperately clinging on to power, and obsessed with a measure that he had no realistic chance of passing into law, since the peers were bound to throw it out. As it eventually emerged, the second Home Rule Bill differed significantly from its predecessor. Eighty Irish MPs would continue to sit at Westminster, but they would only be allowed to vote on Irish issues. Once again there would be a separate Irish parliament, though its composition and its workings were different from what had been proposed in 1886. As before, Gladstone failed to recognize the fears and opposition of Protestant Ulster, which had significantly intensified since the first bill. He introduced his revised scheme in the Commons in February 1893, with another gigantic speech, and even his critics and opponents conceded they were witnessing a remarkable parliamentary spectacle, the like of which they would never see again, and indeed never did. Nevertheless, the Unionist opposition fought the measure clause by clause, and it was only with the support of the Irish Nationalists that it eventually passed its second reading in April by 347 votes to 304. But when the bill reached the House of Lords in the autumn of 1893, it was defeated by 419 votes to 41, the largest majority ever against just a single measure.

That rejection had been a foregone conclusion. Salisbury and Chamberlain were jubilant, and so was the queen: for she was as hostile to Irish Home Rule at the end of her reign as George III had been to Catholic Emancipation towards the end of his. In neither case was such royal obstinacy justified, and even George V, a sovereign not known for his radical views, conceded as much in his grandmother's case: 'What fools we were,' he opined to Ramsay MacDonald, 'not to have accepted Gladstone's Home Rule Bill.' But Gladstone's second failure to give Ireland a measure of freedom meant his last great political crusade was over: there was nothing more he could do, and his increasingly impatient and disaffected colleagues also wanted him gone. Yet he would not depart quietly; indeed, if Gladstone had had his way, he would even then not have departed at all. Anticipating what would happen twenty-five years later, in the aftermath of the rejection of Lloyd George's 'People's Budget' by the upper house, he wanted to dissolve parliament and fight another general election on the single issue of 'Peers versus People'. But while their lordships (like their sovereign) had been intransigent and high-handed in opposing Home Rule, they were also more in harmony with English public opinion on that subject than was Gladstone, and this meant he could not carry his cabinet with him over dissolution. Soon after, he fell out with his colleagues over Lord Spencer's proposal that naval expenditure should be increased. It was a quintessentially Gladstonian issue, for he had always believed that government spending should be held down, especially in the case of the armed services. But the 1890s were not the 1860s: international relations were more tense than they had been thirty years before, which meant more resources had to be committed to the protection of the nation and empire.

This second rebuff by his colleagues meant Gladstone had no choice but to resign for what was to be the last time: thus ended not only his fourth and last administration, but also the longest and most extraordinary career in nineteenth-century British politics. There was an entirely predictable misunderstanding between Gladstone and the queen over his final leave-taking, and with spectacular ungraciousness she refused to offer him a peerage on the grounds that she knew he would not accept it. She also declined to ask Gladstone's views about his successor, and it was no accident that she appointed someone as unlike him as any other Liberal could possibly be: the fifth Earl of Rosebery, who had been Gladstone's last Foreign Secretary, and who was her first and only prime minister to have been born after she had become queen. Privileged, courtly, wealthy and highly intelligent, Rosebery had not only married a Rothschild, but he was also an accomplished historian and man of letters, and a great patron of the turf, whose horses would win the Derby twice while he was prime minister. But such aristocratic grandeur, plutocratic connections and raffish pursuits hardly endeared him to the 'Nonconformist Conscience', and his political views were scarcely those of the majority of Liberals. He was also vain, spoiled and petulant, an impossibly difficult colleague, and it was rumoured that he was homosexual (Oscar Wilde would be sent to prison for 'gross indecency' in May 1895, the very month before Rosebery resigned). To be sure, he had been loyal to Gladstone, and had hosted him at Dalmeny, his great house near Edinburgh on the Firth of Forth, when he had waged his 'Midlothian Campaign'. But Rosebery was more committed to the British Empire than he was to Home Rule, he had never been a member of the House of Commons, two of his colleagues (Earl Spencer and Sir William Harcourt) thought they should have been the next prime minister, and he was the unhappy head of a minority government that would only last for fifteen months.

The cabinet over which he presided was virtually identical to Gladstone's last, with Lord Kimberley moving from the India Office to replace Rosebery at the Foreign Office. Not surprisingly, the government accomplished little. At the Exchequer, Harcourt needed to find more revenue to finance (among other things) the increased military expenditure that Gladstone had recently but vainly opposed; in his budget of 1894 Harcourt not only raised the income tax but also imposed a new form of death duties on landed property. Levied at only

8 per cent on the greatest estates, they were scarcely punitive, but these new exactions caused an enormous outcry among the titled and territorial classes (and three years later Harcourt would have to pay death duties himself when he inherited the Nuneham estate in Oxfordshire from a cousin). In foreign affairs the government sided with Japan in its long-running conflict with China, and helped preserve the independence of the kingdom of Siam, which was sandwiched between the British colonies of Burma and Malaya, from the colonial ambitions of France. But as a self-confessed 'Liberal Imperialist', Rosebery did not share the views of his colleagues who regretted the rapid recent expansion of the empire, and in 1894 he formally annexed Uganda because the British East Africa Company, after only seven years of its existence, had already run into serious financial difficulties. By then, the government was visibly faltering, and in June 1895 it was defeated on another issue relating to defence, concerning its alleged failure to procure for the army a sufficient supply of the smokeless explosive known as cordite. Rosebery thereupon resigned the premiership, and he relinquished the party leadership in the following year; he would live on until 1929, and he would often be courted by Liberal politicians who wanted him in their governments, but he never held public office again.

10

Jubilation and Recessional, 1895–1905

Although the queen regarded Lord Rosebery as being more 'sound' on Ireland and the empire than Gladstone, she was not sorry to see him leave office, and Lord Salisbury formed his third administration at the end of June 1895. Rightly sensing that the political initiative was again with him, Salisbury immediately requested a dissolution of parliament. The ensuing general election was a triumph for the Conservatives and Unionists, and a disaster for the Liberals, still unhappily led by Rosebery. Indeed, the 1895 contest was the greatest success for the right in British politics since the passing of the Great Reform Act, surpassing Peel in 1841 and Disraeli in 1874. Altogether, the Conservatives secured 341 MPs, which gave them an impregnable majority in the Commons, something they had not secured since Disraeli's earlier victory. They were further strengthened by the return of seventy Liberal Unionist MPs, with whom they were more closely allied than they had been during the late 1880s or early 1890s; and together the anti-Home Rulers enjoyed a commanding lead over both the other parties. As before, the Irish Nationalists held steady at around eighty MPs, but the Liberals lost almost one hundred seats and were reduced to a mere 177 MPs. Across the United Kingdom the 'Celtic fringe' remained loyal to the left: the Nationalists were dominant in Ireland, and the Liberals in Wales and (although less completely) Scotland. But in England the anti-Home Rulers outnumbered the Liberals by 342 MPs to a mere 112, which vindicated the decision of the House of Lords to throw out Gladstone's bill. Only on the English periphery did the Liberals survive, in the northeast, the West Country and in Norfolk; but across most of the counties, in the suburbs and in greater London, the Conservatives and Unionists swept the board.

Lord Salisbury's cabinet of nineteen was unusually large, reflecting his Tory parliamentary majority and the fact that the Liberal Unionists were now willing to join what was thus a merged Conservative and Unionist administration. As before, Salisbury combined the posts of prime minister and Foreign Secretary, which he would continue to do until 1900. Among the Conservatives, Disraeli's former colleague R. A. (by now Lord) Cross became Lord Privy Seal, and Salisbury appointed his nephew, A. J. Balfour, as First Lord of the Treasury, who as 'Prince Arthur' would become his acknowledged heir apparent. Sir Michael Hicks Beach became Chancellor of the Exchequer, Sir Matthew White Ridley was made Home Secretary, Lord George Hamilton was appointed Secretary of State for India, and Lord Balfour of Burleigh was Secretary of State for Scotland. Among the Unionists, the eighth Duke of Devonshire, previously Lord Hartington, became Lord President, Lord Lansdowne became Secretary for War, and Joseph Chamberlain was the first businessman to be Colonial Secretary. This was a strong cabinet, and there would be no major changes until Salisbury relinquished the Foreign Office to Lansdowne. It was also, despite three Reform Acts, very aristocratic and patrician: nine ministers (including a duke and two marquises) sat in the Lords; and of the ten commoners, three were close relatives of peers, two were baronets, and two were country gentlemen with substantial estates: Henry Chaplin at the Local Government Board and Walter Long at the Board of Agriculture. With pardonable exaggeration, this fin-de-siècle Salisbury government has been described as the last in the Western world 'to possess all the attributes of aristocracy in working order'.

For someone who, only ten years before, had feared political, social and imperial 'disintegration', and looked upon democratic politics with fear and loathing, Salisbury possessed a remarkable capacity to win a majority of the popular votes at general elections, which he effectively did in 1886 and 1895, and would do so again in 1900. Between 1832 and 1918 he was by far the most successful Conservative leader, and there were many reasons for his seemingly surprising electoral appeal. One was the continued weakness of the Liberals: for whereas Gladstone had united the party in 1868 and 1880 by brilliantly (and opportunistically?) espousing great causes that brought together MPs and peers, and also won him widespread popular support, his later obsession with Home Rule merely served to divide the party and alienate great swaths of English public opinion. A second was the 'Great Depression': for while the Tories had suffered electoral defeat largely because of it in 1880, its longer-term impact was to weaken and sunder the left in Britain (and elsewhere in Europe), and to give a significant boost to the right, as would again be the case during the subsequent economic downturns of the 1920s, 1930s, 1980s and the 2010s. A third was the Conservatives' successful identification with the cause and cult of empire, which Disraeli had anticipated and then established during the 1870s, and which was further strengthened during the next two decades, even though Salisbury was a more reluctant imperialist than Disraeli had been or than Chamberlain would turn out to be. Salisbury was also a strong believer in party organization, and he rightly surmised that there was a great deal of support to be mobilized in what he described as Mr Pooter's world of 'villa Toryism'. Underlying these developments was the general shift in attitudes among the propertied classes, especially landowners, bankers and the inhabitants of suburbia, away from their mid-Victorian Liberal internationalism and towards late-Victorian Conservative imperialism. At the same time there was a significant upsurge in working-class jingoism, of which the prime beneficiaries would be the Tory Party, the landlords of public houses, and managers of music halls up and down the country.

Salisbury was more the gainer than the instigator or the architect of these deep-rooted changes: in particular, the divided and demoralized Liberal opposition could not compete with the Conservatives in terms of party funding or organization, and seemed to have run out of policies and ideas, except for Home Rule. But while that tainted Gladstonian relic still resonated on the 'Celtic fringe', it remained very unpopular in England, where the majority of voters elected the majority of MPs. Salisbury was also a ruthless and formidable political operator, as Lord Randolph Churchill had earlier discovered to his cost, and he was skilful at keeping his cabinet colleagues together, in giving them the leeway to get on with their departmental business, and in consolidating the coalition between the Conservatives and Unionists. From 1895 the politician who derived most advantage from these prime-ministerial attributes and accomplishments would be Joseph Chamberlain. Along with George Goschen at the Admiralty, and C. T. Ritchie at the Board of Trade, he was one of the few authentically middle-class figures in Salisbury's grandee-laden cabinet. Having refused the offer of the

Exchequer, Chamberlain had settled on what had been until then the relatively junior portfolio of the Colonial Office. But no one could have predicted that the man who had begun his parliamentary career as an advanced radical, viscerally hostile to the monarchy and aristocracy, would now embrace loyalism and imperialism with almost messianic zeal and enthusiasm. He would become passionately committed to the expansion and consolidation of the British Empire, and this would make him the most influential figure in Salisbury's administration – but also the most divisive personality in the Balfour government that followed.

WEARY TITANS WAGING WAR

From 1895 to 1902 Britain's relations with the rest of the world would be conducted by Salisbury and then by Lansdowne at the Foreign Office, and for the whole of the period by Chamberlain at the Colonial Office. Although these were years of unprecedented imperial assertiveness and jingoistic display, Salisbury and Chamberlain were both anxious men, who were worried by Britain's increasingly threatened and insecure place in the world. In Salisbury's case this was partly because he was existentially and temperamentally gloomy; but it was also because he had never shared Disraeli's liking for grand gestures in Europe or beyond, even as he reluctantly recognized that it seemed impossible to avoid adding to Britain's overseas possessions – which were already, in his view, too many and too extended. Chamberlain's worries were different, but convergent. He feared that Britain's manufacturing supremacy, which he believed essential for the maintenance of great-power status, was being lost, and he regarded the recent structural changes in the economy, away from industry and entrepreneurship and towards banking and commerce, as a sign of national decadence. At worst, and following Sir John Seeley in *The Expansion of England*, he worried that Britain might replicate the downward trajectory already traced by such once great maritime empires as Venice and the Netherlands. At best, he feared Britain was shouldering imperial burdens that were too heavy for it to bear alone, even as he wished at the same time to add to them. Either way, Chamberlain believed that the future lay with 'great empires' rather than 'little states'; and he was determined to consolidate the United Kingdom, and its diverse and scattered portfolio of maritime colonies and territories, into a secure and unchallengeable global empire, as the essential antidote to what would otherwise be unavoidable national decline and imperial recession.

This meant that for Salisbury, Lansdowne, and especially for Chamberlain, the dominant issues of the day were those of foreign affairs, international relations, war and peace, and imperial conquest and administration. Before 1900 the various differences that arose between Britain and the United States, or with France, Germany, Russia or Portugal, would be more concerned with colonial than with continental matters, and they were often dealt with by Chamberlain rather than Salisbury. One such issue blew up within months of the new government taking office. Despite earlier efforts to settle the remaining boundary questions between Canada and America, relations between the United States and the United Kingdom continued vexed and volatile; and in December 1895 President Grover Cleveland sent to Congress what was in effect an ultimatum to the British government concerning another long-standing border dispute in the western hemisphere, that between British Guiana and Venezuela. Invoking the Monroe Doctrine, which had declared that no European powers should intervene in the affairs of the Americas, Cleveland peremptorily insisted that a United States commission would decide on the contested boundary, and that their decision would be accepted and indeed imposed, by force if needs be, regardless of how the British government might respond. This was a serious challenge to the United Kingdom's transatlantic imperial position and possessions. Cleveland duly appointed his commission, with which the Salisbury administration, having no alternative, deemed it prudent to co-operate. As on previous occasions, the issue was referred to international arbitration; and as it turned out, the principal British claims were confirmed when the award was finally promulgated in October 1899.

This was a relatively minor colonial fracas compared to the events that were unfolding in South Africa. Cecil Rhodes, who was not only a multi-millionaire and head of the British South Africa Company, but had also become prime minister of Cape Colony in 1890, was determined on further expanding British territories in the region. Four years earlier prodigious quantities of gold had been discovered near Johannesburg, which soon transformed the Transvaal from a backward territory into a prosperous state; and to the dismay of Rhodes and Chamberlain this threatened further to upset the unstable relations between the two British colonies and the two Boer republics. From the British point of view, matters were made worse because the Boers felt a closer affinity with the Germany of Kaiser Wilhelm II (who had succeeded on the death of his father, the Emperor Frederick III, in 1888) than with the British Empire of Queen Victoria (although she was, in fact, the Kaiser's grandmother), to which, on the basis of their previously unhappy dealings with London, they had no wish to belong. Meanwhile, more than 40,000 fortune-seeking Britons had poured into the Transvaal, where they were known as Uitlanders (or 'outlanders'), since the Boers denied them equal political rights so as to keep control of their own country. In late December 1895 a friend and associate of Rhodes, Dr Leander Starr Jameson, galloped into the Transvaal at the head of an armed body of men, with the aim of igniting a Uitlander rising that would overthrow the Transvaal government. Rhodes had been behind Jameson's raid, and Chamberlain had known of his efforts to encourage the rising, though he was not directly responsible for it. But the raid was not only an illegal incursion into a foreign sovereign state, with the aim of overthrowing a legitimately elected government: it was also a military and public-relations fiasco, for the Uitlanders did not rise, Jameson surrendered, Rhodes was obliged to resign the premiership of Cape Colony, and the Kaiser sent a telegram to President Kruger of the Transvaal, congratulating him on the failure of Jameson's attempted coup.

Undismayed, Chamberlain sought to force the pace of imperial acquisition elsewhere in Africa. On the western side of the continent, and inland from the British colony of the Gold Coast, he dispatched a military expedition under Colonel Sir Francis Scott to subdue the Ashanti, on the grounds that they were still carrying on the slave trade and human sacrifices that General Wolseley had tried to stop in his punitive raid of 1874. The Ashanti capital, Kumasi, was occupied in January 1896, the king was deposed and exiled to the Seychelles, and the remaining chiefs were placed under British control. In nearby Nigeria, Chamberlain encouraged the establishment of a 'small west African army', with the aim of backing some doubtful treaty claims by force, and fending off French incursions in the area. Meanwhile, on the other side of the continent, the British government established a protectorate over Kenya and assumed control of its administration from the British East African Company. Further north, the commander of the Anglo-Egyptian army, General Sir Herbert Kitchener, began organizing an expedition to conquer the Sudan, thereby avenging the murder of General Gordon at the hands of the Mahdi, and securing the Nile Valley for Britain. Neither Salisbury nor even Chamberlain were initially enthused at this further extension of Britain's imperial responsibilities; but Kitchener advanced by short, relentless stages along the Nile, building a railway as he went, and by 1898 he would have beaten the Sudanese at the Battle of Omdurman, and captured Khartoum. He subsequently faced down the French, who also had designs on the upper Nile, at Fashoda, whereupon Salisbury proclaimed the re-establishment of an Anglo-Egyptian condominium over the Sudan.

It was in this febrile atmosphere of imperial confrontation and colonial expansion that Queen Victoria celebrated sixty years on the throne in 1897. No British monarch had ever attained a Diamond Jubilee before, and the festivities were on a scale far surpassing those held ten years earlier. Chamberlain saw it as an opportunity to raise public awareness of the British Empire, by bringing troops, potentates and premiers from the queen's transoceanic realms to parade on the streets of London, thereby making the empire real, visible, immediate and actual, while across her dominions Victoria's subjects seemed united in similar acts of homage and celebration. The queen drove through cheering crowds to a service of thanksgiving held on the steps of St Paul's Cathedral, while the subsequent naval review at Spithead assembled the largest number of ships from the British fleet ever seen in home waters. Ten years after the Golden Jubilee, the narrative of material progress, democratic advance and imperial expansion was repeated with renewed fervour and conviction (though not in large parts of Ireland, where the Great Famine remained unforgiven and unforgotten). But 1897 was not all national complacency and imperial self-congratulation. The 'Great Depression' had been a blow to the United Kingdom's industrial pride and self-esteem; the international and imperial climate was becoming ever more tense and competitive; and questions were being asked as to whether Britain's imperial prowess and global pre-eminence could last. On the very day that the queen drove in triumph to St Paul's, Rudyard Kipling made this point insistently and unsettlingly in his poem 'Recessional', written specially for the occasion. Far from being a hymn to triumphalism and imperialism, Kipling dwelt instead on the transience of worldly power and the ephemerality of earthly dominion.

These ambiguities and paradoxes were much in evidence throughout the year of the Jubilee. At the conference of colonial premiers Chamberlain made little headway in his calls to convert widespread imperial sentiment into a consolidated imperial structure. He floated the idea of a federal council as a prelude to imperial federation, but with the exceptions of Tasmania and New Zealand, the colonies made plain their preference that things should stay as they were. He achieved no more progress in advocating a single policy of imperial defence, and a unified naval force to carry it out: Cape Colony was willing to pay the cost of a first-class battleship for the Royal Navy, but the Australian colonies wanted more ships in their own waters and under their own command. Nor did Chamberlain obtain support for an imperial *Zollverein*, or customs union, which would have created a free-trade zone encompassing the whole empire, sheltered from the rest of the world by high protective tariffs. Many of the colonies had already embraced protection, and the most they were willing to concede was some more limited form of imperial preference. But while the Jubilee, to Chamberlain's great disappointment, gave no added impetus to the cause of imperial unity, whether political or military or fiscal, it did mark the beginning of the move towards the federation of Australia. The six colonies (New Zealand declined to join) shared language, lineage, laws, institutions and traditions; they were increasingly concerned about French and German penetration into the Pacific; and the severe economic downturn of the 1890s, ending long years of boom time, had been a major jolt. In 1897 a federal convention was established to produce a scheme acceptable to all the colonies, and eventually, in July 1900, the Commonwealth of Australia was brought into being by British legislation, thereby creating, one-third of a century after Canada, the second great unified Britannic community overseas.

The year of the Jubilee also witnessed growing pressure by Chamberlain on the two Boer republics, as he was determined to coerce them into a British-dominated South Africa. The House of Commons had belatedly appointed a select committee of inquiry into the Jameson Raid, but Chamberlain was a member, when he should more properly have been under investigation; it failed to press hard for the truth, and the final report censured Rhodes but acquitted the Colonial Secretary and the Colonial Office. Meanwhile, Chamberlain had appointed the vigorous and aggressive Sir Alfred Milner as the new British High Commissioner at the Cape. Milner had previously worked in Egypt under Sir Evelyn Baring, he sympathized with the Uitlanders' grievances, and he shared Chamberlain's expansionist ambitions. As a result of his more assertive stance, relations with Kruger rapidly deteriorated, and Chamberlain began planning to send more British troops to the Cape, while the Boers and the British disagreed as to what the term 'suzerainty' had meant in the 1881 Convention and whether it still existed. In the autumn of 1899 Kruger issued the British with an ultimatum, demanding they withdraw their troops from the Transvaal frontier where they had recently been deployed, but his terms were not met, and the British and the Boers went to war on 12 October. Until further imperial reinforcements arrived, the Boer forces not only outnumbered the British troops, they were also highly mobile and well equipped with French and German weapons, and they were fighting to preserve their independence. They invaded the Cape and Natal, and laid siege to Kimberley, Mafeking and Ladysmith; and during 'Black Week' in December 1899 the British relieving forces suffered three humiliating reverses at Stromberg, Magersfontein and Colenso. 'We have,' opined Kipling, 'been taught no end of a lesson.'

It had been generally supposed that any confrontation between the world's greatest empire and the two land-bound and far-distant republics would be over rapidly and that the Boers would be crushingly defeated; but the war had unexpectedly begun as a David-and-Goliath contest, and the Boers scored another stunning victory in January 1900 at Spion Kop. In Germany, France and Russia there was widespread *Schadenfreude* at this succession of British defeats and humiliations. The press and public opinion in all these countries were stridently anti-British, and there were calls for armed intervention by the European powers in support of the Boers. In the run-up to the war, and during its first disastrous months, the United Kingdom had been diplomatically isolated, but once the British began to mobilize their vastly greater forces and resources the military tide turned to the empire's advantage. The incompetent military commander, Sir Redvers Buller, was replaced by Field Marshal Lord Roberts, a veteran of the Great Rebellion in India and of many later colonial wars; and his chief of staff was General Lord Kitchener (as he had become), fresh from his recent triumphs in the Sudan. British reinforcements began to arrive, and eventually 450,000 imperial troops faced a mere 60,000 Boer soldiers. The three sieges were successfully raised, and the relief of Mafeking in May 1900, which had held out for 217 days, was greeted in Britain with widespread euphoria amounting to jingoistic hysteria. In the same month British troops occupied Pretoria and Johannesburg, the principal Boer cities in the Transvaal, and they overran the Orange Free State. By September both Boer republics had been annexed by Britain, and Kruger fled to Europe in what would prove to be a vain attempt to secure the military support that had earlier been hinted at but was not in fact forthcoming.

By the autumn of 1900 it seemed as though the United Kingdom had won the Boer War, and Salisbury took the opportunity to obtain an early dissolution of parliament in the hope of exploiting the popular patriotic mood: hence the term 'khaki election', alluding to the colour of the new uniforms worn by the British army in South Africa. The result was virtually identical to that of 1895, and Salisbury became the first party leader to win two consecutive contests since Lord Palmerston, and he had done so on a much narrower franchise. The Conservatives and their Liberal Unionist allies won 402 seats, which was nine fewer than in 1895 but still an impressive showing; while the Liberals won 184 seats, which was seven more than they had managed five years earlier but no significant improvement. The Irish Nationalists remained at their usual number of eighty-odd MPs, while the fledgling Labour Party, founded earlier that year, won two seats. As such, the 1900 election represented the high-water mark of the late Victorian Conservative Party: more support from the middle classes, an even stronger showing in greater London, and also winning a majority of seats in Scotland. Salisbury made some changes to his cabinet: he relinquished the Foreign Office to Lord Lansdowne, replaced him at the War Office with St John Brodrick, brought in Gerald Balfour, his nephew and Arthur's brother, as President of the Board of Trade, and moved C. T. Ritchie to the Home Office. He also gave minor preferment to his eldest son, Lord Cranborne, the future fourth marquis. The social composition of the cabinet remained essentially unaltered, and it contained so many of Salisbury's relatives that it was dubbed the 'Hotel Cecil', after the building recently constructed in the West End on land his family owned.

Yet for all the jubilation that the Boer War was being won, and despite this second Tory electoral triumph, there was an inescapable air of 'recessional' and 'fin-de-siècle' about the years after 1897. Two of the greatest figures of the queen's reign had already died: Tennyson in 1892, and Gladstone six years later, unlamented and still unforgiven by the queen. Even though the foundations had been in some ways fragile and fortuitous, the nineteenth century had 'belonged' to the United Kingdom more than to any other power; but as it ended, there was serious concern, which would turn out to be well founded, that the twentieth century would 'belong' elsewhere. These forebodings about the future were further intensified in January 1901 when Queen Victoria died, after a record-breaking reign of more than sixty-three years. Since average life expectancy remained little better than forty, there were scarcely any inhabitants of the United Kingdom or the British Empire who could remember a time when she had not been on the throne. She had given her name to her age, bestowing on it what was in many ways a misleading unity; her reign had coincided with Britain's economic pre-eminence and global greatness; and during the last two decades of her life she had been venerated and apotheosized as the Gas-Lit Gloriana, the doyenne of European royalty, and the matriarch of the world's largest empire. Like many of her subjects, she had evolved from being something of a Liberal internationalist into something of a Tory jingo, though hers might have been a very different political trajectory had Prince Albert lived another thirty years. And although she was admired and adored by the 1880s and 1890s, she lamented that she could no longer influence public affairs in the way she and her husband had sometimes done during the early years of her reign. Victoria had never wished to be a 'democratical monarch', gaining in ceremonial splendour but losing in any real capacity to determine events; but that was her fate – and in the end also her salvation and apotheosis.

The Great White Queen would be thwarted in another deeply held wish, namely that her deplored and detested eldest son, Albert Edward, the Prince of Wales, should never succeed her. She almost outlived him; but not quite. A philistine, a glutton, a gambler, and a rampant and unscrupulous sexual predator, but also someone of enormous charm and considerable untutored intelligence, the young prince had early on rebelled against his parents' cosy and *gemütlich* 'Balmorality'; and he had also rejected their impossibly demanding educational training, which had been designed to prepare him to be a monarch whose role would be fundamental to government rather than merely ornamental to society. But it was an inadvertently prescient rebellion, for by the time 'Bertie' inherited the throne there was far less for him to do than Victoria and Albert would have wished there to be. The tasks that remained were, in Walter Bagehot's constitutional taxonomy, more dignified than efficient, and Edward VII (he refused to be known, as his mother had wished, as King Albert I) was very good at them. Yet this was not immediately apparent, as his coronation, scheduled for June 1902, was postponed at the last minute because the new monarch, already in his seventh decade, became ill with appendicitis, and had to undergo what was then a major and risky operation. But he was sufficiently recovered to be crowned two months later in Westminster Abbey. Although the service was scaled back in recognition of his temporarily weakened state, it was much grander and more imperial than his mother's coronation, the ceremonial was better planned and performed, and two works, Sir Hubert Parry's anthem 'I was glad' and Edward Elgar's 'Coronation Ode', were inspired compositions.

Despite, or perhaps because of, the autumnal undertones of the Diamond Jubilee and the early military reverses of the Boer War, those years also witnessed the apogee of strident imperial self-consciousness. As A. C. Benson put it, in words he wrote for Elgar's 'Coronation Ode': 'Wider still and wider shall thy bounds be set; God who made thee mighty make thee mightier yet.' In the corridors of power in London, and the pro-consular palaces of the empire, there were some men for whom this was both an exhortation and a mission. As Colonial Secretary, Chamberlain was as determined to raise imperial awareness at home as he was to extend and consolidate imperial dominion overseas, and during his term of office the three most assertive British proconsuls of modern times were extending their sway over palm and pine. In Egypt, Sir Evelyn Baring, now Lord Cromer, had been the de facto ruler of Britain's de facto colony since 1883, and he would remain in charge until 1907. In South Africa the recently ennobled Lord Milner continued in charge of Cape Colony and Natal, and also oversaw the incorporation of the two former Boer republics into the British Empire. And from 1898 to 1905 the Viceroy of India was the proud and imperious Lord Curzon, who passionately believed in the enduring significance and moral righteousness of Britain's imperial mission. He thought it was good for the native inhabitants, and he also recognized that the possession of India and the military force provided by the Indian army were the essential props to Britain's global pre-eminence. 'As long as we rule India,' he presciently observed, 'we are the greatest power in the world. If we lose it, we shall drop straight away to a third-rate power.' Appropriately, the coronation durbar that Curzon arranged in the winter of 1902–03 to proclaim Edward VII as Emperor of India was the greatest imperial spectacle yet.

But even during these climactic years of high imperialism there were (pace A. C. Benson) real limits to the bounds that might be set, for the traditional hostility to additional annexations remained deeply embedded in most parts of Whitehall; and this continued reluctance to take on more territory was especially marked, and successfully maintained, in the Far East. Ever since the Opium Wars some China hands had wanted Britain to take over the Celestial Empire as it had already acquired large parts of the Mughal Empire. But there was no appetite in London for what seemed such a quixotic and irresponsible proposal, and that opposition had hardened since the Great Rebellion of 1857. Maintaining British rule in India was difficult enough; extending it to China was beyond the bounds of reason or resources, and from the Taiping Rebellion onwards British policy was firmly in favour of supporting the established Qing regime. Yet there was a brief moment in the aftermath of China's defeat by Japan in 1895, when it seemed as though it might be partitioned, just as Africa had recently been divided up, and that Britain might proclaim a protectorate along the Yangtze River to safeguard its commercial position in Shanghai. The European powers were competing with each other to secure contracts and concessions, and to finance and construct railways, as the Celestial Empire belatedly sought to modernize, and some British traders urged the government to make a pre-emptive territorial strike. But Britain's influence in China was too limited, its international rivals there were too strong, and the country was too vast and resilient, for this to be possible or practicable. Instead, the Salisbury government settled for a ninety-nine-year lease on the so-called New Territories, adjacent to Hong Kong, which would expire in 1997, the hundredth anniversary of Queen Victoria's Diamond Jubilee.

By the time that Edward VII had been belatedly crowned, the Boer War had been brought to a no-less belated close with the signing of the

Treaty of Vereeniging at the end of May 1902, which confirmed Britain's annexation of the Transvaal and the Orange Free State, in return for the rapid granting of some form of self-government. The conflict had dragged on long beyond the 'khaki election', as the Boer commanders had organized a campaign of guerrilla warfare against the British, and their forces repeatedly raided Cape Colony, destroying railway lines and telegraphs and attacking military posts. In November 1900 Lord Kitchener took over as commander-in-chief of the British army and herded 120,000 Boer women and children into what were called 'concentration camps', where more than 10,000 of them died of disease. By sheer weight of numbers and ruthless determination he eventually forced the Boers to submit, but only after the British had spent well over £200 million, and after 5,800 soldiers had been killed and another 23,000 wounded. Lord Salisbury held on as prime minister until the war was formally concluded, hoping to see Edward VII crowned, but the postponement of the coronation meant he called it a day in July 1902. By then, Salisbury was visibly in decline, and he would die the following year, having been the last British prime minister to sit in the House of Lords. Although Joseph Chamberlain implausibly coveted the succession (he was a Liberal Unionist not a Conservative, and thus ineligible to lead the Tories), it naturally and automatically went to Salisbury's nephew and long-time heir apparent, Arthur Balfour. His cabinet closely resembled that of his uncle (although he did bring in Chamberlain's elder son, Austen, as Postmaster General), so the management of the Hotel Cecil remained in family hands.

As Colonial Secretary, Chamberlain had been more associated in the public mind with what was widely regarded as 'Joe's War' than anyone else; but despite the final, successful outcome, the protracted nature of the conflict, and the additional imperial responsibilities that it eventually entailed, had made him more anxious, not less. At the Colonial Conference held in 1902 to coincide with the coronation of Edward VII, he graphically depicted Britain as 'the weary Titan stagger[ing] under the too vast orb of his fate'. 'We have,' Chamberlain told the assembled premiers, 'borne the burden for many years. We think it is time that our children should assist us to support it.' Here was the latest iteration of the problems that had vexed so many British governments since their unhappy dealings with the American colonists during the 1760s: how was the empire to be held together, and how were imperial subjects to be persuaded to help pay for the cost of their imperial defence, rather than leave it all to the mother country? As in 1897, Chamberlain found little backing among the delegates for his ideas of imperial unification, and scarcely more enthusiasm for colonial contributions towards the cost of the Royal Navy. To be sure, the dominions (as the settlement colonies were increasingly being termed) had recruited 30,000 men to fight in the Boer War, they were proud of their Britannic identity, and would eagerly celebrate Empire Day, marking Queen Victoria's birthday (from 1904). But while they were happily tied to the mother country by immigration, sentiment, trade, investment and culture, they did not want closer political integration, and Chamberlain was rebuffed in his efforts to consolidate the empire and get the colonies to make a serious contribution to the costs of their defence.

The problems facing the British Empire that Seeley had diagnosed twenty years before had thus become even more pressing, and not just in terms of its still-unconsolidated structure, but also because it seemed increasingly vulnerable and over-extended in what had recently become a much more hostile international climate. France was still seeking redress for the British occupation of Egypt and continued to resent the humiliation of Fashoda. The German Navy Laws of 1898 and 1900 had authorized the construction of a great battle fleet, the prime purpose of which could only be to challenge Britain's maritime supremacy. In the aftermath of the Spanish-American War of 1898, and under the aggressive leadership of President Theodore Roosevelt, the United States was asserting itself in the Caribbean and across the Pacific. Russia was a perennial threat to Britain's position in South Asia, where Persia, Afghanistan and Tibet remained contested buffer zones, while the impending completion of the trans-Siberian railway in 1904 might also enhance Russia's potential as a Pacific power. Moreover, Japan's defeat of China in 1895 had signalled the rise of a new and genuinely Asiatic force in that region, which would be consolidated by its devastating defeat of Russia ten years later. The result, as Lord Salisbury had observed in 1902, was 'some great change in public affairs in which the forces which contend for mastery among us will be differently ranged and balanced'. Indeed, the British Empire appeared threatened as never before in his lifetime. 'The large aggregation of human force which lies around our Empire,' he concluded, 'seems to draw more closely together, and to assume almost unconsciously a more and more aggressive aspect.' He left the prime ministership, as he had lived it, a very pessimistic man.

Under these circumstances, the United Kingdom's traditional and much-vaunted policy of 'splendid isolation' no longer seemed so superb, or even sensible. Indeed, Britain's search for allies had already started while the Boer War was being fought, and despite the Kruger telegram, the Navy Laws, and Kaiser Wilhelm II's strange love-hate relationship with his grandmother's country, it had begun with Germany, which had, after all, been the nation's friend for much of the nineteenth century. In 1898 Joseph Chamberlain, venturing from imperial into foreign policy, floated the idea of an Anglo-German alliance, initially in private, subsequently in public. Salisbury, who was still at the Foreign Office, was dubious, but in November the following year Chamberlain tried again, by appealing for some broader arrangement between the United Kingdom, the United States and Germany, which would link 'the Teutonic race and the two great branches of the Anglo-Saxon race'; but the speech was coldly received in all three countries (although it was precisely the union envisaged by Cecil Rhodes, who in his will would provide scholarships for inhabitants of the British Empire to study at Oxford University, and for Americans and Germans, too). Nevertheless, during 1901, a succession of schemes for some new Anglo-German relationship were aired by one side or the other, with more or less formality and authority; and Lord Lansdowne, by now Foreign Secretary, was more ready than Salisbury to contemplate an alliance. But Salisbury was still against, and carried his cabinet with him. At the end of the year Lansdowne reopened conversations with the German ambassador in London, but following instructions from Berlin he no longer evinced any interest, and from 1902, further increases in the German naval building programme began to sour the whole tone and tenor of Anglo-German relations.

As the end of the Boer War approached, the British government redoubled its efforts to find friends and allies, not initially in Europe, but rather by trying to lessen tensions and reduce risks on the boundaries of the empire. Hence the Hay-Pauncefote Treaty negotiated in 1901 between the American Secretary of State and the British Minister in Washington, as a result of which the United Kingdom ceded sole control of the proposed Panama Canal to the United States, and also effectively abandoned all claims to be a major naval presence in the Caribbean. And hence, two years later, the final settlement of the disputed Alaskan boundary, which was also (and unlike the earlier Venezuelan settlement) concluded in favour of the United States rather than the British Empire. These two agreements not only reflected the prevailing assumption that war between the United Kingdom and the United States had become unthinkable for the British because it would be unwinnable; they also represented a significant withdrawal of imperial power and influence in the western hemisphere. Even more important, because it spelt the formal end of 'splendid isolation', and because it was a treaty made with a non-European nation, was the alliance Britain concluded with Japan in 1902. The agreement ensured joint Anglo-Japanese naval supremacy in the Far East over the French and the Russians, and saved Britain from additional expenditure on new warships for its Pacific fleet. But it also recognized that the Royal Navy was no longer paramount and unchallenged in the Pacific: indeed, it would soon be scarcely any presence at all. These accommodations with the Americans and the Japanese were, indeed, significant imperial recessionals, as the government sought to lessen its responsibilities and reduce its vulnerabilities in parts of the world where in earlier times it had seemed as though Britain had been pre-eminent.

A NATION ILL AT EASE WITH ITSELF

Despite the eventually victorious outcome, it was widely recognized that the Boer War had been the most humiliating imperial conflict for the British since the American colonists had won their independence, as what belatedly became a large imperial force had taken three years to subdue an amateur army from two small states with combined (white) populations smaller than those of Flintshire and Denbighshire. 'The war has been the nation's *Recessional*,' noted *The Times* in the first volume of its *History of the War in South Africa*, taking up Kipling's earlier warning, 'after all the pomp and show of the Jubilee'. It also, and very properly, engendered another episode of national self-doubt and soul searching, comparable to that which had taken place during the 1880s, but rendered more pressing by what seemed the close interconnection between a badly waged war abroad and urgent social problems at home. Such was the argument put forward by C. F. G. Masterman and some of his young Liberal friends in 1901 in a book revealingly called *The Heart of the Empire*, and subtitled *Discussions of Problems of Modern City Life in England*. The contributors wrote on what they regarded as such urgent issues as housing, education and temperance, and the final essay was the work of the young historian George Macaulay Trevelyan, who had suggested the title of the book. In his essay he denounced laissez-faire, the modern unregulated economy, and the free play of great material interests, and also the Conservatives and Unionists, whom he believed to be both corrupt and incompetent, in thrall to the brewers and the jingoists, and indifferent to the social problems and social injustices of the time.

Trevelyan may have overdone the party-political polemics and point-scoring, but the belief that all was not well with the faltering heart of the empire was widespread. Although the population of the United Kingdom was still growing, from 37.7 million in 1891 to 41.5 million ten years later, the birth rate had fallen over the thirty years from the 1880s to the 1900s by between one-quarter and one-third, as contraceptive practices spread beyond the middle class to the aspiring working class. This gave rise to a variety of concerns. One was that the population was ageing, and *The Times* warned in 1901 that 'an old man's world would not be a beautiful one', partly because old age was on the whole an impoverished and stricken time for those who lived long enough to have to endure it, but also because an ageing nation would not be a vigorous nation, full of ardent and energetic young men eager to go out into the empire. A second anxiety was that the lowest and poorest social groups were producing proportionately more of the population, which meant that the overall quality of the British people was declining, as too many of them were coming from its 'least successful and progressive elements'. Yet a third worry was that Britons might be swamped by 'inferior' and 'alien' races such as the Irish and the Jews, which would, according to Sidney Webb, result in yet further 'national deterioration', and this also fed alarmist fears that the so-called 'yellow peril', emanating from the Far East, meant that 'the ultimate future of these islands may be to the Chinese' (or, more likely, the Japanese). How, contemporaries wondered, could such a great nation withstand the diminution and the dilution of its population in this way? How would British industry continue to remain competitive, and how indeed would the empire be peopled, consolidated and defended?

This fin-de-siècle fear that the United Kingdom in general, and London in particular, was being 'swamped' by hordes of inferior 'aliens' did not sit easily with the early and mid-nineteenth-century view that the country provided a welcoming home and a safe haven for exiles fleeing persecution elsewhere, and that it was a source of national pride that it offered such sanctuary and succour. Those earlier refugees – from Revolutionary France post-1789, from Latin America during the 1800s and 1810s, and from Europe post-1848 – had been relatively few in number; and that tradition continued, most famously in the case of Emile Zola. He had taken refuge in the Queen's Hotel, Upper Norwood, from October 1898 to June 1899, escaping the imprisonment to which he had been sentenced by the French courts after he had written his open letter, 'J'accuse', in defence of the Jewish army officer Alfred Dreyfus. Such individual instances were still tolerated; but very different were reactions to mass influxes of Irish and Jews. By 1900, thanks to the impact of the Great Famine, there were over 400,000 men, women and children in London alone who were of Irish descent, most of them poor, Catholic and working class. During the last two decades of the nineteenth century they were joined by Jews fleeing the Russian pogroms, especially in Poland, in the aftermath of the assassination of the Tsar Alexander II in 1881. In that year there were 46,000 Jews in London; twenty years later the figure was 135,000. Many settled in the East End, where it was claimed they lived separate, ghettoized lives, and failed to observe the Sabbath on Sunday, while also taking away British jobs from British workers. As a result there were growing demands by the late 1890s for legislation that would for the first time restrict immigration into the United Kingdom, in the hope of preserving the nation's racial purity and vigorous stock.

These anxieties, that British population growth might be on the brink of a debilitating decline comparable to what was already happening across the Channel in France, fed into a second pervasive concern, namely that the economy of the United Kingdom was becoming increasingly unbalanced and uncompetitive. Between 1882 and 1902 the acreage that was given over to growing wheat diminished by 50 per cent, while cheaper supplies poured in from the Canadian and American prairies, along with chilled lamb from the Antipodes and beef from Argentina. From the late 1840s to the early 1900s imports of wheat increased tenfold, and domestic production could feed only one in four of the British people. Hence the growing concerns that, in the event of another European war, a blockade of Britain might reduce the nation to starvation; and hence the growing demands for an ever stronger Royal Navy to keep open the sea lanes for the merchant marine. This was not the only anxiety concerning what seemed to be Britain's increasingly distorted economy. As had been true for much of the nineteenth century, its prime exports remained coal, iron and steel, and cotton (Lancashire's mills were almost entirely dependent on the overseas market). But they were insufficient to pay for the food and goods that Britain imported, with the result that the balance of trade ran an annual deficit of £100 million a year during the 1900s, although this was more than made up by the 'invisible' earnings derived from banking, shipping and insurance. Nevertheless, Joseph Chamberlain was not alone in fearing that a British economy increasingly based on financial services would be significantly weaker than when it had been more securely based on manufacturing industry – and it would not just be the British economy, but the British nation and the British Empire that would, as a result, become weaker, too.

Underlying this concern was the further anxiety that the United Kingdom was continuing to lose its industrial pre-eminence to the manufacturing might and entrepreneurial inventiveness of the United States and Germany, those two, great land-based empires with (as Seeley had feared twenty years before) larger populations, larger markets and greater natural resources. To be sure, the volume of British foreign trade, both imports and exports, was over six times higher by 1910 than it had been during the mid-nineteenth century; but the growing power of America and Germany was one reason why the United Kingdom's share of world exports of manufactures had declined from over 40 per cent in the late nineteenth century to around 30 per cent in the Edwardian period. By the 1900s, Britain was no longer the first industrial nation or the workshop of the world in terms of current manufacturing pre-eminence, but only in the historic and nostalgic sense of having started off before any other country. By 1913 the United Kingdom would be making less than half as much steel as Germany, and less than one-quarter as much as the United States. In 1901 E. E. Williams returned to the subject of Britain's inadequate response to foreign competition in an article entitled 'Made in Germany – Five Years After', where he contended that his earlier forecasts had been confirmed, and that Britain was also being increasingly challenged by the United States. The following year W. T. Stead published *The Americanization of the World, or The Trend of the Twentieth Century*, which left no doubt as to which would be the greater Anglo-Saxon power in years to come. It was a prescient and disturbing prediction.

The combination of declining population growth, fear of racial deterioration and continuing economic retardation fed into a further anxiety, namely the continued existence of widespread poverty, as revealed by an inquiry that was begun in 1896 by Seebohm Rowntree, scion of one of the high-minded Quaker chocolate-making dynasties. Five years later, while the Boer War was still being fought out, he published *Poverty: A Study of Town Life*, based on his recent researches into York, the cathedral city and railway centre where the family factory was also located. Rowntree reported that 10 per cent of York's population were in 'primary' poverty, where income was insufficient to purchase the bare minimum of food and shelter, mainly because wages were too low, but also because of illness, old age, the death of the chief breadwinner, or the presence of too many young children. A further 18 per cent were in 'secondary' poverty, where family income was notionally sufficient to keep body and soul together, but in practice was spent unwisely, especially on drink. These precise figures were open to question, and Rowntree was unsure whether such destitution was better explained in terms of systemic problems or individual failings. But his findings that almost one-third of York's population were living in some form of penury and indigence made it clear that poverty amidst plenty was as much a problem in the county towns as it was in the great industrial cities. They corroborated Charles Booth's earlier figures, published in *Life and Labour of the People of London* (1889–1902), and they effectively answered those critics who had alleged that deprivation and misery were problems uniquely confined to the great metropolis.

One fashionable explanation of such widespread poverty was that too many people at the lower end of the social scale were spending too much money on the wrong things, and this not only made them poorer than they otherwise would have been, but also involved them in activities that were both physically damaging and morally degrading. Pubs might be important social centres, providing rooms for trades union and friendly society meetings, as well as hosting a variety of entertainments; but spending on drink was widely regarded as a major cause of ill-health, family breakdown and poverty, and high-minded Liberal abstainers (such as Rowntree) sought to regulate the drink trade and the brewers, who not only purveyed these deplorable beverages and damaging liquids, but were also closely allied to the Conservative Party. In fact, working-class drunkenness seems to have diminished during the late nineteenth century, but working-class betting on the horses had not, and its consequences seemed at least as reprehensible as the demon drink. Indeed, according to Rowntree, gambling 'ruins sport, checks industrial development, postpones social progress, destroys the character of the gambler, ruins yearly thousands of homes [and] lowers the whole tone of social life'. Equally deplorable was the squandering of money on the music halls: they were closely linked to the drink trade and prostitution, and many of the acts were deemed to be lewd and vulgar, with often-simulated nudity, obscene jokes and double entendres. Thus regarded, the drinking, gambling, music-hall-going working classes were financially indigent and morally delinquent, ill-fed and ill-housed, and so it was scarcely surprising that they were also physically unhealthy and unprepossessing.

Was it, then, any wonder that such weaklings had performed so badly during the course of the Boer War? This was a question that was being widely asked, especially amidst the growing fears that the British population was being blighted by so-called 'racial decay'. Three thousand British soldiers had been invalided back home from South Africa because they suffered with acutely bad teeth; Rowntree's survey included figures which showed that nearly half of would-be army recruits in York, Leeds and Sheffield had been found medically unacceptable; and in 1902 Major General J. F. Maurice, a veteran of several African campaigns, asserted that 60 per cent of the male population were unfit for military duties, a figure apparently confirmed the following year by the Director General of the Army Medical Service. In response to these alarms, Balfour set up an Inter-Departmental Committee on Physical Deterioration, which reported in 1904. Reassuringly, it found Maurice's statistics to be misleading, and that there was no serious sign of racial degeneration. But the inquiry collected a great deal of evidence of the widespread ill-health and deteriorated environments characteristic of working-class life, resulting from the effects of urban overcrowding, air pollution, working conditions, venereal disease and physical defects. The committee was also clear that these problems invariably began early, that many infants and schoolchildren were inadequately cared for, and that this condemned them to a later life of chronic ill-health. It urged that there should be regular medical inspections of schoolboys and schoolgirls, and that local authorities should assume responsibility for feeding those who were obviously undernourished and hungry.

Underlying these concerns about poverty and ill-health were two deeper anxieties. The first was that the seemingly irreversible physical and moral degeneration had taken place because the balance between rural and urban living had been completely upset. By 1901 more than three-quarters of the people in England and Wales were housed in an urban environment (the proportion was lower in Scotland and Ireland), and fewer Britons than ever before had personal memories of country life, or any substantial knowledge of rural living to compare with their urban experience. 'Our children,' lamented Keir Hardie, 'grow up in great cities divorced from the great forces of Mother Nature.' This view was shared by those on the political right, who believed that the countryside, both at home and on the imperial frontier, was the place where decent values and manly virtues were to be inculcated, and also on the left, where strenuous country walks and Alpine mountaineering were advocated as the necessary antidote to the debilitating and corrupting nature of urban living. But the greatest danger from these developments was, as usual, presented by the working classes, to many of whom these healthy antidotes were unavailable (many of them could not even afford bicycles as a means of escape). Crammed in the towns and cities, and lacking the settled sense of place, identity and continuity that the countryside provided, it was widely feared that these alienated, uneducated, impoverished urban masses were shallow, unstable and excitable people, open to exploitation by charlatans and demagogues: hence, in the aftermath of the Boer War, the new verb 'to maffick', meaning to indulge in extravagant and irrational displays of emotional and aggressive crowd behaviour.

These urban-based, degenerate proletarians also seemed to be increasingly godless, as a result of the continuing decline in religious observance, a general watering down of religious practices, and the further undermining of Christian dogma by the claims and advances of scientific inquiry. Between 1886 and 1902 the population of inner London rose by well over half a million people; but church and chapel attendance fell by 164,000; and the Church of England suffered most from this decline, with only three London worshippers in 1902 for every four in 1886. Moreover, Sunday was no longer a day given over as exclusively to religion as hitherto. In 1896 parliament authorized the opening of national museums and art galleries on the Sabbath, and soon after the British Museum, the National Gallery and the South Kensington museums opened their doors to the public on Sunday afternoons. Sunday walks in local parks to listen to a brass band also became a favourite weekend pastime, and Sunday newspapers sold double the number of dailies. Nor was religious observance a high priority of the country-house weekends that became commonplace for the upper classes during the Edwardian era, which were increasingly characterized by excessively good living and frequent adultery, in both of which activities the monarch himself still set the pace. The United Kingdom remained outwardly a Christian nation (albeit in varied Anglican, Nonconformist and Catholic guises), but regular public religious practices were visibly eroding (although the furore over Balfour's Education Bill would show that religion remained a powerful and controversial *political* topic). In the autumn of 1904 the *Daily Telegraph* promoted a discussion of the question 'Do we believe?' Those who did stressed the great importance of faith and of Christ's teachings, while unbelievers contended that in an age of science and reason, people knew where they stood without the need for religious faith or superstition of any kind.

As the lower classes became more rootless and godless, so it seemed as though their social superiors were becoming more decadent and corrupt. In Sherlock Holmes, Conan Doyle had created a resonantly hybrid figure, in one guise a reassuring Nietzschean superman of action, but in another a Wildean decadent, dependent on cocaine, wearing make-up, and often living in a state of lethargy, boredom and ennui. In 1894 John Lane launched *The Yellow Book*, the house magazine of the decadent group, whose spirit was powerfully captured by Aubrey Beardsley (or Aubrey Weirdsley, as *Punch* called him), with his disturbing, erotic pen and ink drawings conveying intimations of cruelty and vice through their sinuous lines. In 1895 the Hungarian Max Nordau published *Degeneration*, denouncing such decadent aesthetes as portending the end of European civilization, and four years later the American Thorstein Veblen produced *The Theory of the Leisure Class*, which criticized the new, super-rich for being in thrall to the material indulgences of 'conspicuous consumption'.

His strictures found many echoes in the United Kingdom, where restless plutocrats and rootless financiers, buying up country houses and town mansions, gobbling up titles and honours, and soon to be at the centre of the court of Edward VII, brought the stench of corruption into the nation's public life, and lacked any serious ancestral attachment to the land or the country. The Liberal peerage creations of July 1895, and Balfour's resignation honours ten years later, were both attacked for 'the furious ennoblement of mere financiers', and although such titles were never directly sold, it was widely recognized that donations to party funds were rewarded in this way. Many of these figures were also Jewish, which further reinforced the growing anti-Semitism of the time (Edward himself was sometimes described as the 'king of the Jews').

A further sign of the decline of 'decency' and the corruption of the national culture was the rise of the mass circulation newspaper, beginning with Alfred Harmsworth's *Daily Mail* in 1896, which was specifically targeted at the middle and lower-middle classes. It was produced by mechanical typesetting and on new high-speed presses, and it soon attached numerous profitable advertisements, some of them full-page spreads. In its first decade the *Mail* enjoyed the highest circulation ever attained by any British daily newspaper, and during the Boer War excitement of 1900 it was selling almost a million copies each day. Harmsworth would go on to acquire the *Evening News*, *The Observer* and, eventually, *The Times*. Other proprietors followed where he had pioneered, and the *Mail* was soon joined by the *Sketch*, the *Herald* and the *Express*, which all catered to the new readership that Harmsworth seemed to have found, with the result that between 1896 and 1906 the total number of daily newspaper readers doubled. These papers were generally conservative and often jingoistic in their politics, they purveyed news in a sensationalist style, they no longer printed the speeches of major politicians in full, and they sought to bribe readers by a variety of stunts, gimmicks and offers. They were well-judged products for a semi-literate readership and a semi-literate electorate, they sought to influence events as much as to report them, and they were equally deplored by high-minded figures at both ends of the political spectrum. Lord Salisbury dismissed the *Mail* as 'a paper written by office boys for office boys', while George Macaulay Trevelyan condemned the 'white peril' of cheap journalism, appealing as it did to 'the uneducated mass of all classes'.

There were, then, many reasons why Britain's rulers and intelligentsia feared that national malaise was widespread by the early twentieth century; and it was against this broader background of concern and discontent that Balfour set up a Royal Commission, chaired by the Liberal peer, Lord Elgin, a former Viceroy of India, to look into the military failings that had been so pronounced during the Boer War, especially during the early stages. It reported in the summer of 1903, and described errors and muddles reminiscent of the opening phases of the Crimean War. 'No plan of campaign,' it declared, 'ever existed for operations in South Africa.' There were scarcely any maps of the ground over which the initial engagements would be fought, there were insufficient supplies of ammunition and khaki uniforms essential for fighting on the veldt, the intelligence and staff work were inadequate, and the Royal Army Medical Corps broke down under pressure. In London and South Africa relations between politicians and military commanders were unsatisfactory: as Secretary of State for War, Lord Lansdowne was supposed to be in charge, but he was no match for Lord Wolseley, the commander-in-chief; and interaction between the governors of the Cape and Natal and the generals leading the troops were no better. Regular soldiers seemed unable to think or act for themselves, while the calibre of the officer class also left a great deal to be desired. In modern warfare professional competence mattered as much as bravery, but the education most officers received still elevated 'character' over intelligence: 'More officers of the studious type are needed,' noted one commentator, but 'the right type cannot be got from the public schools'.

The outcry and embarrassment over these military and political failings gave rise to a more general campaign to improve what was termed 'National Efficiency'. Among those who belonged to the movement were the journalist Arnold White, whose book *Efficiency and Empire* (1901) argued that the failings of the Boer War ought to shame the ruling elite into modernizing itself (perhaps on the model of Wilhelmine Germany), or it would be swept away. Soon after, a group of like-minded figures, including R. B. Haldane, Edward Grey and Sidney Webb, founded a dining club appropriately named the 'Co-Efficients', which aspired to 'permeate' the Edwardian state and reshape its policy agenda. For Fabians like Sidney and Beatrice Webb, 'National Efficiency' was a way of advancing their collectivist schemes of social reconstruction. Many Liberals saw it as a way of escaping the clutches of the Newcastle

Programme, by embracing a project of institutional modernization that was simultaneously progressive and patriotic. At the same time 'National Efficiency' also became associated with a circle of young admirers forming around Lord Milner, including J. L. Garvin and Leopold Amery. As such, 'National Efficiency' was a cross-party movement, whose advocates insisted that the old battles between Conservatives and Liberals were much less important than waging the new but more pressing battle of expertise against incompetence. There was even talk that Lord Rosebery might be persuaded to form a cross-party 'National Government' dedicated to the cause of 'National Efficiency'. But like so many attempts to get this latter-day Achilles to leave his tent, it came to nothing, and the movement achieved little, beyond giving further expression to the mood of national malaise.

There were other, more focused organizations concerned with similar issues, the most important and influential of which was the Navy League, established in December 1894, soon after the cabinet crisis over the naval estimates that had precipitated Gladstone's final departure from 10 Downing Street. It was supported by journalists such as Henry Spencer Wilkinson, Herbert Wrigley Wilson, and also by Arnold White, by such establishment figures as the Dukes of Devonshire and Westminster and Field Marshal Earl Roberts, and by a host of naval personnel, of whom Admiral Lord Charles Beresford MP was the most conspicuous. The League established branches throughout the United Kingdom and the empire, and sought to raise popular awareness of the importance of the Royal Navy for Britain's survival (since most of its food was imported), for keeping the British Empire together (command of the waves was essential to sustain the world's largest maritime empire), and for winning what would soon be the escalating arms race with Germany (which was about navies rather than armies). In addition to the many meetings and publications that it sponsored, the Navy League invented and promoted the annual celebration of Trafalgar Day, beginning in October 1896 when Nelson's monument in Trafalgar Square was festooned with flowers and flags, and similar observances were marked around the towns and cities of Britain and the empire. The League also maintained pressure on successive governments to keep building more ships, and to improve the quality of leadership and the structure of command at the Admiralty. At the turn of the century the League's membership totalled 14,000, which was significantly less than the German Navy League, but by 1912 it would grow to 100,000.

A very different response to the widespread anxieties of the time was a re-emphasis on rediscovering and preserving the wholesome and settled values of the countryside, and this took many forms. In 1895 a group of what would now be termed liberal environmentalists, including Octavia Hill and Canon Hardwicke Rawnsley, set up the National Trust for Places of Historic Interest or Natural Beauty to preserve scenic landscapes and open spaces in England and Wales from the ravages of urban expansion and development. Two years later Edward Hudson, the owner of Lindisfarne Castle in Northumberland and of several Lutyens houses in the home counties, established *Country Life*, which soon became the preferred journal for those devoted to rural recreations and residences. At the very end of the reign of the Queen-Empress, the *Victoria Histories of the Counties of England* were inaugurated, to offer a longer perspective on life in the shires from prehistoric times to the present day, which would be published in a large, lavish, multi-volume series. More proactively, Ebenezer Howard produced *Garden Cities of Tomorrow* in 1901, which argued that it was possible to reconcile essential rural values and environments with contemporary urban living, and two years later work began on creating just such a place at Letchworth in Hertfordshire. At almost the same time Cecil Sharp was collecting English folk songs in Somerset, as was his friend Ralph Vaughan Williams in Somerset, thereby beginning the folk-music revival, the aim of which was to rescue and record authentic melodies and indigenous ballads, which it was believed went back to time immemorial, from the threatening assault of music-hall songs.

The late Victorian and early Edwardian years may have seemed to some to be a time of the great queen's apotheosis, followed by the beginnings of 'la belle époque', but that was not how it appeared at the time, and least of all to those in government, regardless of their party affiliation, as they contemplated the growing catalogue of domestic and international problems they faced, and the spiralling costs of dealing with them. By the mid-1890s, as evidenced by Gladstone's vain protest against the increase in the navy estimates, the majority of the Liberal Party had effectively abandoned its old beliefs in retrenchment, cheap government and low taxation, in the face of mounting welfare and defence expenditure; while under the Conservatives and Unionists, the costs of the Boer War and increased defence expenditure (the naval estimates went up from £13.8 million in 1890 to £29.2 million ten years later) meant that the revenue from direct taxation became more important than that from indirect levies. Between the mid-1870s and the late 1890s the standard rate of income tax was increased from a low point of 2d in the £ to being as high as 8d in the £; and by 1903 it was up to an unprecedented 1s 3d. At the same time local-government spending was also mounting, as municipal authorities assumed ever wider responsibilities in areas of health and education, along with the greater costs of road-making, sewage disposal, hospital provision and policing. The result was that the combined expenditure of central and local government increased from £131 million in 1890 to £281 million ten years later, and national and local indebtedness went up in proportion.

UNIONISM DOMINANT AND DECLINING

The uninterrupted decade of power enjoyed by the Conservative and Unionist Party under Salisbury and Balfour was unprecedented since the government of Lord Liverpool between 1812 and 1827, and it would not be until the second half of the twentieth century that anything like it would happen again. Although preoccupied with foreign and imperial affairs, Salisbury was willing to continue the policy of cautious reform that had characterized his earlier administration from 1886 to 1892, even as he was also determined to safeguard the Union with Ireland; while Joseph Chamberlain was not only concerned with the problems of international rivalry and imperial consolidation, but was also well aware of the increasingly close connection between the condition of the empire and that of the United Kingdom. In countries as varied and distant as imperial Germany and New Zealand, experiments in welfare legislation, such as national insurance and old-age pensions, were being embraced; while the romantic revival of ethno-linguistic nationalism, of which there was growing evidence in Ireland, Scotland and Wales, but not in England, was merely one variant on a much broader phenomenon across much of Europe. Organized labour may (or, alternatively, may not) have wanted more social reform; but politicians were increasingly convinced that it did. Yet welfare reforms, like the waging of the Boer War and the construction of new battleships, were expensive, which invariably begged the question: how were they to be afforded, and would such an unprecedentedly expensive welfare-warfare state be financially sustainable? In 1897, when there was a bumper budget surplus, this did not seem a problem; but two years later it had been replaced by an impending deficit of £4 million, and thereafter the deficit increased.

The seemingly inexorable rise in public expenditure, especially on defence and the empire, helps explain why domestic legislation was so limited during this Tory-dominated decade. The Agricultural Land Rating Act of 1896 provided Exchequer grants that reduced the ratable value of land that was cultivated by one-half for poor rates and by three-quarters for general rates. The intention was to pass on the resulting savings to hard-pressed farmers, but there was criticism from Liberals that the real beneficiaries were the (Conservative-supporting) landowners, and many of the Tory Party's recently acquired urban supporters also disliked the measure. In the same year the government had to withdraw an educational bill, after it had run into fierce opposition from Liberals and Nonconformists in the Commons, and in 1898 the Benefices Act, which sought to help the Church of England by giving the clergy more autonomy, and limiting patrons' rights of presentation, also proved to be unexpectedly controversial (early warnings, which Balfour would ignore, that religion in politics still mattered). By contrast, traditional Tories generally welcomed the Local Government Act of 1899, which split London into twenty-eight newly created municipal boroughs, in a move designed to counter what seemed to be the incorrigible progressiveness that had characterized the London County Council ever since the earlier Salisbury government had set it up little more than a decade before. But there was opposition criticism that this was a petty and vindictive measure, and also of the 'sham municipalities' that had been called into being, with their titled mayors, invented coats of arms, and expensive and over-ornamented town halls.

In much of this domestic legislation as in imperial affairs, the driving force was Chamberlain, even though his colonial portfolio should have precluded his involvement with home issues; but as in his attempts to expand and consolidate the empire, he met with mixed results. In 1897 he piloted through a Workmen's Compensation Act, which introduced the general principle that accidents in the workplace must be paid for by the employer, rather than wait on a costly and protracted lawsuit, as had been the previous practice. Initially, the measure did not apply to seamen, domestic servants, or agricultural labourers; but by subsequent Acts passed during the next ten years, they were all brought in. The previous year Chamberlain had set up a Committee on Old Age Pensions, with Lord Rothschild as chairman. It examined more than one hundred schemes, but failed to recommend any of them. Chamberlain thereupon set up his own parliamentary select committee with Henry Chaplin as its chairman and Lloyd George as one if its members; it reported in 1899 and recommended that 5s a week should be awarded, under strict conditions, to the needy and deserving poor aged over sixty-five. A further committee was then established to ascertain the cost of such a provision; but by the time it reported in 1900, the escalating expenses of the Boer War meant that the proposal was shelved, and the issue would not be revived until Lloyd George took it up on behalf of the Liberal government almost a decade later.

As a Liberal, and subsequently a Unionist, Chamberlain's attempts to mobilize central government, and to deploy the resources of the state, to improve the conditions and prospects of ordinary people had again met with limited success. But Salisbury's administration did hold firm in its defence of the Union, at a time when Liberals such as Lloyd George began to advocate an extension of Gladstone's proposals for Ireland into a more comprehensive constitutional settlement for all the four constituent nations of the United Kingdom, known as 'Home Rule All Round'. The Royal Commission that Gladstone had set up to investigate the land and agriculture of Wales reported in 1896, but was simply disregarded. In the case of Scotland the government set up the Highland Congested Districts Board in 1897, but it only made limited amounts of land available to crofters. In Ireland minor adjustments were made in 1896 to the Land Purchase Act that had been passed five years before. But while Salisbury's government contemplated with relative equanimity the gradual dismantling of the territorial basis of the Anglo-Irish Ascendancy, it was also determined, as Gerald Balfour explained, to 'kill Home Rule with kindness', preferring, in an almost Gladstonian way, conciliation to coercion in a 'constructive' attempt to maintain the Union. Investment in railways increased, and subsidies were provided to support many craft industries. Most importantly, the Local Government Act of 1898 replaced the old Grand Juries, which had been dominated by the landowners and their relatives, with county, rural and urban district councils, which owed much to the English local-government reforms of a decade before. Most of the seats were won by Catholic Nationalists, thus ending centuries of patrician Protestant dominance in Ireland's local affairs.

During these years of Unionist rule, Salisbury and his colleagues could do as much or as little as they wanted, for the Liberals presented no serious parliamentary opposition. Rosebery hung on to the party leadership after the electoral debacle of 1895, but in the following year Gladstone came out of retirement, for positively the last time, to denounce the Turkish massacre of Armenians in Constantinople. Rosebery, who was more pro-Turk than Gladstone, thereupon resigned, and the Liberal leadership was divided between Harcourt in the Commons (who had succeeded Gladstone in 1894) and Kimberley in the Lords (who now replaced Rosebery). But Harcourt was almost as difficult and touchy as Rosebery had been, and in December 1898 he himself resigned as Commons leader over a minor issue, and was replaced by Sir Henry Campbell-Bannerman. To these damaging personal antagonisms was soon added a deep party split over the Boer War. The so-called 'Liberal imperialists', among them Rosebery, Asquith, R. B. Haldane and Sir Edward Grey were in favour, but they were opposed by the 'Little Englanders' or 'Pro-Boers', among them Lloyd George and, eventually, Campbell-Bannerman himself, and in June 1901 C-B (as he was known) denounced the use of concentration camps by the British military as carrying on the war by what he called 'methods of barbarism'. But by then the Liberals had lost the 'khaki election' under Campbell-Bannerman's leadership, and his prospects seemed no better than his party's. The Liberal imperialists continued to distrust him, and in September 1905 Asquith, Grey and Haldane agreed that they would not serve under Campbell-Bannerman in any future administration unless he went to the Lords and left Asquith in control in the Commons.

The Liberals mounted no threat to Salisbury's government, and nor did the Irish Nationalists. Their cause had suffered two massive defeats: the death of Parnell in 1891, which deprived them of their one outstanding leader, and the devastating rejection of Gladstone's second Home Rule Bill by the House of Lords two years later. Thereafter, the Nationalists split: the majority worked closely with the Catholic Church in Ireland and the Liberal Party, while a minority, no less attached to Home Rule, wanted greater freedom of political action. Only in 1900 were the two wings of the party reunited under the leadership of John

Redmond who, like Parnell before him, came from a well-established gentry family, this time in County Wexford. But by then it seemed as though the Unionist policy of 'killing Home Rule with kindness' was succeeding – so much so that at the general election of 1900 it was Herbert Gladstone, the son who had revealed his father's conversion to Home Rule in December 1885, and by now Liberal Chief Whip, who effectively dropped it from his party's programme. In Ireland the Nationalists were overwhelmingly Catholic, but during the 1890s an alternative vision was offered, stressing indigenous culture rather than ultramontane religion, as evidenced by the establishment of the Gaelic League in 1893. But Protestant Unionism was also becoming an increasingly powerful force, in Cork and parts of Dublin, and especially in Ulster, which was the most economically advanced part of Ireland, and where landowners, shipbuilders and members of the industrial working class were united in being vehemently opposed to Home Rule. One hundred years since the Act of Union there was still widespread opposition to it; but there was by this time a diversity of political aspiration which, when it finally assumed more energized and antagonistic forms, would make it more difficult than ever to solve the Irish question by providing one single, simple legislative and constitutional answer.

Nor were the attempts to represent ordinary men in politics amounting to much. The British working class remained fractured along ethnic, religious, status, occupational, geographical and gender lines; the links between those who worked in factories and down the mines and such cerebral, middle-class, Fabian socialists as Sidney Webb, Graham Wallas and George Bernard Shaw remained tenuous; and the relations between the recently established Independent Labour Party and the trades union movement and the Liberals were as contentious as they were ill-organized. At the 1895 general election the ILP put up a mere twenty-eight candidates, all of whom finished bottom of the poll, and Keir Hardie lost his seat in West Ham South. The earlier self-confidence and euphoria soon evaporated, as membership fell after 1895, recovered slightly until 1898, and then entered a steady decline that lasted until the end of the century. Indeed, between 1896 and 1899 nearly one-half of all ILP branches disappeared: in the era of the Diamond Jubilee and the Boer War, it was popular Toryism that seemed more appealing to many members of the working class. In 1900 there was an attempted new beginning, when representatives of the ILP, the Fabian

Society and the Social Democratic Federation joined forces to establish the Labour Representation Committee. Ramsay MacDonald was its secretary, but its initial base was predominantly middle class, trades union interest and support were negligible, and at the 'khaki election' it fielded only fifteen candidates. Hardie was elected for Merthyr Tydfil and Richard Bell for Derby; but relations between the two were not close and soon after, Bell defected to the Liberals.

With the Liberals divided, the Irish Nationalists apparently anaesthetized (an appropriate early twentieth-century metaphor, as Edward VII had discovered), and the ILP and the LRC barely to be seen, it was scarcely surprising that the Conservatives and their Unionist allies scored another triumph at the election of 1900. One measure of their success was that across the United Kingdom as a whole there were 161 uncontested Unionist victories, compared with 122 in 1895. As many of its traditional wealthy backers continued to desert the party, the Liberals' finances were becoming increasingly straitened, and in many parts of the country they simply could not afford to field any parliamentary candidates. And while Chamberlain's denunciations of all Liberals as unpatriotic ('every vote given against the government,' he claimed, 'is a vote given to the Boers') was regarded by the imperialist wing of the party as unfair and unjust, there was a widespread perception among the electorate that a party so divided, demoralized and under-funded should not be trusted to govern the nation and run the empire in a patriotic spirit. But even with a second victory at the polls, the Conservatives and Unionists also faced their own problems. Salisbury may have sought and won the 'khaki' election, but he loathed the jingoism that had brought him his victory: 'The recent reform bills,' he told a friend, 'digging down deeper and deeper into the population' had descended to 'a layer of pure combativeness', which meant the country had 'evil times before it'. His gloom was compounded by the fact that ever since 1895 there had been some grumblings among MPs about Salisbury's cliquish management of his government, and they urged that younger talent should be brought in. But the juniors he now recruited seemed mainly to consist of his relatives: truly, the 'Hotel Cecil' seemed 'Unlimited'.

In fact, its limitations soon became apparent, as the Salisbury-Balfour government fell foul of two issues that would do it lasting damage, and help revive the left. In 1901 the High Court ruled, in the case of 'Taff

Vale Railway Co. versus the Amalgamated Society of Railway Servants', that a trades union could be sued by employers (or, indeed, by anyone else) and its funds confiscated as damages if the case went against it. The judgement overturned the previous protection of their funds that the unions had enjoyed under Gladstone's Trade Union Act of 1871, and the immediate result was that the Amalgamated Society lost £32,000 in costs and damages. This outcome inflamed trades union opinion, and aroused labouring support for the ILP and the LRC, to an extent that had been neither possible nor predicted hitherto. Soon after, a second issue arose that further energized the fledgling Labour Party. Because of a shortage of South African workers, the Conservative government bowed to pressure from Lord Milner and allowed the importation of 50,000 Chinese 'coolie' labourers to be sent down the Transvaal mines. They were paid virtually nothing, and forced to live in closed compounds, where drug addiction, prostitution and promiscuity were rife. Nonconformists denounced such behaviour as immoral, humanitarians called it slavery, and trades unionists complained that the wages of African and European workers were being undercut. Like 'Taff Vale', the issue of so-called 'Chinese slavery' would haunt the Conservatives and Unionists until the general election of 1906.

To make matters worse the first major piece of legislation that Balfour passed in 1902, having succeeded his uncle as prime minister, was an Education Act which proved exceptionally controversial (as had his earlier attempt to deal with this matter in the abortive bill of 1896). The aim was to rationalize the organization and financing of primary and secondary schooling, on the grounds that (as Balfour put it) 'the existing educational system is chaotic, is ineffectual . . . [and] makes us the laughing stock of every advanced nation'. The Act abolished the school boards established in 1870 and brought the 'voluntary' (i.e. religious-based) schools under the authority of local government, which meant that ratepayers' money would for the first time be used to pay Anglican and Catholic teachers. This was wholly acceptable to Protestants and Catholics alike, but the Gladstonian, Liberal, 'Nonconformist conscience' was outraged that public rates would in future be used to support denominational schools, especially in rural districts, where there was often no Nonconformist alternative. Despite their lingering differences over the Boer War, most Liberals could agree on opposing the measure, and Lloyd George first attracted widespread public attention by urging local authorities to ignore the legislation. Dr John Clifford, a Baptist minister, established a National Passive Resistance Committee against the payment of rates that would be used to subsidize faith schools, and two hundred men went to prison for refusing to pay 'school taxes'. In cabinet Joseph Chamberlain, who had grown up a Liberal and a Nonconformist (even though he had largely abandoned both creeds), attempted to modify the legislation but failed; and thus began his own alienation from Balfour and his government.

A second piece of legislation was equally well intentioned but also led to trouble, namely the Irish Land Act passed in 1903 by George Wyndham. Wyndham had been Chief Secretary for Ireland since 1900, when he succeeded Gerald Balfour, but so quiescent had the country by then become that the post was no longer deemed worthy of cabinet rank. Nevertheless, on succeeding Salisbury, Arthur Balfour brought Wyndham into the cabinet, and in the following year he passed the most comprehensive piece of land legislation yet devised, which for the first time successfully encouraged the sale of entire estates, by making more money available from the Exchequer to tenants on more advantageous terms, and by providing an extra bonus for landlords. Thus began, on a significant scale, the deliberate dismantling of the territorial base of the Protestant ascendancy, with the avowed aim of appeasing nationalist grievances against landlords; but many Irish landowners felt they had been betrayed by the party that should have supported them. 'Governments who claim to be Conservative,' thundered Lord Muskerry, 'have been anything but Conservative as regards their Irish policy.' The following year Wyndham carelessly allowed his name to be associated with a controversial plan, devised by his Permanent Under-Secretary, Sir Antony MacDonnell, who was himself both Irish and Catholic. The aim was to grant an additional measure of Irish devolution, beyond that already enacted in 1898, by creating a central government structure for certain administrative purposes. But such was the outrage of Irish Unionists, who regarded this as a second betrayal, that Wyndham was forced to resign in March 1905, and he would never hold office again.

By this time Balfour's government was running into very serious difficulties. This was partly because the Liberals were reviving, the LRC was again advancing, and the two parties were beginning to collaborate against the Tories. Balfour's refusal to introduce legislation overturning the Taff Vale judgement and restoring trades union rights drove many hitherto unenthusiastic unions to support the LRC, on the grounds that such legislation was essential; and by 1903, 127 unions had affiliated to it with a combined membership in excess of 800,000. The Liberals were also sensing that the political tide was turning, and rich businessmen began to return to the fold, which greatly improved the party's chances of contesting more seats at the next election. As Balfour's government declined in popularity, the Liberals and the LRC began to co-operate, albeit cautiously, by not fielding candidates against each other at by-elections. In the spring of 1903 Ramsay MacDonald (on behalf of the LRC) and Herbert Gladstone (for the Liberals) began informal conversations, and they agreed to use their influence during the forthcoming election to prevent the running of 'wrecking candidates' whose third-party intervention would risk handing over a seat to the Unionists. This secret arrangement encouraged broader co-operation between the Liberals and the LRC, and was cemented by the Caxton Hall Agreement of 1905. The divided left was getting its act together, and at just the time when the Conservative and Unionist Alliance was falling apart.

Once again, as Gladstone had discovered during the mid-1880s, the problem was Chamberlain, who had wanted to be as creative a figure in Westminster as he had earlier been in Birmingham, but who had so often turned out to be a thwarted and destructive force instead. By now in his early seventies, he must have known that the leadership of both main political parties had eluded him, and that time was running out. He had failed to pass the major social reforms he had wanted to implement under both Gladstone and Balfour, and had become increasingly critical of the latter for what he regarded as his mishandling of the Education Act. Nor, despite his best efforts in 1897 and 1902, had he persuaded the colonies to embrace his plans for imperial restructuring. Yet still he hankered after carrying a sweeping political scheme, which would simultaneously implement major social reforms at home, especially in the realm of old-age pensions, without raising direct taxes, and also bring about imperial consolidation overseas. In 1897 Canada had introduced preferential tariffs on imports from the United Kingdom, and the other colonies had been keen to follow suit if Britain would abandon free trade and grant them preferences in return. The Boer War had delayed Chamberlain's response, even as the widespread foreign hostility to Britain manifested during that conflict had made him still more anxious to promote imperial unity. But although in 1902 the government had imposed a corn registration duty to help meet the cost of the war, it was abandoned in the budget of 1903, to Chamberlain's regret, for he had hoped that its remission to the colonies might have been the first steps towards some form of comprehensive imperial preference.

During the winter of 1902–03 Chamberlain visited South Africa to see at first hand the reconstruction of the conquered Boer republics. While there he pondered anew the issues of free trade, imperial unity and social reform, and began to develop some ideas that might link them all together. Having been, as he saw it, rebuffed on his return by the government's abandonment of the corn duty, he delivered a speech in Birmingham in May 1903, in which he sketched out a scheme of what he termed 'tariff reform' as the first step towards the larger project of promoting imperial unity. This, he insisted, was the biggest challenge of the day: was the empire 'making for union or are we drifting to separation?' His speech was strong on generalities but weak on detail, and Chamberlain had only intended to open discussion in the hope the government would eventually commit to tariff reform at some point before the next general election. But his words attracted a great deal of attention, and the tariff question suddenly became the burning political issue of the moment, with ministers under strong pressure to declare themselves immediately for or against. The result was a three-way split between the Chamberlainites, the free traders and the followers of Balfour. The prime minister vainly strove to keep his cabinet and party together by not committing himself to anything specific that might offend either side. But as the Tories and Unionists descended into fratricidal chaos, this equivocation was not good enough for Chamberlain, and nor was it for his opponents. He resigned from the cabinet in September 1903 and so did the most ardent supporters of free trade, led by the Duke of Devonshire.

These issues of tariffs and imperial unity would primarily involve the 'Greater Britain' (as Sir Charles Dilke would have put it) of the United Kingdom and the settler colonies of Canada, Australia, New Zealand and the soon-to-be unified British South Africa. But neither at home nor in the 'white' empire would Chamberlain attract the support he needed. Meanwhile, there were also some early indications that the inhabitants of the 'non-white' empire might also be unwilling to endure indefinite imperial subordination. In 1888 the young Mohandas Gandhi had arrived in London from India to study law at the Middle Temple; he qualified three years later, and subsequently headed to South Africa, where he was appalled by the attitude of white people to coloured people, especially his fellow South Asians. In 1889 a group of South Asians had formed the British Committee of the Indian National Congress, which campaigned over the next twenty-five years for greater participation by South Asians in the government of their country. In 1897 Henry Sylvester Williams travelled to London from Trinidad to study law at Gray's Inn, and in the same year he established the African Association. In 1902 the Association organized the first Pan-African Congress, held in London, at which the African-American W. E. B. du Bois famously proclaimed that 'the problem of the twentieth century is the problem of the colour line'. These men and these events were scarcely noticed by the imperial authorities in London, who would be increasingly preoccupied and divided by the issue of tariff reform; but they were the first signs and portents that British rule in the 'tropical' empire would not go uncontested for much longer.

TORIES, TARIFFS AND TRAVAILS

By late 1903 the Conservative and Unionist Party was bitterly divided on the issue of tariff reform. Balfour tried to paper over the cracks, and to preserve some semblance of party unity, by appointing Chamberlain's son, Austen, as Chancellor of the Exchequer, and the Duke of Devonshire's nephew and heir, Victor Cavendish, to be his deputy as Financial Secretary to the Treasury. But his efforts were unavailing. The Tory free traders believed in that cause with an almost religious fervour, as did the Liberals; and Chamberlain launched his Tariff Reform League and set up a Tariff Reform Commission, consisting of sympathetic economists, to organize and make the case against them. Between October and December 1903 Chamberlain campaigned up and down the country, delivering major speeches in the great industrial cities, urging the need to defend British manufacturing from foreign competition, insisting that the revenues derived from protection would be sufficient to finance the costs of welfare and defence, and proclaiming the greater cause of imperial consolidation. So fervent and aggressive were Chamberlain's followers that they drove at least a dozen Tory free traders out of the party; one of them was a young, brash and ambitious MP named Winston Churchill, who early in 1904 broke with the Conservatives, crossed the floor of the Commons, and joined the Liberals.

Like Rosebery before him, Balfour was presiding over a divided and demoralized 'fag-end' government, whose electoral prospects were decidedly gloomy; and like Rosebery again, he seemed irresolute, indecisive, and unable to assert his authority over his party. Nevertheless, Balfour did carry some measures, and implement some reforms, during these twilight months. In 1904 the government passed a Licensing Act, which pleased the temperance lobby but upset the brewers because it led to the closure of many pubs, yet it also pleased the brewers and upset the temperance lobby because the owners would be generously compensated for their losses. In 1905 the Unemployed Workmen Act set up 'Distress Committees' to investigate why and where men were out of jobs, and to provide them with some relief; but the powers of these committees were limited, and they depended on voluntary funding to assist the unemployed. Neither of these measures was of particular significance, and the Licensing Act was quintessentially Balfourian in its equivocated ambiguity. Much more sinister and alarming was another piece of legislation, also carried in 1905, which was largely in response to the mounting pressure from Conservative backbenchers and the popular press against the recent influx of Jewish immigrants; they were stereotyped and scapegoated as 'undesirable and destitute', and their growing presence reinforced the widespread fear that the nation was being 'swamped' by people of alien race and inferior quality. The resulting Aliens Act ended the automatic right of asylum for immigrants, which had been one of the most generous aspects of Britain's nineteenth-century political culture (though in fact the subsequent Liberal government would rarely invoke its provisions).

In the aftermath of the Boer War, and of Lord Elgin's exposure of Britain's widespread military failings, and notwithstanding the distractions and divisions caused by tariff reform, Balfour was determined to shake up both the British army and the Royal Navy. He wanted to render them fit for the purpose of fighting the war with Germany, which, from 1904 onwards, was widely recognized as the most likely opponent the British nation and empire might soon have to face. He established, chaired and subsequently strengthened the Committee of Imperial Defence, to develop, oversee and co-ordinate grand strategy, appointing the relevant cabinet ministers and heads of the army and navy as members, and providing it with its own staff and secretariat. He abolished what had become the anachronistic post of commander-in-chief and established a modern General Staff and Army Board, reforms that had been called for since the early 1890s. And he appointed the dynamic and irascible Admiral Sir John Fisher as First Sea Lord to modernize the navy, which Fisher duly did by scrapping many older vessels of negligible fighting value, commissioning the construction of the first 'Dreadnought' battleship, and filling key posts with hand-picked officers of high quality. Fisher also brought home five capital ships from the Far East as part of the closure of the Pacific Station, and the West Indian and North American naval squadrons were also withdrawn. This was, indeed, recessional, as Britain effectively abandoned its claim to a maritime presence, let along maritime supremacy, on the western side of the Atlantic to the United States, and in the Pacific to the United States and Japan. Instead, the Royal Navy was concentrated on the Channel, Atlantic, Mediterranean and Eastern commands: a sure sign that fighting war in Europe against Germany was a more urgent strategic priority than defending the far-flung empire of the settlement colonies and Africa, though India remained a different and more important proposition.

Having recently settled outstanding issues with the United States and concluded an alliance with Japan, Balfour's government was equally concerned, during what would be its last remaining months, to try to clarify relations with France, Germany and Russia. The French were still unhappy at the British occupation of Egypt and the rebuff they had suffered at Fashoda, while Kaiser Wilhelm II's erratic influence on German foreign policy since his dismissal of Bismarck in 1890 accentuated British anxieties about the economic and naval challenges emanating from the *Reich*. These were serious obstacles to a rapprochement with either nation, but it seemed easier to attempt to mend fences with France than with Germany. In May 1903 King Edward VII paid a highly successful state visit to Paris, which gave added impetus to the negotiations that were already taking place between the British and French governments. In the following year the two nations agreed an Entente, whereby France recognized Britain's claims to Egypt and the Nile, while Britain accepted France's possession of Morocco and Madagascar, and further settlements were reached concerning Newfoundland, West Africa, the New Hebrides and Siam. Meanwhile, the British government also sought to lessen tensions with Russia (and also with China), and to rein in further imperial adventures in Asia, by repudiating the unauthorized efforts made by Sir Francis Younghusband to acquire Tibet for Britain.

These were significant military and diplomatic achievements and initiatives for a government and a party that had been playing out their own version of 'Disintegration' since Chamberlain's resignation in the autumn of 1903. He continued to insist that tariff reform would save domestic industry, provide abundant jobs and work, and ensure the consolidation of the British Empire. But his free trade opponents, both within the Tory Party and among the Liberals, argued that tariff reform meant taxes on food and a consequent increase in the cost of living that would hit ordinary people hard. Moreover, it was far from clear that the empire could be consolidated into a single, unified, self-sufficient economic organization when the majority of Britain's exports still went elsewhere, especially to Europe and the United States. In January 1905 Balfour made another effort to bring the warring factions together, by proposing that duties should be levied for the purposes of international negotiation and retaliation, and that another imperial conference should be called to discuss 'closer commercial union with the colonies'. Chamberlain accepted these proposals, but only after two months' delay, which effectively killed any hope of reunifying the party. By this time Balfour's parliamentary authority had all but vanished, and his position in the country was undermined by the fact that since 1900 the government had lost twenty-six by-elections to the Liberals and Labour. By November, Chamberlain and the tariff reformers seemed to be gaining more ground, as they captured the leadership of the National Union of Conservative Associations. Mistakenly believing that the Liberals were as divided as his own party, and would be unwilling or unable to form a government, Balfour suddenly resigned on 4 December 1905. But after almost two decades out of power, the Liberals were determined to take office, and they did so on the following day.

11

General Election, 1905–06

Asquith, Grey and Haldane had earlier agreed they would not serve under Campbell-Bannerman unless he went to the Lords, but he stayed in the Commons as prime minister, having successfully faced down his erstwhile challengers. The government he formed in December 1905 was in a minority in the lower house, but it was of considerable ability, it was fired up by a deep hostility to Joseph Chamberlain and tariff reform, and despite the recent divisions over the Boer War, the Liberals were determined to end the long years of Conservative dominance. Although there had been many Whig defections since Gladstone had espoused Home Rule in 1885, and notwithstanding Rosebery's continuing and petulant aloofness, Campbell-Bannerman's cabinet contained a significant number of landed grandees, among them the Marquis of Crewe as Lord President, the Marquis of Ripon as Lord Privy Seal, the Earl of Elgin as Colonial Secretary, Lord Tweedmouth as First Lord of the Admiralty, and Earl Carrington as President of the Board of Agriculture. The country gentry were represented by Herbert Gladstone as Home Secretary, Sir Edward Grey as Foreign Secretary, Sir John Sinclair as Secretary of State for Scotland and Sydney Buxton as Paymaster General. This was at least as landed a cabinet as Gladstone's third and fourth administrations had been, but none of the peers were heavyweight figures with significant staying power. Much of the driving force in the government would instead come from such middle-class men as Asquith as Chancellor of the Exchequer, John Morley as Secretary of State for India, R. B. Haldane as Secretary for War, Lloyd George as President of the Board of Trade, and John Burns as President of the Local Government Board (the first man from a working-class background to sit in a British cabinet).

Sensing that the political tide was now running the Liberals' way, and that the Conservatives and Unionists were hopelessly split, tired of office and badly led, Campbell-Bannerman obtained an immediate dissolution of parliament, and the general election was held in January 1906. This time the Liberals could afford to field a full slate of candidates, and they were further assisted by the Lib-Lab pact, which helped them in such places as the industrial northwest, where the Tories had recently done well but only because the Liberal and Labour candidates had divided the opposition vote. The Liberals said little about Irish Home Rule or social reform, but they made much of the Taff Vale judgement and the issue of so-called 'Chinese slavery'; above all they campaigned on the enduring benefits of free trade with a fervour that would have done credit to Cobden and Bright. It had, they insisted, been the foundation of British economic and global greatness since 1846, and it had also secured low prices and cheap imports, thus making possible the rising general prosperity that had been especially marked since the 1870s. By contrast, the Liberals argued, Chamberlain's policy of imperial preference and tariff reform had divided the empire instead of uniting it, while his proposed taxes on food would hit the poor much harder than the rich, and as such they represented a serious threat to the recently risen living standards of ordinary people. Insofar as they mentioned such issues as welfare and defence, the Liberals urged that they should be financed by (progressive) direct taxes on incomes and land, rather than (regressive) indirect taxes on food. In response, the Conservatives mounted a fatally divided campaign: Chamberlain and his friends hammered away for tariff reform, but they were contradicted by Tories who preferred free trade, while Balfour's elaborate equivocations played no better on the hustings than they had recently done in the cabinet or the Commons.

The result, which had been widely expected, was a massive Liberal landslide that completely reversed the electoral verdicts of 1895 and 1900, and delivered a victory surpassing those won by Gladstone in 1868 and 1880. To be sure, the Liberals obtained only 49 per cent of the popular vote, compared to the Conservatives' 44 per cent, but under the 'first past the post' system this converted into a huge Commons majority, as the Liberals won 377 seats, which gave them a lead of 84 over all other parties combined. On the opposite side of the lower house were a mere 132 Conservatives and 25 Unionists, and of that combined total of 157 some two-thirds were supporters of Chamberlain. Tariff reform had conspicuously failed to rouse or carry the country for the causes of imperial consolidation and social reform, but it seemed set fair to take over the Tory Party. The Irish Nationalists, by now reunified under the leadership of John Redmond, won eighty-three seats, but the most portentous result was that fifty-three Labour MPs were elected: twenty-nine under the auspices of the LRC, which now renamed itself the Labour Party, and another twenty-four Lib-Lab MPs who were as yet unaffiliated, and most of whom were officials of the coalminers' unions. These returns further strengthened the Liberals' position, as the Irish Nationalists, the Lib-Lab and Labour MPs were often minded to support them. For the Tories this was a devastating defeat: they lost all six Manchester and Salford seats (thanks to the Lib-Lab pact); both Arthur Balfour and his brother Gerald were rejected by their constituencies; the number of MPs from traditional landed backgrounds was significantly reduced; and more than two hundred men were elected to the Commons for the first time. The result was a major change, both in the personnel and the atmosphere of politics: even the normally detached and philosophical Arthur Balfour admitted that 'the election of 1906 inaugurates a new era'.

Looking back over the same period, Virginia Woolf would later put the same point even more emphatically: 'On or about 1910,' she would famously opine in her essay 'Mr Bennett and Mrs Brown' (1924), 'human character changed.' 'Masters and servants, husbands and wives, parents and children' were behaving in different ways towards each other and, she went on, 'when human relationships change, there is at the same time a change in religion, conduct, politics and literature.' This was undoubtedly overstating the case for rhetorical effect, and Woolf's sense of epochal transition may also have owed something to the fact that her father, Leslie Stephen, had died six years before in 1904, an event that had reduced her for a time to a state of complete mental collapse. Yet there can be no doubt that there was a widespread feeling, by the mid-1900s, that change was not just confined to politics and government in the aftermath of the Liberal landslide, but was indeed much more pervasive and widespread. One indication of this broader shift in mores and mentalities is that references to 'character' and 'perseverance', those quintessential watchwords of Victorian morality to which Leslie Stephen himself had attached so much importance, suddenly began to decline in the pages of local and national newspapers. The speeches of politicians and the deeds of military men were still reported at length, but column inches were also being increasingly devoted to describing the deeds and doings of those actors, singers and dancers who were giving sell-out performances in theatres and music halls across the country. Celebrity was superseding character, trains were being mentioned more than horses, electricity more than steam, and the telephone more than the telegraph. This may not have amounted to a revolution, but Balfour was surely correct in recognizing the advent of a new era.

EMPIRE AND INSECURITY

Nevertheless, in terms of the management of Britain's foreign and imperial affairs, there was little immediate sign of any such abrupt discontinuity on the part of the new Liberal government. From 1905 until 1916 the Foreign Secretary would be Sir Edward Grey, who would hold the office continuously for longer than anyone else, before or since. He was a Whig baronet and a kinsman of the second Earl Grey who had passed the Great Reform Act, and like him, Edward Grey was a Northumberland landowner with an estate at Fallodon. He preferred fishing and birdwatching to reading Foreign Office telegrams, he was not well travelled and had no command of foreign languages. But Grey was widely esteemed for his integrity and public spirit, and like many Foreign Secretaries he would enjoy considerable freedom in pursuing policies that would be virtually indistinguishable from those of his predecessors, Lords Salisbury and Lansdowne. At the Colonial Office the appointment of Lord Elgin, after Alfred Lyttelton (who had briefly replaced Chamberlain when he resigned in 1903), signified a return to the sort of traditional notable who had held the post for most of the nineteenth century. Elgin's father had been Governor General of Canada, where he had introduced responsible government, he himself had been Viceroy of India from 1894 to 1898, following Lansdowne and preceding Curzon, and he had subsequently chaired the inquiry into the military failings of the Boer War. He was a safe and experienced pair of hands, but he was frequently upstaged by his brash and bumptious under-secretary, the young and ambitious Winston Churchill, who was holding his first ministerial appointment, and who would take every opportunity to steal the parliamentary limelight as the Colonial Office's sole representative in the House of Commons.

The British Empire for which Elgin and Churchill bore the ultimate responsibility on behalf of the new Liberal government, and whose foreign policy Sir Edward Grey directly controlled from London, was populated in the early 1900s by approximately four hundred million people, a staggering swath of humanity under one imperial authority. Only one-tenth of them resided in the United Kingdom itself, nearly three hundred million subjects lived in the Indian Empire, there were six million elsewhere in Asia, forty-three million in Africa, fewer than ten million in the Americas, and barely five million in the Antipodes. This meant the British ruled somewhere between one-quarter and one-fifth of the sixty million square miles that constituted the habitable land surface of the globe; and the protracted but ultimately successful suborning and incorporation of the Boer Republics of the Transvaal and the Orange Free State represented a final step in the consolidation of another imperial dominion in southern Africa. In terms of the lists beloved of contemporary authors of geographical textbooks, the United Kingdom held sway over 'one continent, a hundred peninsulas, five hundred promontories, a thousand lakes, two thousand rivers [and] ten thousand islands'. From one, anxious perspective, the British Empire might indeed amount to no more than the dispersed and vulnerable trans-oceanic agglomeration that Joseph Chamberlain had vainly sought to rationalize and consolidate; but from another viewpoint, it was a multi-faith, multi-ethnic, multi-lingual maritime imperium without precedent in human history, and no other contemporary power could equal the range and riches of its diverse realms.

From an Anglo-Saxon perspective the most important constituent parts of the British Empire were Canada, Australia and New Zealand, along with Cape Colony, Natal and the two former Boer republics in South Africa. During the nineteenth century, Britain was not the only country to settle in so-called 'empty' parts of the distant globe: Americans headed west from the Atlantic coast to the Pacific, and likewise Russians migrated east from Europe to Siberia. But there was no equivalent to these overseas Britannic dominions, separated from the 'mother country' by thousands of miles of ocean, and which had consolidated markedly since the 1870s. They were overwhelmingly British in language, law, culture and customs, they traded primarily with the imperial metropolis, and they enjoyed complete self-government in domestic matters, subject to the ultimate authority of the Westminster parliament and the Privy Council. But the increased British investment that had helped pull them out of their late nineteenth-century economic downturns had further strengthened, and partially redefined, the connection between the imperial metropolis and the settler colonies, with the result that in terms of trade and finance, they were becoming *more* subordinate to the mother country as places of investment and as suppliers of food and raw materials. So while Canada, Australia and New Zealand began to develop their own nationalist sentiments, they also saw themselves as being primarily Britannic settlements and offshoots overseas. (South Africa was more complicated: the Cape and Natal were at best only partially Britannic, while the two former Boer republics were never fully reconciled to British suzerainty.) Yet they were unwilling to contribute much to the costs of imperial defence or join in any scheme of imperial federation. And fearing the 'yellow peril' in the aftermath of Japan's victory over Russia in 1905, both Canada and Australia were determined to remain white man's countries (and South Africa even more so, although New Zealand treated its indigenous population more tolerantly, albeit in the aftermath of the brutal Maori Land Wars).

By contrast, British rule in India, both direct and indirect, remained authoritarian, and was an unabashed autocracy tempered by the rule of law. It enjoyed the overwhelming support of the Indian princes, but was increasingly resented by the educated middle classes, some of them London-trained lawyers, who lived in Calcutta, Bombay and Madras. The entire Raj was administered by fewer than one thousand members of the Indian Civil Service, and more than 95 per cent of them were British. As Disraeli and Curzon both recognized, it was the capacity to deploy the Indian Army, which was financed by the Indian taxpayer, that made the United Kingdom a great military power in the east, at no cost to the British exchequer; and the revenues of the Indian government also served to guarantee the extensive British investments in the subcontinent, especially in railways and public utilities. Between 1898 and 1905 the authoritarian Lord Curzon pursued a policy of sound government and administrative reform. But he did so at the cost of alienating educated Indian opinion, while his partition of Bengal on the grounds that it was just too large a jurisdiction had given particular offence. This had greatly strengthened support for the Indian National Congress, whose members deplored the contrast between the 'despotism' of the Raj and the responsible government evolving in the dominions. Curzon's Liberal successor as Viceroy, Lord Minto, was more sympathetic to the views of the Congress, as was John Morley, the Secretary of State for India. But they did not believe that South Asians were capable of ruling themselves, and there was no prospect that India would become another self-governing dominion, let alone a democracy.

The remaining realms of Britain's formal empire consisted of crown colonies of conquest and annexation, and naval bases and military installations. The colonies included Ceylon, Burma, Malaya and parts of Borneo east of Suez; Nigeria, the Gold Coast, the Gambia, Sierra Leone, the Anglo-Egyptian Sudan, Kenya, Uganda, Nyasaland, Rhodesia, Bechuanaland, Basutoland and Swaziland in Africa; Jamaica, Trinidad, British Guiana, British Honduras, the Leeward Islands, the Windward Islands and the Bahamas in the Caribbean; and Fiji, the Gilbert and Ellice Islands, the Solomons and lesser groups in the Pacific. The African colonies were, along with the Pacific islands, the most recent acquisitions, where explorers and adventurers, missionaries and traders, officials and proconsuls, were gradually extending British influence, constructing docks, roads, railways and churches, sinking mines and establishing plantations. But the military and administrative resources that Britain could make available for these tasks during the era of the 'New Imperialism' remained distinctly limited. Hence the initial outsourcing of expansion and administration to private companies in large parts of Africa, while in many areas of their 'tropical' empire imperial rule was economically and indirectly exercised through indigenous hierarchies of authority. The global network of naval bases and military installations that were essential for the maintenance of British maritime power included Gibraltar, Malta and Cyprus in the Mediterranean, Aden, Singapore and Hong Kong east of Suez, as well as Vancouver Island in Canada, Sydney and Wellington in the Antipodes, Mombasa in Kenya and the Cape of Good Hope in South Africa, and Colombo in Ceylon. No other empire could rival this network of overseas outposts, where coal and supplies were readily available to the Royal Navy and the British merchant marine, which enabled them to link and maintain the greatest seaborne empire the world had ever seen.

These were the formal constituent parts of the British world-system as it had evolved by the early twentieth century; but although it had expanded dramatically since the 1880s, it had not done so according to any systematic plan or preordained design, and rarely, except in the case of Joseph Chamberlain, with enthusiastic official approval. At the same time there were other parts of the globe, which were ostensibly independent, over which the British exercised varying degrees of influence via trade, commerce, investment and the leverage that they brought with them. The pre-eminent example of the deployment of such scarcely veiled power was Egypt, which had been ruled by Lord Cromer as Consul General since 1883. In all but name Egypt was a part of the British Empire (and would formally become so in 1914), since the United Kingdom had an essential strategic interest, namely the Suez Canal, which was the vital lifeline between the imperial metropolis and India. Further east, Britain exercised significant influence over the kingdom of Siam, in part because it lay between the imperial outposts of Burma and Malaya, but also to prevent the French from further expanding their rule in Indochina; while in China itself the United Kingdom was the pre-eminent (though by no means unchallenged) trading and investing power in Shanghai and along the Yangtze. The largest area of informal influence was in Latin America, where the British had been involved since the fall of the Spanish and Portuguese empires in the early 1820s. The Monroe Doctrine precluded formal annexation in the region by any European power thereafter; but the United Kingdom soon became the largest investor in Latin American railways, utilities and government bonds (both the Rothschilds and Barings were heavily involved), and by the early twentieth century was importing massive quantities of guano from Peru, nitrates from Chile and chilled beef from Argentina.

As such, the British Empire, both formal and informal, was also an extraordinary economic, technological and social phenomenon. On the eve of the First World War, the United Kingdom would have exported in total almost £4,000 million worth of capital overseas, thanks to the balance of payments surplus created by its huge 'invisible earnings', and of that prodigious sum, nearly £1,500 million was invested in the empire, of which £1,000 million was in Canada, Australia, New Zealand and

South Africa, more than £250 million had gone to India, and approximately £150 million was invested in the remaining colonies. Much of it went into dominion and municipal bonds, which helped fund and finance the infrastructures of such great imperial cities as Toronto, Bombay, Calcutta, Melbourne and Sydney. Much went into transport: the Canadian transcontinental railroads had been financed from London, and by 1910 India had 32,000 miles of track, with another 1,500 under construction. Massive irrigation projects were funded, as at Krian in Malaya in 1895, and later at Gezira in the Sudan, while the construction of the Aswan dam on the Nile between 1898 and 1902 was widely regarded as a miracle of modern engineering, and represented a first attempt to control the river's unpredictable and often destructive annual flood. Large areas of the Burma Delta were given over to rice, of forested Malaya to rubber, of the Gold Coast to cocoa and of Uganda to cotton, while the docks and harbours constructed in such port cities as Singapore, Sydney, Montreal, Cape Town and Auckland were essential for the trade that was the lifeblood of empire. Part cause, part consequence of this global investment was that one-third of all British exports (mostly coal and manufactured goods) went to the empire, and a quarter of British imports (mostly food and raw materials) came from the empire.

As Sir John Seeley had noted a generation before in *The Expansion of England*, holding maritime empires together was intrinsically a more challenging task than consolidating land-based dominions; and it was scarcely a coincidence that the United Kingdom was not only the world's foremost imperial power, but also strove, successfully for now, to remain its foremost naval power, too. Although the Royal Navy was increasingly concentrated in home waters in response to the growing German threat, it was still the strongest fighting force afloat, and remained larger than the next two biggest navies combined. In terms of its shipbuilding industry and merchant marine, the United Kingdom's pre-eminence was as marked in the age of steam as it had been in the age of sail. Throughout the Edwardian era the United Kingdom would build 60 per cent of all ships launched, and such iconic British firms as Cunard, P&O, Union Castle and Royal Mail Lines were carrying about one-half of the world's seaborne trade, including nearly all of it within the formal and informal British Empire. By the early twentieth century these maritime links on the surfaces of the seas were complemented by a new communications network beneath. The first such submarine cable to be successfully laid had linked the United Kingdom and Newfoundland in the mid-1860s (thereby giving Brunel's ill-fated *Great Eastern* steamship a brief and belated moment of success and glory). Many subsequent cables were laid to link Britain to Latin America, Africa, India and the Antipodes, and in 1902 a trans-Pacific cable was laid, from Vancouver to Fiji and thence to Australia and New Zealand. These 'Deep-Sea Cables', encircling the globe, would be celebrated by Rudyard Kipling (who was fascinated by engineering and technology) in his poem of that name; and they would soon be followed by an imperial wireless network, linking Britain with Cyprus, Aden, Bombay, Malaya, Hong Kong and Australia.

The empire, or at least the 'white' dominions of settlement, was also held together by continued emigration, which reached its peak in the years from 1900 to 1914, when 6.7 million people left the British Isles and, for the first time, a majority went to the colonies rather than the United States. This was good news in that such an outflow of those presumed to be young and enterprising people was a significant and timely reinforcement to the imperial Britannic communities; but it worried those who feared that the nation's best blood was going overseas, to be replaced by immigrants of inferior quality arriving in Britain from eastern Europe. There was also traffic the other way, as many imperial emigrants, and Britons born overseas, returned home, either because they had always intended to do so or because they found life on the frontiers of empire less congenial than they had hoped. There was in addition a growing intra-dominion movement of people, especially among members of the professional classes, between one colony of settlement and another. One example was William Maxwell Aitken, a son of the Canadian manse who, having made a fortune in Montreal by somewhat dubious means, departed for Britain in 1910 in search of a bigger stage on which to play. Another was the New Zealand-born scientist Ernest Rutherford, who was a professor at McGill University in Canada from 1898 to 1907, then took a senior post at Manchester, and eventually became Cavendish Professor of Physics at Cambridge, and a Nobel Prizewinner. Even more imperially travelled, and appropriately so, was Kipling: born in Bombay in 1865 and educated in Britain, he spent the years from 1883 to 1892 in India and the United Kingdom, then lived in the United States, became a regular visitor to South Africa, finally settling in Sussex in 1902.

Yet despite the economic, technological and social links that bound it together, and notwithstanding the jingoistic euphoria evinced by the Diamond Jubilee and the Boer War from which the Conservatives and Unionists had undoubtedly benefited at the 1900 election, the empire as a cause and as a creed was never all that popular in Britain itself, even in the age of 'High Imperialism'. Lord Meath, who had established Empire Day as an annual observance, deeply regretted that the history of British overseas expansion was scarcely taught in elementary or secondary schools. The words and music of Benson and Elgar's 'Coronation Ode' may have become the transcendent anthem of the twentieth-century empire, but Elgar himself evinced little enthusiasm for (or for visiting) Britain's overseas dominions, while Benson admitted that the very idea of empire left him 'cold', and probably most other Britons, too: 'How,' he wondered, 'can little, limited minds think about the colonies, and India, and the world at large, and all that it means?' There was also undoubtedly an anti-imperial reaction after the excesses of the Boer War, as the electorate resoundingly rejected Chamberlain's schemes of imperial consolidation. He would be incapacitated by a stroke later in 1906, and although the tariff reformers would remain a powerful force within the Conservative Party, their cause effectively ceased to be practicable politics for the foreseeable future. The previous year Lord Milner had retired from South Africa on health grounds, and soon after would be subjected to a motion of censure in the Commons. Lord Curzon returned from India the following year, defeated and disappointed that the Liberal government refused to extend his term as Viceroy. And Lord Cromer would resign his post in Egypt in 1907, ostensibly on health grounds, but in reality because his authoritarian approach grated with Campbell-Bannerman and his colleagues.

Indeed, it was for Milner, Curzon, Cromer and Chamberlain a constant cause of regret and anxiety that they could never make the cause of empire as central to British politics, or as appealing to the British people, as they believed it ought to be, and after the Boer War that task became even more difficult. In 1899 Campbell-Bannerman had deplored 'the vulgar and bastard imperialism of . . . provocation and aggression . . . of grabbing everything even if we have no use for it ourselves'. Sir Charles Dilke reaffirmed the need to recall 'the true as against the bastard imperialism', by which he meant the doctrine of self-government rather than of authoritarianism and annexation. F. W.

Hirst, journalist and editor of *The Economist*, denounced British aggression on the North-West Frontier of India, the reconquest of the Sudan, and the conflict with the Boers as 'unjust and uncalled-for wars, the product of crude, boyish ambitions and unworthy policy'. The most trenchant critique of all was mounted by J. A. Hobson, in his book *Imperialism: A Study* (1902), who argued that Europe's recent expansion into Africa, notoriously exemplified by British attacks on the Boers, was largely motivated by greed for profit. Finance, he insisted, was the governor of the imperial engine. This was an over-simplification, but what Hobson lacked in evidential conviction he more than made up for in moral outrage, harking back to the free-trade and anti-imperial rhetoric of Cobden and Bright. Hobson assailed the bankers, financiers and capitalists who he believed manipulated the new imperialism, and also the parasitic interests who were their hangers on, among them the military, arms manufacturers, bondholders, aristocrats and missionaries, all of whom had again succumbed to the 'primitive lusts of struggle, domination and acquisitiveness'.

Whether the empire was praised or blamed, by the early twentieth century the United Kingdom's engagement with the rest of the world, via both its formal realms and its informal influence, was more extended and varied than it had ever been before. To that extent, at least, A. C. Benson's words about 'wider still and wider' did not err or mislead. Yet in many ways Britain's closest overseas relations remained, as they had been throughout the nineteenth century, with the other advanced nations of the northern hemisphere, especially those in Europe, and also the United States. Despite Hobson's claims to the contrary, the majority of Britain's overseas investment (£2,500 million out of £4,000 million) was *outside* the formal empire, and so was most of Britain's trade (75 per cent of imports and 62 per cent of exports). This meant that the commercial relations between the imperial metropolis and its colonies were neither close enough, nor exclusive enough, to make Chamberlain's dream of a consolidated and self-sufficient empire an economic reality, let alone a political one. Despite the tariffs that most European nations had imposed during the last quarter of the nineteenth century, the continent continued to offer a major market for British exports, and cultural links with France, Italy and Germany also remained strong (Elgar, for example, was very closely indebted to the German symphonic tradition, despite mistaken assertions that his compositions were quintessentially English). British emigration to the United States, and British investments there, still surpassed those flows to any other single country, and by the early twentieth century, America was also a major exporter of foodstuffs to Britain and of manufactured goods to the empire. Pace Chamberlain, the United Kingdom was too much engaged with the wider world beyond the empire, especially in the northern hemisphere, for tariff reform to work.

As the twentieth century dawned, Europe and the United States were not only the United Kingdom's pre-eminent trading areas: they were also becoming places of increasingly challenging and threatening engagement. On the one hand, there were many proponents, on both sides of the Atlantic and North Sea, for schemes of closer union between the Teutonic and Anglo-Saxon races, or of the English-speaking peoples, in the belief that trade, culture and ethnicity meant that such closer connections were providentially ordained and must be made to happen. But the United Kingdom was also beginning to make commitments, initially in the case of France, to military involvement on the continent, the like of which had not been seen since the Napoleonic Wars, and which would ultimately prove astonishingly costly in terms of men and money in the two world wars that would be fought against Germany. At the same time, the successes of the United States during the Spanish-American War of 1898, and its subsequent annexation of Puerto Rico and the Philippines, had prompted Kipling to urge the Great Republic to 'take up the white man's burden', perhaps even superseding Britain as the greatest force in the West, and become a fully fledged imperial power – something which, during the assertive presidency of Theodore Roosevelt, it came closer to doing. The combined implications of an increased British military commitment to the continent, in alliance with France and against Germany, and the growing power and assertiveness of the friendly but challenging United States, would not in the long term bode well for Britain and the continued existence of its empire.

But by the mid-1900s the United Kingdom was not only more deeply enmeshed in Europe, and more economically and militarily challenged by the United States, than at any time during the previous century: it was also suffering, as many earlier opponents of additional imperial annexation had feared, from having too many territories in too many parts of the world that it was obliged to govern and administer and above all (if it came to another global war) *to defend*. In short, the British Empire was increasingly afflicted and threatened by what would later be termed 'imperial overstretch', for as the empire's bounds became wider still and wider, the task of defending them became correspondingly greater still and greater, even as its military commitment to the European continent simultaneously increased. There were, then, many reasons why the Liberals who took charge in December 1905 were apprehensive about the threats and limitations to what had earlier seemed Britain's apparently unchallenged global hegemony, and they rightly wondered, in this more menacing and competitive era, whether its world position and possessions could be successfully defended and maintained. For the time being the Liberal government would be successful in further reducing tensions and confrontations on the periphery of their empire, in order to focus on the threat increasingly being presented in Europe by Germany. But the fact remained that the United Kingdom's far-flung world system of formal and informal empire was already dangerously over-extended, and it would become even more so in the aftermath of the First World War. As the cataclysmic events of the early 1940s would later prove, it would be impossible for Britain to defend or recover many of its imperial possessions without essential American help and assistance.

Some of these apprehensions were well reflected in the 'invasion scares' that were as widespread in the 1900s as they had been a century before. During the 1790s and 1800s the fear had been of assault by Revolutionary and Napoleonic France across the English Channel, and there had been further anxieties at mid-century; now the cause of national concern was the predatory predilections imputed to imperial Germany across the North Sea. Such worries had first emerged at the time of the Franco-Prussian War, in the aftermath of which George Chesney had written a fictional account of the Prussian army's invasion of England entitled *The Battle of Dorking* (1871). But it was Britain's diplomatic isolation during the Boer War, combined with the subsequent revelations concerning the empire's military unpreparedness, and the growing fear of the German menace, that gave rise to the explosion in 'invasion scare' literature during the 1900s. The prototypical novel was Erskine Childers's *The Riddle of the Sands* (1903), perhaps the first work of espionage fiction, linking the earlier writings of Wilkie Collins with the later 'shockers' of John Buchan. This was followed by William

Le Queux's *The Invasion of 1910* (1906), Patrick Vaux and Lionel Yexley's *When the Eagle Flies Seaward* (1907), and Henry Curties's *When England Slept* (1909). In every case the enemy was invariably Germany, and the genre became so widespread and formulaic that the young P. G. Wodehouse sent it up in his brilliant parody, *The Swoop! or How Clarence Saved England: A Tale of the Great Invasion* (1909), in which a heroic boy scout single-handedly saved his country from the simultaneous invasion of eight foreign armies, joined by the mighty Swiss navy carrying out an audacious attack on Lyme Regis. As with his later Berlin broadcasts, Wodehouse's light-hearted intervention into the serious business of war and national survival won him few admirers.

Fear of threats from without was accompanied by a growing anxiety concerning subversion from within, from spies stealing military and diplomatic secrets to anarchists wanting to blow up important buildings in London. Such figures represented an additional threat from foreign interlopers, in addition to the influx of Jews from eastern Europe. Several of Conan Doyle's later Sherlock Holmes stories explored these themes, among them 'The Bruce-Partington Plans' and 'The Naval Treaty', as did Joseph Conrad in *The Secret Agent* (1907), which centred around a terrorist bomb plot in London based on a real incident that had taken place in Greenwich in 1894. This helps explain the creation of the Secret Service Bureau in 1909, a joint venture by the Admiralty and the War Office, which was initially preoccupied by the German threat, and was the precursor of MI5 (dealing with domestic subversion) and MI6 (concerned with foreign intelligence). Conrad was Polish by birth, had reached the United Kingdom via France and lengthy service in the British merchant marine, and *The Secret Agent* was the only one of his famous novels that he set in England. Others drew on his experiences sailing the oceans of the world, but they may also be read as powerful contemporary meditations on the contradictions and limitations of Britain's global reach and imperial power. *Heart of Darkness* (1899) engaged with the issues of imperialism and racism, 'civilization' and 'savagery'. *Lord Jim* (1900) took many of the standard ingredients of the imperial adventure novels of Rider Haggard and Robert Louis Stevenson, and turned them on their heads. And *Nostromo* (1904) depicted the excesses of multinational capitalism in a story set in the fictional South American republic of Costaguana, for which the author drew on his earlier experiences when he had sailed to

Mexico. But Conrad was far from being the only writer of fiction whose work provides powerful insights into the boastful yet anxious imperial Britain of the 1890s and the 1900s.

NATION AND PEOPLE

A little earlier, in 1895, another novelist, H. G. Wells had published one of his most innovative science-fiction books, entitled *The Time Machine*, which imagined that human beings might travel backwards and forwards across the centuries, to encounter life as it had been lived before their day and as it would be lived in the future. Suppose, then, that a British family living at the beginning of the nineteenth century, a husband and wife with their children, had been able to travel forward through time, and arrived in the United Kingdom just over one hundred years later, during the early years of the first decade of the twentieth century. What would have appeared different to them, and what would have looked most the same? By what continuities would they have been most reassured, and by what changes would they have been most surprised? As they stepped gingerly from their time machine, which had conveniently landed them in the heart of London, they would soon have discovered that Britain's position in the world was familiar, yet not quite the same. They would find out that most of South Asia was now ruled directly or indirectly from London, that the East India Company had long since been abolished, and that vast swaths of the African continent, along with many islands in the Pacific, had recently been annexed. They would read that the colonies in British North America had been confederated into a transcontinental Canadian nation, that the penal colony of Botany Bay had morphed and grown into the Australian Federation, and that an even more tightly bound British dominion was being contemplated for South Africa in the aftermath of the Boer War. In the light of the many continental coalitions through which they had lived in the 1800s, the time travellers might not be surprised to learn that Britain had recently renewed its European commitment after a long period of detachment from direct involvement in continental affairs; but they might have been taken aback to discover that the recent military alliance had not been with Germany but with France.

Depending on their political views and places of origin, Wells's time travellers might have been surprised, relieved or disappointed to learn that the Union between Great Britain and Ireland, which had been so controversial and so novel in their own day, still held. The merging of both crowns and both parliaments had lasted for a century, and there was no historical precedent for such a sustained attempted to integrate the two nations and peoples across the Irish Sea. They might also have noted the important part played by people of Anglo-Irish origins in Britain's nineteenth-century affairs, ranging from the Duke of Wellington and Lord Castlereagh at the beginning, via Fergus O'Connor and Lord Palmerston at mid-century, to Charles Stuart Parnell and Oscar Wilde at the end. But they might also have learned about the extraordinary demographic catastrophe of the Great Famine, about the resulting massive Irish Catholic emigration to the United States, which superseded and overwhelmed the predominantly Protestant Irish emigration of their own time, and about the deep and abiding legacy of bitterness and hostility that the famine had left behind. This in turn might have helped them understand how and why British attempts to assimilate nineteenth-century Ireland had been constantly thwarted and disappointed: by the unbridgeable gulf between Catholics and Protestants, by hostility and resentment on one side and ignorance and incomprehension on the other, by the repeated failures of successive London governments to make coercion or conciliation (or some amalgam of the two policies) work, and by the inability of British statesmen who thought they had answers to the Irish question to carry the necessary measures in London – in the case of the Younger Pitt and Catholic Emancipation, and Gladstone and Home Rule. And they might finally have been surprised by the growth in the intransigence of Protestant Ulster since the 1880s, which in the twentieth century would make solving the Irish question more difficult, not less.

Yet despite these concerns about invasion and Ireland, the time travellers from the 1800s would also notice that the British monarchy still thrived and flourished, and that the great-grandson of George III sat securely upon a throne that was less politically influential, but was at least as popular, as their own sovereign had been one hundred years before. The French monarchy, in both its Bourbon and Bonapartist guises, might have gone, but there was no appetite for republicanism in early twentieth-century Britain, even among most members of the fledgling Labour Party. Like many of the continental crowns, along with such Asiatic thrones as Japan and Siam, the British monarchy was publicly more splendid and ceremonially more spectacular than it had been one hundred years before, and as the sovereign who reigned over the greatest empire the world had ever seen, it was hardly surprising that he had inherited from his mother a new imperial title to go with it: for Edward VII was the first male British monarch to be both *Rex et Imperator*, king and emperor. He also retained close dynastic links with the other great ruling houses of Europe, especially Germany and Russia, although there would be increasing difficulties with the Kaiser, on account of their long-standing personal animosity and growing national antagonisms, while relations with the Tsar were always more cordial, even as his autocracy was generally disapproved of in Britain. As such, Edward VII was what might be termed a multivalent monarch: for he was simultaneously a focus of national loyalty, a senior member of the crowned European cousinhood, and the cynosure of a global empire. He reconciled these different roles with ease and aplomb, but by 1914 his son and successor, George V, would find it much more difficult.

Despite his mother's forebodings, and King Edward's undeniable moral shortcomings, which would have shocked austere Anglicans and the 'Nonconformist conscience' alike had they known, Wells's time travellers might have been interested to learn that although Edward had not been a good or successful Prince of Wales, he turned out to be a generally good and popular king. Unlike his mother, he willingly settled for a royal role that was ornamental rather than fundamental, 'dignified' rather than 'efficient', and he performed the public functions of the monarchy with style, gusto and elan, restoring (for example) the state opening of parliament, something his mother had generally declined to participate in after Disraeli's death. His horses had won the Derby when he was Prince of Wales in 1896 and 1900, and they would do so again when he was king in 1909. These successes brought him enormous acclaim across a broad spectrum of society, since horse-racing was one of the few sports enjoyed by all classes, from rich owners to middle-class spectators to proletarian gamblers. As king, Edward was also noticeably more liberal in his private political views than the Tory jingo his mother had become, and than the insular conservative his son and successor would turn out to be. Moreover, as Prince of Wales, he had been a member of the Royal Commission on the Housing of the Working Classes that Gladstone had set up in 1884, and he had been a pallbearer at the Grand Old Man's funeral in Westminster Abbey. Queen Victoria had been predictably outraged, and inquired of her son whose advice he had sought, and what were the precedents for behaving as he did. The prince replied rather magnificently that he had consulted no one, and did not believe there were any precedents. As king, his relations with the Liberal governments of Campbell-Bannerman and Asquith were much better than his mother's would ever have been, at least until the problems later associated with the 'People's Budget' and House of Lords reform arose.

These Wellesian visitors from a century before might also have been interested to learn that the hereditary principle survived and flourished in another part of the law-making process. The House of Lords was still entirely unelected, and retained the power to reject any measures sent up from the Commons (as it had done recently and overwhelmingly in the case of Gladstone's second Home Rule Bill) with the customary exception of legislation concerning finance. But since the 1880s, patterns of ennoblement of new peers had begun to shift, away from traditional landowners, and towards bankers, businessmen and brewers such as Lord Rothschild, Lord Cowdray and Lord Iveagh. As a result, the upper house was becoming less aristocratic and more plutocratic, reflecting the recent transformations in the nature of the nation's wealth elite. There had been parallel changes in the lower house, where the earlier preponderance of peers' relatives and country gentlemen had been progressively eroded, and once again the 1880s had been the pivotal decade. Many MPs now came from business or professional backgrounds, a few were working men from the trades unions, and they were all elected on a far wider franchise than had existed in the 1800s; but it still only gave the votes to 60 per cent of adult males, and to no women. Across the intervening century, parliament had become much busier, as the range of domestic legislation increased, and as more time was taken up passing laws for Ireland (often with difficulty) and dealing with the empire. Uniquely among the nations of the world, the outward forms of the United Kingdom's constitutional government, namely 'the crown in parliament', remained essentially unaltered, but the personnel and the activities had changed considerably, as indeed had the Palace of Westminster itself; its mid-Victorian Gothic extravagance would certainly have surprised Wells's visitors, who must have wondered whatever happened to its venerable predecessor.

These visitors from an earlier world would also have noticed that the nature of British political culture had significantly evolved across the nineteenth century. The riots, the protests, the violence, the subversion, the repressive measures and the prosecutions that had been so marked a feature of the 1790s and 1800s had long since all but disappeared, as popular engagement with, and acquiescence in, party politics and Westminster politics increased in the aftermath of each of the three episodes of parliamentary reform. Between 1832 and 1865 voter turnout at general elections varied between 53 and 65 per cent; in the contests of 1868 and 1874 the range was from 66 to 69 per cent; but after 1880 it varied between 72 and 83 per cent, an astonishingly high rate of participation by contemporary continental standards. The franchise might still remain restricted, but by the 1890s national parties, enjoying mass popular membership, and involving both women and men as campaigners and volunteers, were increasingly making and unmaking governments, as was the case in 1892, 1895 and 1906. And with a larger voting public and much more sophisticated party organizations than had earlier been the case, elections were more than ever fought on national issues, candidates and senior political figures undertook a greater number of speaking engagements, and the literature and propaganda produced at headquarters in London was supplemented by posters and pamphlets printed locally. Among politicians and voters alike, it was widely believed that the representational system was more 'democratic' than in fact it was, as they extolled (or on occasions regretted) the advent of 'mass politics' and a 'popular' electorate. This was a very different political culture from that with which Wells's visitors were familiar, and from that which the Barry and Pugin Palace of Westminster had been constructed to embody and proclaim.

There would be many other aspects of the changed appearance of the great metropolis by which the time travellers would be startled, in large part resulting from the staggering increase in its population since the 1800s. In addition to the Houses of Parliament, the city had been transformed by the construction of such public buildings as the Foreign Office in Whitehall (Classical not Gothic), the Law Courts on the Strand (Gothic not Classical), and the neo-Baroque Treasury bordering Parliament Square. Wells's time travellers could also visit a remarkable array of museums and galleries, which had been established since their day, among them the National Gallery on Trafalgar Square and the National Portrait Gallery nearby, as well as the Science Museum, the Natural History Museum and the Victoria and Albert Museum, all in South Kensington. They might marvel at the great metropolitan railway termini, of which Paddington, St Pancras and King's Cross were architecturally the most outstanding, and at the many bridges crossing the Thames, of which Tower Bridge was the most recent; and if they journeyed further east, they could observe how much the London docks had expanded since their day. They could indulge their literary interests in the reading room of the British Museum, attend concerts at the Royal Albert Hall or St George's Hall, shop at such department stores as Whiteleys in Bayswater and Harrods in Knightsbridge, or visit theatres in Shaftesbury Avenue and dine afterwards in restaurants close by. They could also get about with much greater ease than had been the case in the 1800s, as mass transport in the metropolis had been revolutionized: horse-drawn omnibuses and hansom cabs had been available since the 1830s, the Underground from the 1860s, and electric trams beginning in the 1880s.

Even more than had been the case a century before, London was a multi-functional, multi-faceted metropolis, as the capital of England, of the United Kingdom and of the British Empire, and also as the one undisputed 'world city'. As in the 1800s, but on a larger scale, it was the place where the court, the government, the legislature, the civil service and the judiciary were all headquartered. It was the centre of the social, cultural, religious and intellectual life of the nation and empire, as well as being a great port, a major manufacturing centre, and the finance capital of the globe. Small wonder that in 1908 it staged the Olympic Games, when 300,000 spectators watched 1,500 competitors from nineteen different nations. Yet although no other city in the United Kingdom or, indeed, anywhere else, could rival London, Wells's visitors would have been right to venture further afield, to see the industrial economy that was so much more pervasive, consolidated and significant than it had been in the 1800s, concentrated in the Midlands and the north of England, and also well developed in south Wales, on the Clyde valley in Scotland and around Belfast in Northern Ireland. They would have noted that the wind, water and animal power that had prevailed in their day had been largely superseded by coal-fired steam power, that the railways had transformed the nation's transport infrastructure, that sailing ships had been rendered obsolete by steamships with iron hulls, and that the earlier small-scale workshops were being replaced by larger factories. They would have been impressed by the coalmines scattered from Glamorgan, via Northumberland and Durham, to Lanark, by the Staffordshire Potteries, the Lancashire and Yorkshire textiles, the shipbuilding located on the Tees, the Clyde and in Belfast, and the new and developing chemical industry on Merseyside.

Wells's visitors would also have discovered that municipal authorities possessed wide powers and raised considerable revenues via the rates, that mayors and lord mayors were expected to be rich and important figures embodying a sense of local identity, and that continuing civic pride was well proclaimed by the grand town halls in such cities as Sheffield, Glasgow, Cardiff and Belfast that had been constructed during the final decades of the nineteenth century. Many mayors, aldermen and councillors were businessmen with strong local connections, and they often lived in well-developed, high-status suburbs, such as the West End in Glasgow, Headingley in Leeds, Edgbaston in Birmingham and Clifton in Bristol. Indeed, one of the most significant changes that Wells's time travellers from the 1800s would have observed was the proliferation of suburbia, not only in the provincial towns and cities, but above all in London itself; and not just for the comfortably off bourgeoisie and professionals in Hampstead and Highgate, but increasingly for the lower middle classes (a social and occupational category that had scarcely existed in the 1800s) in Lewisham and Camberwell. Yet despite increasing efforts by local authorities to eradicate sub-standard housing, pioneered by Joseph Chamberlain in Birmingham, and the early provision of municipal dwellings, begun by the LCC in 1900, slums and rookeries also remained a marked feature of town life and city living, where the urban poor existed in insanitary, disease-ridden, overcrowded conditions that observers such as Charles Booth and Seebohm Rowntree had so vividly described. And most towns and cities were heavily polluted, by the stench of horse droppings, by the smoke and soot emanating from factory chimneys, domestic hearths and railway engines, and by the unrelenting noise of clopping hooves on cobbles (with which Wells's visitors would have been familiar) and the constant din of factory production (much more pronounced than in their day).

The overwhelming importance of urban life and industrial production would not only have impressed but also surprised the time travellers from one hundred years before: for whereas in their day the towns and cities of the United Kingdom had been largely dominated by the personnel and the politics of the countryside, that pattern had been almost completely reversed across the intervening century. They would have noted that the total acreage under cultivation had significantly diminished since the high point of the Napoleonic Wars, that agriculture on the Celtic fringe and in the wheat-producing areas of England was especially depressed, that farm labourers were only a very small percentage of the national workforce, and that the United Kingdom was no longer self-sufficient in food. From this perspective, the countryside was increasingly marginal to national life, and had become more a place for recreation rather than production, as city-dwellers rode their recently invented bicycles into the shires; and as the invention of the internal combustion engine meant an increasing number of car owners were bringing noise, dust and danger to rural roads, among them the prime minister, Arthur Balfour, and Edward VII himself. In 1896 the speed limit had been set at fourteen miles an hour, but the Motor Car Act of 1903 raised it to twenty miles an hour. The following year there were more than 8,000 cars registered. Many of these could be driven faster than the statutory limit, and as a result, traffic offences were significantly on the rise; indeed, the immediate occasion for the establishment of the Automobile Association in 1905 was the ardour of the Sussex police in trying to curb speeding motorists. To the time travellers from the Revolutionary and Napoleonic Wars this would have seemed a very different sort of organization from the radical and subversive corresponding societies that had been set up in the 1790s for very different purposes.

Although the Wellsian visitors would have recognized some buildings that had survived from the eighteenth century and before, among them ruins like Stonehenge and Fountains Abbey, cathedrals and churches, castles and country mansions, royal palaces and customs houses, Georgian terraces and Hanoverian squares, they would have been amazed by the dramatically changed appearance of the United Kingdom's physical fabric, most of which had been constructed since their day. They would also have been incredulous that between 1801 and 1901 the nation's population had risen from sixteen million in

1801 to almost forty-two million a century later, notwithstanding the dramatic reduction of numbers in Ireland as a result of the Great Famine and the departure of millions of emigrants for the New World. Yet they might also have noted that many aristocratic and landed families who had been prominent during the first decade of the nineteenth century were still significant during the first decade of the twentieth. Dynasties such as the Bedfords, Butes, Derbys, Devonshires, Sutherlands and Westminsters remained prodigiously rich, not on account of their huge land holdings, but because of the income they drew from their urban estates, coalmines, docks and harbours, and stocks and shares, which effectively insulated them against the agricultural depression. This helps explain why many of them remained prominent in national and local politics and high society. But other grandees and gentry, with revenues drawn entirely from their agricultural rentals, suffered diminutions in income from 25 per cent to 30 per cent, which was one reason, along with the Third Reform Act and the reform of local government, why many of them were playing a lesser part in public life than their predecessors. In Ireland the decline of the landed establishment was even more marked, following the Land War, successive Land Acts, local government reform, and the threat of Home Rule, which encouraged many owners to sell out to their tenants on the best terms they could get.

Moreover, since Gladstone had taken up Home Rule in 1885, aristocrats and gentry who had previously espoused Whiggery or Liberalism had in the main gone over to the Conservatives, which meant they could be vulnerable to a Liberal government with a large Commons majority; and this was also increasingly true of the proliferating numbers of plutocrats, businessmen and professionals, whose numbers were markedly increasing at this time (although some would return to the Liberal fold for the 1906 election). Just as the richest landowners were often involved in a variety of non-agricultural ventures, so many of the upper middle class invested some, but by no means the whole, of their wealth in land, enjoying the pleasures of rural living, leisure and pursuits, without the corresponding burdens of owning country estates that yielded little in rental. Wells's visitors might, for example, have been impressed by the palatial country residences the Rothschilds (of whom scarcely anyone in Britain had heard in 1800) had constructed in Buckinghamshire. They might have noted the new fashion among plutocrats such as Sir Thomas

Lipton for steam-powered, ocean-going yachts, which he vainly raced in the hope of winning the America's Cup, or the new craze for shooting thousands of game birds, which Edward VII and the Prince of Wales (and future George V) fully shared. Or they might have been present, in 1903, at the christening of a baby named Kenneth Mackenzie Clark, whose Scottish father had inherited a fortune his forebears had made in the textile trade, and who lived a life of philistine repose on an estate in Suffolk. Kenneth Clark would later rebel against his parents, embracing art history and connoisseurship, and insisting that the first decade of the twentieth century marked a high point of civilization, in Britain and in Europe more generally.

Lower down the social scale, Wells's time travellers would have encountered those members of the middle class who lived in the suburbs of London and the great provincial towns and cities, who had increased significantly in numbers from the 1880s to the 1900s. The well to do of Birmingham, many of whom lived in Edgbaston, its most fashionable suburb, might stand proxy for this group. The most famous residents, who lived in the largest villas, were that knot of Nonconformist families who frequently intermarried, among them Chamberlains, Kenricks, Nettlefolds, Cadburys, Crosskeys and Beales. Some were involved in local businesses, others were lawyers, and many served on the City Council. Joseph Chamberlain was the most famous, but he burst the bounds of Birmingham to become a major figure in British politics. So, in another way, did the Chocolate manufacturer George Cadbury, who remained a Liberal after 1885, and in 1901 purchased the *Daily News* to campaign in favour of old-age pensions, and against the Boer War. More local in his loyalties was the solicitor Charles Gabriel Beale, who had a variety of local business interests in banking and railways, was three times Lord Mayor of Birmingham between 1897 and 1899 and in 1905, and subsequently the first Vice Chancellor of the University of Birmingham from 1900 to 1912, of which Joseph Chamberlain was the founder and first Chancellor. Such figures were commonplace among the late Victorian provincial middle classes. Gurney Benham of Colchester was another example; he inherited the family printing business, was editor of the *Essex County Standard* for fifty-nine years, a local and prolific antiquary, served three times as Mayor of Colchester, and reinvented the annual Oyster Feast as a major piece of civil ceremonial.

It would not have taken long for Well's visitors from the 1800s to appreciate that the social hierarchy of the turn-of-the century United Kingdom had become much more elaborate and finely graded than it had been one hundred years before, as the developing industrial and service economy had created many new occupations, and many new jobs. In particular, they would have noticed the marked proliferation of the lower middle class of 'white-collar workers', among them those clerks, small-scale shopkeepers and schoolteachers who tended to live in such London suburbs as Balham, Hammersmith or Leyton, often on the edge of older suburbs of higher social status. They also proliferated in the provinces, and Alfred Roberts, father of Margaret Thatcher, was a quintessential example, eventually owning two grocer's shops in the Lincolnshire market town of Grantham. Such petty-bourgeois neighbourhoods were immediately recognizable, and easily distinguished from the suburbs of the more prosperous members of the middle class, because they generally consisted of terraced houses rather than detached or semi-detached villas, which became more spacious and better equipped by the 1890s, with sufficient bedrooms for the children, and even an indoor bathroom and toilet. One sign of the respectability to which the lower middle classes aspired was the keeping of a servant, even if it was only a single, inexperienced, over-worked, live-in teenage girl or, for the less well off, an occasional domestic help coming in from outside. It was this rapidly expanding social group that Alfred Harmsworth targeted with the *Daily Mail*, as did his competitors; by the turn of the century the lower middle classes were becoming an increasingly reliable source of Conservative and Unionist support.

The temporal voyagers from the 1800s would have been surprised to discover how varied were the middle classes both in the range of their occupations and the range of their incomes. 'Gentlemen farmers', in the fringes of county society, and small-scale cultivators of a hundred acres or less, were as much members of the middle class as Charles Gabriel Beale, or Gurney Benham – or the eponymous Robert Thorne in Shan Bullock's novel, *The Story of a London Clerk* (1907). As for their incomes: at the top were the millionaire dynasties such as the Rothschilds in banking, the Willses in tobacco, and the Guinnesses in brewing, and such prodigiously rich individuals as Harmsworth, Cassel and Cowdray, who were wealthier than all but the most moneyed aristocrats. Successful professional men were also living well, although not at the same high level: R. B. Haldane was earning between £15,000 and £20,000 a year as a barrister before he went into politics, while a Permanent Secretary in Whitehall was being paid between £2,000 and £2,500 a year. But most professional men made much less: barristers' earnings averaged below £400, and general practitioners' less than £400, while many clergymen, on scarcely £200 a year, were getting less than skilled industrial workers in full employment. Even lower down the middle-class income scale was the elementary schoolteacher, on barely £150 a year (the lowest level of taxable income in the 1900s), often living, like many members of the lower middle class, in little more than genteel poverty, aspiring to a middle-class lifestyle but lacking the requisite means to achieve it, while small shopkeepers and clerks suffered the added disadvantage of job insecurity. There were no old-age pensions until 1908, and even then, they were only 5s a week for those aged over seventy.

In the same way that Wells's visitors from an earlier time would have found many different and often subtle gradations demarcating different layers and levels of the British middle classes, so they would have found the same in the case of the working classes, who formed the vast majority of the population. But there was also one single and crucial distinction, between those who were 'respectable' and those who were not, which paralleled, though did not completely coincide with, the difference between those men who had the vote and those who did not, and those men who were usually in work and those who were not. By the late nineteenth century the spread of mechanization meant the erosion of many traditional craft skills and also the reorganization of many enterprises, such as footwear, food processing, cigarette manufacturing and newspaper production along factory lines. Nevertheless, by American standards, British industry remained small scale: in 1898–99 the average workshop employed fewer than thirty male workers, and ten years later there were only one hundred firms with 3,000 people on their payrolls, and these employees represented a mere 5 per cent of the total labour force. But de-skilling was one reason why trades union membership increased from the late 1880s. It was highest among metalworkers (especially in shipping), coalminers and cotton operatives, and it was significant, but weaker, among printers, railwaymen and sections of the building industry. Yet whole areas of working-class life had scarcely been touched by trades unionism, since in many county, market and seaside towns (as in Robert Tressell's 1914 novel, *The Ragged Trousered Philanthropists*) there were no large employers whom the workers could identify as a clear class enemy.

Among such people there was little sense of class solidarity or brotherhood, of the sort that bourgeois intellectuals, from Marx to the Fabians, believed, wished and hoped might have existed. For the manual working class, which comprised over three-fifths of the labour force, were living and working in very disparate circumstances, and the differences between the skilled, semi-skilled and unskilled were of great significance. Indeed, Charles Booth, in his vast and famous survey of *Life and Labour of the People of London*, broke the manual workers down into six separate 'classes' or categories. Engineers who had served an apprenticeship earned almost double the wages of the labourers who worked alongside them, and were highly conscious of their superiority in terms of status and respectability. Likewise foremen and supervisors, who formed the non-commissioned officers of the United Kingdom's industrial army, possessed an even more complex social identity, since they were often manual workers by origin but aspired to the lifestyle of the lower middle class. In coalmining and iron-making, by contrast, stratification was based on age and strength rather than skill, which offered every worker a chance to earn high wages at some point in his career. In the same way agricultural labourers might live in model dwellings on a great aristocratic estate, or they might endure near-poverty in insanitary rural squalor. Many better-paid, urban-based workers in regular employment, earning perhaps 45s per week, which would be double that of their unskilled colleagues, took advantage of cheap workmen's trains and the expanding network of trams, which offered escape from the overcrowding, squalor and expense of inner-city living. Their households might even contain a servant, since they could keep an orphan girl from the workhouse at practically no cost.

Beneath the respectable working class were those whom Marx had earlier described as the 'lumpenproletariat', who were known to turn-of-the-century commentators as the 'residuum', or the 'underclass', or the 'people of the abyss' (as Jack London described them in 1903), whose early nineteenth-century forebears might have been familiar to Wells's recent arrivals. At best, they were casual workers, such as dockers, suffering from chronic underemployment; at worst, they were vagrants, mendicants and paupers, on whose fringes were to be found the semi-criminal and criminal classes, including pickpockets and prostitutes. Unlike the better-off and more securely employed workers higher up the social and occupational scale, casual labourers could not afford to move far from their place of (intermittent) employment: dockers, for example, had to live close to the wharfs, so they could hear by word of mouth when a ship would need loading or unloading. As a result, they often lived in overcrowded accommodation where avaricious landlords charged high rents, and such tenants were consigned, in the words of John Burns, to the lower depths of a 'jerry-built hell'. Well over three million people were living in overcrowded conditions, and millions more fared little better. The extremely difficult existence endured even further down the social scale was vividly portrayed by Arthur Morrison in his novel *A Child of the Jago* (1896), set in a thinly disguised version of the East End slums between Shoreditch High Street and Bethnal Green Road, which explored the vain attempts of young Dicky Perrott to escape the life of poverty and crime to which he was condemned by the degraded environment in which he grew up.

CULTURE AND SOCIETY

The travellers from the 1800s, perhaps clutching their copy of Wells's *Time Machine*, would certainly have been busy, checking out the constitutional arrangements of British government, and getting a sense of its varied industrial, urban and social structures one hundred years on from their own era, visiting the four constituent parts of the United Kingdom, and hearing about the realms and dominions beyond the seas. They would have found many significant continuities, not least that Britain remained a very unequal society, but they would have been at least as much impressed by the extraordinary transformative changes that had taken place in all these aspects and areas of national life across the nineteenth century. If they were statistically inclined, they would have discovered that both the birth rate and the death rate had been declining since the 1870s, that life expectancy had risen across the nineteenth century, albeit only slightly and slowly from less than forty to more than forty, and that family size had been significantly reduced, not only since the beginning of the nineteenth century, but from an average of six for couples who married in the 1860s to four for the pair who wed in the 1900s. They would also notice a corresponding change in attitudes to childhood, at least on the part of the middling classes, namely the recognition that children were not miniature adults but had their own age-specific personalities and needs. Increasingly, childhood was redefined to signify dependence on parents, economic and sexual inactivity, and an absence of legal and political rights, but there was also a growing awareness that children needed to be protected and cared for, and that the state and voluntary organizations had an obligation to do so.

Wells's journeyers might also have learned that the London Society for the Prevention of Cruelty to Children was founded in 1884 (sixty years since the foundation of the society to prevent the mistreatment of animals) by Lord Shaftesbury, the Reverend Edward Rudolf and Benjamin Waugh, a Congregationalist minister, and that it had subsequently developed local branches across the country. From the 1890s, agents of the LSPCC had begun entering working-class homes, and they were soon investigating 10,000 cases a year, instigating numerous prosecutions of parents, only a minority of whom were really poor. In 1885 the age of consent was raised for girls from thirteen to sixteen, and in 1903 (and again in 1908) parliament passed legislation prohibiting incest, as a further way of protecting children. The LSPCC's hand was further strengthened by a succession of new laws, such as the Prevention of Cruelty to Children Act of 1889, which made it an offence carrying a fine or imprisonment to neglect, ill-treat or abandon children in ways that would cause additional suffering, and it placed more restrictions on child employment. A further Act passed five years later obliged Poor Law Guardians to accept children brought to them for protection, and the Poor Law Acts of 1889 and 1899 gave them authority to terminate parents' rights over abused children. By this time prison was no longer regarded as appropriate punishment for young offenders, hospitals were established specifically to deal with young boys and girls, and in 1907 the Child Study Society was founded as a forum for a new type of professional, the child psychologist.

To be sure, childhood experiences continued to vary, by class and region. Infant mortality was one of the few indexes of well-being that did not fall during the nineteenth century: in 1899 it stood at 163 deaths per 1,000 births, the highest figure since records began sixty years earlier. But by the 1900s these deaths were increasingly concentrated among those at the lower end of the social scale, where insanitary and unhealthy environments, combined with parental negligence and ignorance, made the life chances for newborns far more challenging than elsewhere, and infant deaths were also more marked in those industrialized parts of the United Kingdom where married mothers went out to work. For those babies who survived into childhood in the roughest working-class areas, caning and slapping, and much worse, were considered salutary disciplines; and Arthur Morrison's Jago-dwelling boy was probably typical of his milieu in becoming accustomed early on in life to physical violence, not only from his parents, but also from other children. The greatest change in attitudes and circumstances took place among the middle classes, where emotional intimacy was increasingly displayed between parents and children, which led to an idealization of childhood as a magic phase of life, and a time of heightened emotional sensitivity. This child-centred view of the world was proclaimed on the stage and on the page in such works as J. M. Barrie's *Peter Pan* (1904), Mrs [sic] Clifford Mills's *Where the Rainbow Ends* (1911), Beatrix Potter's *The Tale of Peter Rabbit* (1902), Edith Nesbit's *The Railway Children* (1906) and Kenneth Grahame's *The Wind in the Willows* (1908), and by the advent of such new toys as teddy bears and model train sets, sold at Hamleys, which had opened a branch on Regent Street in 1881. But the aristocracy and gentry still saw little of their children, who were handed over to nannies, nursemaids and governesses, and then, in the case of boys but not girls, sent away to boarding school.

The visitors from the 1800s might have been further impressed by the greater provision of state-supported education, especially at the elementary level, where schooling had been free since 1891. In 1880 school attendance had been made compulsory until the age of ten, in 1893 until eleven, and in 1899 twelve. The following year local authorities were empowered to raise the leaving age to fourteen, although few did so. Initially, these increases in the school-attendance age had been opposed by many working-class parents, who expected their children to go out to work to augment the family income. Moreover, there were loopholes in the legislation that meant many boys and girls left school earlier, and the practice of attending school part-time, while doing a job part-time, was also widespread. In the first decade of the twentieth century the number of schoolchildren who were working for wages in shops, agriculture or domestic service, or in factories and workshops, was somewhere between one quarter and one half of a million, and in an effort to reduce costs many employers began to substitute teenage for adult labour. Thirty per cent of textile workers were lads under twenty, 95 per cent of Birmingham boys aged between fifteen and twenty had jobs, and there was a growing use of young female labour in the potteries, shoemaking and hosiery. Perhaps most surprising to the time travellers would have been the simultaneous growth in a range of semi-militaristic organizations, which appealed to the sons of the respectable working classes and the lower middle classes, among them the Boys' Brigade (1883), the Church Lads' Brigade (1891) and, most famously, Robert Baden-Powell's post Boer-War Boy Scouts (1910).

In the immediate aftermath of Forster's Education Act of 1870, the most that working- and lower middle-class children might hope to learn in their elementary school had been a basic grounding in the 'three R's': reading, writing and arithmetic. But by the end of the nineteenth century the curriculum had widened, at least in the best schools, to include sports, physical education, and trips to local museums and historic sites. By 1900 4.6 million children were being educated in 31,234 elementary schools in England and Wales, the highest numbers ever, but there was scarcely any prospect of pupils from humble backgrounds going on to secondary let alone higher education. The best they could hope for in later life was to attend lectures sponsored by the Workers' Educational Association, established in 1903 by a group of bishops, academics and self-educated labouring men. In 1895 a Royal Commission, chaired by James Bryce, concluded that there had been insufficient progress in the provision of secondary education since the Taunton Commission had reported in the 1860s, and that it must be significantly improved if Britain was to remain competitive in trade and in war into the new century. In terms of material prosperity, intellectual activity, and the happiness and moral strength of the nation, Bryce concluded, 'the extension and re-organization of secondary education seem entitled to a place among the first subjects with which social legislation ought to deal'. The result was the creation of the Board of Education in 1900, with the Duke of Devonshire as its first president. But by 1905 there were still only 575 secondary schools in England and

Wales, attended by a mere 95,000 pupils. The overwhelming majority were children of the professional and managerial classes – not surprisingly, since secondary schools charged fees for virtually all of their pupils.

Even more exclusive were the public schools, of which there were approximately one hundred by the end of the nineteenth century, ranging from such authentically ancient foundations as Eton and Harrow, via more recent establishments like Wellington and Haileybury, to relatively 'minor' schools such as Malvern or Oundle. Aristocrats and gentry still preferred to send their sons to Eton, although the majority of pupils in most of the public schools were sons of the new plutocracy, or of the richest members of the entrepreneurial and professional classes. By the early twentieth century it cost in the region of £300 a year to maintain a boy at a major public school, and £200 a year at a lesser establishment. In these total environments, boys fagged, bullied and bonded with each other, and were subjected to severe discipline in order to control the rebellious urges and sexual desires characteristic of that 'difficult' phase of life now known as adolescence. Education was still primarily based on Classics and the humanities, with little attention given to mathematics or the sciences, the aim being to produce young men of good, Christian character, who could play games and who knew how to lead their social (and racial?) inferiors in politics, the armed forces, the civil service and in the government of the empire. Such values would be celebrated in Henry Newbolt's public-school verse, and would be gently lampooned in many of P. G. Wodehouse's stories. But by the end of the nineteenth century, criticism was mounting of the all-pervasive cult of 'Muscular Christianity', as reflected in an article entitled 'Our Gentlemanly Failures', published in the *Fortnightly Review* in the year of the Diamond Jubilee. 'Grown and growing up,' it argued, 'we see them everywhere: bright-eyed, clean-limbed, high-minded, ready for anything, and fit for nothing.'

Wells's time travellers from one hundred years before would also have discovered that the late nineteenth and early twentieth centuries witnessed a revolution in higher education, albeit one that made little impact on the majority of the British population. In Oxford and Cambridge, college teaching was transformed in quality and quantity under the influence of such notable figures as Benjamin Jowett at Balliol and Oscar Browning at King's. Jowett placed great emphasis on public and imperial service (Grey, Elgin, Curzon and Milner were all Balliol men), but Browning was more interested in cultivating personal relations (especially via the secret society known as the Apostles, where he influenced figures such as John Maynard Keynes and Lytton Strachey). The foundation of the London School of Economics by Beatrice and Sidney Webb in 1895 was an innovative venture to promote the study of the social sciences; the University of London was drastically overhauled in 1900, when it became a teaching as well as an examining body; while the establishment of the Imperial College of Science in 1909 was an attempt to create Britain's answer to Charlottenburg, the great German Technical High School in Berlin. During the same period civic universities were also established at Birmingham, Manchester, Liverpool, Leeds, Sheffield and Bristol, while in Scotland the four ancient universities of Edinburgh, Glasgow, St Andrews and Aberdeen underwent significant expansion. In 1893 the University of Wales received its royal charter, affiliating colleges in Cardiff, Swansea, Aberystwyth and Bangor; and in Ireland legislation passed in 1908 would create the Queen's University in Belfast, and the National University of Ireland. The NUI offered a social, political and religious alternative to Trinity College Dublin, which remained the bastion of the Anglo-Irish Protestant Ascendancy.

This new wave of university expansion was the most significant that had yet occurred in the United Kingdom, and the number of full-time students more than doubled between the 1880s and the 1900s. But the proportion of Britons attending university, both full-time and part-time, remained tiny, and was not much above 1 per cent of the population, which may have been slightly less than in France or Germany, and was definitely less than in the United States. Much of the funding for the new civic universities came from local businessmen, such as the Wills (tobacco) and Fry (chocolate) families at Bristol, and the Muspratt brothers (chemicals) at Liverpool, or from tuition fees. The range of subjects taught at Oxford and Cambridge broadened significantly during the last third of the nineteenth century, encompassing natural sciences, moral sciences, the law and modern history. Oxford remained wedded to Classics and the arts, but Cambridge retained its traditional strength in mathematics, and became a world leader in the natural sciences, especially physics. In the new civic universities, the sciences and engineering were important subjects from the outset, in part as a result of the funding from local industrialists; but at least one-third of the students were reading arts subjects. Yet wherever their subject strengths lay, the world of higher education still impinged scarcely at all on the work or expenditure of the government or on the life of the nation. The Treasury had paid a small grant-in-aid since 1889, but in 1903 it only amounted to £57,000, scarcely 5,000 British undergraduates were receiving their degrees each year, and Manchester was the only civic university with more than a paltry 1,000 full-time students.

Of all the aspects of this minor but undeniable revolution in education, Wells's time travellers might have been most surprised and impressed by the fact that, by the Edwardian era, one-fifth of all undergraduates were women, of whom the majority were preparing for a career in school teaching through the study of such subjects as history, English and modern languages. Here, indeed, was an indication of another significant, but limited, revolution, namely the changed position of some (but by no means all) women by the late nineteenth and early twentieth centuries. The earlier ideology of 'separate spheres', according to which women stayed at home and cultivated domesticity, whereas men went out into the world, did a job and made money, had never been as all-pervasive in practice as contemporary prescriptive literature had suggested, and in any case by the late nineteenth century it was declining in importance. For in the United Kingdom as a whole almost one-third of the total labour force were women, according to the censuses, and they were to be found overwhelmingly in three occupations: domestic service (one and three quarter million), textiles (three quarters of a million) and clothing (one third of a million), which together amounted to well over three million women in paid employment. Moreover, this was undoubtedly an underestimate, since many working-class mothers often took on part-time, casual work, as 'chars' or in laundries, to supplement their husbands' low income or compensate for unemployment, while thousands of even poorer women were 'outworkers', employed in sweated trades, such as tailoring or box-making. Others turned to prostitution, perhaps as a rational career choice, but more often in sheer desperation, as the only way they could earn a living, as controversially suggested by George Bernard Shaw in *Mrs Warren's Profession* (1892).

From the 1880s to the 1900s there was also a significant increase in the demand for women from the lower middle classes, especially as teachers, nurses and office workers. Following Forster's Education Act, there was a sharp rise in the demand for female teachers, and by the 1901 census, they constituted three-quarters of the labour force in the classroom. Somewhere between the role of servant and teacher was the position of governess, the opportunities and risks of which were well explored in the Conan Doyle short story 'The Copper Beeches' (1892), in which Violet Hunter asks Sherlock Holmes whether she should accept such a post, given the strange conditions attached to it (she was compelled to cut off her lustrous tresses). Nursing was another occupation in which opportunities expanded, and by 1901 there were more than 60,000 of them. Beginning in 1887, nurses could register themselves voluntarily as appropriately trained and qualified practitioners, and the royal patronage of Princess Helena, daughter of Queen Victoria, and subsequently of Queen Alexandra, further raised the prestige of the profession, while the construction of new hospitals and the expansion of the empire increased demand at home and abroad. The advent of the typewriter, which was in widespread use by the early 1890s, led to a further growth in women's work. Because of its vague similarity to the sewing machine, it drew an increasing number of females into what had previously been male-dominated clerical occupations. The job of telephonist, which also came into being at this time, was dominated by women from the outset. By 1900 women made up 20 per cent of all white-collar workers, which gave them a freedom they were denied as domestic servants or as housebound wives. Hence the popular myth (among men) of the 'fast' female office employee, embodied in such books as Clara Del Rio's *Confessions of a Type-Writer* (1893).

For women from the better-off working classes and from the lower middle classes, employment opportunities were definitely growing, and Wells's visitors might have been equally impressed by the simultaneous improvement in educational opportunities, at least for middle- and upper-class girls. By 1895 there were more than 15,000 private girls' schools in the United Kingdom, of which thirty-six owed their existence to the flourishing Girls' Public Day School Company that had been established in 1872. This may help explain why, by 1901, 6,400 women were actresses, 3,700 were artists, and 22,600 were musicians, ranging from the composer Ethel Smyth to a number of humble music teachers. Smyth had studied at the Leipzig Conservatoire, but by her time there was a significant opening up in higher education for women. Girton and

Newnham Colleges in Cambridge, and Lady Margaret Hall and Somerville College in Oxford, were by this time well established, although female graduates could not formally take their degrees until 1920 in Oxford and 1947 in Cambridge. The University of London, by contrast, had allowed women to graduate from 1878, Scottish universities followed suit in 1892, and Trinity College Dublin (with some restrictions) in 1904, while all the new civic red-bricks admitted women on equal terms with men from the very beginning. By 1900 there were more than 2,500 full-time female students attending English, Welsh and Scottish universities, and the number was rising steadily, year on year. This in turn led to a gradual infiltration into what had previously been exclusively male professions. During the 1890s the British Medical Association had admitted women, and the first female factory inspectors had been appointed; and by 1911 women would constitute 6 per cent of the membership of the higher professions. That may seem a risibly low figure today; but at the time it signalled a very significant – and portentous – change.

There were also many ways in which women of all social levels were engaged in public and political life. The rates of female trades union membership, although relatively low in absolute terms, were high in Britain compared to most other countries, and showed the same pattern of growth as male membership. There were just over 10,000 women trade unionists in 1876; twenty years later the number had increased tenfold, the overwhelming majority working in cotton textiles, the one large sector of female employment that was unionized. Women were also increasingly active in local and national politics, as voters in municipal elections or elected members of School Boards and Poor Law Boards of Guardians. These rights were extended to the parish, rural district and urban district councils that were established by the Local Government Act of 1894, although it would be another thirteen years before females were allowed to sit on county and borough councils. Women were also involving themselves in national politics and party organizations. The (Conservative) Primrose League had been established by them in 1883, and the Women's Liberal Federation followed soon after. By the end of the century there were 50,000 members of the Women's Liberal Federation, and more than ten times that number belonged to the Primrose League. Women were also prominent among the founding Fabians, Beatrice Webb being the most famous example; they were also important in the establishment of the ILP, among them Annie Besant, Eleanor Marx and Julia Varley. Other women were influential in particular political campaigns, such as Josephine Butler, who led the crusade to repeal the Contagious Diseases Acts. Middle-class women were also thought to be naturally suited to philanthropy, because of their domestic orientation and nurturing instincts, and by the mid-1890s half a million were working continuously in voluntary activities.

Despite all these recent developments, women continued to labour under many disadvantages, both legal and customary. Pubs and working-men's clubs were generally off limits to them. Those who were in employment almost invariably earned less than men, even when they were doing the same job, as in the case of schoolteaching. Female clerks drew less than one-third of the average male salary, and those working in the textile industry little more than one-half, while shop assistants received two-thirds of what men were getting. The notion of a 'woman's rate' for the job, which would invariably be less than a man's, was widespread, while shop assistants and barmaids, whether married or not, struggled to get work once they were in their thirties. Moreover, women were often systematically excluded from well-paid jobs, not only by employers, but also by the male-dominated unions. In many sectors of the economy a formal marriage bar operated (whereby a woman who married had to resign her job, whereas a man who married did not), as it did at Huntley and Palmers' biscuit factory in Reading, and throughout the civil service. Many middle-class women, who were the first of their ilk to obtain a university education, chose to put their careers before marriage: writing in the *Nineteenth Century* in 1895, Alice Gordon noted that of the 1,486 women who had received a university education, only 208 had married, although it was unclear whether those who had not done so relished their independence or regretted their solitariness. It also remained the case that no women in the United Kingdom could vote in parliamentary elections, Gladstone having firmly resisted an attempt to amend the Reform Bill in 1884 in favour of female suffrage. Yet some British women had already been given the vote in the Isle of Man in 1881, all females were enfranchised in New Zealand in 1893, and the vote was progressively extended to women in the Australian colonies and the ensuing federation between 1895 and 1911.

There were varied reactions by contemporaries to these changes and non-changes in the position of women, by which Wells's visitors from an earlier time might have been both intrigued and confused. Some women were happy as things were, and at the slow rate of change. The headmistress of Manchester High School, Sarah Burstall, believed that since most girls would get married, they should learn housewifely duties at school. Male-dominated trade unions disapproved of female competition in the labour market, an ILP pamphlet from 1900 declared that true freedom for women lay in not earning 'any wages under *any* conditions'. The trade unionist Mary MacArthur declared herself 'sufficiently old-fashioned' to believe that a woman's place was in the home; Ramsay MacDonald's wife Margaret argued that a married woman should turn her back on paid work, thereby having more time to 'give thought and companionship to her husband'. From a different political perspective, there were those who argued that the essential task of women, in the age of high imperialism, was to be loyal wives and nurturing mothers to the husbands and sons who would guard and govern the empire. Such was the case made by the National Association for the Promotion of Housewifery, which was active during the 1870s and 1880s and had been founded by Lord Meath's wife. Hence, too, the establishment of the Mothers' Union in 1886, which would soon enjoy the largest membership of any women's voluntary association, and which campaigned for a better deal for women within the framework of the existing family. Many women's magazines proliferating from the mid-1880s took the same view, among them *Woman* (1890), whose ambiguous motto was 'Forward, but not too fast'.

But there were also commentators, of both sexes, who thought that female stereotypes remained too conventional and constraining, that opportunities for women were still insufficient, and that change was too slow and must accelerate. In opposition to the policy pursued at Manchester High School, the schools that were part of the Girls' Public Day School Company stoutly resisted pressure from the Board of Education to introduce training in housewifery into the syllabus, on the grounds that it limited female horizons and constrained their ambitions (although they did send the girls home in the afternoon, to avoid strain and so they could spend more time with their mothers). At the same time, and welcoming change even as more was demanded, there was a growing discussion of what was increasingly being called the 'New Woman' phenomenon. The phrase was first coined by the Anglo-Irish writer 'Sarah Grand' (Frances Elizabeth Bellenden Clarke) in 1895, to describe and celebrate those of her sex who sought to exercise economic, social or personal control over their own lives, independent of men, and who protested against the fate of many women trapped in loveless and unequal marriages. In advocating such 'advanced' female aspirations to individual, professional and sexual independence, and also in denouncing the double standard, Grand was taking up issues already voiced and advocated by writers such as the Australian-born and Ireland-dwelling 'George Egerton' (Mary Chavelita Dunne) in *Keynotes* (1893), her collection of short stories. Another energetic propagandist for what some deplored and others acclaimed as the spirit of feminine 'rebellion' was the socialist Edward Carpenter; he advocated a looser marriage bond in which both men and women could explore their sexuality in extra-marital relationships, and predicted that women would no longer be contented with wedded subordination and serfdom, but aspired to be 'the equal, the mate and the comrade of Man'.

It was scarcely coincidence that by the late 1890s a new word, 'feminism', was being used in Britain to describe those who advocated such enhanced rights for women. But when it came to the issue of female enfranchisement, as distinct from female emancipation, both sexes were divided in their attitudes. In the case of male politicians, Tories such as Disraeli, Salisbury and Balfour tended (perhaps surprisingly) to be in favour, whereas Liberals such as Gladstone and Asquith were not (perhaps equally surprisingly). Even those who wrote in praise of the 'New Woman' took different sides: 'Sarah Grand' was for, but 'George Egerton' was against. The result was that even as the 'New Woman' was much talked about during the 1890s, the suffrage movement made little headway for most of that decade. But in the year of the Diamond Jubilee, several provincial organizations came together to form the National Union of Women's Suffrage Societies, which sought to revive the campaign for the parliamentary vote that had flagged since the passing of the Third Reform Act. For the duration of the Boer War parliament and the country were preoccupied with other matters, and no bills were brought before the Commons between 1897 and 1903. But in that latter year Emmeline Pankhurst, her daughter Christabel and four others established the breakaway Women's Social and Political Union in Manchester, which sought to campaign more actively to gain the franchise. In 1904 the Commons passed a pro-suffrage resolution by 184 votes to 70, which was the strongest support yet given to votes for women in the lower house. Soon after, the Pankhursts began to embrace an aggressive style of heckling and interruptions at public meetings, which they first put into practice during the general election of 1906, and within a few years such behaviour would bring them what they wanted, namely massive publicity.

Wells's time travellers would surely have been surprised by these developments, and they might have been equally impressed by the growth in leisure time and activities for men and women at most social levels. The bank holiday legislation of 1871 and 1875 had established days off, which were for secular rather than religious purposes; the last quarter of the nineteenth century witnessed further statutory reductions in the hours of work of certain professions; and by the 1890s most workers could hope to enjoy free time on Saturday afternoons, while large employers were organizing annual outings for their employees or giving them two weeks' (unpaid) summer holiday. With more time to spare and more money to spend, there was a significant proliferation in working-class recreational activities beyond the pub, the horses and the music hall. Some leisure pursuits were solitary, such as breeding canaries or growing leeks; some were more social, such as playing in a brass band or joining one of the choral societies that were such a prominent feature in the Midlands and the north of England. Professional football developed as a spectator sport because factory operatives were able to watch their teams on Saturday afternoons, and the modern features of the professional game date from the 1880s. Day excursions to the seaside were also becoming commonplace, and so were longer holidays. Towns such as Blackpool and Southend were among the fastest growing in the country, with their cheap hotels, their guest houses and landladies, and they offered a whole new range of entertainments, among them piers, illuminations, funfairs, donkey rides and candy floss. At the same time the further expansion of free public libraries, in part funded by the Carnegie Bequest, offered scope for more serious and self-improving pursuits on the part of respectable workers and the lower middle classes.

Middle-class recreations also expanded, among them cricket, which was one of the few sports where 'players' and 'gentlemen' met on the same field. This was the era of W. G. Grace and the consolidation of the county championship, but cricket was also played internationally, with visiting 'test' sides from the empire, and at a local level, as in the case of the Lancashire League, established in 1892. Rugby, by contrast, was played in two class-stratified and place-specific variants: rugby union, associated with the public schools and the south of England (but in Wales was much more working class), and rugby league, which was more proletarian and flourished in the industrial north. Lawn tennis assumed its modern form, with the establishment of the Wimbledon championships in 1877. Like bicycling, tennis was an activity that women could enjoy as well as men, and cycles and racquets may have contributed as much to advance female causes as any amount of feminist propaganda or suffragette campaigning. Golf, by contrast, which also became popular, was generally restricted to men. Meanwhile, the facilities of many towns also improved, with the more widespread provision of orchestras and museums, usually funded by a combination of municipal expenditure and private philanthropy. Again, the middle classes were the prime beneficiaries, as they were of such 'exclusive' holiday resorts as Eastbourne, Worthing, Bournemouth and Torquay, with their winter gardens, theatres, promenades, bandstands and palm-court orchestras. There were also unprecedented opportunities to venture further afield, thanks to the improvement in transport facilities on the continent and in the Mediterranean, and the tours arranged by such travel agents as Thomas Cook and Sir Henry Lunn: to the great cities and capitals of the continent, to the Alps, the Rhine and the Danube, and even as far as Egypt for winter cruises on the Nile, and for extended stays in the grand hotels recently constructed on the river's edge.

Among the plutocracy, gentry and aristocracy the pattern of recreation combined traditional leisure pursuits with new and more geographically extended pastimes, provided they could afford them. Traditional fox-hunting was threatened by the downturn in the incomes of many landowners, which reduced their capacity to participate, by the importation of barbed wire from the United States, which made jumping hedges and fences more dangerous, and by the Land Wars in Ireland, where opponents of the Protestant ascendancy sabotaged hunting as another form of protest. There were also complaints that upstart plutocrats were buying their way in to a form of recreation that had previously been the preserve of the traditional titled and territorial classes; while the new fashion for slaughtering thousands of birds, on organized shoots, was something that only the most moneyed landowners could afford, but which the new rich, and their even wealthier friends from the United States, were increasingly enjoying. The advent of faster and (until the *Titanic*) safer transatlantic liners brought the old-world nobility and the new-world rich together as never before, as evidenced by the increase in the number of Anglo-American marriages, and as described in the novels of Henry James and Edith Wharton. And for those who wished to journey even further afield, there was big-game hunting in East Africa, South Africa and the Indian Empire. The artist who most vividly depicted the men and women belonging to this rich, raffish and luxurious world was, appropriately, the Anglo-American John Singer Sargent. Wells's time travellers might have noticed that his paintings, with their bravura brushwork, and vivid depiction of clothes and jewellery, owed much to the artist from their own day who had also depicted an upper class in flux, namely Sir Thomas Lawrence.

The visitors from the 1800s would certainly have noticed that recreation, leisure pursuits and entertainment played a much greater part in the lives of a broader spectrum of society than they had a century before. During the last two decades of the nineteenth century the number of authors, editors and journalists rose by 81 per cent, and that of musicians by 69 per cent, while between 1881 and 1911 the number of actors and actresses almost trebled. Newspaper circulation doubled between 1896 and 1906, and would double again by 1914, and the annual issue of book titles also doubled between 1901 and 1913. There were some complaints that this mass culture was a debased culture, but serious writers like Wells, Arnold Bennett and John Galsworthy all sold well, while publishers such as Collins, Oxford University Press and Joseph Dent reissued classic novels and Shakespeare's plays. (On the other hand, works by E. M. Forster and Joseph Conrad did not sell well.) By the 1900s the United Kingdom was no longer the 'land without music': Arthur Sullivan had died in 1900 and Charles Stanford was past his prime, but Hubert Parry was at the height of his renown, Edward Elgar was transitioning from oratorios to symphonies and concertos, and the young Ralph Vaughan Williams was making his mark. Even as the music hall flourished, the theatre was becoming ever more respectable and demanding. Henrik Ibsen's plays, exploring such 'taboo' subjects as infidelity and venereal disease, were produced on the London stage in the 1890s, Henry Irving and Beerbohm Tree were pioneering actor-managers, and Irving's knighthood in 1895 recognized that the theatre had become respectable. On the other hand, the belated appointment in 1896 of Alfred Austin in succession to Tennyson as Poet Laureate was widely regarded as quixotic, a view vindicated by his hastily written poem in praise of the Jameson Raid.

Wells's visitors would also notice that science and technology were transforming many aspects of national life in ways that would have been unimaginable a century before. The shift from sail to steam was not the only revolution in maritime transport: the development of the steam turbine by Charles Parsons would greatly improve the speed of battleships and liners alike by the 1900s. By then the telephone, patented by the Edinburgh-born American, Alexander Graham Bell, in Philadelphia in 1876, had become widespread in Britain, under the auspices of the Post Office. London was linked with Birmingham in 1890, and by the turn of the century most large public offices and commercial establishments had installed a telephone. The number of private subscribers quadrupled during the 1890s, and would double again between 1900 and 1905. Even more extraordinary for contemporaries was the advent of the wireless, effectively dating from 1896, when the Italian inventor Guglielmo Marconi, rebuffed in his own country, brought his experimental apparatus to the United Kingdom. Soon after he took out his first patent, established his own company, and in 1901 successfully transmitted a message across the Atlantic from Cornwall to Newfoundland. Both the Admiralty and the merchant marine soon realized the potential of the wireless as an efficient and reliable form of offshore communication, and it shortly complemented the undersea cables as a way of linking the empire together, although its potential for domestic broadcasting was as yet unrecognized. But the most important single scientific development was the spread of electric lighting, made possible by the invention of an effective light bulb by Thomas Edison in 1881. By the end of the century many British towns were lit by electricity, trams and substantial sections of the London Underground were electrified, and electric illumination shed a new form of glamour over theatres, shop windows, and seaside piers and promenades.

These inventions, along with those of the internal combustion engine (largely pioneered in Germany in the 1880s) and the heavier than air flying machines (the Wright brothers first flew at Kitty Hawk, North

Carolina, in 1903), heralded the end of the carboniferous age, when coal – and when Britain – had dominated the modernizing world. But as the examples of the internal combustion engine, the wireless and manned flight suggest, while the United Kingdom had led and dominated the first industrial revolution, it would neither lead nor dominate the second. Hence the growing concerns, from the 1880s onwards, that Britain was not only losing out to Germany and the United States in such traditional industries as iron and steel production, but also that they were pioneering a new economy based on electricity and petrochemicals. And hence, too, the growing fears that the deterioration in the United Kingdom's international industrial competitiveness was in part because its scientific research was insufficiently rigorous, inventive and professionalized, and that relations between science, technology and industry needed to be much closer and more mutually reinforcing than in fact they were. The travellers in Wells's time machine could scarcely have failed to pick up this issue, in which Wells himself was involved, both as a propagandist and critic. In fact, early twentieth-century Britain excelled at *pure* science, as recognized by the award of Nobel Prizes in physics to Lord Rayleigh, J. J. Thomson and Ernest Rutherford in 1904, 1906 and 1908 respectively. But there were concerns that the country was far less successful at *applied* science, perhaps because government support was insufficient. Thus began a debate that has continued virtually uninterrupted down to our own times, and as in other realms of national life it was (and still is) Germany and the United States who were (and are) the competitors and comparators.

In following these issues, Well's visitors could hardly have failed to recognize that the sum total of human knowledge had proliferated and become increasingly specialized, compared to the 1800s when a few exceptional individuals such as Joseph Banks could know virtually everything about virtually everything. A century later, research and scholarship were no longer the preserve of gentlemanly amateurs, or of fellows of Oxbridge colleges in holy orders, but were increasingly being carried out by professional experts and academic specialists, often based in Britain's expanding university system. The publication of the *English Historical Review*, which began in 1886, was one example of this trend. The establishment of the Economics Tripos at Cambridge University by Alfred Marshall in 1903 was another. The appointment, four years later, of L. T. Hobhouse to the first British chair in sociology, at the University of London, was yet a third. Such university-based academics were eager to make the case that the humanities and the social sciences were as specialized, scholarly and significant as the sciences themselves, and it was that impulse which lay behind the establishment of the British Academy in 1902. But neither the humanities nor the sciences were fully in thrall to this new academic rigour. The young George Macaulay Trevelyan left Cambridge in protest against what he regarded as the misguided cult of scientific history limited to academic professionals, and the improper disparagement of literary history written for a broad public audience; while many eminent scientists, among them Lord Rayleigh and Sir Oliver Lodge, thought they could establish by scientific methods the existence of a non-religious spiritual world, and were ardent supporters of the Society for Psychical Research, formed in 1882.

In the same year that the British Academy was inaugurated, Edward VII established the Order of Merit, which was something entirely new in the British honours system. There had been desultory discussion of such an Order during the Napoleonic Wars, and Prince Albert had later taken an interest. But once again it was the early twentieth-century fear of German competition that provided the immediate incentive, for the model for this new British Order was the German *Pour Le Mérite*, which had been established by Frederick II of Prussia in 1740 to recognize outstanding achievement in both military and civilian life. The British Order of Merit was similarly divided, and the initial recipients provide a revealing snapshot of who was deemed to be especially deserving of recognition in the British Empire in 1902. Five of the initial twelve appointees came from the armed services: Field Marshals Earl Roberts, Viscount Wolseley and Lord Kitchener, and Admirals Sir Henry Keppell and Sir Edward Hobart Seymour. Four more were distinguished scientists: Lords Rayleigh, Kelvin and Lister, and Sir William Huggins. And the final three represented the arts and the humanities: the biographer and Liberal politician John Morley, the historian W. E. H. Lecky, and the painter and sculptor George Frederic Watts. The relative balance between the military and non-military figures, and between the sciences, humanities and arts are an instructive indication as to the public and cultural priorities of early twentieth-century Britain. Since Wells's voyagers had come from an earlier era when military affairs were paramount, they might have been less surprised by the balance, even if they were confused and disoriented by much else that they had seen and heard and learned on their journeys.

Sooner or later, Wells's time travellers would surely have been completely exhausted, and would be eager to leave Britain and its empire in the first decade of the twentieth century, and return to their own times. But they might have wanted to take back some mementoes of their extraordinary journey, and nothing would have been better or more appropriate than the fictional and non-fiction works of Wells himself, which offered so many vivid and varied insights into the contemporary scene, while also venturing perceptive predictions and anxious warnings as to the scientific and technological future. His early comic novel *The Wheels of Chance* (1896) depicted lower middle-class men and women cycling out to the countryside. *The War of the Worlds* (1898), an innovative work of science fiction, reflected the prevailing sense of contemporary national anxiety and international tension. *Anticipations* (1901) was a critique of the debased nature of contemporary democracy. *A Modern Utopia* (1905) looked at the movement for 'National Efficiency', which would be promoted by a new governing elite called 'the Samurai'. In *Kipps* (1905) the eponymous hero was a shop assistant, enduring long hours and low wages in a regimented atmosphere. *Tono-Bungay* (1909), a semi-autobiographical novel, offered a powerful critique of the squalor, waste and muddle that were the defining characteristics of modern capitalism. *Ann Veronica* (1909) was a sympathetic portrait of an independent young woman, who opted for free love, which was banned from many libraries. *The History of Mr Polly* (1910) explored the tragi-comic ups and downs of lower middle-class provincial life. *An Englishman Looks at the World* (1914) warned that manned flight meant the United Kingdom was no longer protected by the sea from military assault. 'Will the Empire live?' Wells wondered in its pages. 'What will hold such an Empire as the British together?' By then, he would not be alone in asking that question.

TIME OF HOPE OR A FALSE DAWN?

The landslide victory of 1906 was the greatest for the Liberal Party in more than a generation, as the defeats, divisions and disappointments that had so hampered and debilitated it since 1885 seemed finally to have been left behind. Although they inherited problems aplenty, and faced serious challenges at home and abroad, the electoral triumph won by Campbell-Bannerman and his colleagues ushered in what seemed a new time of hope, the final triumph of the nineteenth-century tradition of liberal humanism and cosmopolitan internationalism against the forces of conservative jingoism and political obscurantism. This optimism was especially felt by a young group of intellectuals and creative figures, who were among the most ardent of the 'New Liberals', and for whom it did indeed seem as though a bright new dawn had arrived. Among them were the politicians C. F. G. Masterman and Noel Buxton, the philosopher Bertrand Russell, the classicist Gilbert Murray, the economists William Beveridge and John Maynard Keynes, the composer Ralph Vaughan Williams, and the historians H. A. L. Fisher, G. P. Gooch and George Macaulay Trevelyan. There were, as Trevelyan observed, sensing the mood, 'occasions in history when new principles of government are being formed', and when 'men are moved by appeals to the imagination', and 1906 seemed to him and his contemporaries just such a time. Hence his Garibaldi trilogy (1907–11) celebrating the virtues of Liberalism and internationalism, and hence Vaughan Williams's settings of Walt Whitman's verse in his *Sea Symphony* (1903–09) which, in a different medium, proclaimed very much the same values.

But these ardent hopes for change, which were shared by many in the Liberal Party, would be set against the forces of inertia that had become more resistant and persistent as the nineteenth century wore on. No other major European nation, be it France, Germany, Spain, Russia, Italy or Austria-Hungary, could boast the continued existence of a single, sovereign legislature as the United Kingdom had done between 1800 and 1906, and for much of that time it had shown itself to be extraordinarily adaptable and flexible. Yet by the 1890s and 1900s there was growing concern that Westminster was losing its capacity to deliver timely and imaginative constitutional change in the way it so often had between 1829 and 1885. The relations between the four constituent nations of the United Kingdom were in many ways unsatisfactory, as were those between the United Kingdom and the rest of the British Empire. Instead of being the most advanced democratic system in Europe, as it had been at mid-century, by the 1900s the House of Commons was elected on one of the narrowest franchises in the whole of Europe, and no women had the vote; while the House of Lords was an almost entirely unelected, oligarchic and self-perpetuating chamber. This in turn meant that the popular mandate of the Liberal landslide was insufficient by itself to ensure that major changes could be implemented. For as Arthur Balfour made plain in the aftermath of the Conservative wipe-out, the Liberals might now command the Commons, but the Tories retained their impregnable majority in the Lords, and he insisted that they would 'continue to control the destinies of this great empire' from what seemed to be their lofty and unassailable citadel in the upper house.

In terms of making progress and making changes, the Liberal government would have it easiest in the realms of foreign and imperial policy. At the Foreign Office, Sir Edward Grey had inherited the Anglo-French entente from his Conservative predecessor, and in August 1907 he added to it an agreement with Russia which, like the earlier accommodations with the United States, Japan and France, also sought to lessen tensions on the periphery of empire. Accordingly, the two governments gave up their rival ambitions in Tibet and recognized Chinese sovereignty. Russia accepted that Afghanistan was a British sphere of influence, although at the same time Britain agreed not to interfere in its domestic affairs. Persia was divided into three spheres: a British zone in the southeast adjacent to the Indian frontier, a large Russian zone in the north, and a neutral zone in between; and by a separate note, Russia also recognized British predominance in the Persian Gulf. From a diplomatic perspective this treaty was a great success for the Foreign Office, as the long-feared threats that Russia had mounted to Britain's position in India had been effectively allayed. But there were also downsides: the alliance with autocratic tsardom did not play well with radical opinion in Britain, and the fact that the British and Russian royal families were closely related made that worse not better. Moreover, the creation of the triple Franco-British-Russian entente completed what could easily and plausibly be seen in Berlin as the hostile encirclement of Germany. That was certainly how the increasingly paranoid and unstable Kaiser saw things, and this in turn prompted him to make threatening and destabilizing gestures overseas, as when he sent a German gunboat to the Moroccan port of Agadir in July 1911 in response to the French government's earlier decision to dispatch a substantial number of troops to put down domestic discontent and unrest in a region where their paramountcy had recently been internationally recognized.

In imperial matters the main task facing the new government would be the settlement of South Africa in the aftermath of the Treaty of Vereeniging. The Liberals would grant self-government to the Transvaal (1906) and the Orange Free State (1907), and bind them together with Cape Colony and Natal in the Union of South Africa (1910). In order to retain what it believed was the recently acquired goodwill of the Boers, the cabinet would reluctantly acquiesce to a Union franchise, which was confined to whites and excluded all Africans (although it did succeed in retaining full British control over Swaziland, Basutoland and Bechuanaland). The government hoped it had created another strong, pro-British dominion in control of the strategic tip of Africa; but it would be the Boers rather than the British who would in fact be in charge. In the case of India, John Morley and Lord Minto would implement the limited reforms that bear their joint names in 1909. The legislative councils were enlarged to accommodate twice as many Indian members, who represented class and communal interests, and who were henceforward allowed to debate budgetary matters; while the first Indian was also appointed to the Viceroy's executive council. Meanwhile the Liberal government soon came to recognize that the 'dominions' of Canada, Australia, New Zealand and South Africa had to be treated with more tact and consideration than Joseph Chamberlain had displayed. At the 1907 colonial conference, plans for an imperial council and a permanent secretariat to service it were thrown out; and four years later Asquith, by then prime minister, would declare that the establishment of any formal links between the dominions and the United Kingdom would be 'absolutely fatal to our present system of responsible government'.

In domestic affairs the new government found its hands tied almost from the outset, as the Tory leadership exploited its massive majority in the upper house to emasculate the Liberals' legislative programme, just as Balfour had threatened. They did not reject measures directly benefiting the working classes, such as the Trade Disputes Act of 1906 and the Factory and Workshop Act of the following year. Instead, they focused their opposition on measures that were not generally popular or only of marginal interest, throwing out a Plural Voting Bill and so modifying an Education Bill that the government was forced to withdraw it. Towards the end of his prime ministership a frustrated Campbell-Bannerman warned the Lords that 'a way must be found, a way will be found, by which the will of the people, expressed through their elected representatives in this House, will be made to prevail'. After Asquith succeeded Campbell-Bannerman as premier in 1908, Lloyd George followed him as Chancellor, and carried the legislation setting up old-age pensions, while Churchill, now President of the Board of Trade, established Labour Exchanges. But the Lords threw out a Licensing Bill, modified a second Education Bill so much that it was again withdrawn, and in 1909 rejected Lloyd George's so-called 'People's Budget'. The result was a confrontation between the Commons and the Lords, the people and the peers, which harked back to the earlier struggles over the first and third reform bills. It was also the battle that Gladstone had wanted to fight when their lordships had thrown out his second Home Rule Bill. The eventual result of the peers' rejection of Lloyd George's budget would be that the reform of the upper house became a necessary and a high priority, and the whole matter of Irish Home Rule would also be reopened. Yet neither issue would be satisfactorily settled during the twentieth century, and even now they both remain fundamentally unresolved. In these ways, and in others, too, the nineteenth century has not yet finished with us, and nor as a result have we yet finished with it.

Epilogue

On 21 October 1905, which was less than two months before Balfour resigned the premiership, the United Kingdom had celebrated the one-hundredth anniversary of the Battle of Trafalgar, that defining naval victory which had ushered in a century of British maritime supremacy. Nelson's column was bedecked with flags and streamers like a giant maypole, the Trafalgar Dinner at the Royal Naval College, Greenwich was the grandest ever held, the Navy League was at its most celebratory and assertive, Henry Wood composed his 'Fantasia on British Sea Songs' for the Promenade Concerts, and the following year the first 'Dreadnought' battleship would be launched. But almost ten years later, on 18 June 1915, the British army would be mired in mud on the Western Front, and there would be only muted observances to mark the centenary of the Battle of Waterloo, which had been Britain's second crushing victory (albeit with essential allied help) that had brought to a triumphant end the Revolutionary and Napoleonic Wars. Yet how circumstances had altered across those intervening hundred years, as Britain's allies and opponents had completely changed places. In 1805 France had been (as so often before) Britain's enemy; by 1905 it had recently (and unusually) become Britain's ally, which made the Trafalgar celebrations more than a touch sensitive. In 1815 Prussia had been Britain's friend, and would remain so for much of the nineteenth century; but by 1915 it had become Britain's foe, and under varying forms of leadership it would continue as such for most of the next thirty years. So much for those vain hopes of Queen Victoria and Prince Albert, Cecil Rhodes and Joseph Chamberlain, that the two great Teutonic nations, linked rather than sundered by the North Sea, and sharing the Protestant religion and much more besides, might live in amity and harmony to their lasting benefit.

Across the nineteenth century, and far into the twentieth, the peoples of the British Isles and the British Empire were exhilarated yet also intimidated by those victories of Trafalgar and Waterloo. On the one hand, they were both the necessary precondition for and the incontrovertible evidence of Britain being a providentially blessed realm and imperium, as the nineteenth century seemed to show; but they also set impossibly high standards of endeavour and achievement that subsequent generations, facing their own trials and challenges, aspired to meet but doubted they could. As the *Anglo-Saxon Review* noted in June 1900 when the relief of Mafeking was greeted with unprecedented jingoism: 'We accepted Waterloo and Trafalgar more calmly than this deliverance of a few hundred Colonial volunteers besieged in a Bechuanaland village.' Indeed, the frenzied response to that news was as much a sign of relief that another imperial humiliation had been averted as it was exaggerated jubilation that a minor military engagement had been won. Throughout the First World War the hope would remain that two great British victories, one by sea, the other on land, might once again bring the hostilities to a triumphant conclusion. But despite such high political and public expectations, such successes would never materialize. Instead of another Nelson and another Trafalgar, there would be Jellicoe and Jutland; and instead of a second Wellington and a second Waterloo, there would be Haig and the Somme. It would be the same during the Second World War, when Churchill would pester and harass his admirals and his generals unrelentingly so that they might deliver for Britain victories on a scale and of a significance that would merit comparison with those that Nelson and Wellington had won. But the biggest land battles would be between the Germans and the Russians on the Eastern Front, while the largest naval engagements would be between the Americans and Japanese in the Pacific.

One further attempt to commemorate the 'national sacrament' of the centenary of the Battle of Trafalgar had been the publication of a multi-authored, nine-hundred-page compilation entitled *The Empire and the Century*, in which a variety of authors, among them Disraeli's biographer W. F. Monypenny, wrote enthusiastically and appreciatively about the evolution and extension of Britain's overseas dominions. But some contributors were less confident and optimistic, among them the journalist (and future biographer of Joseph Chamberlain) J. L. Garvin, who posed the disquieting question: 'Will the Empire last the century?' 'Despite the optimism which is the fashion of the hour,' he opined, 'national instinct recognizes that the answer is no foregone affirmative.' Even amidst the anniversary euphoria of 1905, Garvin was not alone in that opinion. In the same year an anonymous, futuristic pamphlet (Wells was not alone in being in the prediction business) was published entitled *The Decline and Fall of the British Empire*, which had allegedly been written in Japan in 2005. It enumerated eight causes of Britain's rapid recession across the twentieth century: the prevalence of town over country life, the weakening of British interest in the sea, the growth of luxury (a very Gibbonian touch), the decline of taste, the debilitation of the physique and health of the people, the deterioration of religious and intellectual life, excessive taxation and municipal extravagance, and the inability of the British to defend themselves and their empire. From one perspective these causes of imperial decay owed much to the contemporary concerns of the 1900s; but in retrospect they were also prescient and perceptive in their predictive power, as Britain's imperial enterprise would in the end prove to be indefensible and unsustainable. Indeed, with the handing back of Hong Kong to the Chinese in 1997, the very centenary of Victoria's Diamond Jubilee, the British Empire would truly be gone, over, vanished and done with.

Many of the reasons why the British Empire (and other European domains, too) disappeared were specific to the twentieth century, in particular the two world wars and the international critique of colonialism that became so pervasive after 1945. But in the British case nationalist campaigning and agitation had already begun in the 1880s, in Ireland, Egypt and India. Among Conservatives and Unionists it would long be fashionable to disregard or denounce such agitators as being either not serious and not worth bothering with, or not to be negotiated with if they were. Yet by the 1900s, London as the imperial metropolis and the world city had played host to many indigenous colonial subjects, who obtained the sort of education that would equip them with the tools which would later allow them to engage directly and devastatingly with their British rulers, and eventually enable them to lead successful nationalist movements against the empire. Mohandas Gandhi was the most famous of them. But he would soon be followed by Muhammad al Jinnah, who studied law at Lincoln's Inn from 1892 to 1895 and was a protégé of Dadabhai Naoroji; and by Jawaharlal

Nehru who, after Harrow and Trinity College, Cambridge, went on to the Inner Temple between 1910 and 1912 (and who would later claim that he had been inspired to work for Indian independence having read Trevelyan's Garibaldi books). All three men came to appreciate the importance of the British constitution and the rule of law, and learned how to turn their knowledge against their imperial masters. The result of their successful agitations would be, as Curzon had feared, that the independence of India meant Britain would indeed fall into the ranks of a much-diminished force in the world.

Yet during the first half of the twentieth century, the preservation of that empire and the maintenance of Britain as a great power would be the shared views of most British policymakers and men of government, among them Winston Churchill, whose youth had been nurtured in what he nostalgically but mistakenly recalled as the 'august, unchallenged, tranquil glow of the Victorian era', whose first government job had been Under-Secretary for the Colonies, and who regarded his life's work as being to keep Britain great and its empire strong. Despite the heroic resistance and magnificent defiance that Churchill mounted between 1940 and 1945, he increasingly came to believe, in his saddened old age, that his endeavours had been in vain. This was not the whole truth of things, but it was substantially correct. Yet no one could have succeeded where Churchill failed because there was so much about Britain's nineteenth-century economic, naval and imperial pre-eminence that was accidental, ephemeral and therefore could not last – being the first nation to industrialize, and also benefiting from a relatively quiescent Europe, United States, Russia, Middle East and Far East. This meant that for a relatively brief span of time a relatively small European nation came to wield an influence over the affairs and the peoples of the world out of all proportion to its size, population and resources. But once other countries caught up economically, and once aggressive nationalism asserted or reasserted itself in many parts of the world, the writing was on the wall for Britain as global hegemon. From this perspective, what is the more remarkable is not that the United Kingdom's power and reach eventually declined, but rather that they lasted for as long as they did.

Until 1945, and even for another decade or so thereafter, the political will to sustain the United Kingdom's assertive position in the world still existed, and nor should this come as any surprise. For just as nineteenth-century Britain had been ruled by eighteenth-century men until 1868, when Derby was replaced by Disraeli, so twentieth-century Britain was ruled by nineteenth-century men until 1963, when Macmillan was superseded by Douglas-Home, who in turn would be followed by Harold Wilson. Indeed, there were many ways in which the late-Victorian United Kingdom lasted and lingered until the mid-1960s: in its great-power pretensions, global empire and imperial monarchy, in its heavy-industrial economy, moral code and gender relations, and in its outward conformity to Christian ethics. The nineteenth century cast a long shadow. Only since the 1960s has Britain significantly de-Victorianized, de-imperialized and downsized, and begun to come to terms with that 'recessional' that Kipling so prophetically foretold in 1897. Despite some lamentations of 'national decline', and despite Margaret Thatcher's espousal of 'Victorian values' and determination to make her nation great again, that process of uncoupling ourselves from our nineteenth-century past, or from the many exaggerated and mythologized versions of that past, continues apace. This in turn may explain why in recent decades it has become fashionable to denounce Victorian Britain more energetically and systematically than Lytton Strachey ever did, for espousing a set of assumptions that seem at best alien, at worst deplorable, and for being (among other things) sexist, misogynist, homophobic, racist, classist and imperialist. The British nineteenth century was not what would now be regarded as a politically correct culture, country or civilization.

Beyond any doubt the decades from the 1800s to the 1900s witnessed many extraordinary and traumatic challenges and wrenching and disorientating changes, as expressed and mediated through (among other things) the poetry of Wordsworth and Tennyson, the paintings of Turner and Landseer, the novels of Dickens and Eliot, and even Gilbert and Sullivan's comic operas and Oscar Wilde's brilliantly brittle plays. How far did the men (and apart from Queen Victoria, they were all men) who were ostensibly in charge of the affairs of the United Kingdom and the British Empire understand what was going on and know what they were doing? To be sure, many of those who occupied 10 Downing Street were prime ministers of exceptional experience and ability, and they were often deeply learned in the classics, history, theology and languages. But for all their impressive erudition and undeniable qualities, which would put most politicians today to shame, none was ever fully comprehending of or in complete command of events. Wellington's paranoia over parliamentary reform was preposterous, Peel was a failure as Conservative Party leader, Palmerston's foreign policy depended too much on a bluff that was eventually called, Gladstone understood Ireland far less well than he thought he did, while Salisbury (and Chamberlain and Queen Victoria) were deeply misguided in their intransigent opposition to Home Rule. And there were also more than enough military disasters, from Corunna to the Crimea to the Boer War, to make plain that failure, incompetence and cock-up were also an integral part of the British way of doing, and of misdoing, things. Much of what seems to us as arrogance, intolerance and bigotry was often born out of fear and ignorance, and of personal self-doubt and national anxiety, for both of which there was often ample justification.

More often than not, the men of government and, indeed, the men of God, of business, of the professions, and of the military, were riding tigers that they could never fully control, which were heading in directions that they could often not discern, and to destinations that they either did not know or did not like. How else (for example) to resolve the paradox that scarcely any major politician in London wanted to take on yet more imperial responsibilities, yet that was precisely and frequently what in fact happened? And that sense that so much was going on, which was by turns worrying and exhilarating, baffling and contradictory, and presenting opportunities a-plenty but also corresponding challenges, has never been better caught than in the words of Charles Dickens, written about the 1790s, but penned in the 1850s, which form the first epigraph to this book. Truly, for the nineteenth-century British, theirs was the best and the worst of times, and often something much more indeterminate and in between. Yet while economic and technological advance gave some of the inhabitants of the United Kingdom a pre-eminent sense of limitless possibilities, the constraints under which most people lived out their private and public lives also demand our attention. The bounds to most people's existences were set by their poverty, their lack of education, and the fact that they were lucky to survive into their forties, while politicians, businessmen and military commanders often endured rejection, failure and defeat – and with good reason. Hence, then, the second epigraph of this book, from Karl Marx, another incomparable observer of the nineteenth-century scene, that men (and we would now add women) might make their own histories, but not under conditions of their own choosing. In many, varied and contradictory ways, the British experience from the 1800s to the 1900s, both at home and abroad, and at all social levels, offers constant validation of Marx's dictum.

Not surprisingly, then, the nineteenth-century legacy, in the United Kingdom itself and in those many other parts of the world once ruled from London, remains simultaneously tainted and contested but also appreciated and acclaimed. In some parts of the globe the British are still unforgiven and unforgotten, for the violence, racism and exploitation of their colonial regimes, the bitter legacies of which remain to this day, especially in South Asia and across much of Africa. Elsewhere, in many of the former colonies of settlement, the impact of British rule seems to have been more benign: parliamentary democracy, the rule of law, museums and universities, hospitals and railways. As for the United Kingdom itself: there is much of the nineteenth century that remains part of our world, and to our benefit. Although Britain has de-imperialized, de-Victorianized and downsized cumulatively and irreversibly since the 1960s, a great deal of the infrastructure of our present-day living – the roads and railways, the sewers and drains, the civic buildings and suburban houses, the department stores and town halls, the museums and galleries – was constructed between 1800 and 1906. Whatever some might now deem to have been their faults and failings, the Britons of those times were prodigiously energetic, and they built to last. They also left behind an extraordinary cultural legacy, in art and architecture, science and engineering, theology and political philosophy, drama and (eventually and belatedly) music, and above all in that most quintessential of all nineteenth-century creations, namely the novel. And it is impossible to read *Hansard's Parliamentary Debates* from that time without being impressed and intimidated by the remarkable qualities of mind and spirit, head and heart, eloquence and erudition, that MPs and peers alike brought to bear on the conduct of their nation's public business and on debating the great issues of their time.

Beyond any doubt, many nineteenth-century Britons were extraordinarily vigorous, industrious and creative, even as they were also in many ways a flawed and fallible people. Not surprisingly, they have been both celebrated and criticized by posterity, and there is ample justification for these very different points of view. Yet neither perspective quite gets to the root of the matter, since handing out praise or blame is of little help in reaching a more nuanced historical understanding of such an exceptionally complex period of the past. Sir John Seeley put it well when he observed that the 'special characteristic' of his times was an 'unusual moral earnestness' combined with 'unprecedented perplexity and uncertainty'. Indeed, many nineteenth-century Britons were only too well aware that their nation's pre-eminence rested on transient and insecure foundations. 'Assyria, Greece, Rome, Carthage, what are they?' asked Lord Byron in *Childe Harold* in 1818. Byron was scarcely renowned for his 'moral earnestness', but he was well aware that past empires had not only risen but had also fallen. And Kipling's words, so presciently set out in *Recessional* in 1897, were another sobering reminder that while global greatness and imperial dominion might seem to be divinely sanctioned and to represent the permanent order of things, the more humbling and troubling reality is that they do not endure and they do not last:

Far-called, our navies melt away;
On dune and headland sinks the fire:
Lo, all our pomp of yesterday
Is one with Nineveh and Tyre!
Judge of the Nations, spare us yet,
Lest we forget – lest we forget!

A Note on Further Reading

The earliest works on the history of the nineteenth-century United Kingdom were written by men who were themselves the products of its final decades. The most explosive of these early writings was Lytton Strachey's *Eminent Victorians* (1918), with its iconoclastic and irreverent studies of Cardinal Manning, Florence Nightingale, Thomas Arnold and General Gordon. Two other high-minded late-Victorians responded to Strachey's deflating sarcasm with much more positive and empathetic interpretations: G. M. Trevelyan, *British History in the Nineteenth Century* (1922), celebrated the nation's peaceful evolution towards parliamentary democracy (even as he had reservations about the final outcome), while G. M. Young, *Victorian England: Portrait of an Age* (1936), provided an incomparable evocation of those years from the standpoint of the highly educated, moderately liberal, professional middle class. On a much larger scale, Elie Halevy produced his (alas, unfinished) *History of the English People in the Nineteenth Century* (6 vols, 1924–48), and Sir Llewellyn Woodward, *The Age of Reform, 1815–1870* (1938), and Sir Robert Ensor, *England, 1870–1914* (1936), divided the century between them in what were then the final volumes of the *Oxford History of England.*

But it was only after the Second World War that the post-Waterloo years became a major field of investigation, and the subsequent burgeoning scholarship has been synthesized in a succession of general accounts. David Thomson, *England in the Nineteenth Century (1815–1914)* (1950), was first among single-volume surveys. He was followed by A. Wood, *Nineteenth-Century Britain, 1815–1914* (1960); Norman McCord, *British History, 1815–1906* (1991); W. D. Rubinstein, *Britain's Century: A Political and Social History, 1815–1905* (1998); Norman Davies, *The Isles: A History* (1999); H. C. G. Matthew (ed.), *The Nineteenth Century: The British Isles, 1815–1901* (2000); Hugh Cunningham, *Challenge of Democracy, Britain 1832–1918* (2001); and Eric J. Evans, *The Shaping of Modern Britain: Identity, Industry and Empire, 1780–1914* (2011). Many surveys, following the precedent set by Woodward and Ensor, have treated the period in two volumes, among them: A. Briggs, *The Age of Improvement, 1783–1867* (1959), and Donald Read, *England, 1868–1914: The Age of Urban*

Democracy (1979); Derek Beales, *From Castlereagh to Gladstone, 1815–1885* (1969), and Henry Pelling, *Modern Britain, 1885–1955* (1960); N. Gash, *Aristocracy and People: Britain, 1815–65* (1979), and E. J. Feuchtwanger, *Democracy and Empire: Britain, 1865–1914* (1985); Eric J. Evans, *The Forging of the Modern State: Early Industrial Britain, 1783–1870* (1983), and K. G. Robbins, *The Eclipse of a Great Power: Modern Britain, 1870–1992* (1982).

Meanwhile, American historians of the United Kingdom have been producing their own accounts of the nineteenth century in which, unlike their British counterparts, they were challenged to explain and evoke the history of what was, when viewed from the United States, essentially a foreign country that had to be rendered both comprehensible and appealing for a transatlantic audience. The two earliest versions were the single-volume surveys of Walter L. Arnstein, *Britain Yesterday and Today: 1830 to the Present* (1966), and R. K. Webb, *Modern Britain: From the Eighteenth Century to the Present* (1968), both of which still repay reading. By contrast, T. W. Heyck, *The Peoples of the British Isles: A New History*, vol. II, *From 1688 to 1870* (1992), and vol. III, *From 1870 to the Present* (1992), provided more extended treatment. The most recent such surveys, reflecting developments in global, imperial and 'four nations' history, are Stephanie Barczewski, John Eglin, Stephan Heathorn, Michael Silvestri and Michelle Tusan, *Britain since 1688: A Nation in the World* (2015), and Susan Kingsley Kent, *A New History of Britain since 1688: Four Nations and an Empire* (2017), the former stronger on high politics, the latter focused much more on identities, especially class, gender and race.

But for any historian of nineteenth-century Britain the three most indispensable books are those that form part of the *New Oxford History of England*: Boyd Hilton, *A Mad, Bad and Dangerous People? England, 1783–1846* (2006), K. Theodore Hoppen, *The Mid-Victorian Generation, 1846–1886* (1998), and G. R. Searle, *A New England? Peace and War, 1886–1918* (2004). In each case, the authors summarize and synthesize a lifetime's learning, which places any subsequent historian of the British nineteenth century admiringly in their debt. Hilton is predictably strong on high politics and theology, Hoppen on Ireland and culture, Searle on politics and society. All of them contain exhaustive bibliographies and detailed chronological tables and, in the case of Hoppen and Searle, lists of the personnel of each successive cabinet. Of the three, Hilton is the most unrepentantly 'Little England' in his approach, Hoppen the best on 'four-nations' issues, and Searle the strongest on international relations. None of them fully manages to reconcile their treatment of historical processes and contingent events, or to devise expositional structures that satisfactorily incorporate narrative and analysis – criticisms which, having myself struggled to address these same issues in this volume, I offer more sympathetically than censoriously.

Three recent developments, reflecting broader public issues and concerns as well as changing patterns in historical inquiry, have stimulated new approaches of significant relevance to the study of the nineteenth-century British past. The first, owing much to what has been celebrated in some quarters (and criticized in others) as the coming of unprecedented globalization, has been the rise of what has been varyingly described as world or global or transnational history. Long before the latest phase of globalization took hold, Eric Hobsbawm was an early pioneer in such a wide-ranging approach to the 'long' nineteenth century, with his magisterial trilogy on *The Age of Revolution, 1789–1848* (1962), *The Age of Capital, 1848–1875* (1977), and *The Age of Empire, 1875–1914* (1987), and so, from a different perspective, was Paul Kennedy in *The Rise and Fall of the Great Powers: Economic Change and Military Conflict from 1500 to 2000* (1988). The two most recent works, which owe much to recent global developments, and which provide the widest geographical context for understanding British nineteenth-century history, are C. A. Bayly, *The Birth of the Modern World, 1780–1914* (2004), and Jürgen Osterhammel, *The Transformation of the World: A Global History of the Nineteenth Century* (2014).

A second development has been the increasingly vexed relations between the United Kingdom and the European Union, culminating in the vote in 2016 in favour of Brexit. During the referendum campaign historians were divided as to whether Britain should remain or leave, and several books have recently appeared offering different versions of the United Kingdom's connections to or divergences from the rest of Europe. Richard J. Evans, *The Pursuit of Power: Europe, 1815–1914* (2016), argues persuasively that many of the themes common to the history of the nineteenth-century continent apply with equal plausibility to Britain itself. Brendan Simms, *Britain's Europe: A Thousand Years of Conflict and Cooperation* (2016), argues more circumspectly that relations between England (or Britain) and the continent have been an uncertain and constantly changing amalgam of connection and contact, criticism and confrontation, but that the United Kingdom's economic well-being and national security have always depended on the stability of the European mainland. Simms's vision of a united Europe did not include Britain, and Robert Tombs took a similar but stronger view in *The English and their History* (2014), which celebrated English (rather than British) difference and exceptionalism with a fervour that even G. M. Trevelyan might not have equalled.

The third development, which owes much to the settlement of the 'troubles' in Northern Ireland, the granting of devolution in Wales and Scotland, and the recent referendum on Scottish independence, has been the rise of separate histories of what was once described (and dismissed) as 'the Celtic fringe'. For Ireland, R. F. Foster, *Modern Ireland, 1600–1972* (1988), and K. Theodore Hoppen, *Ireland since 1800* (1989), are both essential, as is R. Bourke and I. McBride (eds), *The Princeton History of Modern Ireland* (2015). For Scotland, Bruce Lenman, *Integration, Enlightenment and Industrialization: Scotland 1746–1832* (1981), and Sydney and Olive Checkland, *Industry and Ethos:*

Scotland, 1832–1914 (1984), were pioneering works, to which should be added T. M. Devine, *The Scottish Nation, 1700–2000* (1999), and T. M. Devine and Jenny Wormald (eds), *The Oxford Handbook of Modern Scottish History* (2012). For Wales, there is Kenneth O. Morgan, *Rebirth of a Nation: Wales, 1880–1980* (1981), and also Matthew Cragoe, *Culture, Politics and National Identity in Wales, 1832–86* (2004). The attempts to manage the vexed identity politics of the nineteenth-century United Kingdom are well discussed in Alvin Jackson, *The Two Unions: Ireland, Scotland and the Survival of the United Kingdom, 1707–2007* (2012), and K. Theodore Hoppen, *Governing Hibernia: British Politicians and Ireland, 1800–1921* (2016).

The best introduction to the history of the British Empire during this period is Andrew Porter (ed.), *The Oxford History of the British Empire*, vol. III, *The Nineteenth Century* (1999), along with Jeremy Black, *The British Seaborne Empire* (2004), and Ronald Hyam, *Britain's Imperial Century, 1815–1914: A Study of Empire and Expansion* (3rd edn, 2002). To these should be added James Bellich, *Replenishing the Earth: The Settler Revolution and the Rise of the Anglo-World, 1783–1939* (2009), and John Darwin, *The Empire Project: The Rise and Fall of the British World-System, 1830–1970* (2009), and also his *Unfinished Empire: The Global Expansion of Britain* (2012). The particular contribution of the Scottish to the British Empire is dealt with in John M. MacKenzie and T. M. Devine (eds), *Scotland and the British Empire* (2011), and John M. MacKenzie, *Scotland and the British Empire* (2016), and of the Irish in Stephen Howe, *Ireland and Empire: Colonial Legacies in Irish History and Culture* (2000), and Kevin Kenny (ed.), *Ireland and the British Empire* (2004). For two views of the empire from the colonies of settlement, see P. A. Buckner (ed.), *Canada and the British Empire* (2010), and D. M. Schreuder and Stuart Ward (eds), *Australia's Empire* (2008). The sense that the whole imperial enterprise rested on insecure structures and impermanent foundations is explored in Antoinette M. Burton, *The Trouble with Empire: Challenges to Modern British Imperialism* (2015) and Stephanie Barczewski, *Heroic Failure and the British* (2016).

The domestic history of the nineteenth-century United Kingdom has been very well treated, and the literature on politics is overwhelming. A good source book remains H. J. Hanham (ed.), *The Nineteenth-Century Constitution: Documents and Commentary* (1969). For general surveys, see M. Bentley, *Politics without Democracy, 1815–1914* (1985), M. Bentley and J. Stevenson (eds), *High and Low Politics in Modern Britain* (1983), and Jonathan Parry, *The Rise and Fall of Liberal Government in Victorian Britain* (1993). The first half of the century is dealt with, from a variety of standpoints, in Linda Colley, *Britons: Forging the Nation, 1707–1837* (1992), J. C. D. Clark, *English Society, 1688–1832* (1985), Arthur Burns and Joanna Innes (eds), *Rethinking the Age of Reform: Britain, 1780–1850* (2003), Malcolm Chase, *Chartism: A New History* (2007), and David Fisher, *The History of Parliament: The House of*

Commons, 1820–1832 (8 vols, 2009). For the period from the 1850s onwards, see: Eugenio Biagini and Alastair Reid (eds), *Currents of Radicalism: Popular Radicalism, Organized Labour and Party Politics in Britain, 1850–1914* (1991); Patrick Joyce, *Visions of the People: Industrial England and the Question of Class, 1848–1914* (1991); Jon Lawrence, *Speaking for the People: Party, Language and Popular Politics in England, 1867–1914* (1998); and James Thompson, *British Political Culture and the Idea of 'Public Opinion', 1867–1914* (2013). Angus Howkins, *Victorian Political Culture: 'Habits of Heart and Mind'* (2015), is a fine synthesis of much of this recent work.

The economic history of nineteenth-century Britain is covered by Joel Mokyr, *The Enlightened Economy: An Economic History of Britain, 1700–1850* (2009); Martin Daunton, *Progress and Poverty: An Economic and Social History of Britain, 1700-1850* (1995), and idem, *Wealth and Welfare: An Economic and Social History of Britain, 1851–1951* (2007); and Roderick Floud, Jane Humphries and Paul Johnson (eds), *The Cambridge Economic History of Modern Britain* (2 vols, 2014). For social history, see F. M. L. Thompson (ed.), *The Cambridge Social History of Britain, 1750–1950* (3 vols, 1990), much of which he effectively anticipated in idem, *The Rise of Respectable Society: A Social History of Victorian Britain, 1830–1900* (1988), and also Jose Harris, *Private Lives, Public Spirit: A Social History of Britain, 1870–1914* (1993). For urban history, see *The Cambridge Urban History of Britain*, Peter Clark (ed.), vol. II, *1540–1840* (2000); Martin Daunton (ed.), vol. III, *1840–1950* (2000). Two recent books offer alternative accounts of the development of modern English/British society: W. G. Runciman, *Very Different but Much the Same: The Evolution of English Society since 1714* (2015), contends that little changed, whereas James Vernon, *Distant Strangers: How Britain Became Modern* (2014), urges that, on the contrary, a great deal did.

The history of what might broadly be termed nineteenth-century culture has also been very fully covered. On religion, recent work includes Hugh McLeod, *Religion and Society in England, 1850–1914* (1996), D. Hempton, *Religion and Political Culture in Britain and Ireland: From the Glorious Revolution to the Decline of Empire* (1996), Andrew Porter, *Religion versus Empire? British Protestant Missionaries and Overseas Expansion, 1700–1914* (2004), and Hilary Carey, *God's Empire: Religion and Colonialism in the British World, 1801–1908* (2011). 'Thought' is well served by Stefan Collini, *Public Moralists: Political Thought and Intellectual Life in Britain, 1850–1930* (1993); S. Collini, R. Whatmore and B. Young (eds), *History, Religion and Culture: British Intellectual History, 1750–1950* (2000), and Gareth Stedman Jones and Geoffrey Claeys (eds), *The Cambridge History of Nineteenth-Century Political Thought* (2013). The arts are best approached through Boris Ford (ed.), *The Cambridge Guide to the Arts in Britain*, vol. VI, *Romantics to Early Victorians* (1988); vol. VII, *The Later Victorian Age* (1989). For science and technology, see James A. Secord,

Visions of Science: Books and Readers at the Dawn of the Victorian Age (2014), Douglas R. Burgess, Jr., *Engines of Empire: Steamships and the Victorian Imagination* (2016), and Crosbie W. Smith, *The Science of Energy: A Cultural History of Energy Physics in Victorian Britain* (1998).

For historians of gender, the Gas-Lit Gloriana has proved a fertile subject, on which see Dorothy Thompson, *Queen Victoria: Gender and Power* (1990), and Susan Kingsley Kent, *Queen Victoria: Gender and Empire* (2015). For those lower down the social scale, see Frank Prochaska, *Women and Philanthropy in Nineteenth-Century England* (1980), Judith R. Walkowitz, *Prostitution and Victorian Society: Women, Class and the State* (1980), Leonore Davidoff and Catherine Hall, *Family Fortunes: Men and Women of the English Middle Class, 1780–1850* (1987), and Patricia Hollis, *Ladies Elect: Women in English Local Government, 1865–1914* (1987). More recent books include Sonya Rose, *Limited Livelihoods: Gender and Class in Nineteenth-Century England* (1992); Deborah Valenze, *The First Industrial Woman* (1995); Anna Clark, *The Struggle for the Breeches: Gender and the Making of the British Working Class* (1997); and Kathryn Gleadle, *British Women in the Nineteenth Century* (2001). The issues of gender and empire are explored in Angela Woollacott (ed.), *Gender and Empire* (2006), and Philippa Levine (ed.), *Gender and Empire* (2004). For men, see Michael Roper and John Tosh (eds), *Manful Assertions: Masculinities in Britain since 1800* (1991); John Tosh, *A Man's Place: Masculinity and the Middle-Class Home in Victorian England* (1999); and idem, *Manliness and Masculinities in Nineteenth-Century Britain: Essays on Gender, Family and Empire* (2005).

Among works of reference, the indispensable sources remain Chris Cook and Brendan Keith, *British Historical Facts, 1830–1900* (1975), and B. R. Mitchell, *British Historical Statistics* (1988). For bibliographies, the essential starting points are Lucy M. Brown and Ian R. Christie (eds), *Bibliography of British History, 1789–1851* (1977), and H. J. Hanham (ed.), *Bibliography of British History, 1815–1914* (1976). From 1976 until 2002 the Royal Historical Society published its annual *Bibliography of British and Irish History*, since when it has been available online (www.history.ac.uk/projects/bbih). The most useful guide to this vast accumulation of scholarly literature is Chris Williams (ed.), *A Companion to Nineteenth-Century Britain* (2004). The lives of many nineteenth-century Britons are recorded in the *Oxford Dictionary of National Biography* (www.oxforddnb.com), and many images of them are available on the website of the National Portrait Gallery (www.npg.org.uk). Converting the monetary values of the nineteenth-century British pound into their modern equivalents is exceptionally difficult, since different commodities have followed very different trajectories (compare the price of wheat and that of Old Master paintings). Again, there are many websites that make it possible to calculate historical currency conversions, but all need to be used with caution.

Index